HE'LL NEVER STOP THAT TRAIN IN TIME!

THE EVOLUTION OF THE TRIPLE VALVE AND DEVELOPMENT OF RAILWAY AIR BRAKING - AN AUSTRALASIAN PERSPECTIVE

FERGUS MOFFAT

ISBN-13: 978-0-6485290-0-2
Library of Congress Control Number: 2014901298
Lightning Source Australia Pty Ltd, Global Print on Demand platform
Melbourne, Australia

CONTENTS

ACKNOWLEDGMENTS

MANY PEOPLE OF significant expertise—from both the past and present—in the Australian and New Zealand railway air brake and operations establishment have assisted me in the preparation of this book in a multitude of ways. They have generously shared their time to discuss air braking matters, have answered my questions, and—in many instances—have assisted with resources and illustrative material that I have used to produce a significant portion of the content and many of the diagrams that appear in this book. In numerous cases they have also vigorously discussed and debated certain issues with me and will possibly not agree with some of my comments. I am grateful to them nonetheless for their contribution and for their willingness to provide their views. This book could not have been compiled without their involvement.

They include Martin Baggott, John Brooks, Kevin Heslop, Martin Janssens, Mike Kendrick (NZ), Alex Kennedy, Ian Livingstone, John Scott, George Shepherd, Barry Payne, Chris Formenton, Les Smith, Phil Dunne, Cliff Stockman, Peter Stokes, and Albert Taylor (NSWGR, retired). My friends Milt Deno (Florida, USA) and John Layton had a level of influence that would probably surprise them.

AUTHOR'S NOTE

I HAVE 'BORROWED' the cover artwork and title for this book from a Westinghouse Air Brake Company advertisement that appeared in the Pittsburgh Press in December 1957 and in the industry journal *Railway Age* at about the same time. The illustration depicts locomotive engineer Dan Tait (or Tate) making what is said to be the first-ever emergency air brake stop in history—using the straight-air brake then under test on his train—although the artist might have taken some license with aspects of detail. The artwork is credited to *Steelways* magazine - the journal of the American Iron & Steel Institute. The artist is unknown, although the author wonders if this could be the painting mentioned in 'The Air Brake' in chapter 7.

The actual location of the incident depicted in this painting is outside the southern portal of the Grant's Hill tunnel (officially the Pittsburgh & Steubenville Extension Railroad Tunnel—but known also as the Panhandle Tunnel—opened in 1863) below Pittsburgh's Central Business District. This location now lies beneath Forbes Avenue where the latter passes the old jail (closed in 1995, and now converted into the juvenile and family sections of the Common Pleas Court). The tunnel was constructed by the cut-and-cover method and was 930 feet long and about 40 feet below the surface when opened. Between 1983 and 1985 the right-of-way was rehabilitated to serve as an underground route and station area (Midtown or 'Steel Plaza' Station) for the Port Authority of Allegheny County's Light Rail Transit subway (the 'T').

"He'll never stop that train in time!"

The dramatic incident that started George Westinghouse, inventor of the air brake, on the road to the Hall of Fame for Great Americans.

THE PAINTING SHOWS A FAMOUS EVENT in railroad history. Twenty-four-year-old George Westinghouse was demonstrating a new brake for trains. He said he could do the job with compressed *air*, instead of a long chain that stretched between cars. Few people thought that an "air brake" would work.

The test train clanked out of Grant's Hill Tunnel in Pittsburgh—but there, right on the tracks, was a team of horses and a wagon. The engineman yanked the strange new brake lever for a full emergency stop. The brakes clapped shut on the wheels, and the train slid to a stop only four feet from the terrified, bucking horses. Never before had a train stopped so quickly. The dramatic story was told all over the land and young George had won his first round toward acceptance of the Westinghouse Air Brake.

His genius was apparent even at 15, when he invented the rotary engine. At 22, he perfected a device to put derailed cars back on the track. Against heavy opposition, he proved the value of alternating current and, with the aid of transformers, made it possible to transmit high voltage electrical power over long distances. He invented dozens of railroad switch and signal devices and pioneered distribution of natural gas for heating purposes. In a 48-year period, George Westinghouse received 361 patents, an average of *one every six weeks*. He caused an uproar by creating the Saturday half-holiday for his workers. In his lifetime, he was director of 60 corporations, president of 30.

On December 1, 1957, a bronze bust of George Westinghouse will be unveiled and permanently enshrined in the Hall of Fame for Great Americans. The choice was made by a select committee of 125 distinguished Americans representing every state in the Union. Only 86 Americans have been so honored, among them such men as George Washington, Abraham Lincoln, Benjamin Franklin, Samuel Morse, Eli Whitney, and Robert Fulton.

The first company formed by George Westinghouse was the Westinghouse Air Brake Company. It exists today, as a greatly expanded enterprise that plays an important part in the industrial scene. The company is still the world's largest manufacturer of railroad brake and signal equipment. But, as you can see from the listing of the various divisions, it has moved into new and growing areas of our economy, inspired by George Westinghouse, one of the truly great creative geniuses of all time.

FIGURE 1: COVER ART: THE FIRST EMERGENCY AIR BRAKE STOP – *PITTSBURGH PRESS*, DECEMBER 1957

A GREAT MANY engineers who take an intelligent interest in the working of the locomotive, whose technicalities they have thoroughly mastered, display no desire whatever to understand the air brake, and are perfectly contented with its action so long as it will stop the train. The air pump, so wonderfully interesting to those who understand its movements, receives no more attention than is necessary to keep it going so that the required air pressure is maintained…

A belief prevails among men who labour principally with their hands, that laziness is exclusively physical. This is a mistake. It is a psychological fact, well known to metaphysicians, that mental laziness is prevalent enough to dwarf the minds of half the human race. Men who would willingly work with their hands during half their leisure time to keep their engines in proper condition for running, have to be driven, by fear or jealousy, before they will force their mental faculties to do trifling labour in a new channel.

Any engineer of ordinary intelligence, who will spend one hour a day for two weeks studying up the Westinghouse instruction book, will understand the brake so well, from the pump to the hind end of the train, that any imperfection happening to its working will be as readily located as an ordinary defect in a locomotive. Yet it is an intensely hard matter to induce men running passenger engines to go through this trifling mental exercise. The consequence is, that the brake sometimes becomes inoperative from causes so slight that men should be ashamed to report them; and they would be so if they only comprehended how small a mole-heap became their mountain…

- Angus Sinclair D.E.
Locomotive Engine: Running and Management, Seventeenth Edition, 1890, Chapter XXI
The Scientific Press, Robert Drummond & Company, Brooklyn, NY
Member: The Brotherhood of Locomotive Engineers, American Railway Master Mechanic's Association, American Society of Mechanical Engineers, etc.

FOREWORD

THE CRITICAL ROLE of air braking in railway operations—being the means by which locomotive drivers control their trains under a variety of operating conditions—is obvious. It is not, though, just the ability to control train speed that is important; the reliable and effective conduct of train braking such as to minimise unwanted shock loadings along the train is also a vital requirement. The need to understand railway braking systems and their capability and limitations is fundamental to safe, efficient railway operations. These equipment arrangements need to be carefully specified to suit not just the rollingstock upon which they are installed but also to comprehend planned operations, particularly in support of the optimisation of route capacity. The question of a driver's need to possess a sufficiently-detailed technical knowledge of the braking system they use—in order to be considered as competent—has often been debated but it is clear that as a minimum a sound understanding of the equipment being used by drivers should allow for better judgement in air brake manipulation, thus promoting safety and efficiency goals.

There is always (and there should always be) a thirst for knowledge and this book will contribute significantly towards satisfying that need. It is vastly preferable to provide technically accurate information than to allow informal hearsay to influence operational process and to populate the pages of history. The dramatic developments in rail capacity over the last 160 years, and the corresponding need for braking system technology to keep pace with these developments provide an interesting and dynamic background to the combination of historical, operational and technical description covered in this book.

The author balances a considerable historical and operational narrative with a comprehensive description of brake equipment used in Australia, New Zealand and—where relevant to the heritage of this equipment—overseas. These detailed technical descriptions require some concentration to properly absorb, but the effort involved carries the reward of an enhanced understanding of the technology, the story of its development, and the intended manner of its use. He also includes some frank opinions that, for obvious reasons, reflect the perspective of an operating person. This is important, since it is—in the final analysis—locomotive and train drivers who are left to manipulate the brake to best effect, thereby providing the required effective outcome in train operations. It is critical that the driver's needs are properly analysed and considered in all development and applications of railway brake systems. The author's measured—sometimes, perhaps, even defiant—opinions are what sets this work apart from other rail-specific histories and works of technical reference.

Fergus Moffat has produced a thorough and comprehensive work that is well-developed, with credible substantiation of argument. This, along with his conviction and passion, make for an inspiring and thought-provoking read. This may well be quite the first work of railway history and reference of its type, and I commend it to all who have a technical and historical interest in railway operations, particularly air brake systems and their use.

David Ferris, ARMIT (Mech Eng), MIE Aust (CP Eng)
David Ferris & Associates
Melbourne VIC
2015

DAVID FERRIS IS a railway mechanical engineer who has a senior rail industry management background. Latterly, he has carried out general consulting in the field of railway rollingstock including rail safety audits, asset evaluation, maintenance planning and strategic asset management both in Australia and overseas.

PREFACE

MY AIM IN producing this book is to provide an interesting and informative reference that chronicles the evolution generally—and within Australasia particularly—of that most arcane of railway subjects; the art and science of air braking. This has, for all of its history, been a dynamic yet disruptive technology, the advancement of which has tended toward intermittent if not spectacular increments, and for this reason I hope that the book serves as an historical, general information reference that explains the progress of this continuum.

I have not intended to provide a treatise on railway safety – this theme having previously and adequately been embraced by others. Nor have I sought to present a long-winded instructional dissertation, although in some sections certain operating details *are* included where I believe the subject matter warrants the narrative. Further detailed information regarding operation of the equipment described in this book is available in manufacturers' instructional pamphlets and numerous other industry publications. Operational information on the various auxiliary pneumatic devices—for example, brake cylinder relay valves, slack adjusters, and empty/load devices—that are often component parts of some of the air brake schedules and systems described herein is also freely available from a range of sources both within and outside the rail industry. Rather, this work is offered as a descriptive summary intended to stimulate interest in the technical discipline of railway air braking and to provide the reader with an historical account of its creation, the progressive development of its two principal mechanisms—the *operating valve* (generally referred-to within Australia and New Zealand as the "triple valve") and the locomotive *brake valve*—and some detail regarding the application of these devices. Where the narrative includes a general discussion about a wagon triple-, control-, or distributing valve (this often being in the international context), please note that the generic term *operating valve* is used rather than *brake valve* due to this latter having prior connotation as the train or locomotive driver's brake operating controller.

The triple valve was fundamental to both the invention and development of a reliable railway power brake system; however, full advantage could never have been taken were it not for the parallel development of an efficient means of controlling it - the aforementioned driver's air brake valve. This study will also look at progress made in this regard since the beginning of railway air braking.

I hope that rail industry personnel at various levels will find this book useful – even if only as a collection point for disparate pieces of information otherwise gathering dust in forgotten cupboards or packed somewhere in anticipation of disposal. Moreover, I hope that it will help the interested layperson to gain a practical insight into a technical speciality that is perceived by many people (including within senior railway management) as something of a black art. To all of these ends, I have deliberately utilised diagrams and illustrations as much as possible, the better to illuminate and enliven the subject as well as to both complement and complete the historical record. Countless hours have been expended over the last 12 years digitally drawing, cleaning up, and otherwise rendering more comprehendible numerous diagrams, (including some back-of-an-envelope creations) collected over the years. I hope they are informative.

Some might consider an attempt to present a comprehensive discourse on a subject as diverse as railway air braking—with its many permutations and multitude of applications—as at best bold and at worst foolhardy, and I am under no illusions as to the magnitude of the task. I welcome feedback from interested or knowledgeable readers who may have a contribution in technical accuracy or detail of an omission that would improve the book should there be an opportunity for a second edition. And one last note. Throughout this book I make reference to the Australian Westinghouse company. Over recent decades—until its purchase by Knorr-Bremse—the company was the Westinghouse Brake & Signal Co. (Australia) Ltd. or Westinghouse Brakes (Australasia) Pty. Ltd.; however, for practical reasons, I have generally contracted the name to "Westinghouse Australia."

Fergus Moffat
Melbourne, Australia, 2015
femoffat@gmail.com

1.

THE INVENTION OF RAILWAY AIR BRAKING

If I understand you, young fellow, you propose to stop a railroad train with wind. I'm too busy to listen to such nonsense…

—"Commodore" Cornelius Vanderbilt, president, New York Central Railroad, to George Westinghouse.

THE STORY OF railway air braking in Australia and New Zealand is necessarily an American one. It is generally well-known that the continuous, end-to-end air braking system, nowadays a standard part of everyday train operations around the world, was a product of the prolific US inventor, George Westinghouse Jr., although he did not so much *invent* railway air brakes as create the first practical scheme for their use. The story behind this innovation, though, and the development of its major element—the triple valve—may be less familiar.

FIGURE 2: GEORGE WESTINGHOUSE, AN EARLY CAREER IMAGE

The grandson of John Hendrick Wistinghaus (a German immigrant), Westinghouse was born in October 1846 at Central Bridge in Schoharie County, upper New York State. The American railroad industry at the time was nascent; the railroads were only beginning to break out of their cultural British chains and about to challenge the eastern canal and barging business. After working in his father's agricultural machinery factory in Schenectady, Westinghouse enlisted in the Union army in 1862 at age fifteen, serving briefly during the Civil War as a private in the 12th Regiment New York National Guard and in the 16th Regiment New York Cavalry. During this service his unit was operating in the District of Columbia and in northern Virginia on the hunt for Confederate guerrillas but not seeing combat (something for which he would later carry guilt). A year spent feeding and caring for horses was not much to the young George's liking, so shortly after a brother was killed-in-action in Mississippi[1], George followed his other sibling into the Union navy in 1864 as an acting ensign. Commissioned as an acting 3rd assistant engineer aboard the *USS Muscoota*, he was soon transferred to another warship, the *USS Stars and Stripes.*[2] After the war ended in April 1865, he attended William Seward's[3] Union College for only three months, dropping out soon after obtaining his first patent in October 1865, at age twenty, for a rotary steam engine. In fact, Westinghouse was asked to leave the college due to reported "inadaptability," with one instructor even labelling him "dull and backward."

Westinghouse grew up in an American era that witnessed the rapid spread of twin steel ribbons across the nation. Railroads, largely confined at the time to territory east of the Mississippi River, had played a significant role in the Civil War, and George Mortimer Pullman had been building his specialised passenger cars since its end. The first transcontinental route was completed in 1869, and within 30 years four more would follow. In 1888—the first year in which such statistics were compiled—2 070 railroaders were killed at work and more than 20 000 injured – a large proportion of them brakemen.

At best, handbrakes were slow to take hold. On a downgrade they often let slack pile up unevenly along the length of a train, bringing on the embarrassment—or, all too often, the calamity—of a break-in-two. The even application of brakes throughout a train called for skill and long practice. Railroad owners devised a sobering punishment for a brakie who tightened down too hard, skidding and flattening the wheels. The price of the wheels, usually about $45 apiece, was deducted from his paycheck – and at that time [about 1873] *a brakeman's pay was exactly $45 a month.*[4]

Having observed the laborious restoration to the rails of some freight cars following a derailment, Westinghouse invented a device to expedite the procedure and started a business to manufacture it.[5] He also patented a new style of railroad switch frog that was reversible; so that when worn out it could be turned over and reused. Moreover, he designed this frog in cast steel rather than iron, and hence he is credited with one of the first successful steel castings ever produced in America – an entirely appropriate outcome for a man who would become an eminent Pittsburgher.[6] These two attempts at patented railroad products ultimately—although practical successes—turned out to be commercial failures. In his paper *George Westinghouse: The Mystery* (Pittsburgh Quarterly; Spring/Summer 2006), William S. Dietrich III states 'The railroads adopted both [products] but sidestepped his modest patent protection, yielding him little return. But he was a quick learner and would not make that mistake twice'. Significantly though, quite apart from his regular rail travel on behalf of his father's business, these two projects were probably the genesis of what would become Westinghouse's lifelong focus on railway technology.

The concept of retarding train speed by applying a braking effort to the coupled vehicles was not an unexplored science. Despite an early patent by Robert Stephenson for a locomotive steam brake

1. See further details in the Westinghouse biography in chapter 7.

2. The propulsion machinery on both the *USS Muscoota* (a 1 370-ton side-paddle, iron-hulled "double-ender" steam gunboat) and the *USS Stars and Stripes* (a 407-ton screw steamer acquired by the US Navy during the war for service as both a gunboat and a tug) must surely have intrigued Westinghouse. Serving aboard these vessels, he became aware of the loss-of-efficiency inherent in reciprocating steam engines and began to conceive of the steam turbine.

3. President Abraham Lincoln's Secretary of State.

4. Keith Wheeler/TIME-LIFE Books, The Old West: The Railroaders, Chapter 5 (New York, TIME-LIFE Books, 1973 and 1975).

5. Note that William Stroudley (1833–89), an eminent British locomotive engineer, invented a "re-railing ramp" around 1860 while working as a foreman fitter at the Great Northern Railway in Peterborough, Scotland.

6. Pittsburgh, from the early days, had been a centre of the North American iron and steel industry. See chapter 7 for more about George Westinghouse's life and achievements.

(6484/1833), the concept was originally one that tended to be associated with trailing rollingstock rather than the locomotive itself. In 1836 Henry Booth[7] invented a "counter-pressure" brake for the Liverpool & Manchester Railway. This device was, in effect, a dynamic brake for steam locomotives that retarded the locomotive by using the cylinders as air compressors and converting the kinetic energy of the locomotive movement into heat; the steam resulting from the evaporation of water during this process serving to cool the cylinders. In 1840 H.M. Grover[8] is credited with conceiving an electro-magnetic brake that proposed the use of an *"electric current induced in the rails or in special conductors laid parallel to the rails* [the electromagnets being "suspended" adjacent to the wheel surface, and—when energised—would "grip" said wheel] *that would retard the forward motion of the train."* (This man was surely ahead of his time.)

Four years later, James Nasmyth[9] and Charles May patented the first vacuum-operated brake. One John Wilson is credited with having devised a continuous brake to enable loads of soft coal to be lowered down a wagonway to the Union Canal in Scotland. Then, in 1848, Robert Heath invented a chain-activated, weight-and-lever-operated automatic continuous brake. In 1867, A. Chambers modified what is known as the Fay-type of chain brake, which was evaluated on the North London Railway. This contraption—not easily controllable—utilised a chain coupled between the trailing wagons and running the length of the train, terminating at the locomotive. When a brake application was desired, the fireman engaged a mechanism against a locomotive wheel that wound the chain up onto a drum and drew it taut. Of course the train had to be moving for this brake to be applied; the logic being that if the train were already stopped, then no brake would be required. This, of course, failed to take account of any potential need to *secure* a stopped train.

In 1853 William C. Creamer, a railway supplier who owned a foundry in Brooklyn, New York City, patented a spring-operated system with a trigger rope running the length of the train and able to be operated by brakemen and locomotive engineers alike. But it was only an emergency brake; the effect being all or nothing. And since the springs were attached to the handbrake shafts, it required them to be rewound after the train stopped, something not all brakemen could be relied-upon to do (passengers were advised not to pull the rope). Creamer reportedly sold about one thousand of these contraptions[10] while trains remained of modest proportions, including one system to the South-Eastern Railway in

7. (1788–1869) Liverpool, England. A corn merchant, businessman, and engineer, Booth was one of the Liverpool and Manchester Railway's twelve initial directors as well as company secretary and treasurer (equivalent to the post of chief executive officer) until 1846, and he was a leading proponent of the concept of working all British railways to one standard time. On the formation of the London & North Western Railway (LNWR) in July 1846, he became a director, serving until 1859. He was also the LNWR's first secretary, serving between 1846 and 1848. Booth proposed the use of the recently invented multi-tubular boiler for Stephenson's Rocket steam locomotive built in late 1829 for the L&M Railway. He is acknowledged as having assisted Robert Stephenson in the construction of Rocket, generally accepted as being the first steam locomotive to combine the essential features of the orthodox type.

8. A clerk from Boveney, Buckinghamshire, England. Patent 5999, May 7, 1840, for an improved method of retarding and stopping railway trains. Per the *Metropolitan Magazine*, vol. 28.

9. Also the inventor of the steam hammer and the pile-driver. Nasmyth's company was the forerunner to locomotive builder Nasmyth, Wilson & Co.

10. The Creamer Safety Brake, a device for stopping runaway passenger cars in an emergency, placed a heavy spring inside a drum between the brake wheel and the brake shaft. The brakemen wound the wheel (like winding a clock); a pawl held the tension within the drum. A rope ran the length of the train connected to each pawl in such a way that a strong tug on the rope would release all the pawls, and the spring tension would then set the brakes on all the cars, relieving the brakemen of this task (and therefore posing a threat to their career!).

England. The scheme was eventually discredited following a fatal train wreck in 1871;[11] much criticism *"...being directed at Creamer and the railroad for depending on so absurd a system as a collection of clock springs to stop an express train."*

For some light relief—and because we might acknowledge the need for these proposals to have been considered first, before being disposed of so that more profound ideas would be entertained (and could begin to influence the concept of train braking)—we might take note of the following rather aberrant notions:

- The use of *"...large retractable scoops that could be opened to provide substantial wind resistance..."* (horizontal umbrellas, in other words!).
- The location of permanent trackside "braking facilities" by which high-pressure water jets — mounted at permanent trackside locations — would be directed "at suitably-shaped blades constructed on the vehicles."
- From a proposal in a *Railway Gazette* issue of the 1880s, to drop an anchor from the hind-end of the train to *"engage with the ties* [so that] *a train might easily be brought to a stop within fifteen or twenty feet...if something did not give way."* From that trailing thought, you have to give the proposer credit for thinking of all the possibilities (and perhaps for coining what would become a global euphemism for stopping in a hurry).

Patent Office records reveal that in the forty years following 1830, hundreds of would-be inventors conceived various devices to improve train brakes. Indeed, experimentation with compressed air began in England during the 1840s (D. Crawford's and S.C. Lister's patents awarded in 1843 and 1848, respectively). The development of patented railway brakes, though, is a story in itself, and these few paragraphs will have to suffice as an introduction to the notion of power brake systems for locomotives and trains. There are, however, various intriguing historical references to a Westinghouse *vacuum* brake. In an excellent treatise on the subject of early railway accident history[12] that author records in regard to the Newark Brake Trials in England,[13] *"Of the brakes experimented with, three were American in their origin – Westinghouse's automatic and vacuum, and Smith's vacuum."* These references to a Westinghouse vacuum brake deserve some exploration.

Prior to the American Civil War, John Y. Smith, a native of Cumberland who had been involved in engineering in the English Midlands, including with the St. Helens & Runcom Gap Railway, had emigrated to America where he formed a partnership with Henry Kirke Porter and opened the Smith & Porter Machine Works[14] in Pittsburgh, repairing and building industrial equipment and, later, railway locomotives – most of

11. In subzero temperatures on the night of February 6, 1871, on the New York Central & Hudson River Railroad, Extra No. 3—a 25-car southbound freight train with a derailed car caused by a broken axle—stopped on the Wappingers Creek Bridge outside New Hamburg, New York. In stopping, several other cars also became derailed, one of them—an oil tank car—obstructing the opposite running track. Before the site could be protected, a late-running *Pacific Express* bound from New York to Chicago approached at speed, its crew sighting the obstruction only six hundred feet distant. The Creamer brake on the passenger train was "sprung" and the engine reversed, but a collision could not be avoided. The impact and ensuing conflagration collapsed the bridge, pitching the locomotive and tender, baggage cars, and one passenger car into the frozen river. Of the remaining seven cars of the passenger train, two, plus several of the freight train, were destroyed by fire. As the remaining portions of the bridge were consumed in the conflagration, the superstructure fell upon the cars already in the river. Passengers and crew uncoupled the last three cars of the express and pushed them off the bridge to safety, a feat that would have been impossible had the Creamer brake taken hold as intended. Twenty-two people lost their lives and many were badly injured. The dead included the passenger train engineer and an out-of-work locomotive engineer riding aboard the engine, as well as the head-end brakeman. By the following year, the New York Central RR had commenced to equip all of its passenger trains with the Westinghouse brake.

12. Charles Francis Adams Jr., *Notes on Railroad Accidents* (New York: G. P. Putnam's Sons, 1879), chapters XX and XIX. (NB: an amusing, educational, and erudite comparison of British and American rail safety issues of the time).

13. Held on 9 and 10 June 1875, at Thurgarton, near Newark-on-Trent in Nottinghamshire, on a level section of the Midland Railway's Nottingham-to-Lincoln line, later to be part of the LNER.

14. Later to become the Porter Locomotive Works, then the H.K. Porter Company.

them being industrial models. Cognizant of the nascent market for power brakes—and probably well-aware of Westinghouse's efforts to satisfy that market—Smith sought a piece of the action and selected the vacuum brake system. Whether he knew it or not, his apparatus copied (*"part-for-part"*) that of Nehemiah Hodge of North Adams, Massachusetts, who in 1860 had obtained the first American patent for a non-automatic vacuum brake. Smith took out patents for improvements to the scheme, and for a time his system was used widely throughout the eastern United States. He used an ejector on the locomotive and *"air pumps in the vans."* The brakes were applied by collapsible India rubber bags and pistons located below each vehicle. There were two train pipes, one of which went straight to the end of the train so that the vacuum was created on all vehicles more-or-less simultaneously. This brake also met with considerable success in Britain.

Being less costly than the Westinghouse air brake (and perceived by many in the industry as being simpler in concept and construction), it provided sufficient competition for George Westinghouse to decide in 1875 or 1876 to purchase the rights to Smith's patents and seek to satisfy the British market by producing vacuum brakes of that design for several years under his own name. Westinghouse also purchased Hodge's patent and had it re-issued in 1879. Paralleling Creamer's experience in the United States, though, the non-automatic Smith brake eventually fell into disrepute in Britain following a series of rail accidents; the final straw being the 1889 Great Northern Railway disaster at Armagh in Ireland.[15]

About a year after John Smith's brake appeared, Frederick W. Eames received his first patent for a vacuum brake system (No. 153,814 dated 4 August 1874). The primary difference between Eames' brake and Smith's appears to be that the latter used a piston mounted on the car, while Eames used diaphragms mounted separately on each bogie (not to mention "cup-shaped" brake cylinders). He established the Eames Vacuum Brake Company in February 1876 and began manufacturing the brake apparatus in his father's machine shop on Beebee's Island at Watertown in upstate New York.

The Eames vacuum brake sold well in the United States and South America, and Eames, aware of British government pressure for their railways to adopt some form of continuous power brake, decided to cash in. He refurbished a locomotive and shipped it to England to demonstrate his equipment there, including at the 1881 Exhibition of Lifesaving Appliances. Transatlantic sales, however, were less-than-spectacular, perhaps because the British were perfecting their own vacuum brake and had developed an automatic version as early as 1879. Realising the limitations of his own equipment, Eames returned to the United States to develop his own automatic version (which he achieved in 1883), became involved in a commercial dispute concerning his factory and business, and was shot and killed by a disaffected employee in April 1883.[16]

The Eames Company was aware that the vacuum brake would likely not prevail in the United States, and two years later, their new Chief Mechanical Engineer, A.P. Massey, devised an air brake resulting in a lawsuit from Westinghouse. Supported by an American railroad industry that feared a Westinghouse monopoly of this new technology, the company persisted and in 1890 reorganised as the New York Air Brake Company. By now, the earliest Westinghouse patents had expired, and the new company was free to replicate Westinghouse designs. Litigation with the Westinghouse Air Brake Company was finally resolved in 1912 when the two companies agreed on a cross-licensing accord covering all air brake patents.

15. On 12 June 1889, an excursion train stalled while negotiating a heavy grade. When the crew divided the train, the rear portion had inadequate braking power and ran back down the hill, colliding with a following train. Eighty-eight people—most of them children—were killed, and 170 were injured. This accident spurred the UK Parliament to pass the Regulation of Railways Act 1889, mandating improved brake and signal systems. The development of "automatic" functionality for the vacuum brake was a direct result.

16. See chapter 7 for more information about Eames.

The Watertown location on the Black River, thirty miles from the Canadian border, is still the American manufacturing headquarters of the New York Air Brake Company, now owned by Knorr-Bremse. We shall see later how ironic was this alliance.

A description of the Newark Brake Trials will serve to provide a background to railway air braking and to place Westinghouse's eventual achievements in context. The Newark trials might be acknowledged as having their genesis in a comment made by Captain H. W. Tyler, Chief Inspecting Officer of the British Board of Trade in his report on the Shipton-on-Cherwell accident[17] wherein he expressed the opinion that:

If…the train had been fitted with continuous brakes throughout its whole length of about 150 yards, which brakes could have been put on by the driver merely turning a handle and opening a valve, as in the Westinghouse or Smith's vacuum brake, or pulling a cord as in Clark's brake, much more simple operations than the reversal of an engine, there is no reason whatever why the train should not have been brought to rest without any casualty.

FIGURE 3: THE SHIPTON-ON-CHERWELL DERAILMENT SCENE. *Illustrated London News, 1874*

17. Derailment of the GWR London (Paddington)-to-Birkenhead express at Shipton-on-Cherwell on Christmas Eve 1874. In frigid weather an obsolete wrought-iron tyre on the wheel of a leading coach fractured, and severe braking action effected on the locomotive sent the derailed train down an embankment near the Hampton Gay paper mill on the Oxford Canal (at this location, incorporated as part of the River Cherwell), killing 34 and injuring 69.

THE TRIALS WERE a requirement arising from the Royal Commission on Railway Accidents in 1874 and were described at the time as the most important railway tests since Rainhill.[18] Captain D. S. Galton[19] of the Railway Department Board of Trade was placed in charge with the assistance of the Royal Engineers. Eight types of brake systems competed, and a train consisting of a locomotive and thirteen cars was operated for each test run. About seventy runs were made over six consecutive days and the results recorded and tabulated. The competitors were:

- the Clark-Webb chain brake in use on the London & North Western Railway;
- the Westinghouse vacuum brake in use on the London, Brighton & South Coast Railway;
- the Clark's hydraulic, Barker's hydraulic, and Westinghouse automatic air brakes, all used on the Midland Railway;
- the Steel and McInnes "pneumatic reaction" brake[20] used on the Caledonian Railway;
- the Smith's vacuum brake used by the Great Northern Railway; and
- the Smith's vacuum brake in use by the North Eastern Railway.

The trials later carried out by Galton, Stroudley and colleagues, provided empirical evidence for—among other things—phenomena we might be tempted to dismiss today as so elementary as to be comical:

He [the president] *was very glad to find in the paper a confirmation of the old idea, referred-to so forcibly by Mr Fay, that the retardation by braking wheels without skidding was far more effectual than the retardation by skidding.*[21]

Up to this time, "*...little effort* [had] *been directed toward a practical application of the knowledge of the fluctuation of the coefficient of friction with the speed, so definitely discovered in these experiments.*"[22] Indeed; there was a significant body of opinion within the railway industry of the time (although not among enginemen), that a railway wheel skidding on a rail provided the greatest retardation force against velocity. Prout comments:[23]

---

18. The Rainhill Trials in October 1829 in Rainhill, Merseyside (between Liverpool and Manchester), were an important early competition organised by the Liverpool & Manchester Railway to decide whether stationary steam engines or locomotives would be used to haul their trains. The trials were arranged as an open contest by which all the candidates would demonstrate their locomotives in action, with the company deciding according to the candidates' ability to satisfy a series of conditions; a prize of £500 being offered to the winner. Ten locomotives were entered, but only five actually began the tests. George Stephenson's Rocket was the only locomotive to complete the trials and, therefore, won the contest. The Stephenson Company was thus awarded the contract to supply locomotives to the L&MR.

19. Captain Douglas Strutt Galton (1822–99)—described as *"a man of wide experience, a very able and perceptive man whose words were not lightly overlooked"*—held various positions under the Commissioners of Railways and in the Railway Department of the Board of Trade between 1847 and 1854 and became chief of the Railway Department, 1854–59. Commissioned in 1840, Galton worked for the Ordnance Survey (1846–47), the War Office (1860–75), and was in charge at the Newark Brake Trials between 1875 and 1880. He was knighted for his services in 1887. He also organized further brake trials during 1878 under the auspices of the Institution of Mechanical Engineers and in association with William Stroudley (locomotive superintendent of the Westinghouse-sympathetic London, Brighton & South Coast Railway) and George Westinghouse. These tests were conducted to determine the coefficient of friction between the brake-shoe and the wheel, and between the wheel and rail at different velocities when the wheels were revolving and when stopped in their rotation and caused to slide. Among other things, the results showed conclusively that skidding did not enhance braking performance (Jeffrey Wells, *Back Track* magazine, vol. 13, no. 11, 1999). These latter trials culminated in the "Galton-Westinghouse Law of Friction."

20. This apparatus was a form of continuous straight-air brake that utilised cylinders and pistons located vertically at the end of each car and connected by links, chains, or rods to apply and release the brake blocks. An application valve and whistle (to identify which valve had been opened or of an inconsistency in pressure in the equipment on that car) were fitted at each passenger compartment. The driver operated the system by releasing pressure at his two-way valve, or the guard or passengers could apply the brake from their compartment. Because it used air pressure applied to both sides of each car piston (with an attendant maintenance effort that can be imagined), the scheme used an enormous amount of air and took a long time to exhaust at the driver's valve during an application. Although a worthy early attempt to harness pneumatics for a power brake, it was not "fail-safe" and was only ever experimental.

21. On the *Effect of Brake s Upon Railway Trains* (second paper); Capt. Douglas Galton CB Hon, DCL FRS, MIME. Excerpt from the minutes, Institution of Mechanical Engineers, Manchester, 24 October 1878; John Robinson Esq., President, in the Chair. (Printed on behalf of the Westinghouse Air Brake Co., Pittsburgh, PA by Bartlett & Co., New York).

22. From the Preface to the paper referred-to above, provided by the Westinghouse Air Brake Co.

23. Henry Prout. Refer to the George Westinghouse biography in chapter 7.

It will astonish some to know that such a fallacy persisted as late as 1878. But that it did then persist is certain. In the discussion of one of Galton's papers a speaker said that the result mentioned in the paper as to skidding was certainly somewhat surprising after the deductions drawn by the Royal Commission on Railway Accidents to the effect that when wheels were skidded they retarded the force of the train more than when revolving. During the further discussion the Royal Commission found supporters, ***but one speaker said that it had been well known by every practical engine driver for the last twenty-five years that the skidding of wheels was a great mistake.***

In Volume 8 (1927/1928) of the Transactions of the Newcomen Society, T. Rowatt has this to say on the outcome and the subject of Railway Brakes and provides some fascinating insight into how and why the British Isles (and thus—we might consider—also its narrow-gauge customers in Australia) became a vacuum brake stronghold:

Every competitor tried to obtain conditions as favourable as possible for itself and many were dissatisfied, but on the whole, the trials were fair. Of the competing brakes, the two mechanical brakes, Fay's and Clark's, were out-classed. The two hydraulic brakes were inconvenient, and the Westinghouse vacuum was, more or less, a failure. This left the Smith vacuum, the Westinghouse pressure, and the Steel & McInnes pressure; and of these, the Westinghouse was decidedly the best. So much was this the case that it seems to the author that had Westinghouse acted sensibly he should have swept all other brakes off the field, instead of which the supporters of other brakes struggled, night and day, to bring them to the level of success of the Westinghouse, with the final result that the vacuum held the field on British steam trains to the exclusion of the pressure brake.[24]

During the early decades of public railway transport in England, America, and other developed nations, the stopping of trains had hardly been accorded the same consideration by interested parties as was the actual running of them. In an authoritative and instructional volume[25] first published in Chicago in 1899, master machinist C.L. McShane allows that:

When running at high speed in the ante-brake days, the writer used to feel that the [locomotive] *engineer was like the man who pulled the trigger of a rifle to send the ball into space. He could start it into speed, but he must wait for air friction, or the hitting of some object to stop it.*

With no reliable system for regulating the movement of trains in single line territory, *"running by the smoke or headlight"* was—at least in the United States—the order of the day (or night!), and it must have required extraordinary qualities of courage (or perhaps optimism?) and vigilance on the part of enginemen to traverse the land under those circumstances.

In the main, stopping or slowing a moving train required coordinated action between the locomotive crew applying whatever means of retardation existed on their locomotive—usually just the tender hand brake and possibly reversing the engine—and other on-train personnel perhaps applying such vehicle handbrakes as may be fitted. In later years, steam brakes were fitted to locomotives, and in some countries—most notably the Americas—the craft of Brakeman emerged. The hazardous occupation of these hardy individuals was to make their way through or over the cars when signalled by the engine whistle to manually apply handbrakes on each vehicle and to release them when appropriate. When not required to be out on the cars, these brave souls were accommodated in the guard's van (caboose) or locomotive – and sometimes atop the cars or locomotive tender in rudimentary

24. This last claim was not strictly correct. Several British railway companies were Westinghouse air brake customers. On UK railways prior to re-grouping in 1923, the proportion of vacuum to air brakes was about 3:2. After 1923 it was decided to adopt—with certain exceptions—the vacuum brake as standard, the decision based on political and financial grounds rather than consideration of the operating characteristics. Prior to 1914, although the air brake was successful on passenger trains, difficulties were encountered on freight trains owing to the slow propagation that caused violent longitudinal shocks. The British draw-gear of the time was inadequate in absorbing the energy of run-in when brake applications were initiated. This was not the experience in America, where centre couplers were standard and gave adequate absorption and fewer operational problems.

25. Charles Leo McShane, *The Locomotive Up To Date* (Chicago, IL: Griffin & Winters, 1899). Subsequent editions available.

accommodation, rearward-facing for a view back along the train. It was generally accepted that one brakeman could manage only three cars, although some freight railroads required more from them.

The brakeman's job was surely the most perilous in an industry replete with hazards. In 1881 at least ten brakemen lost their lives every day on US railroads, and many more were maimed.[26] Different railroads employed different numbers of brakemen per train, and fewer on a freight than on a passenger run. Most brakemen on North American trains were immigrants—initially Irish—as others tended to shun the dangers and low wages.

How effective was hand braking? The average burly brakeman could apply around 150 lbs of force to the brake wheel, the wheel itself multiplying this to 900 lbs. The levers and linkage, depending on the arrangement, could compound this to place the blocks against the wheels with 4,500 lbs of retarding force - on only one bogie! Having accomplished this, the brakeman then ran through or across the car to the next brake wheel, and so on. A long-rod-and-linkage arrangement was eventually developed by Willard J. Nicholls—car shop foreman of the Hartford & New Haven Railroad[27]—to brake the distant bogie as well, this being the genesis of today's "foundation" brake rigging. Nicholls never patented his invention, and others are known to have utilised and developed its valuable features, appropriating the invention as their own. This invention doubled the output of a brakeman.

But what resulted once the whistle shrilled to "down brakes"? One biographer of George Westinghouse writes:

"...the old 'armstrong' method would halt a passenger train travelling at 30 mph in 1,600 feet."[28] This appears at first glance as a self-serving exaggeration if one is aware of actual braking tests conducted around the mid-1800s, indicating that brakemen could achieve stops of far shorter measure; an 1855 report, for example, recorded an alert test crew stopping a 119-ton train running at 30 mph in 450 feet. However, given these tests were 'staged stops', it is apparent that the crews were alert and ready to spring into action. During normal over-the-road operations, passenger train brakemen were just as likely to be otherwise engaged about the train any time an unexpected whistle signal called them to the 'wheels' (note that passenger train brakemen had additional duties such as tending stoves, positioning clerestory ventilator windows, trimming lamps, reversing seats). In such cases, if it should take the brakemen, say, 15 seconds to get to their nearest brake wheels, a train making 30 mph might have traveled 500 feet before those brakes

FIGURE 4: THE BRAKEMAN. THESE IMAGES DEPICT THE DRAMA AND THE DANGER OF THE RAILROAD BRAKEMAN'S TRADITIONAL ROLE. *Celebrated railroad artist Ted Xaras; O. V. Schulbert (Image courtesy of Harper's Magazine); Jack Delano, 1943 (The Thorpe History Archive); B&O Railroad Museum.*

26. John H. White Jr., *The American Railroad Passenger Car, Part 2*, Johns Hopkins Studies in the History of Technology (The Johns Hopkins University Press, 1985).

27. Angus Sinclair, *Development of the Locomotive Engine* (Cambridge, MA: The MIT Press, 1907).

28. Prout, H.G., *A Life of George Westinghouse* (NY: Charles Scribner's Sons, 1922). Refer to the George Westinghouse biography in chapter 7.

were against the car wheels, stretching the stop to around 800 feet. Consider, though, if these brakemen were to suffer any further hindrance in getting to the job, and it took them twenty or thirty seconds to get their 'club' into the brake wheels, then 1 500 or so feet might not be so fanciful.

The year 1866 was a seminal one in George Westinghouse's life. He was just twenty years old when—traveling one day on a passenger train between Schenectady and Troy, New York—he was delayed by a track blockage caused by the head-on collision of two freight trains on the track ahead. Westinghouse later learned that although the two locomotive engineers had been aware of their proximity to each other, the response of the brakemen to the "down brakes" whistle signal had been insufficient to prevent the collision. He decided there must be a way to provide a reliable braking system for trains that would function simply and that could be operated easily by the locomotive engineer. After a couple of abortive attempts at conceiving the solution (one involved a motion brake system that applied on each car in response to buff or 'run-in' slack action, while the other attempted to use live steam piped to car brake cylinders), Westinghouse had another idea.

FIGURE 5: DRILLING THE MONT CENIS RAIL TUNNEL

He had read in a news magazine of the compressed air-powered drilling machines being used in the excavation of a 13-kilometre rail tunnel under Mont Cenis in the Franco/Italian Alps (now known as the Fréjus Tunnel). Reasoning that a force capable of being transmitted several thousands of feet by rubber hose to operate rock drilling machinery could also be used to stop trains, he set about developing his concept. At his farewell address as president of the American Society of Mechanical Engineers in December 1910, Westinghouse would reveal how two unrelated occurrences (the first having been the delay to his train caused by the train collision previously described) led to his conviction that compressed air would be the ideal medium to operate his power brake. One day, he recounted, two young ladies who were aware of him and his mechanical interest summoned the courage to ask to see him, whereupon they canvassed his subscription to an engineering magazine. In an issue of this journal, Westinghouse learnt of the compressed air technology referred-to above. He would spend the next nineteen years refining his railroad air brake, although here we might pause for breath.[29]

In extolling Westinghouse as, "*...the preserver of The Travelling Public whose great invention of the Automatic Air Brake has preserved more human lives than any military general or tyrant ever succeeded in destroying...*" the eminent Dr Angus Sinclair[30] might have exaggerated. Without questioning the man's faculties as an engineer and businessman, we might be wary of eulogising Westinghouse. Train braking and the evolution of a successful power brake had persisted through the efforts and innumerable successes and failures of many engineers and thinkers in Britain since at least 1833, and in America soon thereafter. Although few indications exist of Westinghouse having recognised this research and the many resultant experiments, he was nothing if not well-read. It beggars belief that—despite his youth—he would not have been aware of contributions in this area and would not have benefited from some awareness of his predecessors' experiences and from what must have been a considerable body of knowledge even then extant. Several patents for air brake schemes had been awarded well before George Westinghouse put his mind to the matter.

From the earliest days of railway operations, there had been public consternation regarding the quantum leap in transport velocity; from the modest speeds of the horse-drawn wagon and stagecoach to the hitherto undreamt-of swiftness now provided by the steam locomotive and railway train – fifteen, twenty, even thirty-five miles per hour! Given this new form of transport had from its earliest days developed as a common-carrier service, the widespread community concern as to the control of this speed—that is, the ability to stop—was perhaps understandable. One commentary records the prevailing societal trepidation:

Under the circumstances, the criticism [that passengers on these trains could not trust their lives to the safety of this new transport mode] *was a perfectly reasonable one. The danger in going at such a rate of speed, and the impossibility of stopping in time to avoid a sudden danger, would naturally suggest themselves to anyone as insuperable objections to the new system for any practical use. Some means of preserving a sudden and powerful control over a movement of such unheard-of rapidity would, almost as a matter of course, be looked upon as a condition precedent...*

Yet down to the year 1870 the brake system remained very much what it was in 1830. Improvements in detail were effected but the essential principles were the same. In case of any sudden emergency, the men in charge of the locomotive had no direct control over the vehicles in the train...

The need of a continuous train brake, operated from the locomotive and under the immediate control of the engine driver, had been emphasised through years by the almost regular occurrence of accidents of the most appalling character. In answer to this need almost innumerable appliances had been patented and experimented with, both in Europe and in America. Prior to 1869, however, these had been almost exclusively what are known as emergency brakes.

Patent brakes had become a proverb as well as a terror among railroad mechanics, and they had ceased to believe that any really desirable thing of the sort would ever be perfected...[31]

It was, therefore, into this environment of public scepticism that Westinghouse in 1868 submitted a patent application[32] for his Straight-Air brake system (or "atmospheric brake" as it was originally known)

29. See chapter 7 for more details.

30. Angus Sinclair, *Development of the Locomotive Engine* (Cambridge, MA: The MIT Press, 1907).

31. Charles Francis Adams Jr., *Notes on Railroad Accidents* (New York: G. P. Putnam's Sons, 1879), chapters XX and XIX.

32. Patent awarded April 1869.

and persuaded some influential railroad acquaintances on the Pan-Handle Division[33] of the Pennsylvania Railroad to fit and test it at his own expense and risk. This invention, which had been under development for three years, introduced compressed air as a practical power brake medium under the control of the locomotive engineman. To capitalise on its potential profit, Westinghouse and his colleagues organised the Westinghouse Air Brake Company and appointed George Westinghouse as president.

The invention used an increase in positive pressure to make a brake application: the locomotive being fitted with a compressor sufficiently compact to be carried on the side of the boiler, a storage receiver (nowadays known as the main reservoir), a "three-way cock" (the engineer's brake valve), and the required brake rigging. Cars were equipped with one or two "train pipes" (nowadays known as brake pipes), a brake cylinder, and appropriate brake rigging.[34]

Brake application was accomplished by movement of the three-way cock handle to a position that closed the atmospheric (or exhaust) vent and opened the train pipe connection to the stored reservoir supply. The stored pressure would then flow down along the continuous train pipe and into the brake cylinders of the locomotive and then each car in turn. Of course, this required careful manipulation of both throttle and brake since the front cars in the train—receiving their air supply first—would be applying well ahead of the rearmost cars. Release occurred when the engineer moved the three-way valve to the position that exhausted the pressure from the train pipe (and therefore the brake cylinders). All air pressure in the system would then have to flow back to the locomotive and out to atmosphere at the engineer's three-way valve, once again requiring good judgment as the front cars released earlier than those at the rear. Indeed, this was an inherent weakness of the system as, "*...application and release times were very long and slow on trains of over five or six cars...Ten to twelve 50-ft cars were the practical maximum for good control with straight-air brakes.*"[35]

FIGURE 6: THE FIRST EMERGENCY AIR BRAKE STOP. AN ARTISTIC VARIATION ON THE THEME DISPLAYED ON THE COVER OF THIS BOOK, THIS ILLUSTRATION IS PERIOD ARTWORK FROM A Story Of America HISTORY CARD. Panarizon Publishing Corp., New York, artist unknown

On September 18, 1868, a four-car passenger train, equipped experimentally with the Westinghouse straight-air brake, was on its first test run. As the train—a local stopper known as the *Steubenville Accommodation*—emerged from suburban Pittsburgh's Grant's Hill tunnel at about 30 mph, engineer Daniel Tate was confronted with the sight of a wagon stalled on a level crossing ahead; the horses rearing in fright from the whistle and the sight of the oncoming locomotive. Tate slammed the brake valve across, and in what has been recorded as the first effective emergency air brake stop in railway history,[36] the train came to a stand only a few feet from the incoherent wagon driver who, by then, had fallen from his conveyance onto the track. Railroad officials aboard the train were thrown from their seats by the violent stop and, infuriated, staggered off the train intent upon admonishing the engineer - and probably Westinghouse, who was riding aboard the

33. The spelling is per PRR timetables. The Pennsylvania Railroad Company's Pan-Handle Division in Ohio was comprised of right-of-way originally belonging to the Pittsburgh & Steubenville Railroad Company and the Chartiers Valley Railroad Company, chartered in 1852 and 1853 respectively. These routes became the Pittsburgh, Cincinnati, Chicago, & St. Louis Railway in 1890, and later, part of the PRR. The term 'Panhandle' is an informal American geographic term for an elongated arm-like protrusion of a state that extends away from the main body. See *The First Air Brake – A Short History* in chapter7 for further details.

34. For further details, refer to the biography of George Westinghouse in chapter 7.

35. *Engineering and Design of Railway Brake Systems* (Wilmerding, PA: The Air Brake Association, 1975).

36. See engineer Tate's story in *The First Air Brake – A Short History* in chapter 7.

locomotive. The worthy officials, though, being fortuitously on-hand to witness the effect of a disaster averted, soon had a change of heart. The incident had served to validate the nascent technology in the most dramatic way imaginable.

A realistic limit was reached when trains eventually became 25 or more cars in length. At this point severe and almost uncontrollable slack action began to occur; compressive "run-in" during brake applications and tensile "run-out" during release. Overshadowing this, however, was an even greater systemic flaw:

In case of a break-in-two (so common in the link-and-pin coupling days) the rear section [of the train] *had no air brakes and the front section might or might not have much brake, depending on the length of brake pipe from the locomotive to the point of undesired parting. Straight-air brake hose couplings were often equipped with check valves which would close in a break-in-two and retain any pressure which had been developed; if the brakes had been applied when the undesired uncoupling occurred!*[37]

Nonetheless, the Straight-Air brake made an undeniable contribution to a safer rail operations environment, and the invention was crucial as the precursor to a system that revolutionized the railroad industry; providing controlled, on-demand braking and thus improving the safety of train operations and permitting higher train speeds.

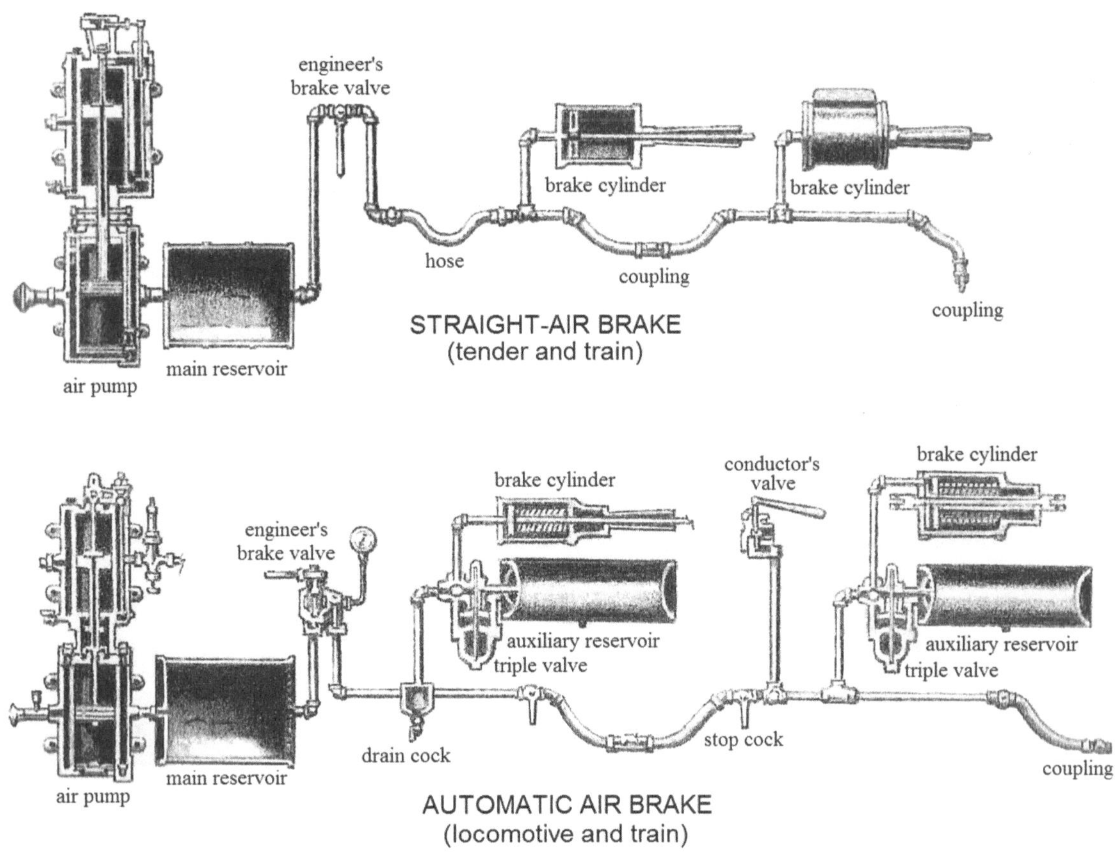

FIGURE 7: EARLY AIR BRAKE SYSTEM COMPARISON. THIS ARTWORK, FROM AN EARLY WESTINGHOUSE AIR BRAKE HANDBOOK, COMPARES THE ORIGINAL STRAIGHT-AIR AND AUTOMATIC AIR BRAKE SYSTEMS, AND DEPICTS—IN SECTIONED VIEW—SYSTEM LAYOUT AND OPERATION. TYPICAL OF AIR BRAKE MANUALS AND PAMPHLETS OF THE PERIOD, IT IS A SUPERB GRAPHIC RENDITION.

37. *Engineering and Design of Railway Brake Systems* (Wilmerding, PA: The Air Brake Association, 1975).

2.

Development Of The Triple Valve In North America

"Let the reader carefully note this, for it was an epoch, not only in the history of the brake, but in the history of land transportation."

—Prout, A Life of George Westinghouse.

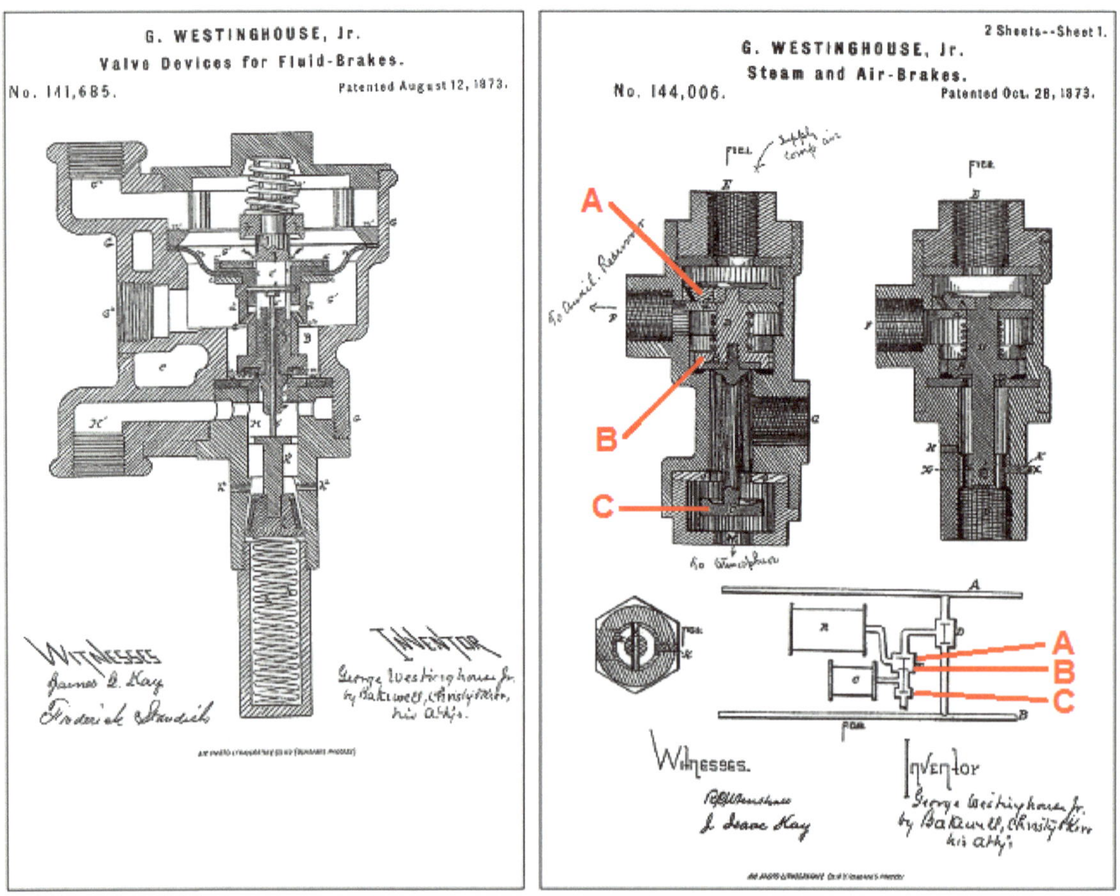

FIGURE 8: EARLY TRIPLE VALVE CONCEPTS

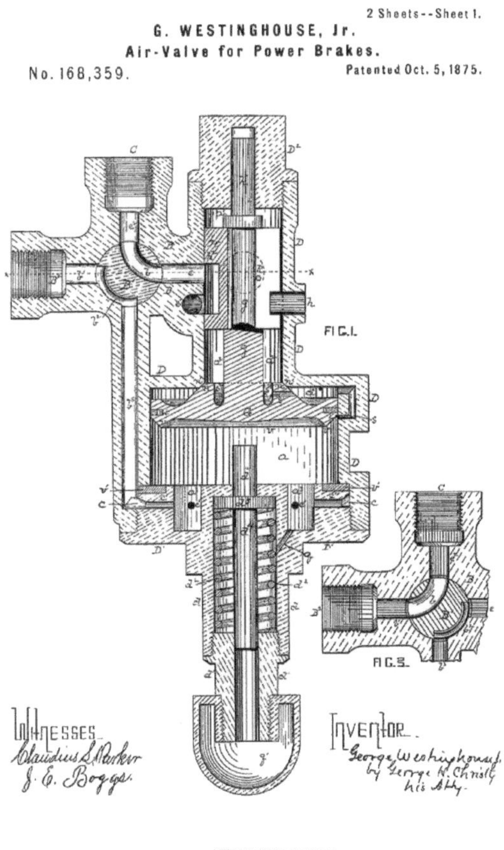

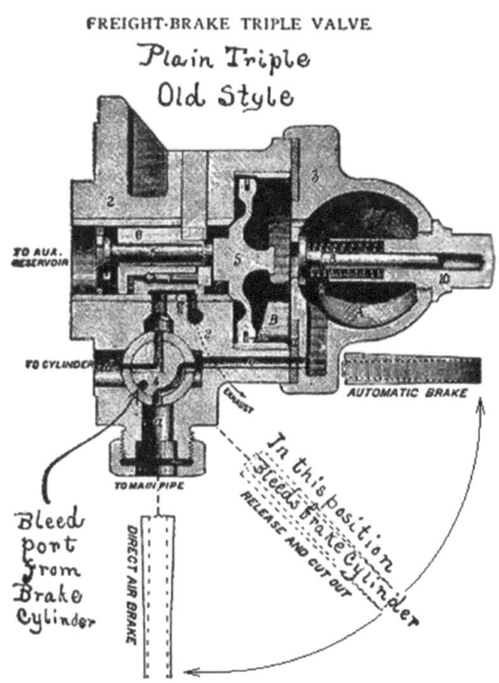

FIGURE 9: (UPPER) AN EARLY PRODUCTION FREIGHT CAR SLIDE VALVE TRIPLE INCORPORATING STRAIGHT-AIR FUNCTIONALITY. (LOWER) IN 1875 WESTINGHOUSE PATENTED AN IMPROVEMENT TO HIS EXPERIMENTAL TRIPLE VALVE BY MODIFYING IT WITH A SLIDE VALVE CONFIGURATION. THIS ITERATION DID NOT PROVE SUCCESSFUL AND WAS ITSELF SOON IMPROVED-UPON.

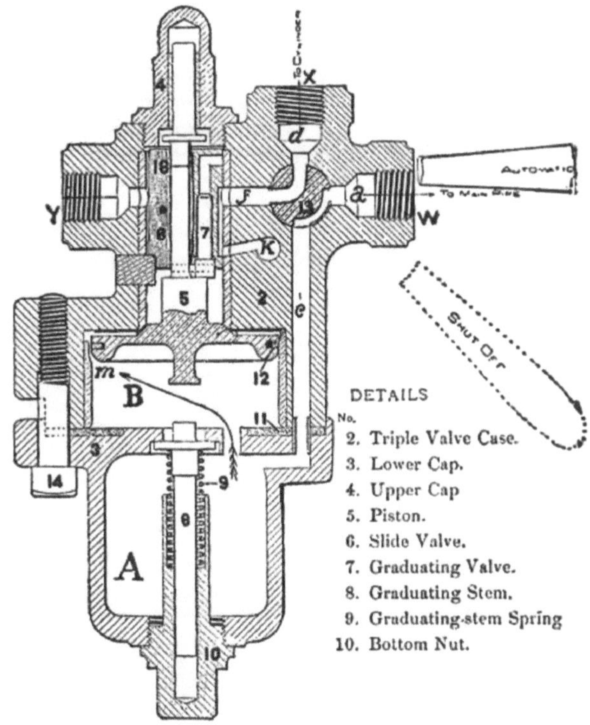

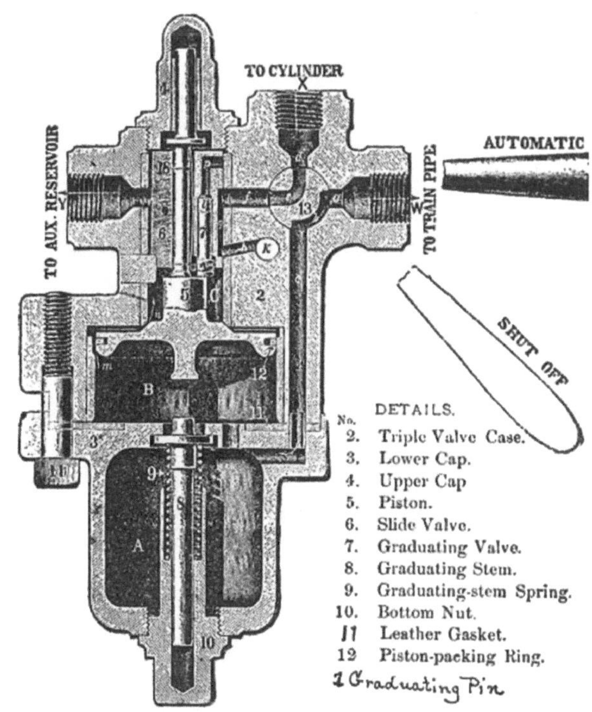

FIGURE 10: THE AMERICAN WESTINGHOUSE PLAIN TRIPLE VALVE. IN AUSTRALIA THIS WOULD BECOME THE ORDINARY TRIPLE VALVE

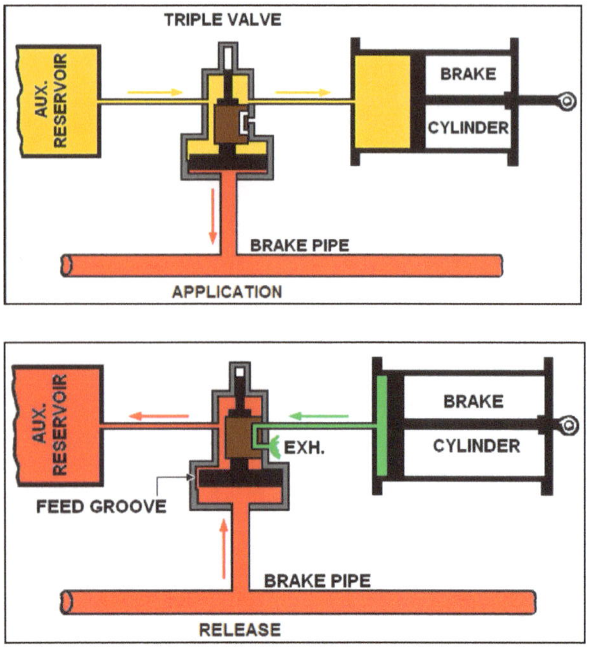

FIGURE 11: THE TRIPLE VALVE SCHEME

The Plain Triple Valve – 1873

WESTINGHOUSE WAS ACUTELY aware of his straight-air brake system's shortcomings, and he soon set about improving its effectiveness. It was clear to him that some degree of 'fail-safe' performance was required, and he and his engineers found the answer by inverting the manner in which the pneumatic result was achieved. The new development employed a brake pipe pressurised in its normal state to hold the brakes released; a reduction in this pressure, whether by accident or design, would now be utilised to effect a brake application. The pressurised brake pipe would provide a supply of air to an auxiliary reservoir located on each car, and this supply would, in turn, be provided on demand to the brake cylinders.

What Westinghouse now needed was a means to control these actions at each individual car, and he delivered it with his crucial invention, the Triple Valve. This new development—known originally as the "plain triple valve"—ensured that in the event of a train separation and concomitant loss of brake pipe pressure, the result would be an immediate brake application individually on all vehicles. Westinghouse had realized his goal of achieving a degree of 'fail-safety' and thus was born our present-day automatic air brake.

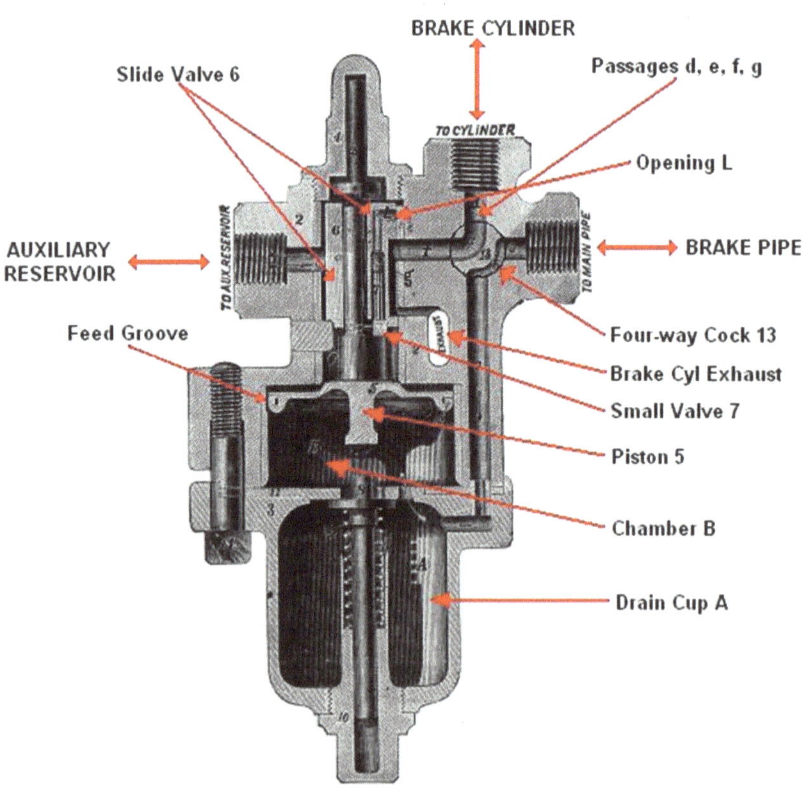

FIGURE 12: THE MODIFIED PLAIN TRIPLE VALVE

The early concept diagrams shown in Figure 8 depict a device that was obviously the foundation of the triple valve, clearly showing the original diaphragm ("flexible annulus")-and-poppet-valve arrangement later discarded in favour of the renowned piston-and-slide-valve assembly. Note that the two devices depicted in the right-hand portion of Figure 8 do not appear to incorporate the capacity for graduating the brake in application. This was a triple valve designed to function only in Emergency braking, with normal Service braking still being achieved via a straight-air brake application. Figure 8 also includes a depiction of the intended two-pipe arrangement described in Westinghouse's biography in chapter 7.

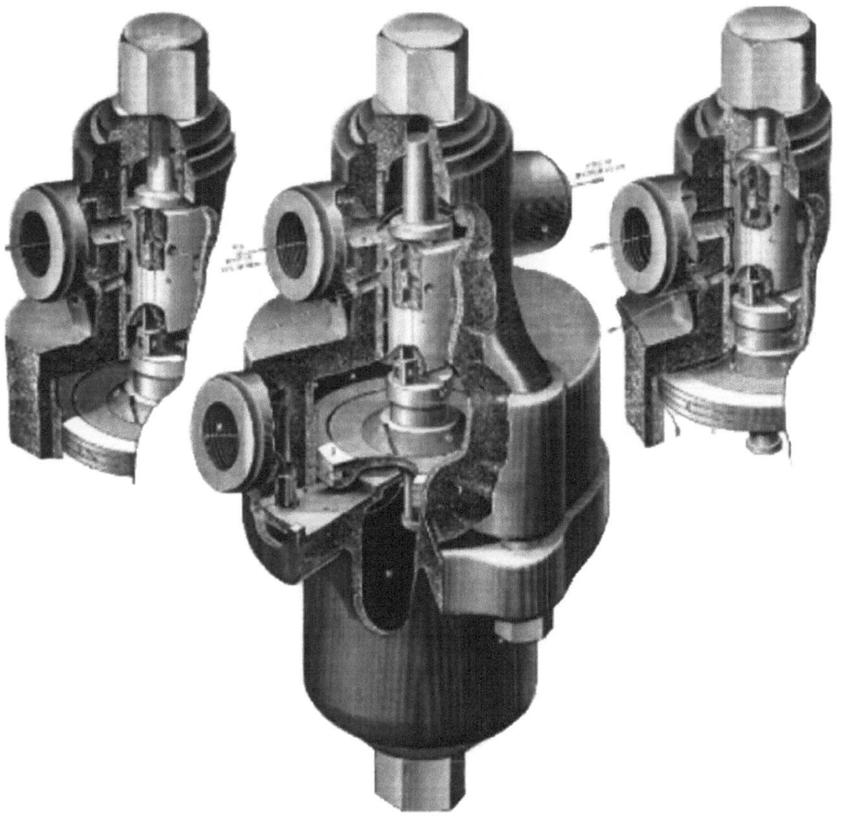

FIGURE 13: THE UK WESTINGHOUSE PLAIN TRIPLE VALVE

Figure 11 is a simplified depiction of the operational intent of Westinghouse's triple valve. Through all of its eventual evolution the device was basically designed to accomplish just TWO things; *apply* and *release* the car brakes in response to a brake pipe pressure signal. To achieve these two things, it was arranged to function thus; in Application, the car reservoir was connected to the empty brake cylinder and was depleted by supplying its (brake pipe) pressure into that brake cylinder. Upon Release, the brake cylinder was permitted to exhaust its pressure and the car reservoir was recharged by the brake pipe.

(Refer to the operating description on the following pages.)

The new brake system, though, had its limitations: *"There was no emergency as such, simply a full application as fast as could be obtained with a wide-open 3-way cock exhausting the brake pipe pressure."*[38] Yet, this brake provided quicker, more flexible control in application and release, which was possible because of the relatively small volume of air required to be taken from or re-supplied to the brake pipe compared to the previous system.

George Westinghouse's original triple valve utilised a poppet valve action, including a rubber diaphragm and leather packing rings. Due to deficiencies inherent in the materials of the day, this design proved problematic because of leakage past the piston and at the valve seat. Thus, in 1873 Westinghouse developed the metal-on-metal piston-and-slide valve assembly—as depicted in Figures 9 and 10—that would predominate until the development (and would remain an essential part) of the AB control valve some sixty years later.

This is the part whose operation gives the brake its automatic action. Those who have opposed this form of brake have made great objection to the complicated nature of the triple valve. But some familiarity with the device shows that it is far from being complex, considering the functions it performs. It is merely a piston-valve carrying a slide valve along with it. The arrangement of the parts of the triple valve is shown [in preceding illustrations].

The triple valve has a piston 5, working in the chamber B, and carrying with it the slide valve 6. Air enters from the brake pipe through the four-way cock 13 into the drain-cup A, and passes to the chamber B, forcing the piston up, and uncovering a small feeding-groove in the upper part of the chamber, which permits air to flow past the piston into the auxiliary reservoir, while, at the same time, there is an open communication from the brake cylinder to the atmosphere through the passages d, e, f, and g. Air will continue to flow into the auxiliary reservoir until it contains the same pressure as the brake pipe.

38. *Engineering and Design of Railway Brake Systems* (Wilmerding, PA: The Air Brake Association, 1975).

Action of the Triple Valve

To apply the brakes with their full force, the compressed air in the brake pipe is permitted to escape, when the greater pressure in the auxiliary reservoir forces the piston 5 down below the feeding-groove, thus preventing the return of air from the reservoir to the brake-pipe. As the piston descends, it moves with it the slide valve 6, so as to permit air to flow directly from the auxiliary reservoir into the brake-cylinder, which forces the piston out, and applies the brakes. The brakes are released by again admitting pressure into the brake pipe from the main reservoir; which pressure, being greater than that of the auxiliary reservoir, forces the piston 5 back to the position shown in the engraving, recharges the reservoir, and at the same time permits the air in the brake cylinders to escape.

To apply the brakes gently, a slight reduction is made to the pressure in the brake pipe, which moves the piston down slowly until it is stopped by the graduating spring 9. At this point, the opening l in the slide valve is opposite the port f, and allows air from the auxiliary reservoir to feed through a hole in the side of the slide valve, and through the opening l into the brake cylinder. The passage l is opened and closed by a small valve 7, which is attached to, and moves with the piston 5, provision being made for a limited motion of these parts without moving the valve 6. When the pressure in the auxiliary reservoir has been reduced by expanding into the brake cylinder until it is the same as the pressure in the brake pipe, the graduating spring pushes the piston up until the small valve 7 closes the feed opening l. This causes whatever pressure is in the brake cylinder to be retained, thus applying the brake with a force proportionate to the reduction of pressure in the brake pipe.[39]

The plain automatic triple valve having only direct release made train handling difficult on long grades. The application would increase with brake pipe leakage and would eventually stall the train. Also some cars might leak off due to difficulty in keeping the piston packing leathers of the day in really tight condition. Thus it was necessary to release and recharge from time to time. Straight-air control was obviously much easier to use since its pressure could be varied up or down to take account of variations in grade as well as curvature and could easily replace brake cylinder leakage. Thus it was common to stop at the top of the grade and change all cars over to direct or straight-air before descending any sizeable grade. For some years a two-pipe system was offered having both a straight-air line as well as an automatic air brake (or 'Main') Pipe. Triple valves were made with an integral double check valve in the brake cylinder passage. However, the two-pipe system with its additional complication never enjoyed very wide acceptance.[40]

Of those early triple valves previously depicted, an interesting feature that incorporated straight-air functionality is the reference to and depiction of an integral *"four-way cock."* This and following references appear to be a mistake or misunderstanding—perhaps a commonly made one—on the part of the responsible writer. The so-called four-way manual valve (nowadays a simple two-way, cut-out cock relocated to the branch pipe) does indeed have four major ports but only three handle positions. If we can accept that by "way" the technical writers mean "position," then it is, in fact, a "three-way" cock. Referring to the previous illustration, operation of the (so-called) four-way cock was described as follows:

When the handle of the four-way cock 13 is turned down, there is a direct communication from brake pipe to the brake cylinder, the triple valve and auxiliary reservoir being cut out; and the apparatus can be worked as a non-automatic brake, by admitting air into the brake pipe and brake cylinder, to apply the brakes. When from any cause it is desirable to have the brake inoperative on any particular car, the four-way cock is turned to an intermediate position, which shuts off the brake cylinder and reservoir, leaving the brake pipe unobstructed to supply air to the remaining vehicles. The drain cup A collects any moisture that may accumulate, and is drained by unscrewing the bottom nut.[41]

This is interesting. In modern-day rail operations, some operators—where slow-speed operations or holding on a grade is required, such as during the loading of a unit train or a ballast-dropping operation—have

39. Angus Sinclair, *The Locomotive Engine: Running and Management*, 17th edition, 1890, Chapter XXI.

40. *Engineering and Design of Railway Brake Systems* (Wilmerding, Pennsylvania, USA: The Air Brake Association, 1975).

41. Angus Sinclair, *The Locomotive Engine Running and Management,* 17th edition, 1890, Chapter XXI.

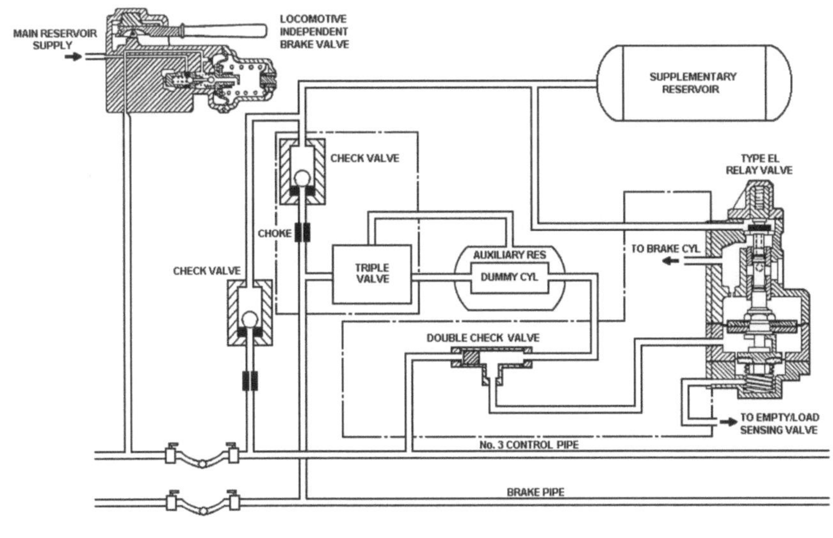

FIGURE 14: EXTENDING LOCOMOTIVE INDEPENDENT BRAKE FUNCTIONALITY. THE INDEPENDENT BRAKE ADAPTOR – SINGLE PIPE BRAKE SYSTEM

arranged for the locomotive independent brake pressure to be separately piped through and to a number of the leading freight cars in the train to provide a controlling brake function supplementary to that of the locomotive (Figure 14).[42] This is not unusual in Australia; certainly, there is little that is new under the sun!

Westinghouse and his engineers had determined that the optimum brake pipe pressure to be carried was 70 psi (accepted as 500 kPa), an operating pressure nowadays little-used in North America but still the standard across Australia. This pressure was sufficient to provide adequate brake cylinder force for the 40-ton cars of the day; was a realistic objective for locomotive compressors given the train lengths then extant; and did not overstress pipe-joints, hoses, couplings, and seals. From early experiments and test installations some fundamental pressure/volume relationships were established that apply to this day. The American auxiliary reservoir volume of the day was approximately 2½ times that of the brake cylinder, thus (using pounds per square inch to keep it simple) a 10 psi brake pipe reduction results in 25 psi (2.5 x 10) in the brake cylinder. Let's assume the driver makes a "split reduction," so a further 5 psi is taken from the brake pipe. In response, the triple valve permits the release of another 12.5 psi (2.5 x 5) of auxiliary reservoir pressure, providing a total of 37.5 psi in the brake cylinder for a total 15 psi brake pipe reduction (note that any total brake pipe reduction greater than around 12 psi or 85 kPa is considered—even today—to be heavy braking). Let's now suppose that we decide to make yet a further 5 psi air brake set. The total brake pipe reduction is now 20 psi, which, for a 70 psi original brake pipe pressure, is Full Service, and a further 12.5 psi of auxiliary reservoir has been supplied, giving a total of 50 psi in the brake cylinder. Since this has left the auxiliary reservoir also at 50 psi, a state of *Full Service equalization* has been attained, and no greater Service brake cylinder pressure is possible. This standard 50 psi maximum design brake cylinder pressure was evolved by the Westinghouse Company from issues germane to economics, the practicalities of air consumption and replenishment times, and the geometry of foundation brake rigging.

Another way of expressing the above convention might be a rationale known as the *"2:5:7 Rule"*. From this code, a 2/7 reduction (20 psi) of the original brake pipe pressure puts 5/7 (50 psi) of that original pressure into the brake cylinder and leaves the same pressure in the auxiliary reservoir. Thus, a Full Service reduction is 2/7 of the brake pipe operating pressure and results in 5/7 of that pressure being admitted to the brake cylinder. This ratio continues to apply even for other brake pipe operating pressures, including 80 psi (New Zealand), 90 psi (Australia's Pilbara region iron ore and most US Class 1 railroads), 100 psi (some US railroads running heavy mineral and grain trains over mountainous terrain), and 110 psi (US passenger train operations). The question might be asked; why then, should not Australia introduce the 90 psi (accepted as 620 kPa) brake pipe pressure into general use since contemporary local equipment is designed for it, and

42. QR employed this feature for a number of years to assist with the loading of coal trains and to provide a static holding capacity on 2% grades in the event that a release and recharge was necessary while standing.

the adoption of railway operating practices paralleling those applying in North America is not a foreign concept here and might offer synergistic benefits?

It should be noted that the above pressure relationships differ slightly in Australia and New Zealand due to different reservoir and cylinder volumes.

System Pressure Relationships - air brake operation: Australia (non-Pilbara region)

	Brake Pipe	Auxiliary Reservoir	Emergency Reservoir	Brake Cylinder
Release	500	500	500	0
Minimum Reduction (50kPa)	450	450	500	125
Full Service BP reduction (143 kPa)	357	357	500	357
Emergency	0	450	450	450

FIGURE 15: BRAKE PIPE-TO-BRAKE CYLINDER PRESSURE RELATIONSHIPS (kPa) WITHIN THE WAGON AIR BRAKE SYSTEM, INCLUDING WHERE EMERGENCY CAPABILITY MIGHT BE PROVIDED

Until 1880 the use of power brakes was confined to passenger service. The straight-air system had been refined to a reasonably advanced degree up to this point; however, the increasing length of passenger trains beyond about ten cars made their handling under braking unwieldy, and with the advent of the triple valve, the conversion of passenger train rollingstock to the new automatic air brake occurred rapidly. The railroads, though, as well as Westinghouse himself, were keen to get the air brake onto freight trains, and tests soon confirmed its value, especially in mountain-grade braking. One important breakthrough, though, would critically affect the migration of air brakes from passenger to freight rollingstock:

Development of the combined cylinder, reservoir and triple valve in 1879 made practical the application of air brakes to freight cars. Separately mounted devices, common on passenger cars, were not economical on freight cars where fairly heavy slack action was frequent and car body deflections were high.[43]

The first commercial application of air brakes to freight cars occurred in this year; to Denver & Rio Grande Western RR rollingstock used over heavy grades; the intent being to avoid the requirement for hand-braking of individual cars in heavy-grade operation. This particular fitment utilised both a brake pipe and straight-air pipe (as previously described), the latter to permit the variation of braking pressure on the steep grades. The Westinghouse Company soon developed a brake cylinder pressure-retaining valve, thus permitting

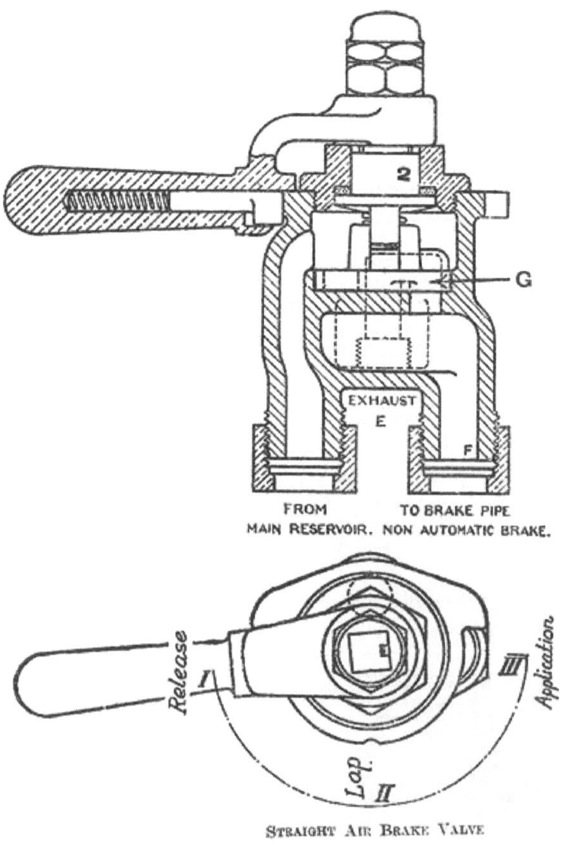

FIGURE 16: EARLY BRITISH STRAIGHT-AIR BRAKE VALVE

43. *Engineering and Design of Railway Brake Systems* (Wilmerding, PA: The Air Brake Association, 1975).

elimination of the straight-air pipe and therefore the requirement for the triple valve to incorporate straight-air functionality.

As the US railroad network expanded, railroads with ever longer and heavier train consists were operating over ever more steeply graded territory and being required to interchange cars with each other. Accidents had been occurring with mounting frequency, and it was obvious that some kind of effective power brake system should be established to a suitable uniform standard. In 1886 the Master Car Builders Association—forerunner of today's Mechanical Division of the Association of American Railroads—designated a committee to conduct a series of competitive tests of the various brake system inventions that were then being offered. Inventors would be required to apply their systems to a fifty-car freight train.

The tests were held the following year on the West Burlington Hill in Iowa (a stiff grade out of the Mississippi River valley), and the Westinghouse Company—having previously submitted their straight-air brake for scientific comparison with others in Great Britain[44]—now offered their recently developed automatic brake. The MCBA committee reported that the operation of this system met the ordinary requirements of Service work. They also reported, however, that the action of the Westinghouse brake system was unsatisfactory in Emergency applications because of the slow transmission of the pneumatic signal from the front to the rear of a long train.

FIGURE 17: THE BURLINGTON TEST TRAIN

The Westinghouse automatic air brake certainly permitted an effective brake application to be established on the entire train; indeed:

...it quickly became possible to exert 3 to 5 times the force input into the brake rigging than with handbrakes.[45]

But there still existed a problematic time difference between the build-up of car brake cylinder pressures at the front and rear of the train – a time difference that had the potential to cause (and did cause) severe compressive slack action; a grave danger to personnel, lading, and rollingstock. Thus, the MCBA decided the test would be repeated the following year, and Westinghouse set about finding a means of

44. Charles Francis Adams Jr., *Notes on Railroad Accidents* (New York: G. P. Putnam's Sons, 1879), chapters XX and XIX.

45. *Engineering and Design of Railway Brake Systems* (Wilmerding, PA: The Air Brake Association, 1975).

increasing the emergency speed-of-action of his system along the length of the train. Of this second series of tests, one report states the following:

Six competitors took part...one brake was operated by electricity alone, a second by compressed air alone, a third by electricity and vacuum, while Westinghouse and one other manufacturer contributed brakes combining compressed air and electricity.[46]

Only an electrically-controlled air brake, Westinghouse thought, would provide the rapidity of serial action in a long train, and his first air brake system reflected this belief. His electric-over-air system of 1883 (electro-pneumatic for Emergency applications only) was relatively simple and had the theoretical benefit of the built-in redundancy of pneumatic backup of any possible electrical failure. Additionally, his recent improvements to the triple valve had more than doubled the speed-of-transmission of an Emergency brake application from the front to the rear of the train from 17 to 6 seconds. This fifty-car train was also fitted with the newly developed Janney automatic coupler and draft gear arrangement replacing the obsolete link-and-pin coupler.

The inventor, though, was not entirely satisfied with the electrical augmentation of his air brake; unsurprisingly, he did not wholly trust the reliability of the electrical equipment and the required connectors. This first electro-pneumatic freight car brake was tested (and approved by the MCBA) but never sold. The main competition in the Burlington tests of 1887, though, came from a young American expatriate based in Berlin, Germany—one Jesse Fairfield Carpenter—who had also developed an electro-pneumatic brake. At a forum organized by the American Society of Mechanical Engineers in 1937 to commemorate the career and achievements of George Westinghouse, Mr W.W. Nichols,[47] who had been the ASME representative engineer at the 1887 tests, stated that the Westinghouse brake was in fact defeated by the Carpenter version and that this had caused some anguish in the Westinghouse camp. Thus, although the problem of excessive slack action had not yet been controlled to the satisfaction of the MCBA, their eventual report—expressing their satisfaction with the Carpenter system—was:... *generally favorable to a brake operated by air, having valves actuated by electricity.*[48]

FIGURE 18: GEORGE WESTINGHOUSE AT WORK IN HIS DRAFTING OFFICE, POSSIBLY WITH WALTER TURNER

This comment amounted to an indictment of brakes operated by air alone, but Westinghouse by now believed there was a solution and was unperturbed. With characteristic industry he set about perfecting his system. Deciding that the electric factor must be eliminated because of its unreliability, he concentrated on further improving the triple valve and also embarked upon the development of a new friction draft gear that, by the early 1900s, would become the principal system on US freight cars.

As for Jesse Carpenter, his German company would evolve following his death in 1901 and—as Knorr-Bremse AG—eventually become a major air brake industry player outside Europe as well, even to the point of taking over the Westinghouse Air Brake brand. An irony, indeed!

46. Francis E. Leupp, *George Westinghouse: His Life and Achievements* (Boston: Little, Brown & Co., 1918), chapter VII.

47. William Wallace Nichols (1860-1948)—who at the time of the forum was Assistant to the Chairman of the Allis-Chalmers Manufacturing Co. Inc., NY—had been in charge of the CB&Q dynamometer car during the tests. A mechanical engineer specializing in steam and hydraulics, he graduated with a Masters degree from the Sheffield Scientific School of Yale University, served as superintendent of telegraph for the Chicago, Burlington & Quincy Railroad and was later Superintendent of the Chicago Telephone Company. He temporarily retired from business to instruct at Yale and supervised the technical training functions of the ASME. He subsequently became assistant manager of the Baltimore Copper Smelting & Refining Company and Vice-President of Allis-Chalmers. In 1914 he was elected Vice-President of The American Manufacturers' Export Association, and through that position, President of The American Industrial Commission in France (1916) and the American Industrial Mission to Mexico (1924) both of which he lead.

48. Francis E. Leupp, *George Westinghouse: His Life and Achievements* (Boston: Little, Brown & Co., 1918), chapter VII.

The Introduction of Quick-Action – 1889

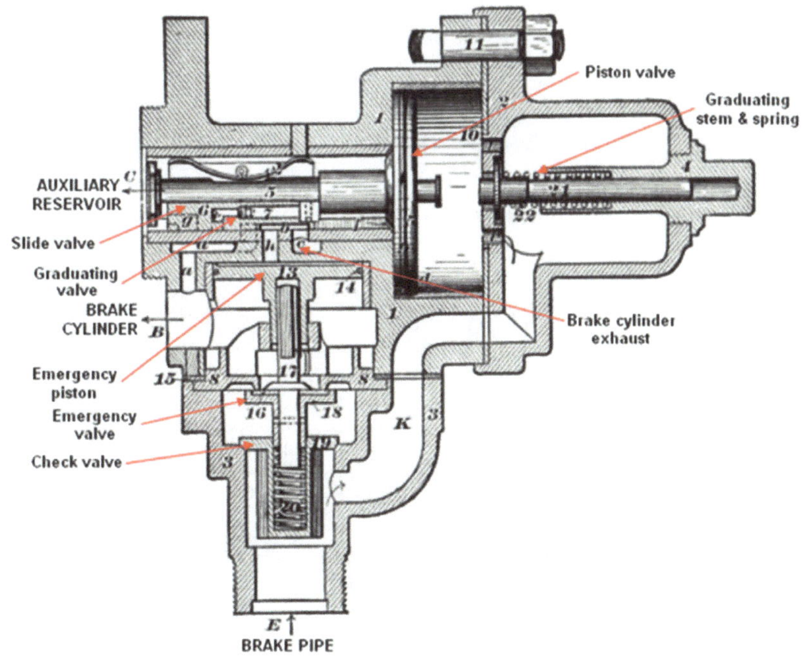

FIGURE 19: H-TYPE QUICK-ACTION TRIPLE VALVE

Freight train lengths continued to increase, and with consists of fifty cars or more, an Emergency brake application was effective so much sooner toward the front part of the train than at the rear that severe compressive slack action was often still occurring. And when it did, derailment and a train wreck often resulted. Westinghouse had to find a way to increase the rapidity of Emergency serial action of his triple valve along the length of the train. Revisiting his drawing board, he and his staff redesigned the device and within three months in 1889 solved the problem.

The triple valve was modified to improve internal air flow, including increasing the brake pipe diameter from 1 to 1¼ inches, and the resulting Quick-Acting version completely reversed the previous MCBA verdict. The propagation of Emergency serial action along a fifty-car train (2 000 ft or about 670 m) had been reduced from 6 to 2½ seconds, and the degree of train handling precision provided to the engineman increased manyfold. Subsequent tests and field trials validated the improvements, and it can be noted that this propagation rate has scarcely been bettered to this day.

The MCBA committee, who had provided observers to these trials, suggested—without overtly endorsing any particular brake system—that a freight train brake needed to fulfil the following conditions:

- A 70-lb (500-kPa) brake pipe pressure and an 8-lb (56-kPa) minimum reduction;
- Shock-free operation on a fifty-car freight train;
- 200-ft stopping distance for an empty fifty-car freight train running at 20 mph (32 km/hr) on level track;
- The maintenance of a constant speed of 15 mph (24 km/hr) down a grade of 53 ft/mile (1 percent) with no more than 5 mph (8 km/hr) variation plus-or-minus during such descent;
- Brakes capable of application, release, and graduation on the entire train by the engineer without the assistance of brakemen or a conductor;
- Hose couplings to be the standard Westinghouse couplings. (NB: The Westinghouse Company also invented the interconnecting clutch type of *palm* or *gladhand* air brake hose coupling—standard even today—whereby the design of the connector provides a nondestructive pull-apart capability.).

Since the Westinghouse air brake system was the only one that could fulfil these requirements, it was—by the action of the MCBA committee—unofficially acknowledged as the pre-eminent system, and US railroads began installing it wholesale.

To improve upon the previous model, the Type-H or Quick-Action Triple valve was designed to take some of the air from the brake pipe during an Emergency application so as to hasten the reduction in brake

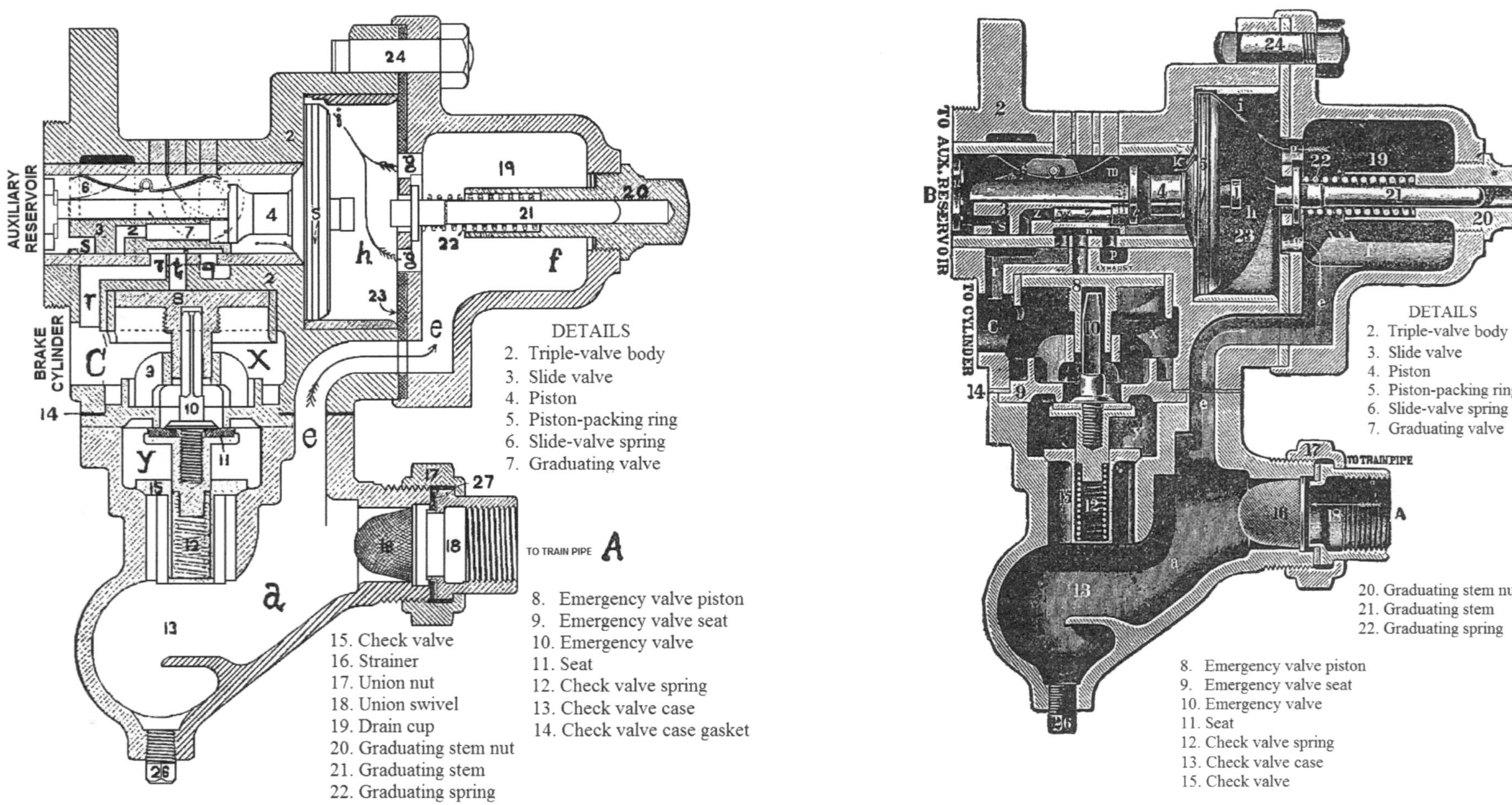

FIGURE 20: AMERICAN WESTINGHOUSE QUICK-ACTION PASSENGER CAR TRIPLE VALVE

pipe pressure and reduce the pressure from the face of the adjacent triple valve piston fast enough to cause it, in turn, to assume its Emergency position. Each triple valve would then render the brake pipe reduction fast enough to cause the next one to go into Emergency quicker, resulting in the brakes applying rapidly-enough throughout the train to prevent the development of severe slack action.

When the brake pipe pressure is reduced fast enough to cause an Emergency brake application, the resultant slide valve movement provides a large opening for the supply of auxiliary reservoir pressure to the brake cylinder and to the triple valve emergency piston, which unseats the emergency valve, permitting brake pipe pressure to also flow into the brake cylinder. This initiates the 'serial action' whereby each triple valve:

...makes a local pressure change [i.e. reduction] *which helps to transmit the function on to the next car. This ensures a much faster propagation of the function through the train than would occur if all the brake pipe air has to flow forward to be exhausted through the* [engineer's] *brake valve. This type of "serial action" enabled the 50 to 80-car trains of the 1890 to 1910 period to be safely and adequately handled if Emergency applications were made. While this type of quick-action emergency transmission is no longer used, it does illustrate the serial action principle.*

Another important air brake principle is illustrated here: when necessary to reduce brake pipe pressure, allow some of this air to flow into the brake cylinder and obtain the dual advantage of faster brake pipe reduction plus higher brake cylinder pressure. The quick-action triple valve produced about 10–12% higher Emergency brake cylinder pressure than for Full Service.[49]

In the United States at the time, other manufacturers thought they could improve upon Westinghouse's design and that such improvements warranted their own patents. Westinghouse's original automatic brake patent of 1872 had expired by 1893, and other companies—including an upstart newcomer called the New York Air Brake Co. (NYAB)—were looking for some of this new air brake business. It was not NYAB, though, who were the focus of the Westinghouse Air Brake Company's attention when, in a legal action brought by them for patent infringement,[50] the judge—in his report—stated (in part):

This patent [for the Westinghouse Triple valve], *although it introduced a novel feature into the art* [of air brake design], *does not seem to have been entirely successful in its practical operation, since in October of the same year an improvement was patented (No. 376,837), with the object of still further increasing the rapidity of action. As observed by the district judge in this connection, "...the success of this improved device (No. 376,837) has demonstrated that the invention, by which the further traverse of the triple-valve piston beyond the extent of the traverse required for the ordinary application of the brakes is made to admit a large volume of train pipe air directly to the brake cylinder, was one of great importance. The proofs show that a quick-action automatic brake, which would give the results which this brake has accomplished, was eagerly sought-after by inventors and car builders, and all had failed until Westinghouse discovered that it could be done by this mode of operation."*

These Quick-Action triple valves could, if required, be combined with (by attaching to one end of) the car auxiliary reservoir, and they became the forerunner of the eminent Type-K triple valves that introduced two innovative features, Quick-Service and Retarded-Recharge (the latter function also known as "retarded release" and "uniform release").

In the United States, Westinghouse had expanded into the railroad signalling industry by organizing the Union Switch & Signal Company, where devices based on his inventions and the patents of others were designed to control the increased train speed and operational flexibility made possible by the invention of the air brake. By 1905 more than two million freight, passenger, mail, and baggage cars as well as 89,000 locomotives were equipped with the Westinghouse Quick-Action automatic brake.

Perhaps we might use as a concluding comment on the advent of Quick-Action, yet another snippet gleaned from Prout's biography; that of G.W. Rhodes, Superintendent of Motive Power at the Chicago, Burlington & Quincy RR and chairman of the Master Car Builders Committee at the time of the Burlington trials:

49. *Engineering and Design of Railway Brake Systems* (Wilmerding, PA: The Air Brake Association, 1975).

50. US Supreme Court, *Boyden Power-Brake Co. v. Westinghouse*, 170 US 537 (1898). http://laws.findlaw.com/us/170/537.html.

MrWestinghouse's wonderful optimism and confidence in the principle of air power after the early failures were very marked. After the first collapse, in rapid succession, three different triples were invented. Finally came the gem, and then, greatest of all acts, exhibited all over the country in operation on a fifty-car train. It made stops with no shock in the fiftieth car, not enough jar to upset a glass of water, a marvellous condition when one considers the wreck and disaster that used to take place in the fiftieth car.

The Introduction of Quick-Service and Retarded Recharge – 1908

KC Freight Car Brake

FIGURE 21: K-TYPE TRIPLE INTEGRATED WITH CYLINDER AND PISTON

By the early 1900s—some thirty years after the introduction of the original straight-air brake—the trains were still getting longer, and the limits of effective operation of the current air brake equipment, especially with respect to the Quick-Action triple valves, were again being reached. The propagation[51] of the brake pipe reduction along the train during Service applications was again too slow for current train consists, and harsh slack action resulted for the same reason as had happened before the adoption of the Quick-Action Emergency triple; increased train length meant the brake application was transmitted slower than the natural run-of-slack.

Quick-Service

The practical solution was to extend the Emergency *quick-action* principle to include Service applications, hence the term "Quick-Service." The Type-H triple was redesigned with an additional position to which the slide valve would be moved during a normal Service brake pipe reduction, and that provided a positive additional reduction in brake pipe pressure locally as each car sped up the transmission of the Service reduction through the train. Graduating valve movement at almost the same time opened the slide valve service port to permit auxiliary reservoir air to also flow to the brake cylinder. From these two actions, a minimum brake pipe reduction would produce a very effective 6 to 7 psi (42–50 kPa) brake cylinder pressure, which was sufficient to cause all brakes to apply within a few seconds on an eighty- to one hundred-car train.

Retarded Recharge/Uniform Release

A debilitating aspect of air brake releases was the fact that the greater recharge pressure at the front of the train meant that head-end triple valves tended to release well ahead of those further to the rear. As well as being physically "first in the queue," the wholesale use by the leading triple valves of the brake pipe recharge air pressure left less of that pressure available for the rearward valves. Thus, slow-speed releases were problematic, risking violent slack run-out and a parted train. The type-K triple valve addressed this problem and greatly assisted locomotive drivers with their train handling. Here's how it worked:

51. Here, the reader is asked to take time to comprehend the contextual distinction between the terms *transmission* and *propagation*. 'Transmission' refers to the rate at which the fluid content of the brake pipe is capable of sending a pressure-variation signal along its length; this generally being a function of the laws of physics and the design of the brake pipe as applied to rollingstock. 'Propagation' refers to the rate by which the normal functioning of advanced triple and control valves (those providing Quick-Action/Service or Accelerated Application/Release) can improve the overall transmission time by conducting the pressure-variation signal more rapidly along the brake pipe, using each triple valve to "trigger" the next, which enables the pressure signal to run ahead of the natural transmission time. These terms are often used interchangeably, but in the interests of technical accuracy ought not to be.

During a brake release, brake pipe pressure initially increases relatively quickly near the point of recharge, traditionally the locomotive at the front of the train. Because auxiliary reservoir pressure is low (the auxiliary reservoirs having discharged their pressure into the brake cylinders) a large pressure differential exists across the piston of the head-end triple valves. The slide-valve assembly is thus driven to an extreme Release position, this movement being controlled and dampened by contact with the *retarded recharge spring*. At this position, the slide valve only partially opens both the brake cylinder exhaust port (providing a slower rate of release) and the auxiliary reservoir charging port (providing a slower rate of recharge of auxiliary reservoirs). Further back in the train, the situation differs. Here, the brake pipe recharge pressure is less and the triple valve pistons are subject to a less extreme pressure differential. This results in these rearward slide valves providing greater openings for brake cylinder exhaust and auxiliary reservoir recharge. David G. Blaine[52] put it in these terms:

This allows more brake pipe air to flow toward the rear by preventing the cars in the front of the train from "robbing all the available air." In the rear of the train, brake pipe pressure rises more slowly and does not create the differential necessary to compress the retarded recharge spring. Thus charging or recharging is much more uniform throughout the train.[53]

It could therefore be stated that "retarded recharge promotes uniform release," the benefits of which cannot be overstated. The Type-K triple valve increased the rate of brake pipe propagation to 125 ft/sec (38 m/sec) during a Service application and 680 ft/sec (207 m/sec) in Emergency.

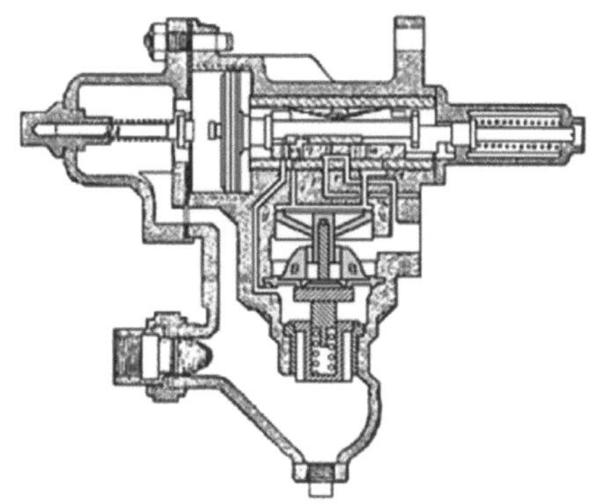

THE WESTINGHOUSE TYPE-K TRIPLE VALVE

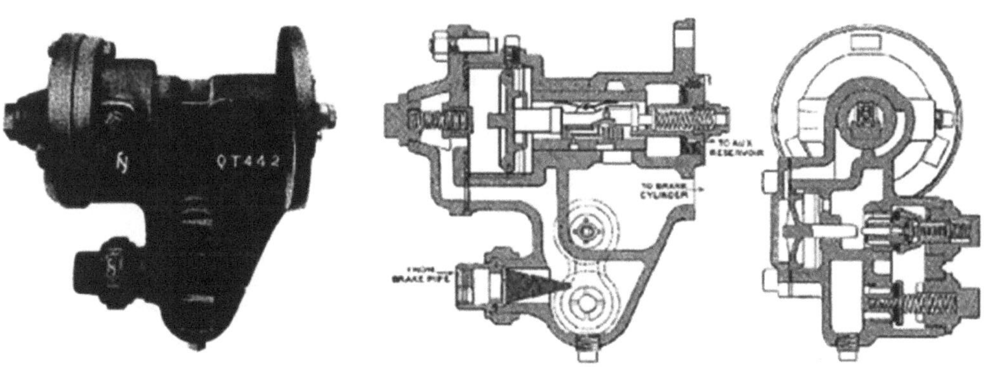

THE NEW YORK AIR BRAKE K-6 TRIPLE VALVE

FIGURE 22: K-TYPE TRIPLE VALVE — WESTINGHOUSE AND NYAB VERSIONS

52. Refer to chapter 7, *Biographies*, for further information.

53. David G. Blaine, *Modern Freight Car Air Brakes* (Simmons-Boardman Publishing Corp., 1979), 55–65.

The Automatic Straight Air Brake Company

In 1910 the Interstate Commerce Commission ordered that from September of that year, 85 percent of the cars in a train must have air brakes controllable by the locomotive engineer. Anyone with an entrepreneurial eye could see a business opportunity here, and by 1915 another would-be air brake start-up—the Automatic Straight Air Brake Company—began promoting a graduated release air brake for which they claimed certain desirable features not incorporated in other brakes then in use. These were claimed to be:

- an Emergency application being obtainable at any time;
- graduated release with brake cylinder pressure uniform throughout the train and maintained against leakage; and
- the conduct of a terminal brake test being impossible until the train was fully charged, thus minimising the possibility of a train departing with an inadequate brake capability.

The Westinghouse Air Brake Company duly rebuffed these claims. Moreover, the Association of American Railroads contended that brake systems then in use were adequate and moved to adopt the K triple and associated equipment as the standard for all new freight rollingstock built after January 1919.

By sad coincidence, there occurred—in July 1919—a devastating rear-end collision between two passenger trains on the New York Central Railroad at Dunkirk, New York, resulting in the death of twelve people, including the engine crew of a following train and injury to 178 others.[54] The cause was found to be loss of braking capability on the second train due to the brake pipe angle cock between the locomotive tender and the first passenger car having been closed, probably by the action—suspected to be intentional—of a trespasser riding the coupler. An official of the Automatic Straight Air Brake Company took advantage of the moment to comment as follows in a letter to the editor of *Railway Age* magazine:

Nearly forty-seven years have passed since the automatic air brake came into use as a means for controlling the motion of railway trains, and for over thirty years it has been in general use on both passenger and freight trains, yet it still retains that fundamental element of danger - the closed angle cock.

With the automatic brakes now in general use [the worthy official is, of course, referring to Westinghouse equipment]*, the train may be charged, the brakes tested and everything put in working order before leaving the terminal, and yet before or after the train departs, an angle cock may be closed between the engine and tender, or elsewhere in the train, without the knowledge of the engineman or of any member of the train crew and thus deprive them, at the critical moment, of the only means on which they rely for safely controlling the train. This may be caused accidentally through something falling upon the angle cock, by a tramp accidentally or maliciously, or by an employee forgetting to open a closed angle cock. There have been many cases of the last.*

54. At 2:20 on the morning of July 1, 1919, eight passengers, the baggageman, the engineer and the fireman of train No. 7, and a trespasser were killed; seventeen passengers were injured; and many others sustained minor injuries when the 2nd division of westbound passenger train No. 41—standing at the Dunkirk station—was struck in the rear by westbound passenger train No.7 at about 45 mph, wrecking the rear sleeping car of No. 41. The wooden baggage car of No. 7 was crushed, and the steel car behind it was turned across the track and its side torn open. The passenger fatalities occurred in this car and the rear sleeper of No. 41. The trespasser was riding between the tender and the leading baggage car. Train 41 had been standing at the station about 47 minutes because of a hot journal. It was properly protected by a Home signal 2 000 feet back and a Distant signal 4 125 feet farther back; and the flagman was back 1 700 feet with a red light. The engineer of No. 7, F. L. Clifford, had sounded the whistle for hand brakes, indicating that he had attempted to reduce speed but that his air brakes had not acted properly. Alive when taken from his engine, Clifford confirmed that the brakes did not hold, and an examination of the wreck disclosed that the rear tender angle cock was shut. It is believed that the trespasser, identified as Charles Schiller, had been riding there and had closed it, intentionally or otherwise. Schiller was a resident of Dunkirk, which tends to confirm the supposition that, finding the train was likely to pass through the town without stopping, he decided to apply the brakes and assumed that he could do this by closing the angle cock.

No more convincing evidence of this truth is required than that furnished by the investigation of the recent rear end collision at Dunkirk, resulting in the loss of many lives, serious injury to a score or more of passengers and a heavy loss in property damage, due to an angle cock being closed between the engine and the first car.

It is reported as a result of the investigation that the engineman of the second train shut off steam probably half a mile east of the distant signal, and that he began to sound the whistle signal for brakes at about the time the locomotive passed the distant signal, which was some 6,000 ft. east of the point at which the collision occurred. The evidence indicated quite positively, therefore, that the angle cock on the rear of the tender must have been closed before passing the distant signal, or at any rate but very little later. The engineman probably realized he was powerless to stop his train fully a minute and a half before the collision occurred.

It has remained for the Automatic Straight-air Brake Company to provide against this menace to the safety of the movement of passenger and freight trains by the development of a brake designed on superior principles, in which, when an angle cock is closed, the brakes apply within a few seconds. On the other hand, a sluggish feed valve does not cause the brakes to apply,

The application of the A.S.A. brake depends merely upon a brake pipe reduction being made, and is positive and precise regardless of the rate of reduction, which is such a vital element in the operation of other types of brake equipment. With brake pipe leakage of, say, two pounds per minute, which is a reasonable assumption for a train of that character, it may be positively asserted that had A.S.A. brake equipment been in use on Train No. 7, the train brakes would have been applied before the collision occurred, with the result of, if not entirely preventing the accident, at least greatly mitigating the severity of the collision and reducing the extent of the disaster. Had the angle cock been turned as far back as the water station pans the A.S.A. brake would have stopped the train without any collision whatever. It is obvious that the improved brake is an urgent necessity for the protection of lives and property in the daily operation of the railroads.

The A.S.A. Brake Co. petitioned the ICC alleging that current brake equipments were unsafe, and—responding to the controversy—the ICC, in May 1922, investigated the adequacy of power brake systems then extant and whether improvements were needed. The extensive hearing ran through the end of the year, and in 1923 the ICC tested a new A.S.A. brake on the Norfolk & Western Railroad. The tests exposed application and release failures as well as undesired releases and unacceptable slack action. According to the AAR report to the ICC, the performance of the Automatic Straight-Air Brake product was suboptimum, and the system was undesirable.

The Introduction of Graduated Release – 1908

Early in the development of the triple valve, it had become obvious that the differences between freight and passenger cars (in this context principally size, since passenger cars were often heavier when empty and required larger brake cylinders) required triple valves to be optimised for one form of service or the other. Thus, the Type-H triple valve was remodelled to become slightly larger and was reborn as the Type-P. Used with the newly developed "high-speed reducing valve,"[55] it helped to improve passenger train handling, especially during stops.

The degree of difference between passenger and freight car triples had increased to the point in 1908 that the two styles diverged with the introduction of the Type-L that incorporated graduated release capability. This feature—not practical for freight train use with brake pipe lengths in excess of 1,500 ft (500 m)—has since remained a basic function of all US passenger car control valves. Apart from graduated

55. Introduced around 1891, the "high-speed" brake was applied to fast passenger trains. It consisted of an H-type triple valve with the addition of an automatic brake cylinder reducing valve. By means of this composite device, brake cylinder pressure was not permitted to exceed the ordinary Service application limit, although a considerably higher cylinder pressure was provided at the beginning of an Emergency application, this higher pressure being automatically reduced to a safer limit as train speed was reduced.

release, the L triple valve and associated equipment provided Quick-Service, higher Emergency brake cylinder pressure, and Quick-Recharge. The Type-L was sometimes referred to as the *pipeless triple* since it was removable without breaking any pipe connection.

With the introduction of the PC equipment developed for passenger trains with cars weighing 130,000 lb (59 metric tonnes), the range of operating features of the apparatus was greatly expanded, and the term *control valve* came into use. From this point on, it would remain the American name for operating valves of both passenger and freight cars.

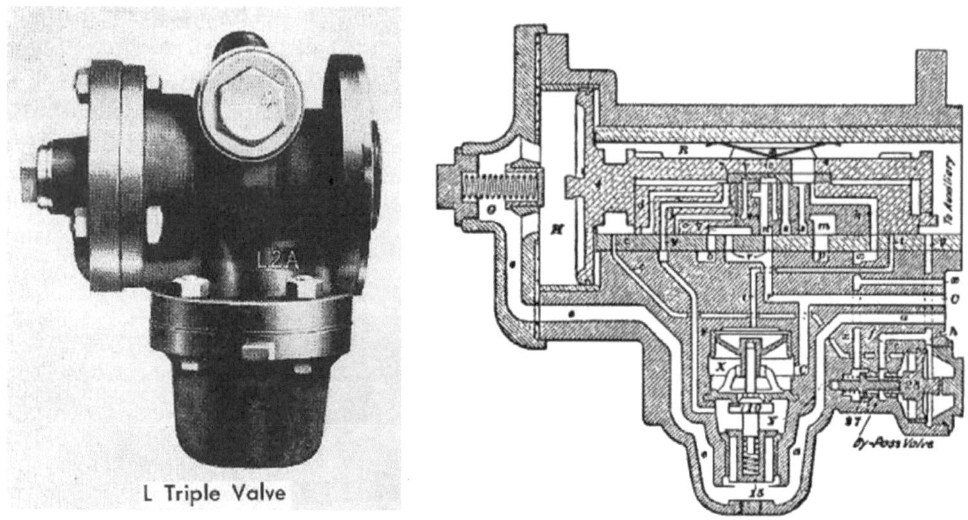

FIGURE 23: THE WESTINGHOUSE L-TYPE TRIPLE VALVE

Graduated versus Direct Release (Europe v. America)

In the United Kingdom and elsewhere in Europe and Asia, the triple valve had evolved somewhat differently. From its earliest days, the Westinghouse Air Brake Co. had been active in Europe—the British Westinghouse subsidiary having been formed in 1899—and European equipments evolved, logically, from the early Westinghouse influence. The first practical, modern air brake system designed on the continent was introduced by an American, Jesse Fairfield Carpenter—some nine years junior to Westinghouse—who in 1879 was residing in Berlin and involved in railway air braking in response to the developing needs of the monolithic Royal Prussian State Railways. In 1883, with the benefit of some patented technology obtained courtesy of German engineer Friedrich Glaser,[56] Carpenter won a competition to design a standard express passenger train brake with a quite sophisticated dual-chamber, gradual-release system incorporating a brake block adjuster as well as automatic brake line coupling. Having concluded a long-term supply contract with the Prussian State Railways that would enable him to build a manufacturing facility, Carpenter poached from them a 24-year-old graduate—who had started his working life as a railway apprentice—to be his chief engineer. After Carpenter's death in 1901, this young man would move to acquire the company; his name was Theodor Georg Knorr.

56. Friedrich Carl Glaser (1843-1910) was the founder, in 1877, and editor of the famed journal *Commercial and Construction Annals* ('Glasser's Annals') and co-founder of the Association of German Machine Engineers; now the German Mechanical Engineering Company (Deutsche Maschinentechnische Gesellschaft). Glaser studied mining and metallurgy, and from 1861 to 1869 worked in the inspection service at the Chemin de Fer du Nord in Paris, excelling from 1870 to 1871 in the field of railway services through successful pioneering work for The Royal Commissariat and as a patent attorney.

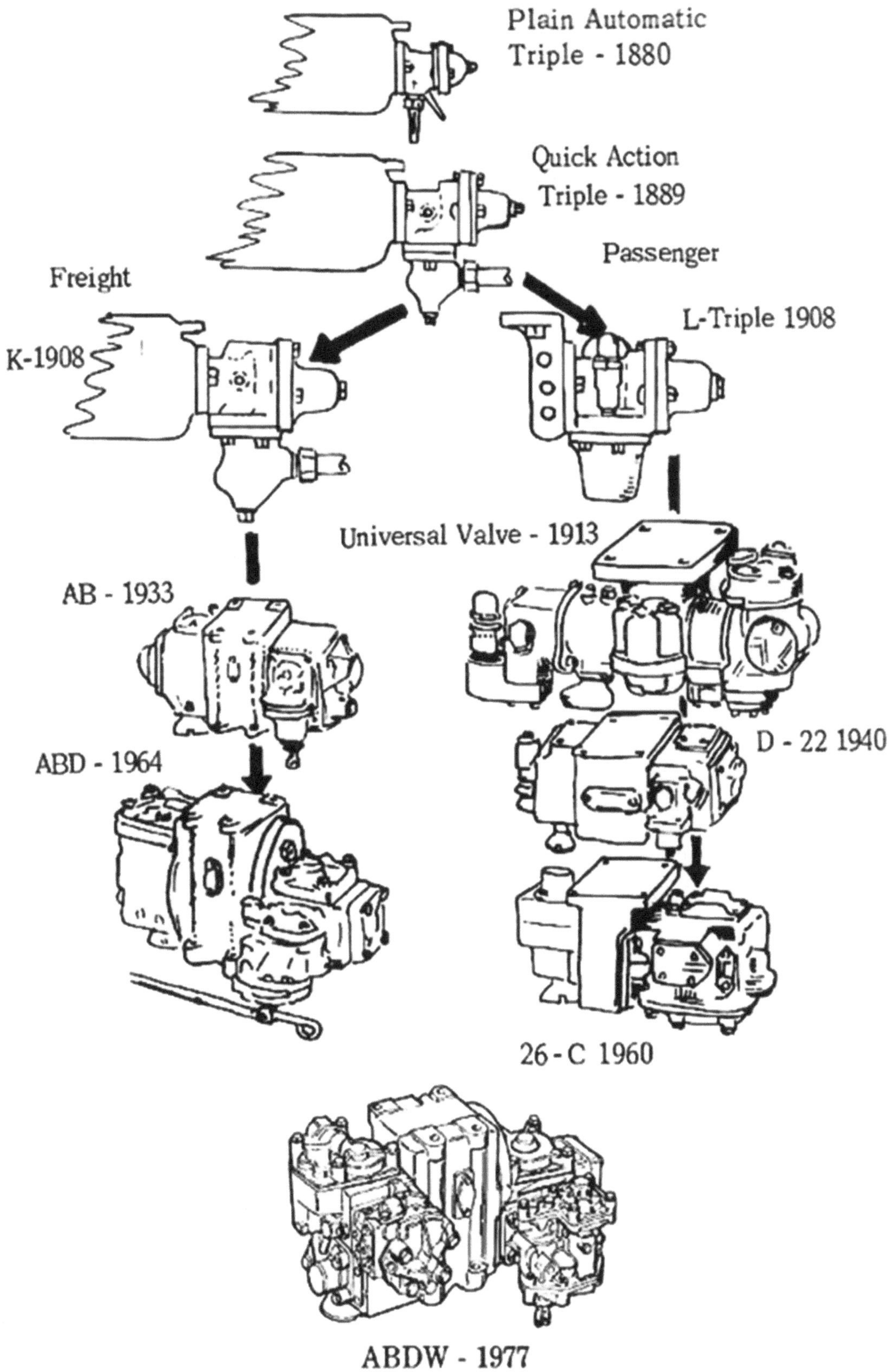

FIGURE 24: THE INCREMENTAL EVOLUTION OF THE AIR BRAKE OPERATING VALVE IN NORTH AMERICA. *The Air Brake Association*.

Carpenter's system overcame competing Westinghouse and vacuum systems due not only to its simplicity, clarity of design, and reliability, but also—decisively—to its "gradual release" capability.[57] This early air brake had been specified by the Prussian State Railways for use only on express passenger trains, and it was undoubtedly this directive that encouraged the early development of graduated release air braking throughout Europe. Carpenter's contract would ultimately not be renewed, however, and the railways of the German Empire—newly enamoured of the fast-acting Emergency feature of the Westinghouse single-chamber brake—would opt for the latter, cementing the Westinghouse presence on the continent. Despite this development, the general preference in Europe—resulting, no doubt, from the different railway operating culture and practices that had evolved on that side of the Atlantic—continued to be for operating valves providing graduated release capability with the so-called inexhaustible brake cylinder supply. Following is a 1968 discussion paper excerpt from the Westinghouse Air Brake Division in the United States:

From the standpoint of certain European countries where the preponderance of rail operations is in mountainous terrain, an inexhaustible brake requiring less-skilled manipulation of the brake valve by the driver [this being a rather bold assertion] *was considered desirable. Contrary to the trend in America toward longer trains, friction draft gear, and all braked vehicles in a train, the European conditions were different. Trains were short. The hook-and-link coupler was still used and a large proportion of the vehicles within a train were unbraked. With these differences in conditions it is understandable that a divergence in braking philosophy began to develop.*

Under these circumstances and conditions, the graduated release brake provided satisfactory operation and became the accepted standard, the current performance specification for rollingstock in interchange service on the continent being issued by the UIC in 1953. The European operating valve eventually became known as a "distributor valve," probably as a description of its functioning to distribute air in various directions or to various devices and possibly to distinguish it from the "triple valve" nomenclature that had become synonymous with Westinghouse. Generally, distributor valves were designed with two-pipe air brake systems in mind (having a continuous brake pipe plus a main reservoir pipe to keep the control reservoir on each distributor charged). One—perhaps somewhat generalised—description of the difference between triple and distributing valves is as follows:

The main stem regulating the inlet valve to the brake cylinder is controlled by one diaphragm in the case of the triple valve and by two in the distributor…

With the distributor it will be seen that two diaphragms are involved and in release, with the brake cylinder at zero, equilibrium occurs between the brake pipe pressure and, instead of the auxiliary reservoir pressure, that of a control reservoir which remains at a virtually constant value during application and release.[58]

This arrangement was such that a partial rise in brake pipe pressure could be controlled, enabling the graduated release of a brake application.

Toward the end of 1925, the AAR—at the behest of the ICC—commenced a four-year comparison of graduated and direct release operating valves. The testing was conducted using a one hundred-car air brake test rack at the University of Purdue School of Engineering in West Lafayette, Indiana (renowned for leadership in railway engineering), and evaluated the type-K triple valve as a reference point, the ASA Brake Co. product described earlier, and the Westinghouse FC-5 graduated release and FC-3 direct release control valves (the FC-3 eventually becoming the landmark AB control valve). From this comprehensive exercise were derived, among other things, the following findings:

The *graduated release* brake system

57. Prof. Manfred Pohl, *Safety First by Road and Rail: The History of Knorr-Bremse AG* (Munich: Piper Verlag GmbH, 2005).

58. H.R. Broadbent, *An Introduction to Railway Braking* (Chapman & Hall, 1969), chapter 3.

- limits train length (by approximately 75 percent of that possible using direct release equipment) due to inherent slow release times;
- prohibits "running release" at reasonably low speeds;
- is slow both to release the brake and reactivate the Quick-Service activity (the latter being important for good control of slack action, especially over undulating terrain);
- tends to impose lower average train operating speeds;
- requires more driver skill for train operation over level and undulating territory (the predominant terrain style traversed by most large US railroads); and
- provides debatable advantage only for train operations down long, steep gradients.

Here is a final comment from the 1968 air brake discussion paper quoted on the previous page:

At the present time, the direct release brake is the only system available which has permitted the operation of ever-increasing train lengths and tonnage at higher average train speeds.

Meanwhile, one might perhaps pause to differentiate between these (predominantly) European *distributor* valves and the operating valves that later came to be fitted to steam locomotives—and eventually to some of the earlier diesels—called *distributing* valves. Triple valves were originally used on locomotives in the United States, providing locomotive brake cylinders with an auxiliary reservoir supply. However, with time it became obvious that greater volumes of air were required to be supplied more rapidly to multiple locomotive and tender brake cylinder arrangements, as well as under varying operating conditions (e.g. shunting/switching), and some form of relay system was required. This resulted in the development of devices of greater complexity that became known as "distributing" valves. Specialised locomotive *control valves* would appear later.

It should be noted that it is still commonplace within Australasia for a triple valve to be used as the locomotive operating valve.

The Introduction of the Control Valve – c1910

Previously-mentioned was the introduction of a specialised triple valve, the Type-L, for passenger rollingstock in 1908. This was utilised in the LN equipment (described below) for passenger car application. As passenger car weights and train speeds increased, it became necessary to provide further specialisation in car air brake systems. The result was the development and use of many devices that delivered continuous

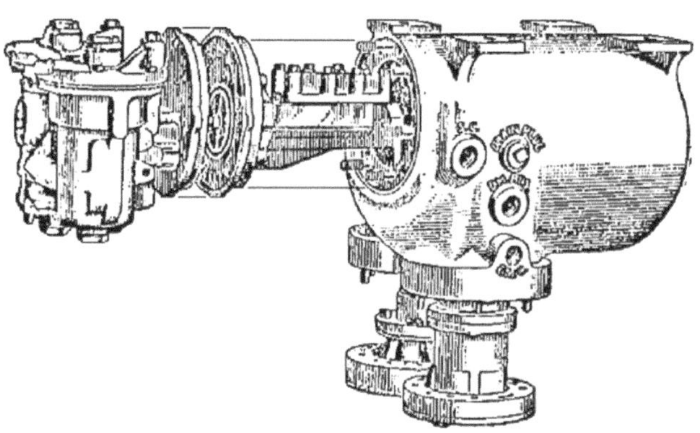

THE EXTERNAL CONSTRUCTION OF THE PC CONTROL VALVE SHOWING ITS RIGHT-HAND SIDE AND THE LINES OF SEPARATION OF THE EQUALISING PORTION. NOTE THAT THE APPLICATION, QUICK-ACTION, AND EMERGENCY PORTIONS EXTEND INTO THE RESERVOIR WHEN IN PLACE.

FIGURE 25: PC CONTROL VALVE

improvement in passenger car brake system performance and train handling (e.g. retainers, electro-pneumatic adaptations, speed governor control, wheelslide control, etc.). This discourse will continue with its focus on the core operating valves and the outcomes and efficiencies they offered but will not include details of the numerous subsidiary devices, such as wheelslide protection systems, that were both concurrently and subsequently developed to provide operating flexibility, but not applied within Australasia.

The LN equipment consisted of the Type-L triple and a specially-sized brake cylinder (Type-N) as well as other parts. This early, specialised system provided five outstanding features vital for effective operation of the passenger trains of the day: graduated release, quick-service and quick-recharge, high emergency brake cylinder pressure, and a service-application safety valve.

The PC equipment (using two No. 3-D and later 3-E ***P***assenger ***C***ontrol valves) was developed, in 1910, for cars weighing 130 000 lbs (59 metric tonnes) or more. Because of several new functions to be performed in the manipulation of the brakes, this fitment utilised—for the first time—a *control valve* instead of a triple. The equipment provided separation of Service and Emergency features and more sensitive and precise brake application and release functions. The control valve can be thought of as corresponding to the distributing valve in the ET locomotive equipment.

U-12-BD Universal Valve

FIGURE 26: UNIVERSAL CONTROL VALVE

The UC equipment (designed to the Westinghouse ***U***niversal ***C***ommon standard for all passenger, subway, and interurban cars and incorporating the Universal[59] control valve) was introduced in 1914 to handle the increasing car weights, lengths, speeds, and faster timetables of American passenger trains. All control valve functions were improved, especially stability and sensitivity, and a degree of flexibility for adaptation to different service conditions added. The UC equipment became a North American industry standard.

D-22-AR Control Valve

FIGURE 27: D-22-AR CONTROL VALVE

Although most Westinghouse Company air brake developments since the Type-L triple had been overseen by a brilliant young British understudy of George Westinghouse, one W.V. Turner,[60] the great man lived to see these developments before his passing on March 12, 1914. Turner, who was chiefly responsible for the K triple and the LN and UC passenger equipments yet to come, had been greatly inspired by Westinghouse. He modestly acknowledged, *"It is truly remarkable that through all subsequent improvements not one of the original functions of the triple valve has been discarded, and many new functions added."* [61]

Introduced in 1940 and designed for the prestige streamliner trains then being launched by various railroads, the D-22-P equipment was more a refinement than

59. 'Universal' in that the device did not have to be resized for different applications. Specific choke sizes were used to control the flow of air to-and-from the brake cylinders. The valve featured separated Service and Emergency operation.

60. Walter Victor Turner, born in 1866, hailed from Epping Forrest, Essex County, England. He was educated at the Textile Technical School, Wakefield, Yorkshire, and came to the United States in 1888. He was granted more than four hundred patents for various inventions and received the Longstreth Medal in 1911, the Elliott Cresson Medal in 1912, and a testimonial as a leading air brake expert by the Air Brake Association in Richmond, Virginia, also in 1912. He died in 1919.

61. Prout, *A Life of George Westinghouse*, p62.

a radical change from previous systems. Control valve functions were yet further improved, and the equipment incorporated features to harmonise with locomotive brake operation, with many parts interchangeable between both freight car and locomotive brakes. It was designed to operate effectively with electro-pneumatic control (instantaneous operation on each car) as well as speed-governor (braking force proportional to speed) and Decelostat (wheelslide) control.

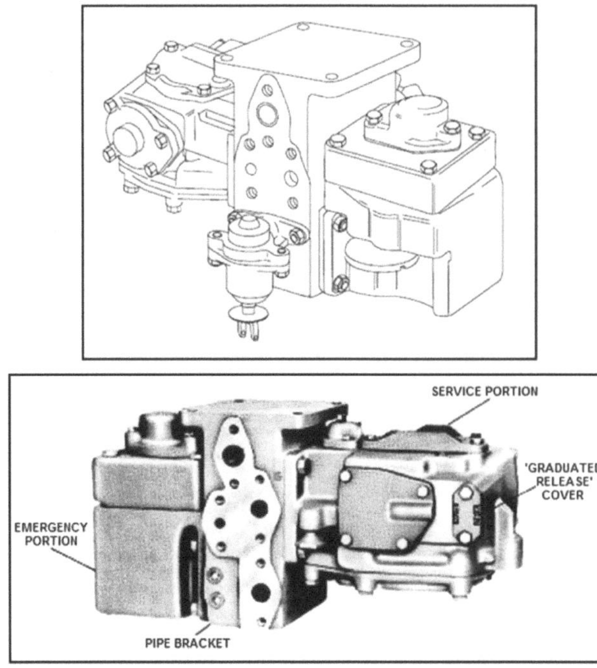

FIGURE 28: 26C CONTROL VALVE – ILLUSTRATION

The 26-L schedule was developed in the United States in the late 1950s and rapidly gained acceptance throughout the air-braked world. No doubt contributing to the swift uptake of this new equipment was the strong US locomotive export industry. This new air brake schedule featured several innovations (e.g. fixed minimum brake pipe reduction, automatic lapping, pressure-maintaining while lapped, reliable diaphragm/spool/poppet valve operation, the ability to incorporate various control functions, and simplified maintenance requirements). The 26C car control valve, introduced in 1960, is a development of the 26F and 26D locomotive control valves also used with the 26-L schedule.

Although its only Australasian use was in New Zealand, the 26C car control valve is state-of-the-art equipment that deserves some attention in this book. The 26C control valve (Figures 28 and 29) is designed for use on passenger cars operated in trains not exceeding 24 x 90-foot-long cars, and in equipment arrangements where brake pipe pressure does not exceed 150 psi (1 030 kPa).

The design emerged from a post-war search by American railroads for a new, improved, lightweight, universal, automatic valve with reduced maintenance requirements, and compatibility and interchangeability with existing LN, PC, UC, and D-22 equipment. The American air brake equipment manufacturers had gained significant development and operational testing experience with the new design concepts of long-travel diaphragms and 'O' rings through their work on the production of a new, universal air brake operating valve for Europe (that became the E-3 triple valve). By comparison with the D-22 valve, which had become large, heavy, and complex, the № 26 equipment—as it was originally known—was a logical solution in simplicity. The 26C was designed to produce extremely fast Service brake response and to be easily modified to suit straight-air pipe electro-pneumatic operation.

The 26C passenger car equipment consists of the control valve (Service and Emergency portions mounted on a common pipe bracket, including a Direct/Grad release cap on the Service portion for compatibility with AB equipment) plus a control and selector volume and a suitably-sized combined auxiliary/emergency and supply reservoir of size determined by total brake cylinder volume. Ancillary components commonly associated with passenger car equipment included the A-2 Quick-Service valve, water-raising equipment, conductor's emergency control, and locomotive air signal functionality. In normal Service applications with other control valves, whereas a specific amount of air is removed from the auxiliary reservoir—in direct proportion to the brake pipe reduction—and is delivered to the brake cylinder or displacement volume (the Quick-Service function adding some brake pipe air), with the 26C the brake pipe reduction is measured and this differential employed to operate a brake cylinder relay valve portion. The 3:1 ratio of control diaphragm-to-brake cylinder diaphragm area determines that the amount of brake cylinder pressure will be three times the effective brake pipe reduction. Since it is relay-type equipment employing both Service (adjustable) and Emergency brake cylinder pressure-limiting valves, only one volume is needed as a combined auxiliary/emergency/supply reservoir for the normal ranges of brake cylinder volumes encountered in contemporary passenger car design. Much of the operating functionality of the 26C was carried through to the 26-type locomotive control valves (the 26D and -F control valve Service portions being interchangeable) and the later ABD freight car control valve and derivatives.

The graduated release function operates as follows:

The Service valve contains the control diaphragm and Service valve assembly, including the brake cylinder diaphragm and the application check valve that admits air to and exhausts it from the brake cylinder. The control diaphragm has brake pipe pressure on its upper and control reservoir pressure on its lower sides. A differential between the two, resulting from a brake pipe reduction, permits the auxiliary reservoir to supply the brake cylinder. A pressure differential caused by an *increase* in brake pipe pressure opens the brake cylinder exhaust connection. Since the control diaphragm is opposed and controlled by brake cylinder pressure acting on a smaller diaphragm, it will assume a Lap position when the opposing forces are equal; the ratio of the two diaphragms ensuring that minor fluctuations in brake pipe pressure will not produce undesired applications.

In graduated release operation, responding to a partial brake pipe recharge, the selector valve (integral to the Service portion and acting as a combined Quick-Service and graduated release valve) vents control reservoir pressure at a controlled rate into the selector volume, thereby reducing the differential across the control diaphragm and exhausting brake cylinder until a Lap condition is achieved. If set for Direct Release operation, control reservoir pressure is connected to brake pipe when the selector valve moves to Release position, destroying the differential and venting the brake cylinder (#16 line) through the Service valve.

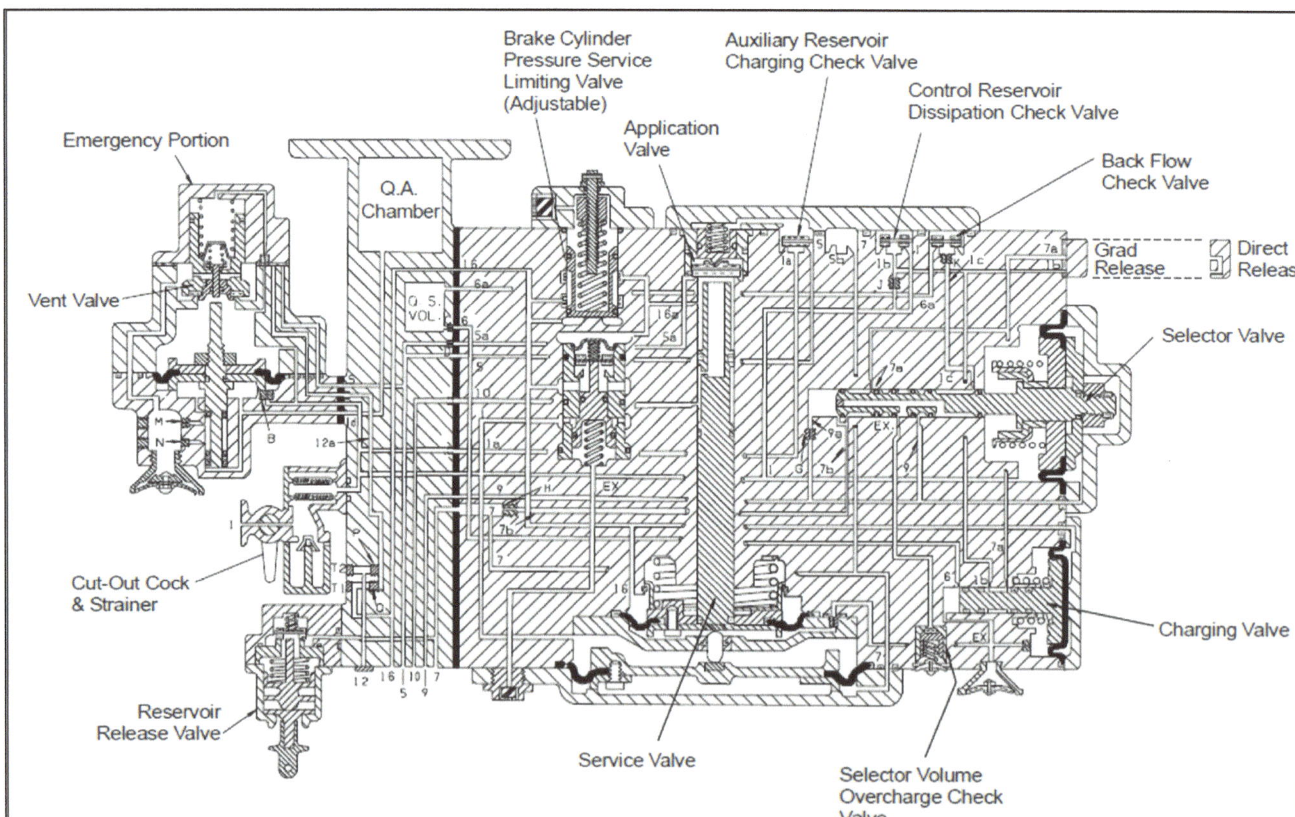

FIGURE 29: 26C CONTROL VALVE - DIAGRAM

The then New Zealand Government Railways was—in 1961—the first customer outside North America to introduce the 26-L schedule, and remains the only operator within the Asia-Pacific region to have regularly utilised "graduated release" on air-braked, locomotive-hauled passenger trains. These were the Hitachi/Nippon-Sharyo-built *Silver Star* overnight luxury sleeper trains (Figure 30) that ran each way six nights per week between Auckland and Wellington from 1971 until 1979, and they sported a modern relay brake system employing the 26C control valves. This service, though, could not compete with air travel, and most of the rollingstock was later sold to the UK leisure and tour company Orient-Express Trains & Cruises, re-gauged to 1 000 mm, reliveried, and remodelled for a five-star luxury operation between Singapore and Bangkok as the *Eastern & Oriental Express*. Having had some experience handling the *Silver Star* in New Zealand, I was privileged on one occasion, while working on a tourist passenger rail project in Southeast Asia, to be invited to ride the head-end of (and to handle) this train within Malaysia, although, regrettably, both the locomotives (26L-equipped GMD[62]) and the passenger cars were configured for direct release operation.

FIGURE 30: NEW ZEALAND RAILWAY'S *SILVER STAR* WAS THE ONLY TRAIN IN AUSTRALASIA EQUIPPED AND OPERATED WITH GRADUATED-RELEASE AIR BRAKE OPERATING VALVES, A CONTEMPORARY STANDARD FOR PASSENGER TRAIN BRAKING IN NORTH AMERICA. THESE IMAGES DEPICT THE TRAIN BOTH IN NEW ZEALAND (LEFT, NEAR AUCKLAND) AND IN ITS SUBSEQUENT *EASTERN & ORIENTAL EXPRESS* GUISE, (ARRIVING AT GEMAS IN MALAYSIA). *Graeme McClare (New Zealand) and Patthanasuwanna P. (Malaysia). Used with permission*

In Queensland, QR has used Metcalf-Oerlikon EST/4d graduated release distributor valves on the SX suburban fleet since its introduction in 1961. These Budd-designed stainless steel cars were intended to be the first EMUs on Brisbane's suburban network; however, electrification did not eventuate at that time, and the SX sets remained locomotive-hauled for their entire working life. The advanced capability, though, that was offered by these operating valves was negated by the fact that QR operated this coaching stock with diesel locomotives equipped with either the older A7-EL or with 26-L equipment not configured for graduated release operation.

The lack of interest within Australia and New Zealand for graduated release capability for passenger trains warrants some critical attention. NZR might be excused, since—until the 1971 introduction of the *Silver Star*—all locomotive–hauled passenger cars had dated to the late 1930s, and when a modern-day opportunity was presented to introduce contemporary-design passenger rollingstock, NZR did in fact equip it with the latest apparatus. It should be acknowledged, though, that over the years, NZR has rebuilt those 1930s cars more than once and could have fitted WG triple valves once they were available; after all, NZR had been using graduated release capable 26-L locomotive brake equipment since 1961.

62. Keretapi Tanah Melayu Berhad (Malaysian National Railways Corp.) 25-Class, EMD GT18LC-2.

A knowledgeable observer might likewise have expected the Australian transcontinental *Indian Pacific* rollingstock, delivered from 1964, to have been so-fitted. Perhaps it was because it would be years before the first Australian state railway introduced the necessary 26-L locomotive brake schedule, and the adoption by the railways of a long-term view was too much to expect. Or perhaps it was because some Australian state railways were notorious, anyway, for not placing the specific choice of air brake equipment high on the list of technical priorities. Whatever the reason, it is a reality that within Australasia, passenger cars are treated as freight cars when it comes to the selection of air brake equipment, and locomotive drivers are obliged—from an air braking perspective—to operate locomotive-hauled passenger trains as if they were freight trains.

Victorian Railways dabbled briefly during the 1970s with graduated release in passenger operation as a result of pressure from the Australian Federated Union of Locomotive Enginemen. In this latter case, the Locomotive Cab Working Party of the Victorian branch of the AFULE had—during a visit to Westinghouse Australia—discussed developments in self-lapping independent brake valves. As a result of these discussions, Westinghouse suggested adapting the SA26 independent valve from the 26-L schedule for use with the A-7-EL equipment then predominant on Victorian Railways, and this hybrid fit-out was tested in operational service with a B-Class locomotive. These discussions with the Victorian AFULE also resulted in some experimentation by VR with graduated release on locomotive-hauled trains. This brief foray utilised the WG1 triple valve (see later section on Australian-built equipment) fitted experimentally to an H-Class passenger consist, plus a 26-L-equipped P-Class locomotive. V/Line management, however, not exactly famous for its early embrace of the latest in air brake equipments, were uneasy. Apparently not trusting their enginemen with the new equipment, they were concerned about the possibility of drivers unintentionally running with the cut-off pilot valve in the wrong position (that is, running a freight train with the brake valve configured for passenger operation). There were also concerns within V/Line regarding the cost of fitting graduated release triple valves to the passenger cars. The program was soon ended, and the use of graduated release on air-braked, locomotive-hauled passenger trains has not featured in Australian rail operations since. For a comparison of railway administration attitudes, see the comments later in this book regarding ARG's relaxed views on running freight trains with locomotive equipment configured for graduated release operation.

American freight **control valve** development began with commencement of the Power Brake Investigation started in 1924 by the ICC and continued through the following year. The result—developed to replace the Type-K triple valve—was the AB valve, originally called the FC3[63] (***F**reight **C**ontrol*, as opposed to the "PC" equipment referred-to previously). This was the designation for the type of brake equipment made standard for new freight cars built after September 1, 1933, rebuilt cars after August 1, 1937, and compulsory in interchange after January 1, 1945. The AB valve:

> *...was the result of several years of trial in the development of a brake to meet as nearly as possible the requirements of the Interstate Commerce Commission. It is an interesting fact that the AB brake is now the basic design not only for freight car brakes but also for passenger cars and locomotives...*[64]

63. David G. Blaine, Fellow ASME (interview by Donald E. Palmer, Henry Christie, Joe Hettingeron; April 3, 2001) on behalf of the ASME Rail Transport Division, History and Heritage Committee. https://community.asme.org/rail_transportation_division/m/mediagallery/1296/download.aspx.

64. C. M. Drennan, *Car Air Brakes,* Chalk Talks, vol. 2 (New York: Simmons-Boardman Books, 1955). Carl Mac Drennan (1886-1971) was originally a carman with the Santa Fe Railroad and commenced railroad air brake instructing in 1908. He wrote and compiled many books and pamphlets on air brakes, and retired as Supervisor of Air Brake Training for the Westinghouse Air Brake Co.

FIGURE 31: AB CONTROL VALVE

The first operating valve certified by the ICC and AAR for 150-car train operation, the AB control valve consisted of two separate portions (***A***utomatic Service portion and ***B***alanced Emergency portion)—each bolted to a common pipe bracket—and boasted the following performance advantages:

- Doubling of the Quick-Service propagation speed through 150 cars to 500 ft/sec (152 m/sec) in Service application and 930 ft/sec (283 m/sec) in Emergency;
- Emergency Quick-Action and Accelerated Emergency Release being faster throughout the train than for any previous equipment;
- 10 lbs (70 kPa) brake cylinder pressure on every car from a minimum brake pipe reduction and faster restoration of brake pipe pressure after release;
- A *Release-Ensuring* feature (brake release after either Service or Emergency application with as little as 1½ psi (10 kPa) of brake pipe rise);
- Brake cylinder release rate almost identical for all cars regardless of their position in the train, and;
- Vastly reduced maintenance requirements.

FIGURE 32: AC CONTROL VALVE - ILLUSTRATION

The AB control valve was safe to go into Emergency at any road speed since the propagation rate was rapid enough to ensure that slack action would not cause a derailment. It became the foundation for all freight control valves to follow:

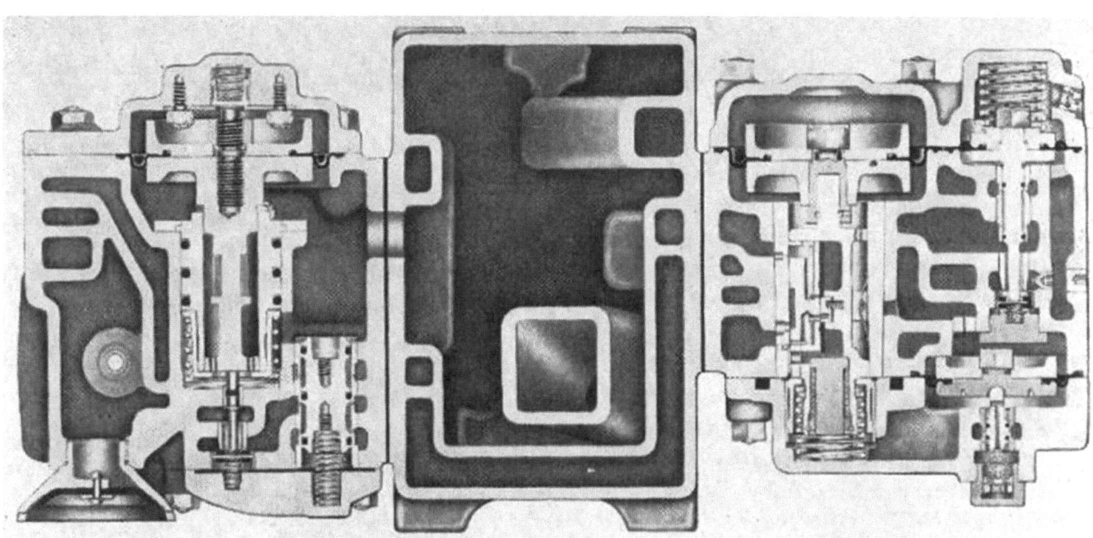

FIGURE 33: AC CONTROL VALVE – CUTAWAY VIEW. Simmons-Boardman Books

ABLC: 1940s. For cars built to a lighter empty-car weight.

AC: 1955. Incorporating removable portions approximately 40 percent lighter in weight, with a special pipe bracket (and partly for this reason, not a commercial success) and the first diaphragm valves, and was a forerunner and prototype of the ABD.
The pipe bracket is in the centre. The Emergency portion is to the left of the pipe bracket and includes the diaphragm piston, slide valve and bushing (centre), and the accelerated release selector valve (right). On the right-hand side of the pipe bracket is the Service portion, with the service diaphragm piston, slide valve and graduating valve (left), the limiting and inshot diaphragm piston valve (upper right), and the release-insuring diaphragm piston and valve (lower right).
The story and even the existence of the AC control valve is not widely known; however, it was a bold step forward. Refer to the end of this subsection for further details on the introduction of this equipment.

ABD: 1963. (Manufactured also by the New York Air Brake Co.) Wabco replaced the standard piston and slide valve in the Service portion and the piston and graduating valve in the Emergency portion with rubber ***D***iaphragm assemblies and re-oriented them vertically. Advertised benefits included more efficient and uniform operation, more sensitivity in Release, and enhanced maintainability. Other internal valvular components were redesigned as spool valves and rubber-seated poppet valves with reduced internal leakage as a major advantage. A new feature appeared: <u>Accelerated Service Release</u>. The *accelerated release valve* operates when brake pipe pressure increases above auxiliary reservoir pressure by approximately 1.5 psi (10 kPa). The control valve feeds emergency reservoir pressure into the brake pipe (and thus also the auxiliary reservoir) until the pressure differential becomes less than 10 kPa. In addition, the design of the control valve provides for Emergency Quick-Action at any time, regardless of the degree of Service application (or release) and for Emergency brake cylinder pressure development. This valve became the alternate AAR standard from 1967.

FIGURE 34: ABDW CONTROL VALVE

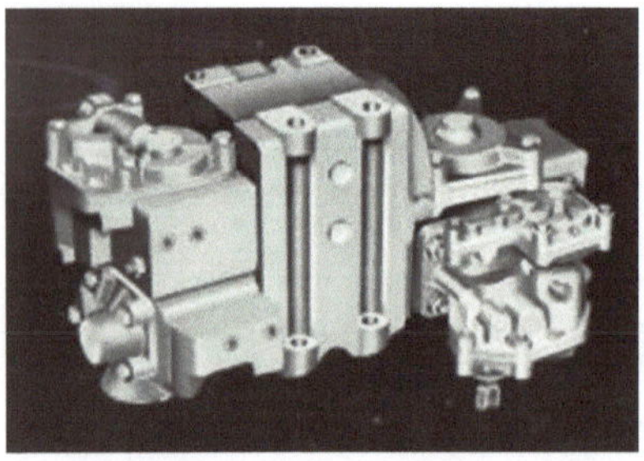

FIGURE 35: ABDX CONTROL VALVE

ABDL: 1975. Simplified Locomotive control valve. This valve utilised a standard ABD Service portion in place of the 25-F Service portion and was overtaken by the 26-F valve.

ABDW: 1977. Introduced the *accelerated application valve* (the "W" feature) for <u>Accelerated Service Application</u> (included as part of the Emergency portion). The AAV functions during a brake application to locally assist the brake pipe reduction by venting the brake pipe in a large number of very small steps or "bites." Reduces brake application time by up to 50 percent. Became the alternate AAR standard from 1977.

ABE: 1980s. Offered by NYAB as a low-air-consumption brake control system designed specifically for use with the truck-mounted unit brake and as an optional, integral empty/load valve.

ABDX: 1989. Incorporated internal improvements and simplified valve structure to improve the initial Quick-Service function and further reduce the potential for internal leakage and provide stability against UDEs (***Un******D***esired—that is, uncommanded and unexpected—***E***mergency applications). Improves brake application times by 30–40 percent over the ABDW. Accelerated Application Valve integrated into the Emergency slide valve. The ABDX-L version is provided for longer cars (i.e. having more than 23 m of brake pipe), the Quick-Service volume being enlarged. Also manufactured by NYAB.

With varying degrees of success, other companies have offered their own control valve designs. All are (or were) ABD-compatible:

Z1AW: Hawker-Siddeley Westcode. Extremely lightweight. Used soft-seated poppet valves rather than slide-and-spool valves and a different design for the Service and Emergency portions. Its development was overtaken by Wabco's introduction of *Accelerated Application* followed by a market recession in the United States that eventually led to the demise of Westcode. This innovative design never received full AAR approval and is no longer manufactured. These control valves were tested in service at Mount Newman Mining Company P/L (now BHP Billiton Iron Ore) during the 1980s.

FIGURE 36: Z1AW CONTROL VALVE

DB-60: (And DB-60L long car version). Produced by NYAB as a result of their acquisition by Knorr-Bremse of Germany. Designed for zero leakage, stability, and sensitivity to small brake pipe reductions and ensured to release from a minimum application. Is fully compatible with all US control valves. Conventional O-rings are replaced with an exclusive K-ring design providing dynamic sealing with none of the weaknesses inherent in other methods. K-rings require no compression when installed, and "float" within the assembly - being effectively "inflated" by the control valve internal air pressure. In 1993 NYAB ceased manufacturing Westinghouse-design control valves in favour of the Improved DB-60.

WFE-W: Produced by Westinghouse Brakes Australia. Evolving from the WF family of triple valves described in this book and developed following a request from the Pilbara iron ore railroads. Not a commercial success and ultimately discontinued. Refer to section 3 for further details. These control valves were also tested in service by the Mount Newman Mining Co.

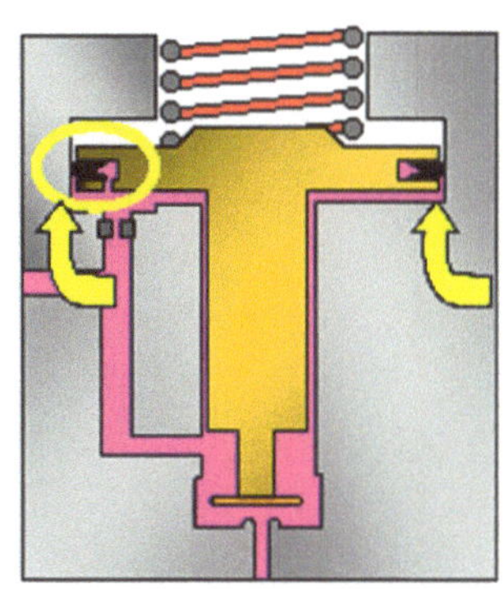

FIGURE 37: DB-60 CONTROL VALVE

Note: The Accelerated Application functionality—the so-called 'W' feature—seen in the control valve nomenclature, ABD ***W***, Z1A ***W***, and WFE- ***W*** was named after Richard Wilson, the Westinghouse engineer who developed it for the ABDW. The term has been adopted by industry as generic terminology.

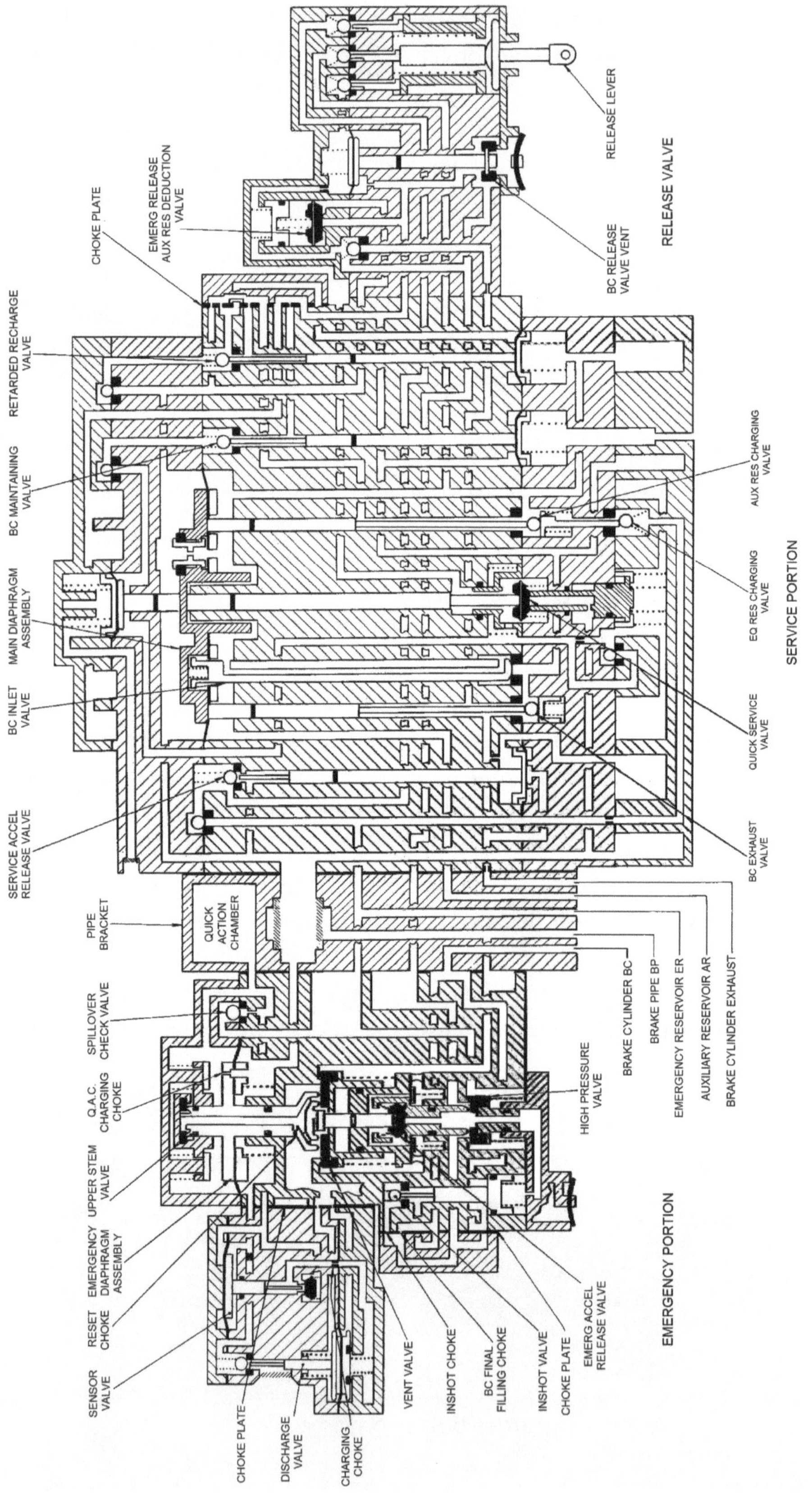

FIGURE 38: Z1AW CONTROL VALVE - DIAGRAM

Development of the AC control valve

As the first modern diaphragm-operated device, the AC was a distinctive step in the evolution of the American control valve. The AC variant is relatively little-known and therefore warrants inclusion in the historical record.

An improved air brake control valve for freight cars designed to operate in complete harmony with the then-standard AB brake, but with removable portions approximately 40 percent lighter in weight, was developed by the Westinghouse Air Brake Company and placed before the AAR for its consideration in 1955.

The AC control valve was designed to provide positive brake operation with extremely rapid serial release action, be simpler in design, and be less costly to maintain. The main parts of the AC valve were made of aluminium. Ring-packed pistons were replaced with diaphragm pistons. Except at two locations, slide valves were replaced with spool valves incorporating rubber sealing rings.

The AC brake valve performed essentially the same functions as the AB valve. The arrangement of the elements, however, was changed and simplified. For instance, in the AB valve the inshot piston and valve and the *quick-service* limiting valve were separate functions. These were combined in the AC valve, but a choke arrangement allowed them to function at substantially the same cylinder pressure as in the AB.

The Emergency section of the AC valve included a piston, slide valve, and graduating valve. The Emergency diaphragm piston moved a small two-port slide valve, and the remaining functions of the large slide valve in the AB equipment were performed by two spool valves packed with the rubber sealing rings.

In the AB valve the air strainer is in the pipe bracket portion, whereas in the AC valve the brake pipe is connected directly to the vent valve in the Emergency portion by a large passage, then through the strainer to other brake pipe passages so that the airflow to the vent valve is unrestricted.

The three-stage build-up of Emergency pressure as effected by the AB valve was changed to two stages in the AC valve. In the initial stage, air from both auxiliary and emergency reservoirs flowed to the brake cylinder by two routes. This accelerated build-up of brake cylinder pressure was terminated when the *inshot limiting* function cut off one route at 13 psi (approximately 90 kPa) cylinder pressure. Thereafter, the build-up continued from both reservoirs through the brake cylinder control choke at about the same rate as with the AB control valve. The maximum cylinder pressure of 60 psi (approximately 414 kPa) remained the same.

Another change that contributed to the rapid serial release of brakes after a severe application was the dumping of emergency reservoir air to the brake-pipe side of the service diaphragm piston instead of into the auxiliary reservoir space as is the case in the AB valve. This contributed to the speed-of-increase of brake pipe pressure restoration in each car, inducing a rapid and positive movement of each Service diaphragm piston and slide valve to the Full-Release position.

On a freight train of 150 cars, brakes at the rear end of the train begin to release nine seconds after those at the head end. This compared with an interval of fifty seconds on a train of the same length equipped with AB valves and required a pressure wave traveling through the brake pipe at about the speed of sound, permitting the release of a brake application at low speed instead of bringing the train to a complete stop to avoid damage. The *accelerated release* feature did for the release function what *quick-service* did for the application.

A novel feature of the AC control valve was the brake cylinder release portion that combined two functions available in the AB valve and added an important third. This comprised a diaphragm-piston-operated release valve, a manually-operated triplex release valve and a manually-operated reapplication valve.

The triplex release valve—operated by a single handle—permitted an overcharge to be released, both auxiliary and emergency reservoirs to be drained completely, or brake cylinder pressure to be released without loss of reservoir pressure when breaking up trains. The selection between the functions was automatic, based on the presence or absence of brake pipe pressure. By use of the manually operated reapplication valve in the absence of brake pipe pressure, the brakes could be reapplied for piston travel readjustment.

When brake pipe pressure was restored following operation of the release valve, the AC valve reset automatically for proper train operation without reapplication of the brakes.

The use of diaphragm pistons instead of sliding pistons fitted with rings resulted in substantial maintenance savings. Piston re-grooving, ring-fitting, and piston chamber re-bushing were completely eliminated. The diaphragms of synthetic rubber provided long piston travel and—being free from friction and leakage—were highly sensitive. The spool valves, packed with rubber O-rings, and the bushings in which they operated were subject to little wear. The relatively large Emergency slide valve bushing was ring-packed and could be readily removed and replaced.

The main diaphragm pistons operated vertically, which improved uniformity and stability. In addition to improving brake operation, these features were also expected to permit longer intervals between required cleaning periods. Wherever controlled rates of airflow were required, replaceable chokes were employed, their size being readily adaptable to various sizes of brake cylinders.

The AC valve for freight service was considered by the Westinghouse Air Brake Company to be a major freight brake control development. For over twenty years, the AB freight brake equipment had permitted the operation of long, high-tonnage trains and was still performing in a highly satisfactory manner. Its performance, as compared to the previous K triple equipment, provided many advantages that were past history and were well-known to air brake specialists of the day. Minor changes had been made in the AB brake from time to time, but the general construction and performance was still essentially the same as when adopted in 1933.

However, although it is possible to incorporate a limited number of improvements into existing devices, in many cases to do so introduces structural handicaps and even functional limitations that prevent the full realisation of the improvements. The development of the AC valve manifested as the incremental outcome of the inability to modernise the AB valve by modifications.

The AC valve employed diaphragms in place of ring-fitted pistons and used O-rings and spool valves in place of slide valves wherever possible. The construction was aimed at simplicity, ease of maintenance, and lightness-of-weight. Performance-wise, the AC valve operated in complete harmony with AB brakes and complied with all the requirements of the AAR specifications for freight brakes.

The use of diaphragm pistons in place of the ring-fitted Service and Emergency valve pistons was designed to guarantee the positive release of valves at the rear of long or moderate-length trains. In addition, an *accelerated quick-release* feature was added to ensure a prompt release of all brakes. An improved brake cylinder release valve was made a part of the Service portion instead of having to be mounted separately, thus permitting the use of only one release rod to the side of the car. Diaphragm pistons also allowed the use of smaller brake cylinders (and reservoirs) when the car weight or type of shoe permitted, thus reducing air consumption. Provision for independent choking also allowed the use of brake cylinders larger than ten inches where required.

Laboratory tests and demonstrations performed on a simulated train are evidence that the design of the AC valves achieved the intended performance specifications.

Although the AC valve was intended for application to freight cars, the Westinghouse Air Brake Company believed that its expected superior performance and lower maintenance cost would make it attractive as a replacement for older operating valves as they came to require major overhaul. For various reasons these expectations were not met. David Blaine, who worked on the project at Westinghouse, once stated;

[The AC control valve] *went over like a lead balloon. I don't think there were more than 300 000 sold. From a practical standpoint its downfall was its special pipe bracket. Many of these valves were used by Erie Mining, hauling Taconite. It operated fine in a unit train.* [65]

65. David G. Blaine, Fellow ASME (interview by Donald E. Palmer, Henry Christie, Joe Hettingeron; April 3, 2001) on behalf of the ASME Rail Transport Division, History and Heritage Committee. https://community.asme.org/rail_transportation_division/m/mediagallery/1296/download.aspx. (accessed June 2013)

ECP: Electronically-Controlled Pneumatic Brakes

ECP is a train power braking system actuated by compressed air and controlled, during brake applications, by electronic signals originating from the locomotive. The brake pipe provides a constant supply of air to the car reservoirs (and thus, brake cylinders).

The key word here is *electronically* rather than *electrically*. Conventional E-P brakes (***E**lectro-**P**neumatic* brake systems fitted to passenger trains and multiple unit sets) utilise several train wires to operate individual valves (or variations in switching of the wires) to control brakes. Most of these systems use a second pneumatic trainline for the main reservoir air supply, and they do not have the built-in two-way communications that ECP systems have. A freight wagon in an ECP train can self-diagnose and report the information to the driver, requiring only the standard brake pipe.

Most expert observers within the international air brake industry believe that fully pneumatic technology has reached the limits of performance.

In conventional brake technology, the pneumatic brake pipe is used as both the communication media and the power source for brake actuation. Although compressed air is an ideal media for brake actuation due to its inherent ability to store large amounts of energy efficiently, it is a poor communications media due to that same capacitance, low signal transmission rate and high signal loss.[66]

Since 1993 the AAR has been developing a performance specification (S-4200-00) for ECP brake systems that will permit complete interchangeability between different manufacturers and equipment. ECP brake systems represent the next generation of conventional freight train braking. The quest for ever-more instantaneous and controllable train braking continues, and the following quote alludes to the reasons for developing these systems:

Conventional air brake systems transmit brake signals near the speed of sound whereas electronic brake systems transmit brake signals at the speed of light.[67]

FIGURE 39 : GENERATOR TAPERED ROLLER BEARING. *Timken Co., courtesy Locomotive Engineer's Journal*

Overlaying electronic control onto a "conventional" pneumatic air brake system separates the brake communication and actuation media and provides a lead-in to full ECP braking for the railroad industry, which is not prepared to scrap air brake systems wholesale and move overnight to the costly new technology. Although—in theory—brake block and wheel wear is expected to decrease with the use of ECP, some industry commentators have urged caution, observing that as drivers become more familiar and competent with the new braking system, the average speed of trains may rise with the result that more and heavier brake applications may become commonplace. This, of course, has cost implications that might cancel out the benefits in this area, and it can be expected that ECP-equipped railways will be considering this aspect.

The ECP system provides near-instantaneous brake control and some features hitherto unattainable such as graduated release (with no train size limit) and constant status feedback of the brakes on every car

66. Proceedings of the 87th Annual Convention and Technical Conference 1966, the Air Brake Association, Chicago, IL, 86–146.

67. Proceedings of the 64th Annual Meeting 2000, International Association of Railway Operating Officers, 44–53.

in the train, combined with the simplicity and sturdiness of pneumatics for the "muscle." The ECP brake uses the present brake pipe as a reservoir-charging pipe only. The brakes can be applied, partially released, and then reapplied as often as necessary to negotiate a changing grade, within the ability of the locomotive compressors to replenish reservoirs at the rear of a long train.

Referring to the communication link to all cars along the train, ECP systems may be radio-based (wireless) or cable-based (wireline). The radio-based option requires the installation of a transmitter and a receiver on the cars and locomotives. A cable-based system provides power and communications to all ECP brake devices in the train via a two-conductor electric trainline running the length of the train. The cable-based system's main challenges will probably be the requirement for battery power and the potential for reliability problems with electrical connections (George Westinghouse would undoubtedly have a wry smile at this!). However, these problems will no doubt be surmounted, and the cable-based ECP schemes might be expected to prevail. Both wireless and cable-based systems have the clear potential to be utilised for Distributed Power operations, and this function may well secure the predominance of the latter.

As the science of the electronic control of air braking continues to evolve, companies not historically associated with air brake systems are likely to play an increasing role. In late 1994, the Timken Company in the United States unveiled its innovative Generator Tapered Roller Bearing, which has an electric generator built into its centre, designed to provide a reliable low-voltage power supply at each car for electronic brake system use. Creating 25 watts at 8 mph and 600–700 watts at 80 mph (Timken expects an electronic braking system to require less than 10 watts), the supply was expected to permit sensors to monitor bearing heat, car load, and brake pressure as well as operate the braking system. Further suggestions were the fitment of safety side-lights to freight cars plus a reliable power supply for an end-of-train monitor. At the time of writing, this invention remains under test by both Timken and the US Federal Railroad Administration and could be described as a product waiting for its time. Timken believes that a demand for the Generator Bearing will arise once electro-pneumatic braking systems are in more regular production and use.

Meanwhile, some suppliers have produced an electronic multifunction operating valve called an *emulating control valve* that may be used in either conventional or ECP braking operation, either wireless or cable. These devices mount directly to a conventional freight car control valve pipe bracket. They are self-contained and utilise—for a power supply—an internal miniature turbo-generator that uses a tiny amount of brake pipe air to generate electrical energy that is stored in ultra-capacitors.

Radio-based ECP braking has been validated in operation in the United States and has been tested at BHP Billiton Iron Ore in Western Australia's Pilbara region (not without some dramatic moments, it must be stated), while cable-based ECP braking arrived in the Hunter Valley with the Wabtec Railway Electronics' product having been installed on the QR National Goninan-GE 5000 Class locomotives and a new fleet of coal wagons and being installed on subsequent fleets. Cable-based ECP braking from Knorr-NYAB also features on the GE Dash 9 locomotives and Chinese-built ore car fleet for Pilbara newcomer, Fortescue Mining Group. On the back of installations in specialised AAR-standard railroads in Canada and South Africa, these sales by the major US manufacturers have placed Australia in the vanguard of global ECP take-up. Undoubtedly, the promoters of practical ECP systems will have to address the cost factor and, as alluded to previously, it is highly probable that cable-based systems will prove reliable enough to overwhelm the wireless option and become the industry preference.

Widespread adoption of ECP braking by the global rail industry will surely signal the next major air braking revolution since the adoption of the automatic brake itself. Ironically, although the technology has been under experimentation in the United States since 1993 and revenue testing since 1995, widespread implementation of ECP within North America has been delayed by a number of constraints. One is the prerequisite that the Federal Railroad Administration address compliance issues and industry concerns regarding the equitable distribution of costs. There is a belief that car owners—who are mostly shippers or leasing companies—will bear most of the costs while the railroads will secure most of the benefits. In addition, the railroads must confront some logistical challenges. Perhaps the first of these is a need to ensure the equipment is targeted in a practicable way for application to those types of operations for which the greatest benefit will be gained with the least interference to current processes. The second might be how to manage what will no doubt be an extended conversion period as all parties struggle with the problematic question of interoperability. Nonetheless, in late 2007, Norfolk Southern became the first US Class 1 railroad to place ECP into revenue service, in this case on 115-car unit coal trains equipped with Knorr-NYAB EP-60 systems operating in south-western Pennsylvania.

In the early years of ECP take-up, it stands to reason that the first large-scale commercial applications of the technology will be the easy ones, involving operators who operate dedicated unit train consists in some sort of high-mileage "merry-go-round" service. Thus, we see the coal haulers in the United States and both the coal and iron-ore haulers in Australia—those who can most immediately put the new technology to efficient use—equipping new rollingstock with ECP brake systems. The remaining industry challenge will be to manage the mix of ECP and non-ECP air-braked cars among the one-and-a-half million vehicles approved for interchange throughout North America.

(Note: The Australian-designed system, originally known as OzECP, is described at the end of Chapter 4.)

How an ECP system works

For the historical record, and because the terminology introduced herein will recur throughout this section and is largely generic within the industry, the following description of the prototypical overlay system developed by TSM[68] is provided [note: I expect that as use of this technology develops and is adapted within Australia, some terminology and acronyms may eventually differ]:

The TSM brake system consists of a Car Control Device (CCD) and a manifold. The manifold contains the solenoid valves and pressure transducers, which are used to fill and vent the brake cylinder, and monitor brake pipe and reservoir pressure. The manifold is mounted between the brake pipe and the Service portion of the control valve. When the ECP system is energised, the manifold cuts off communication between the Service portion and the brake cylinder, but the Service portion continues its function of charging the [auxiliary] *reservoir. If the Emergency portion is retained, then the car can operate in either an ECP-braked train or in a conventionally-braked train. This is known as an* **overlay ECP brake system***. If the Emergency portion is removed and replaced with a blanking plate and the Service portion is removed and replaced with a blanking plate equipped with charging chokes, then the car can only operate in ECP-equipped trains and this is known as a* **stand-alone ECP system***. A stand-alone system has been tested by the TTCI[69] at the Facility for Accelerated Service Testing.*

The CCD is a separate box containing the computer, battery, battery charger, and other electrical components that form the brains of the car brake system. The CCD can be mounted anywhere on the car, but is usually in close proximity to the manifold.

The system is controlled by the Head End Unit (HEU) mounted on the engineer's control stand in the locomotive. The HEU consists of a control box that has push buttons and soft keys.[70] *The Service Application and Release button is mushroom-shaped; pushing the button applies the brake and pulling it releases the brake. Brake applications are made as a percent of Full Service, with this latter being a 100% application, a Minimum Reduction being a 15% application, and Emergency being 120%. The brake can be applied and released in 1% increments from 0 to 100%; or if the button*

68. Technical Services Marketing Inc. of Kansas City, Missouri, installed the first ECP brakes on a Burlington Northern coal train in 1993. Rockwell International Corp. acquired TSM in July 1996, and the operation was sold again to Westinghouse Air Brake Co. in 1998 as part of the Westinghouse purchase of Rockwell's railroad electronics division.

69. Transport Technology Center Inc. (Pueblo, Colorado), a subsidiary of the AAR (refer to chapter 8).

70. See Glossary, chapter 8.

is held, the application or release will be continuous. The other push button is for Emergency application; it being caged to prevent accidental use. The soft keys provide for Minimum Reduction and Full Service applications, and direct release. These keys are also used when initialising the train brake system. The HEU also has a flat screen that is used to inform the engineer of the status of the train brake system. The display includes brake pipe pressure on the last car, the air brake command currently in effect, and any existing error messages.

The car brake systems are connected to each other by a shielded, two-conductor, 8-gauge cable carrying both power and signal. Power at the locomotive is 230 volts dc from a power supply connected to the locomotive batteries. For the Conrail and BNSF tests the cable between cars was connected using a Conomac connector, which is similar to a welding connector. This was considered to be a temporary solution until an AAR-standard component was developed. The Conomac connectors suffered from exposed electrical contacts and the lack of a positive locking feature [note: the AAR eventually settled on an Amphenol brand connector with "pull-apart" capability].

The communications protocol tested is LonWorks by Echelon, this having been selected by AAR committees as the standard for cable-based ECP brake system communication. LonWorks is an off-the-shelf protocol widely used in applications such as controlling building environments and in rail transit.

The system allows the engineer to directly control the brake cylinder pressure on every car in the train. The brake pipe is used only to charge the reservoirs, even during brake applications. This allows the system to maintain full reservoir pressure at all times (except immediately after a brake application, when the reservoirs are drawing on the brake pipe to recharge), and allows the CCDs to maintain brake cylinder pressure even if moderate brake cylinder leakage is present. If a CCD cannot provide the brake cylinder pressure commanded by the HEU, then that CCD will send an error message for display on the HEU. If communications are disrupted the system automatically goes into Emergency without venting the brake pipe. This allows for the maintenance of brake cylinder pressure even with some brake cylinder leakage. Automatic electric Emergency applications also result from loss of brake pipe pressure and reduction of operative brakes to less than 85%.[71]

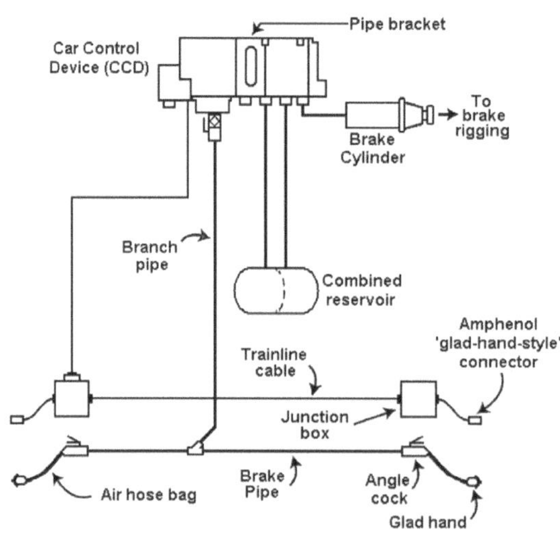

FIGURE 40: TYPICAL WAGON ECP SYSTEM – BASIC SCHEME

ECP system description (from the S-4200-00 family of US specifications)

Due to the novelty of this innovative technology, the opportunity is taken by the inclusion of this section, to examine it in some detail.

71. US Department of Transportation FRA Office of Research and Development, "Electronically Controlled Pneumatic Brake Revenue Service Tests," Final Report (DOT/FRA/ORD-99/07).

In addition to providing a near-instantaneous response to braking commands—including graduated brake release and re-application—primary functions of ECP braking include adjustment of braking level with reference to car loading, continuous fault detection and equipment status monitoring, and an appropriate response to undesired train separation or the malfunction of hoses, cabling, or brake pipe.

Major components (generic)

Car Control Device (CCD): This electronic device is fitted to each car. It replaces the conventional pneumatic control valve Service and Emergency portions during electronic braking, and in a Stand-Alone system it also controls reservoir charging. The device is activated by trainline power; each CCD containing a battery that is charged from the trainline. The CCD will also send a warning signal to the locomotive should any of the components not respond appropriately to a braking command. Each CCD has a unique electronic address.

There are a number of different types of CCD applications:

- *Overlay*. This system can operate in either a conventional or electronically-braked train. Should this system fail, the train can continue in 'conventional' operation using the standard pneumatic control valves once the ECP system is cut out;
- *Emulator*. This is an optional feature by which the CCD can imitate the function of the pneumatic control valve when in a conventionally-braked train. A suitable power source is provided to allow the emulator CCD to function correctly at all train speeds for a minimum of 48 hours;
- *Stand-Alone*. These types of CCDs can only operate in full ECP-equipped trains. They cannot operate within a conventional, pneumatically-braked consist.

Head End Unit (HEU): Mounted within the locomotive, this device controls the ECP brake system. Its specific functions are to:

- provide a data display and the man/machine interface;
- monitor the End-of-Train beacon;
- provide control signals (1) to discontinue trainline power if communication with the EOT is interrupted and (2) to command a momentary trainline power application to restart any CCDs that are shut down;
- permit the conduct of brake system diagnostic tests;
- interface between the ECP system and interacting locomotive signals.

End of Train Device (EOT): This is physically the last network node in the train. Connected to the end of the trainline on the last car in the train, it is activated by trainline power and communicates with the HEU. It contains a brake pipe pressure transducer and a battery that is charged from the trainline cable. The EOT transmits a status message once per second that includes current brake pipe pressure and trainline voltage for cab display by the HEU.

For ECP trains capable of operating in "pneumatic mode" (Overlay or Emulator systems), a pneumatic EOT function is also required. The EOT functions for ECP operations may be combined with pneumatic EOT functions in a dual-mode device. In Distributed Power operations, the ECP equipment on a locomotive at the rear end of a train may perform the EOT function.

Trainline: This is a two-conductor electric wire carried through each car, and therefore extending throughout the length of the train. It carries both power to operate CCDs and the EOT as well as communications network signals (superimposed on the power voltage).

<u>ECP Trainline dc Power Supply</u>: Operates at nominally 230 Vdc to provide electrical power as above. The power supply is mounted within the locomotive and is controlled by a Power Supply Controller. A single power supply is capable of supplying power to a minimum of 160 CCDs and an EOT.

<u>Power Supply Controller (PSC)</u>: This network device interfaces with the trainline communication network and controls the trainline power supply as commanded by the HEU. Multiple power supplies may be enabled by the HEU.

<u>Head-End Trainline Terminator</u>: This is attached to the front of the trainline to provide electrical termination at the Lead locomotive to minimise impedance-related communications problems. The HEU will automatically confirm the presence of head-end trainline termination.

<u>Pneumatic Backup (PB)</u>: This capability, needing no electrical power, is required on each car to apply emergency brake cylinder pressure (assuming reservoirs are adequately charged) in the event of a vented brake pipe, whether or not the CCD is operational. The PB system is also capable of assisting in the propagation of pneumatic pressure signals through the brake pipe.

<u>Car ID Module</u>: Mounted on the car, this stores car-specific data for provision to the CCD such that the CCD always contains the correct characteristics, parameters (constants), and other information for the car or brake set on which it is placed. In this way, such data is mechanically tied to the car and cannot be changed inadvertently in the field, even if CCDs are swapped. Such information includes the following as a minimum:

- Car owner and type;
- Stretched length/Loaded weight/Empty weight;
- Default Full Service Net Braking Ratio;
- Braking characteristics and air reservoir details;
- Presence/type of pneumatic or electronic Empty-Load device;
- Number of operable brakes controlled;
- Number of axles associated with CCD;
- Vehicle orientation reference.

<u>Locomotive ID Module</u>: Mounted on the locomotive, this stores locomotive-specific data for provision to the HEU and PSC such that the HEU and PSC always contain the correct characteristics, parameters (constants), and other information for the locomotive on which it is placed. In this way, such data is mechanically tied to the locomotive and cannot be changed inadvertently in the field, even if the HEU and PSC are replaced. Such information includes the following as a minimum:

- Locomotive road number and type;
- Stretched length/Nominal weight/Number of axles;
- Battery voltage Fault and Fault Clear threshold;
- Brake pipe pressure default set-point;
- Default train Full Service Net Braking Ratio;
- Suppression brake application;
- Nominal wheel diameter;
- Vehicle orientation reference.

ECP system operation

The following is from AAR documentation. I acknowledge that in Australian practice, this procedure may vary. Nonetheless, the process described here will provide a generic understanding of the requirements to enable an ECP system:

1. Setup

- *HEU Power-up*. HEU defaults to Trail status (cannot broadcast any beacons or system commands) and designates the train as "Loaded";
- *HEU to Lead*. Selected by locomotive driver (only one locomotive unit can assume Lead status). Once selected, and prior to completing the Setup process, the Lead HEU broadcasts beacon messages containing a Full Service (100 percent) Train Brake Command, a trainline "Power Off" command, and Initialisation Mode. The HEU then enters the Setup process, allowing the driver—if required—to change the ECP brake system configuration. With the exception of the Empty/Load setting and trainline power mode, Setup changes cannot be made unless the train is stopped or a Full Service brake application has been in effect for at least 120 seconds. Setup changes require operator confirmation to become effective;
- *BPP set-point*. The driver selects the required brake pipe pressure (or accepts the default setting) to establish the ECP brake system reference BP pressure;
- *Empty/Load setting*. For trains without a "local" load-sensing capability, the driver must designate whether the train is empty or loaded prior to moving it (the default setting being "Loaded"). If the cars have on-board load sensors or Empty/Load devices, the CCD will ignore any "Empty/Load" command from the HEU;
- *Train Net Braking Ratio*. The capability for the locomotive driver to vary this may be specified by the individual railroad. This is unlikely to be an option accepted by current Australian operators.

After the above values have been selected, they will be displayed by the HEU, which requests confirmation from the locomotive driver. The desired operating mode is now entered.

2. Operating modes

- *Initialisation Mode*: Initialising the ECP system 'registers' network devices and train conditions and 'sequences' the train. This means that the identity and position of all network devices in the trainline communication network are established and confirmed, and network addresses are assigned to each. Setup data are also downloaded. An End-of-Train device is required to be connected, and when Initialisation Mode is entered a Full Service brake interlock is applied and fault logic is reset. The locomotive driver may cancel this procedure at any time. As part of the initialising procedure, the Lead unit HEU determines the relative position and orientation of each vehicle within the train, although should this process not occur (for example, specific component requirements are not met), normal train operation can still proceed. When the brake pipe pressure read by the EOT exceeds 0.74 percent of the set-point, the system may enter Run Mode;
- *Run Mode*: This is the main operational mode of an ECP brake system. It is selected when the train consist is fully configured and the brake system fully functional. In Run Mode, the train is ready for normal road operations;
- *Switch Mode*: This mode provides for operation of the ECP brake system should the EOT not be communicating with the HEU, or should the train be separated for en-route switching activities. In this case, many of the system fault detection and response procedures are suspended. Basically, Switch Mode operation differs from Run Mode in that the EOT is not required, train configuration may be changed (cars added or removed), and there is a restricted maximum speed limit applied;

- *Cut-Out Mode*: This is used to shut down the entire ECP brake system, either when converting to conventional pneumatic brake operation (overlay- or emulator-equipped system) or when the train is parked or the driver wishes to cut out all CCDs and release the air brakes prior to shunting.

The Wabtec ECP-4200 equipment

In November 2000, Wabtec Railway Electronics installed—on a fleet of privately owned coal cars for operation on CSX Transportation—the first ECP system in the United States to be fully compliant with the AAR specification for ECP brakes (Standard S-4200, *Performance Requirements for ECP Cable-Based Freight Brake Systems*) as it had evolved at the time. In Australia, the application of this equipment is currently limited to the Aurizon (originally QR National) coal fleet operated in captive service between Hunter Valley coal mines and the Kooragang Island coal terminal in the Port of Newcastle. This Australian Wabtec system does not strictly comply with the current iteration of S-4200, but since neither railroad interchange nor single-system interoperability are relevant—as they would be in North America—this fact is of little moment.

Components:

- *Network Interface Unit* (NIU): Also called the Head End Unit (HEU), this is the computer that manages the ECP system. It controls the power supply via Locomotive ID modules (LIDs) for trainline control, monitors the status of all operable cars via CCIDs, and EOT status. It also provides a means to start up and shut down CCIDs and the EOT;
- *Head End Terminator* (HET): Attached to the front of the trainline to provide electrical termination of the trainline at the Lead locomotive, its function is to minimise impedance-related communications faults;
- *Power Supply*: Converts the 110 Vdc supplied from the locomotive to 230 Vdc for ECP trainline power. It is controlled by the NIU performing the function of HEU via the LID;
- *Locomotive Identification Module* (LID): Controls the ECP power supply in response to HEU (NIU) commands and contains and provides locomotive characteristics and identification information for the NIU;
- *Car Control and Identification Device* (CCID): Contains data that are specific to the individual car and are required to determine braking. It controls the ECP manifold in response to NIU (HEU) commands. Powered from the trainline, it has a four-hour battery backup;
- *ECP Manifold*: Replacing the pneumatic control valve Service portion, the manifold is mounted on the pipe bracket and contains magnet valves to charge and vent the brake cylinder plus transducers for brake cylinder, brake pipe, and reservoir pressures. There are also two manual bleed valves to vent brake cylinder and reservoirs, and a pneumatic Emergency lockout valve that ensures that system electronics cannot release brake cylinder pressure during an Emergency application;
- *Manual Release push button*: Two on each Lead car of a married pair, they cause the venting of brake cylinder and reservoir pressures when held depressed for approximately five seconds and then released. When released, the system compares brake pipe and reservoir pressures. If brake pipe is equal to or less than the reservoir, the vent latches, completely venting both brake cylinder and reservoir. If brake pipe remains higher by a fixed value, only the brake cylinder will be vented. Holding the button depressed for 120 seconds will cause the vent request to be ignored. This feature protects against short circuit;

- *End-of-Car Junction Box*: Located on each end of a married pair and the locomotive(s), it permits easy in-service replacement of the Amphenol Freightmate™ intercar connectors. BIW end-sill connectors protect car and locomotive permanent wiring;[72]
- *End-of-Train device* (EOT): Situated on the rear of the last car, it connects to both the brake pipe and the 230 Vdc trainline and provides the last node of the ECP Network. It contains a brake pipe pressure transducer, a motion sensor, and a High Visibility Marker (HVM) otherwise known as the tail-light. The device also includes a backup battery charged from the trainline. The EOT continuously reports brake pipe pressure and trainline voltage to the NIU. It must be connected to enable ECP Run Mode.

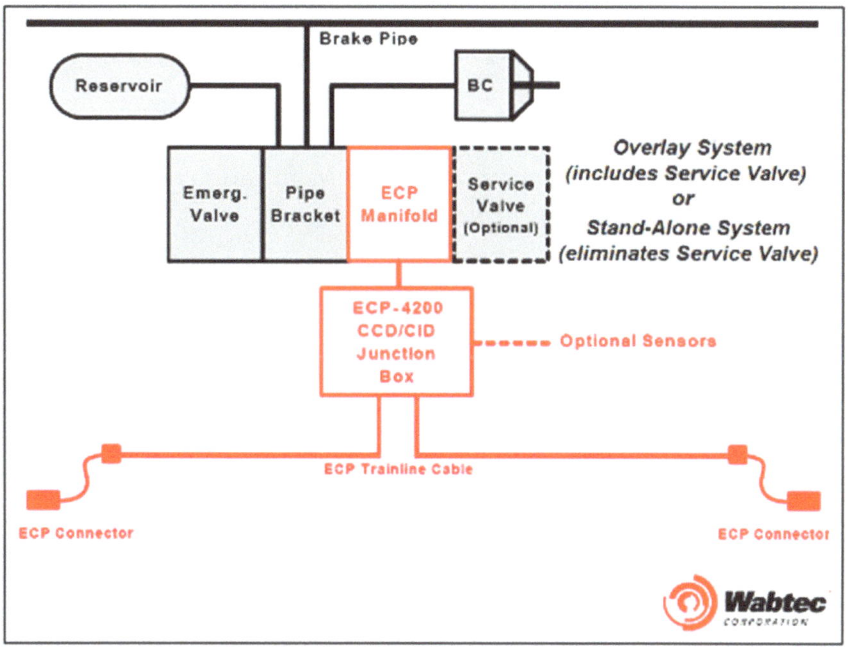

FIGURE 41: WAGON CONFIGURATION WITH ECP-4200 (SHOWN IN RED) – BASIC SCHEME

FIGURE 42: ECP-4200 IN AUSTRALIA (1)

72. Originally Bath Iron Works, BIW is a major American shipyard located on the Kennebec River in Bath, Maine. BIW was purchased in 1995 by General Dynamics and is part of their Marine Systems group.

System operation:

1. HEU-designated NIU searches for HET, Power Supplies, and LIDs;

2. HEU/NIU turns on Power Supply for eight seconds to "wake" up CCIDs and EOT;

3. HEU/NIU sends initialisation beacon signal to acquire all cars in the train as well as the EOT;

4. If the EOT is found, trainline power is applied and CCIDs are searched for and acquired. Run Mode is enabled;

5. Driver moves brake handle. EPIC sends a request to NIU, which puts a "brake request" onto the ECP trainline;

6. The CCIDs read this request and command the ECP Manifold magnet valves to apply brake cylinder pressure. Note: Brake pipe pressure does *not change*.

Operational modes:

- *Run Mode* – Full ECP functionality. Requires an operative EOT;
- *Switch Mode* – Allows ECP operation without the EOT. Used when manoeuvring to cut cars in and out of the train consist. Switching (shunting) speed is limited;
- *Cut-Out Mode* – All CCIDs are cut out. This mode is used when shutting down the ECP system.

ECP incidents:

The introduction of ECP braking into Australia has not been without some moments of drama. Later, in the GE section a major train runaway incident is discussed, but the application of the Wabtec S-4200 system to Australian coal trains in the Hunter Valley has also had a documented 'moment'.

One evening in August 2010, a Pacific National ECP-braked coal train had completed loading at the Drayton Colliery and had drawn up to the Start signal which was at Stop. Shortly after a crew change had occurred, the train commenced an undesired movement resulting in the signal being 'passed at Danger' - an event referred-to as a SPAD. The incident resulted in the entire locomotive consist and the leading two wagons of the train being derailed at the catch-points which protect the running line. This location is on a descending grade of 1-in-50 (2%).

The train consisted of three TT-Class locomotives (Downer EDI Rail-EMD GT46C-ACe) and 88 x 120-tonne coal hopper wagons, was over 1 350 metres in length and had a trailing load of 10 560 tonnes. The TT-Class locomotives manufactured by Downer EDI Rail under licence to EMD were delivered to Pacific National in 2009 equipped with standard pneumatic braking and had been retro-fitted with a wired ECP braking system supplied by Wabtec Railway Electronics, re-entering service in 2010. This ECP braking equipment had been recorded as being compliant with AAR Standard S-4200 as a Risk Control during the risk assessment for introduction of the locomotives and the recently-built wagons.

At the time of the crew-change, the train's ECP braking system was energised, with the driver's brake valve in the Full Service application position. Having boarded the locomotive, the incoming locomotive driver undertook the screen-based process to isolate ECP braking. In the context of this train operation and of any crew control actions that could be expected at this time, there was no valid operational reason for this crewman to have undertaken such an action.

Due to the design of the Wabtec Railway Electronics ECP brake system, the act of de-energising the system caused the train brakes to release. The locomotive Independent brake was unable to hold the train stationary and—despite the efforts of the crew to apply the train brakes by use of the cab-mounted Emergency Brake Cock (providing a direct brake pipe exhaust)—the train commenced moving down the gradient, passing the Start signal and subsequently derailing with the locomotives in danger of overturning.

Basic instruction provided to train crew members has long been that the brakes of a train will always "Fail Safe", which has traditionally meant that a loss of brake pipe air from whatever reason will cause a brake application that will stop the train. In this instance manipulation of the Emergency Brake Cock did not result in a 'safety' brake application because the brake pipe pressure on the standing train was too low to activate the Emergency application feature due to the lack of Vent Valves on the locomotives. The fact that—by design—the train brakes would release if the ECP brake were Cut Out co-incidentally with the Automatic Brake Valve handle being in any position other than Emergency was not understood by Pacific National at the time of the incident. It would be found that there was insufficient specificity in the training documents or manuals provided to Pacific National to instruct train crews that, prior to cutting-out ECP brakes, the Automatic Brake Valve handle must be in the Emergency position (that is to say, if the system were being isolated with the locomotive(s) coupled to a train, there would be an Emergency application in place on that train).

The crew of this train stated that they had been advised during instruction to "play with the ECP screens" when they got a chance, to become more familiar with ECP. They also claimed that training in ECP brake operation had been rushed and disorganised. It appears that this locomotive driver heeded the advice to experiment but chose extremely inappropriate circumstances under which to do so.

The action required to isolate an active ECP system involves undertaking a series of 'key-strokes' on one of the locomotive's computer screens, and at critical points in the process, these key-strokes are required to be acknowledged. The locomotive driver completing this process would need to enter and acknowledge five screen views in order to cut out the brakes, and the system provides a highlighted warning message, "CONFIRM END ECP" after the fourth keystroke.

The system Operators Guide stated, "*Ending ECP Operation: Place the Automatic brake handle into SUPPRESSION*"[position], and the Operations and Maintenance Manual included a warning that "*If the brake pipe is charged, any car that does not have its hand brake applied will immediately release its brakes when ECP is Ended.*" However, there was no specific warning to Pacific National training personnel or train crew members that by following the instructions to place the Automatic Brake Valve in either the Full Service or Suppression positions when ending ECP, the train brakes would release; and procedures and documents used for ECP training of operations personnel were not validated by technical experts prior to release for training or operational use.

It was determined after the derailment that the cutting-out of ECP braking with the Automatic Brake Valve in the Full Service or Suppression Positions as indicated in training materials would cause a brake release on Wabtec Railway Electronics ECP-equipped trains and that cutting-out of ECP brakes *with the Automatic Brake Valve in the Emergency position* **would** leave the train brake applied.

The incident was found to have occurred due to a system failure to propagate an Emergency application of the train brakes when the locomotive driver isolated the ECP Braking. When the crew members realised the train was moving, the driver moved the Automatic Brake Valve handle to the EMERGENCY position and when this had no effect he then used the train's Emergency Cock, also without effect.

It appears that Pacific National, having identified a need for a fleet of standard-gauge high-horsepower locomotives for unit coal train operation, contracted with Downer EDI Rail to lease some of their LDP-Class (Locomotive Demand Power) units. For some reason, these locomotives had not been provided with brake pipe vent valves in their air brake scheme, and it seems that Pacific National either did not comprehend this or did not think it mattered. These locomotives were later purchased outright by PN, and others (the TT-Class) followed, built specifically for PN. The LDP units had been built with contemporary pneumatic air braking and were retro-fitted with Wabtec ECP equipment under a subsequent contract variation.

Brake functionality was tested prior to the first of the units commencing operation with PN, but the tests were conducted only on short wagon groups rather than a full 88- or 92-wagon train, and with a double-unit locomotive consist rather than a 3-unit set as was intended by PN for their coal train operation. This equipment arrangement would later be found to have concealed the fact that without vent valves (which are a contemporary AAR air brake schematic requirement) a 3-unit locomotive consist presented an excessive brake pipe length between the operating cab and the first wagon such that with a depleted brake pipe pressure (as existed on the train at Drayton) an Emergency application signal of sufficient strength could not be generated within the brake pipe and propagated to the first wagon. With ECP braking cut out, this propagation was necessary for the first wagon in the train to trigger the next, and so on, to provide the conventional serial action along the length of the train.

This incident clearly occurred as a result of deficiencies in equipment specification, technical training, and planning for the introduction into service of new equipment and systems. It seems to the author, too, that insufficient consideration might have been accorded to differences between North American and Australian railroad practices—such as a lower brake pipe pressure in Australian operations—and the ECP system could have benefitted from the inclusion of a 'Low Brake Pipe' pressure feature as exists in software in the Locotrol distributed power system.

The incident serves as a salutary lesson on the critical requirement for the customer to be 'across' all air brake issues and for due diligence to be satisfactorily concluded in all aspects of a project of this nature.

The Knorr-NYAB EP-60 equipment

The EP-60 Stand-Alone ECP system was initially developed in 1996 and commenced in revenue testing with iron ore producer Quebec Cartier Mining Company in eastern Canada in early 1998. It has been developed to AAR interoperability and compatibility standards. The equipment is integrated with NYAB's Computer Controlled Brake (CCB) on the locomotive such that EP-60 is an extension of the CCB controller. This gives the CCB locomotive brake controller—through brake 'valve' handle movement (the EBC)—the ability to control the trainline brake command pneumatically or electronically and eliminates the need for a separate brake controller and operator display. In ECP operation, the CCB system provides Automatic and Independent brake handle positions to the EP-60 Head End Unit (HEU) while continuously charging the brake pipe. EP-60 is also fully integrated with the locomotive's Integrated Locomotive Computer (ILC) driver's display, providing train setup functions, operator display feedback, and a link to the event recorder. The Empty/Load changeover process is also carried out from the locomotive and electronically controlled at each car when ECP is operating.

Primary components of EP-60 on the locomotive:

- *Electronic Brake Controller* (EBC) is what has traditionally been known as the driver's brake valve. The 'electrical switch' provides the man/machine interface to control the EP-60 Stand-Alone trainline brake commands;

- *Operator Interface Unit* (OIU) provides the man/machine interface for EP-60 setup, brake command display, operational information, and system alarms;
- *Trainline Communications Controller* (TCC) interfaces between the Lead locomotive and the rest of the train to control the EP-60 brake effort, reporting trainline equipment status and diagnostic information and to provide the gateway to the ECP trainline;
- *Trainline Power Supply* (TPS) converts the locomotive 75 Vdc supply to 230 Vdc trainline power for the EP-60 system and supplies 24 Vdc to the trainline for train sequencing;
- *Locomotive ID Module* (LID) provides the circuitry for locomotive identification and train initialisation (i.e. sequencing, orientation, Remote unit setup, etc.).

The system architecture utilises one communications link from CCB to EP-60 to provide brake control commands and another separate communications link from EP-60 to the ILC to process all setup, display, event recorder, and diagnostic information to Locomotive System Integration (LSI)[73] standards. EP-60 can also link Distributed Power functionality to the trainline to provide Wired DP.

System functionality:

- *Setup*
  To operate EP-60 on an integrated locomotive, first the CCB system and then EP-60 must be set up by the driver via the ILC.
- CCB setup requires the Lead unit to be cut-in, ER setting known, Frt/Pass brake conditioning, Independent brake to Full Application, Automatic brake to Release, BP to be charging, zero road speed, and radio DP not in operation.
- EP-60 setup requires Lead, required Mode, Power, Empty/Load, and Full Service Interlock. With these activities complete, CCB will allow transition from conventional mode to ECP mode, and at this point the EP-60 system can be initialised, also by way of the ILC display.
- Train initialisation requires the establishment/confirmation-of-identity of all trainline devices (locomotives and cars), position/orientation of all EP-60–equipped locomotives and cars, assignment of unique network address, collection of device information, and downloading train configuration information. During this process, an ECP Full Service Interlock exists until EP-60 confirms that the brake pipe is charged and all ECP equipment in the train is fully operative. Once these conditions are met, the Automatic brake handle must command a Full Service brake call to reset the interlock. At this point, EP-60 is fully enabled, and the brake can be released. If required, Wired DP can now be selected and the Remote unit locomotive "setup" performed. All Remote unit setup procedures are completed from the Lead unit; there is no requirement for setup actions at the Remote unit as for Radio DP.

- *Brake controller handle commands*
  The CCB system transmits the brake controller handle position (the brake effort requested) to the EP-60 HEU, which uses the information to control the EP-60 trainline brake command. The Independent brake handle position (thus level of Independent braking effort) will not initially be transmitted on the EP-60 trainline; however, this functionality is a long-term goal, and the communications interface is designed to support such future installation.

73. See Glossary, Chapter 8.

CCB transmits the Independent brake handle position to the HEU in the form of an Independent "% Brake Call."

- *Penalty commands*
The EP-60/CCB Integrated System utilises existing penalty brake application logic within the CCB system. ILC penalties (e.g. Alerter/Vigilance timeout, train overspeed, AWS/ATP activation, etc.) are processed through the CCB to EP-60. The CCB interprets the command, performs suppression tasks, transmits the command to EP-60, and performs penalty reset functions. In conventional mode, the CCB commands a brake pipe reduction at the Service rate and actuates power knock-out (PCS) in response to the penalty command. In EP-60 mode, the trainline penalty command occurs as an EP-60 Trainline Brake Command (TBC) with power knock-out but no reduction in brake pipe pressure. Should the EP-60 TBC acknowledgement that the penalty call is being executed not be received by the CCB, then CCB will respond by initiating a conventional Emergency brake application.

- *Emergency commands*
Both conventional and EP-60–equipped locomotives can operate in the same train. The Lead unit transmits the train commands via the EP-60 trainline and the Trail unit via the Independent Brake No. 20 pipe, Actuation No. 13 pipe, and the MU jumper cable. Wired

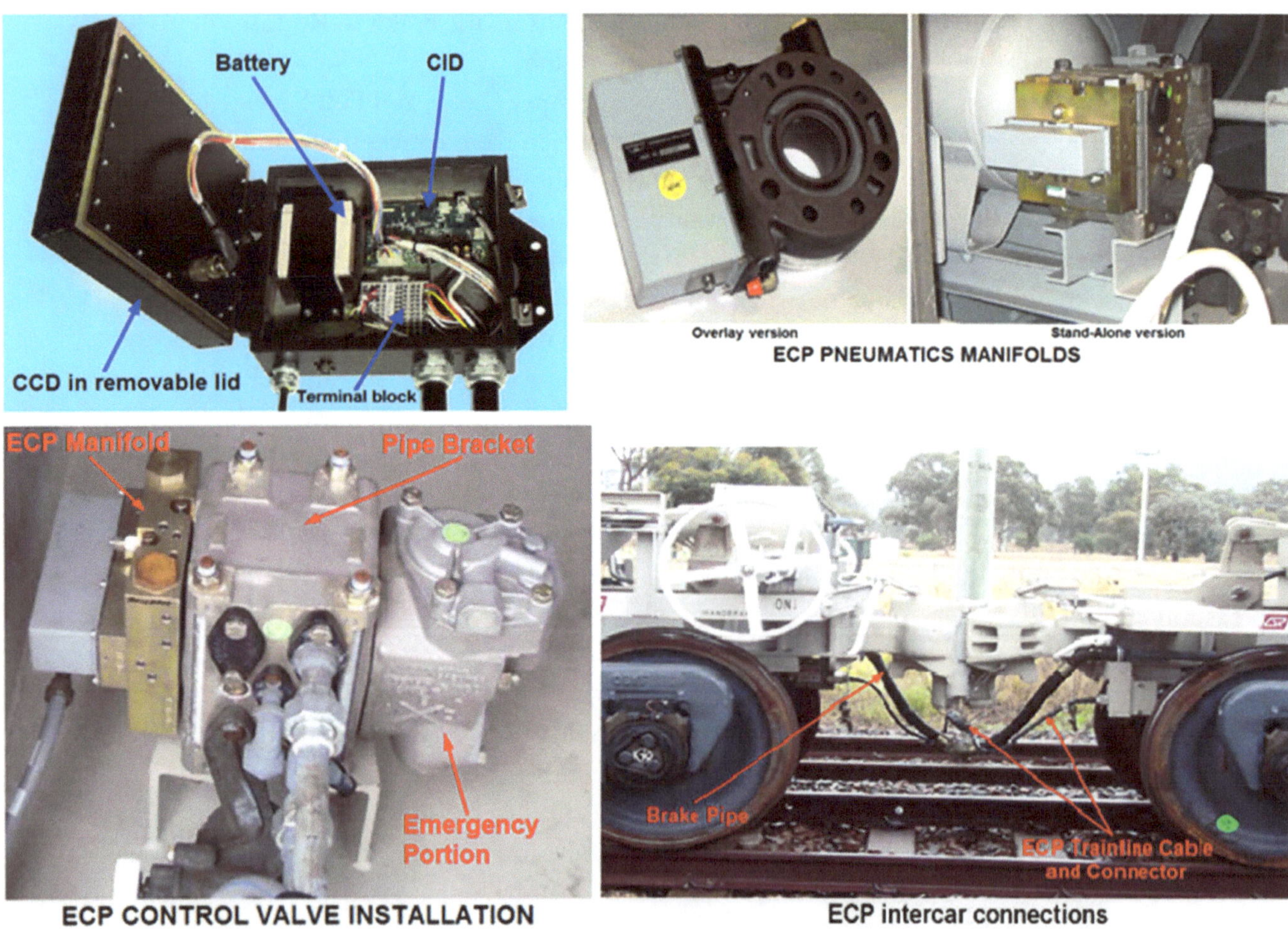

FIGURE 43: ECP-4200 IN AUSTRALIA (2

DP mode requires the control of Remote lead and trail units. In this case, the Remote lead unit includes EP-60 HEU equipment, but any Remote trail units — receiving all commands through the MU cable and air hoses — do not. The intention is to eventually provide for trail locomotives to read and respond to the ECP trainline and eliminate the MU cable and hoses.

- *Wired Distributed Power*

  A fully compatible, AAR-compliant Wired DP system is currently under development at NYAB. In this case, when operating in EP-60 mode, only Wired DP can be operational, with "conventional" Radio DP prohibited. Conversely, when operating in conventional mode with Remote DP active, EP-60 Wired DP will not be possible.

The integrated CCB/EP-60 system has been installed on a fleet of Chinese-built iron ore cars for service in the Pilbara[74] on the Fortescue Metals Group railroad.

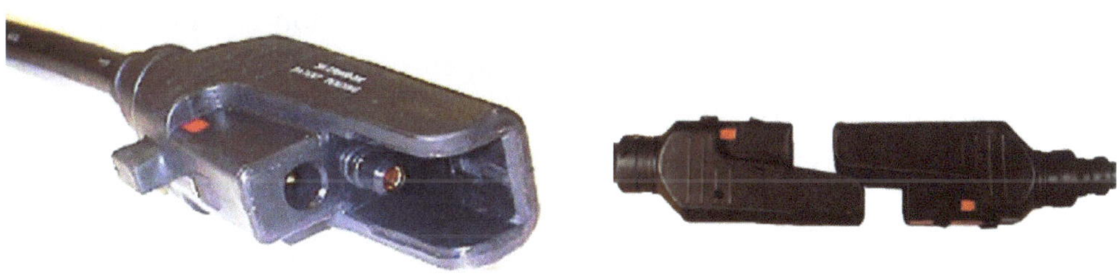

Connector Design adopted by the AAR

FIGURE 44: AMPHENOL *FreightMate*™ ECP TRAINLINE CABLE CONNECTOR

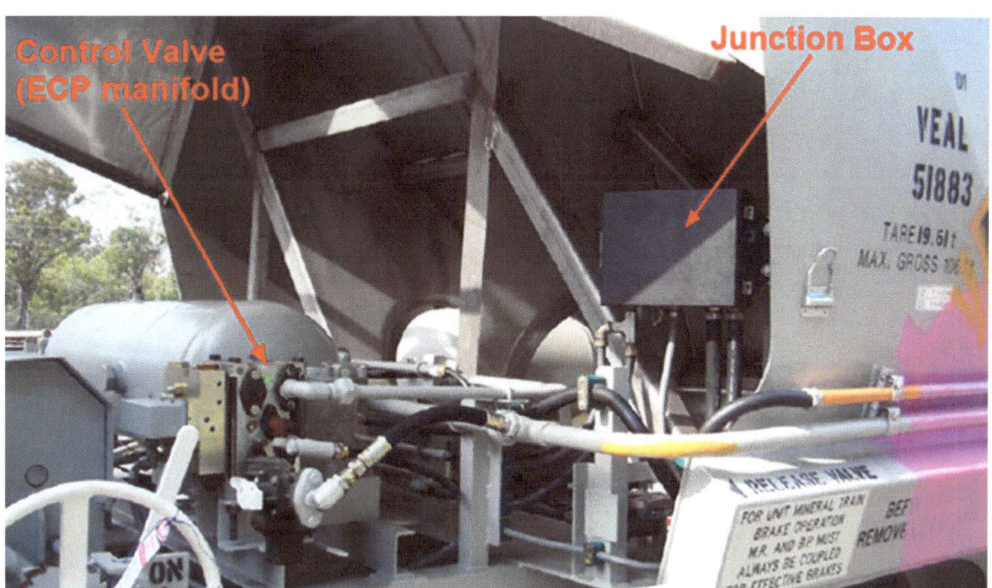

FIGURE 45: ECP FITOUT ON A *QR NATIONAL* COAL WAGON

74. A region in the north-west of Western Australia

The GE Transportation EPx equipment

BACKGROUND

At about the time that BHP Billiton Iron Ore was examining the marketplace for a new-generation train braking system, two North American-manufactured schemes were presented and marketed as developed systems. One was a wire-line system, and the other utilised radio telemetry. The GE-Harris (now GE Transportation) system was marketed as the Legato ECPB system (subsequently changed to "EPx") and was purported to be 'operating' on the Florida East Coast Railroad. This latter claim possibly originated from the fact that Harris Controls in Melbourne, Florida, had long used the adjacent FEC for on-track testing of new systems developed within its Railroad Product Line.

After extensive evaluation, BHPBIO—early in 1998—entered into a contract with GE-Harris to trial their radio-based ECP brake system, including their wireless TrainTalk™ technology. The contract was to test the system to evaluate its suitability for the BHPBIO operation and its provision of operational cost savings and safety benefits. This would be the first implementation of ECP brakes for BHP.

The wireless TrainTalk intra-train communications system uses a direct sequence spread spectrum radio that is said to provide a five-fold increase in bandwidth data transmission capability over cable-based ECP systems and provides a continuous communications link in all conditions, including during train separations. This capacity enables the transmission of data from a variety of other applications such as on-car sensors and diagnostics, EOT functions, and distributed power.

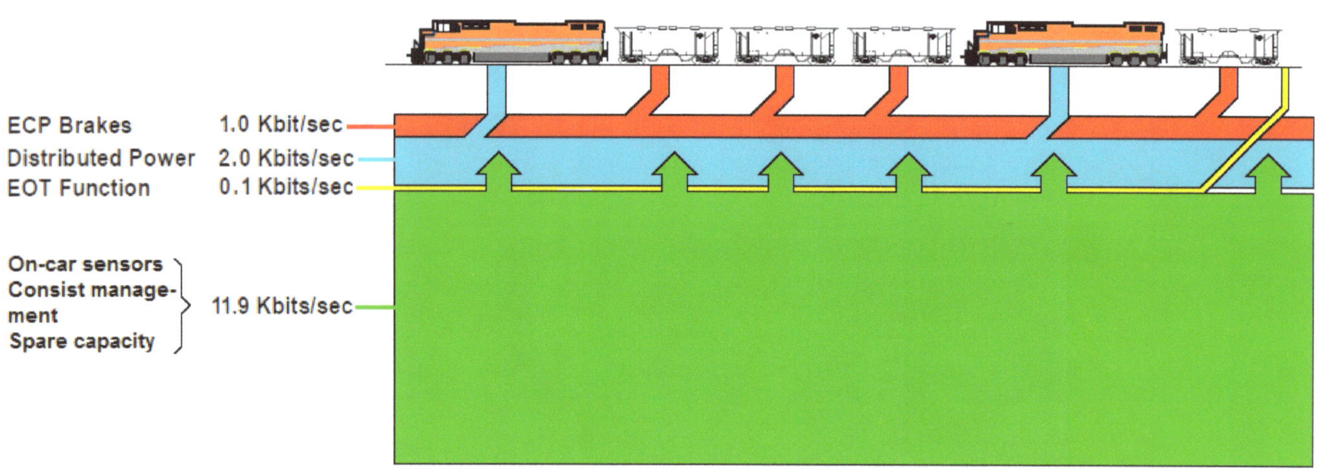

FIGURE 46: GE-HARRIS RAILWAY ELECTRONICS *TrainTalk*™ SYSTEM COMMUNICATIONS CAPACITY – DIAGRAM

The program of works required installation and testing being completed by October 1998 and the six-month trial period to be completed by April 1999. During the program, it was realised that the level of development in the United States on limited train lengths tested at the GE-Harris laboratory and in the field tests on the FEC, when compared to the real-world train lengths of BHPBIO, caused problems that had not been evident to the manufacturers. This resulted in consequent delays to the program at Port Hedland.

The system's development reached a stage where an entire train consist was equipped with the new braking system and a six-month revenue-hauling trial period commenced in August 1999. It was conducted on a conditional basis, since key features such as system "Autolinking" were not functioning correctly insofar as it was requiring 24 minutes to link the consist.

The following section is a perspective on ECP braking generally, and radio-based ECP in particular. It has been adapted from a technical paper written by Tony Lynch, BHP Billiton Iron Ore, and David Peltz, GE-Harris Railway Electronics that discusses the implementation of the EPx radio-based ECP braking system tested at BHPIO. Opinions expressed are those of the writers. The information is used with permission.

Introduction

As part of its continuous improvement philosophy, BHP Billiton Iron Ore was looking for a fast-acting freight braking solution for its iron ore trains. During the early nineties, GE-Harris—who had a long working history with BHPBIO (and predecessor Mount Newman Mining) in the provision of Locotrol distributed power (DP) equipment—submitted proposals for an ECP brake system.

The vast length of BHPBIO trains and their heavy-haul axle loads dramatically affected the in-train forces arising from the intense train dynamics. Along with Queensland Rail, BHPBIO was a long-time, heavy-haul leader in the application and use of DP to facilitate improved handling by drivers of long, heavy mineral trains and reduced consequent effects of in-train forces. Thus, the idea of a fast-acting train brake to complement the advantages of distributed power was appealing. Such a system, it seemed, would increase the railroad's ability to safely maintain higher average speeds, and it was decided that ECP braking could be the answer. GE-Harris was approached to provide a solution, piggybacking off their proven Locotrol system for the radio remote control of distributed motive power.

Radio-based versus Wire-line

Two ECP brake systems were offered to BHPBIO: one a wire-line and the other a radio-based Intra Train Communication (ITC) system. The wire-line system, as the name suggests, requires a physical conductor—running the full length of the train—over which the control commands and data are transmitted to-and-from the cars and the locomotive. The radio-based system employs a Direct-Sequence Spread Spectrum radio[75] to achieve the same purpose. BHPBIO undertook an extensive evaluation and a Failure Modes Effect Analysis on both proposals to determine which promised to be the most cost effective and best suited to their operations.

After much deliberation, BHPBIO elected to trial the radio-based system. Factors that influenced the decision were [1] the required system operating voltage, [2] the capacity to operate over an unlimited train length, [3] the inherent data capacity, [4] the elimination of hardwired-to-smart-car technology, [5] that the equipment was not an overlay on the existing brake system, and [6] that the radio system provided an emulation mode for mixed-fleet running.

- *Operating voltage*

BHPBIO preferred a low-voltage, low-power system so as to allow maintenance personnel to maintain the braking system in a safe environment. The 12 V radio-based system as opposed to the 230 Vdc wire-line system also suited BHPBIO operations.

- *Unlimited train length*

BHPBIO did not want to be restricted in train length. The company regularly runs 240-car trains, and it was their view that any limitation on train length could impact future operational requirements.

75. DSSS is a frequency modulation technique. As with other spread spectrum technologies, the transmitted signal takes up more bandwidth than the information signal that modulates the carrier or broadcast frequency. The "spread spectrum" qualifier derives from the fact that the carrier signals occur over the full bandwidth (spectrum) of a device's transmitting frequency. Certain IEEE 802.11 standards use DSSS signaling.

The length of a train equipped with wire-line ECP is primarily constrained by resistive loss, although inducted noise and impedance mismatching also plays a role. The communication protocol being employed by the wire-line ECP systems is generally a Carrier Sense Multiple Access (CSMA) network[76] protocol. With this, each car contends for the right to transmit on the wire (although critical messages have a reserved transmit time). The term "carrier sense" means that each communicating car listens for signals on the wire for any other car transmitting. This is important because the protocol attempts to avoid simultaneous communications that will result in data collisions and corrupted communications. The protocol at a car attempts to transmit only when data traffic on the wire is idle. The term "multiple access" simply means many cars are contending for use of the wire.

Put another way, every car must be able to hear every other car to prevent communications collisions and corruption. Unfortunately, as the train length increases, the 5.5 V communication signal put out by each car—and which rides on top of the 230 V dc power—diminishes because of the electrical resistance of the wire, to the point where it cannot be distinguished from noise. Boosting this signal partway through the train is not just a simple matter of a repeater or amplifier. A unique network router or bridge had to be installed at the appropriate point in the train to receive the weak signals and process them to re-address and re-transmit the messages on a new and separate network.

The radio-based system, on the other hand, is based on a "token relay network protocol". In this type of network, the cars in the train relay the messages from one car to another in an orderly and structured method, with messages being essentially regenerated each time they are relayed. Sophisticated data handling methods adapted from high security and anti-jam military communications ensure that the messages are accurately relayed. This method results in robust and redundant communications with no limitations on train length. As a practical matter, the radio network management system placed a one thousand-car constraint on the network to optimise addressing space, but this was not seen as an operational limitation.

The high reliability of this method of relaying messages does not come free, and the price paid is the time involved in the structured relaying of messages. This consideration was addressed by using a very sophisticated relaying method, extremely fast electronics, and very high data rates. Because of the advanced design, the radio-based system enjoys the benefits of the token-based network while still delivering messages to all cars in a 240-car train in a fraction of a second.

The chosen radio-based system had a 15 kilobits-per-second bandwidth that allowed expansion into "smart car" technology to provide features and functionality that would assist BHPBIO with future needs. Because the radio-based system was conceived and designed specifically for on-train use, all aspects of the communications means could be selected. A 5x margin above the near-term expected use of the ECP communications and DP systems was selected to allow for future growth in the smart car applications. Currently-fielded ECP braking and DP applications require about 2.5 to 3 kbps of usable data bandwidth. The radio-based system was designed to support up to 15 kbps of usable data bandwidth. The actual transmitted bandwidth of the radio system is 1 megabit per second. All of the differential bandwidth is used for ensuring reliable communications. Figure 46 illustrates the amount of available capacity for new applications.

---

76. Carrier sense multiple access (CSMA) is a *probabilistic media access control* (MAC) *protocol* in which a node verifies the absence of other traffic before transmitting on a shared medium such as an electrical bus, or a band of the electromagnetic spectrum. "Carrier sense" means that a transmitter uses feedback from a receiver to determine whether another transmission is in progress before trying to send; that is, it tries to detect the presence of a carrier wave from another station before attempting to transmit. If a carrier is sensed, the station waits for the transmission in progress to finish before initiating its own transmission. In other words, CSMA is based on the "sense before transmit" (or "listen before talk") principle. Transmissions by one node are generally received by all other stations using the medium.

• *User-available bandwidth*

The wire-line system, on the other hand, is limited to about 3.3 kbps of available user bandwidth—as recommended by the hardware manufacturer—from the total transmitted bandwidth of 10 kbps. Looking at the bandwidth needed for the applications in Figure 46, this means that virtually all of the throughput would be used by a combination "ECP brake and DP" application. There would be nothing for smart car applications. While it is possible to restrict or minimise the amount of data sent to fit within 3.3 kbps, GE-Harris suggested that requiring this was a clear sign that the wire-line communication system was undersized from the start.

• *Avoiding the requirement for inter-car connectors*

The wire-line system presented a potential maintenance problem and failure point at the inter-car connector. It was also considered to be a handicap in cutting cars in and out in marshalling yards. While it is true that ECP brake systems still use gladhands for coupling the car brake pipes, having yet another connection to make and un-make was considered an unnecessary burden. [The author notes that the Freightmate II version of the connector—which hard-mounts the ECP connectors on the brake pipe glad-hand so that only one coupling action is necessary to make both connections—might negate this problem.]

• *Compatibility with Locotrol*

The ease of integration and the compatibility with the distributed power system was considered a benefit. The majority of trains run on the BHPBIO main line are DP trains. The radio-based system would allow distributed power—which is transmitted over the company's wayside radio system—to be carried over the ECP brake system radio link. This would eliminate black spots that can potentially disrupt the service. The radio-based ECP system was delivered with Locotrol compatibility built in. For locomotives not equipped with an integrated driver's display, a dual-screen Consolidated Display Unit (CDU) was provided to minimise cab clutter. The CDU allowed ECP brake operation to be displayed on one screen and standard Locotrol operation on the other, all within one box the size of the current Locotrol control console. The screens could be swapped for redundancy at the user's discretion. If a railroad was not currently using Locotrol, then the CDU provided redundant screens for ECP operation.

The migration of Locotrol onto the intra-train communication (ITC) system was made as seamless and flexible as the customer desired. The 2.4 GHz TrainTalk radio-based system works in parallel with the Locotrol UHF radios. In this way, a railroad may continue to use its existing Locotrol radio equipment and cut over to the radio-based ITC system at a time of its choosing. Both Locotrol and the radio-based ECP systems incorporated the interfaces to account for the elimination of brake pipe reductions during ECP brake operations because brake pipe reductions and recharges are normally used by Locotrol for certain communications redundancy.

With the radio-based ITC system installed, a railroad may switch to this full-time communication system for Locotrol as well. The radio-based ITC system needs no repeaters because of its sophisticated token relay method of communication. In other words, there would be no "shadow zones" or need for Locotrol radio repeaters. A significant advantage of this approach was the retention of most of the already-installed Locotrol hardware. The only portion of Locotrol that would be unused was the UHF radio. In this way, a railroad may continue to enjoy the proven performance of its existing Locotrol equipment while gaining the benefits of full-time communications and ECP braking performance.

• *Ability to skip around defective cars*

Inherent in the radio-based system is the ability to skip around defective cars and provide continuity of service, whereas—with the wire-line system—if the wire is broken, ECP brake system functionality beyond that point is lost.

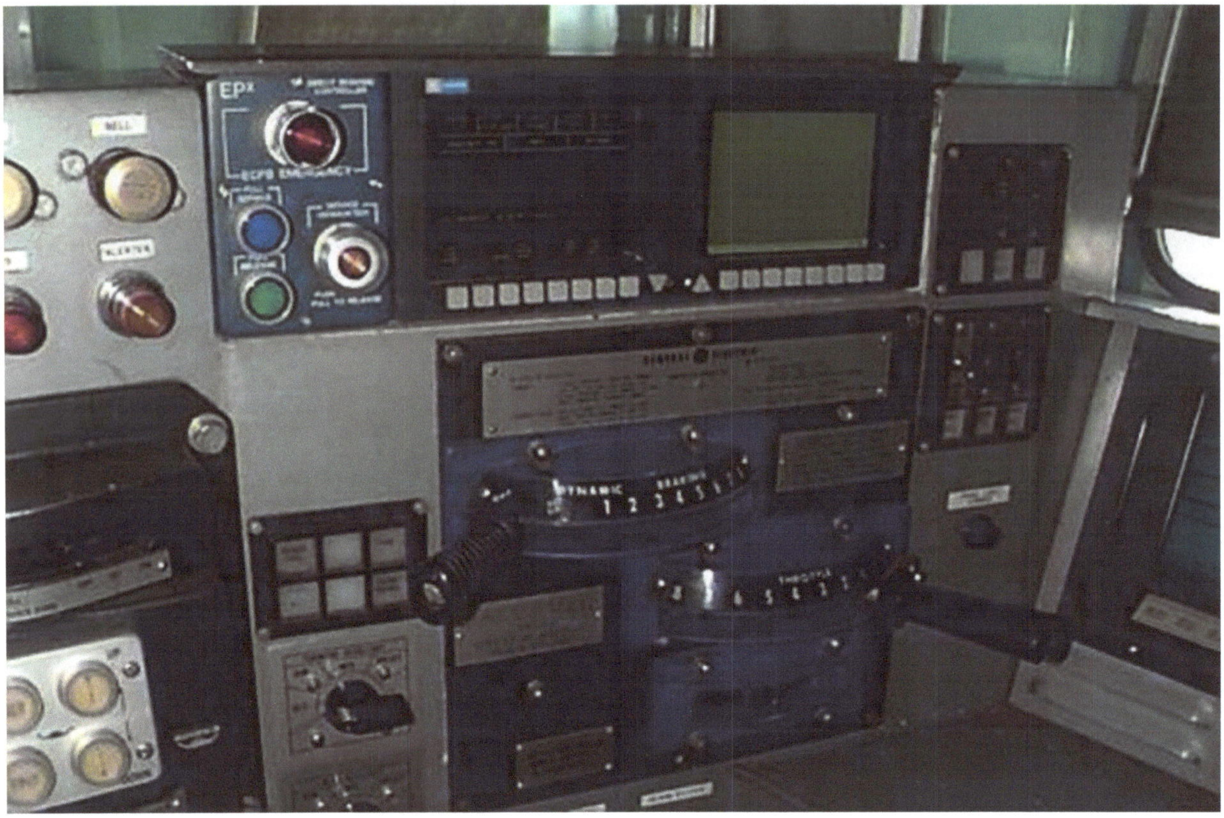

FIGURE 47: EPx *CONSOLIDATED DISPLAY UNIT* INSTALLED IN A BHPBIO LOCOMOTIVE CAB. THE COLOURED BUTTONS ARE USED FOR AIR BRAKE OPERATION AND ARE THE SAME AS THOSE USED ON A LOCOTROL AIR BRAKE CONSOLE. *Photo J.Ward, Western Australia*

All ECP brake systems are designed with fail-safe communications. A regular communications beacon or "heartbeat" must be received by each car, otherwise that car will initiate communications failure procedures to stop the train, a communications loss of several seconds will initiate a penalty brake application. Ultimately, the challenge for the radio-based ECP communication system was for it to be reliable enough to allow the train to keep running.

GE contends that in the wire-line system, every connector between cars becomes a potential single point of failure that may stop the train. The radio-based ECP communication system bypasses this potentiality - the radio signals will ignore failed cars or relay nodes. Fifteen percent of the cars in an ECP train are allowed to fail before the ECP system must enforce a penalty brake application.

- *Seamless transition into fleet*

The radio-based system can operate in conventional pneumatic emulation mode, which allows it to replicate the operation of a conventional ABDX pneumatic control valve and permit an ECP-equipped car to run in a conventional train. As such, mixed car fleets can be run during the period taken to equip the full fleet.

That the power source is on each car is key to this kind of installation. Therefore, an individual car or several cars may have ECP brakes installed and then be returned to service in conventional pneumatic mode. When a full consist of cars has been thus equipped, it is a simple matter to just push the button on the Head End Unit to enter the ECP mode.

For a retrofit, wire-line systems mean that the entire train will be out of service during installation because the wire is required for power throughout the train. This out-of-service time is a significant problem for organisations such as Pilbara iron ore companies or eastern-states coal-haulers with long consists.

Some wire-line vendors attempt to address this by offering an "overlay" system where the conventional control valve Service portions are retained and mounted alongside the new ECP valves on a special manifold. Contrary to the goal of reduced maintenance, this places two completely separate and different brake control valves on every car.

- *Ease of migration to Smart Car technology*

Specific on-board features, such as *load-weighing* (which will also facilitate proportional braking of individual cars), *bearing condition monitoring* (which will provide real-time monitoring of bearings as opposed to wayside hot box detectors that provide after-the-event warnings), *hand brake detection,* and the potential for eliminating the need for wayside systems for on-board reporting provide favourable options for operators of large trains.

An additional feature that BHPBIO explored was the ability for the radio-based system to communicate with the PMAC[77] system that controls the Car Dumper operations for the emptying of ore cars at the port terminal. At present, all rakes of cars being processed through the car dumpers require the attachment of a special compressor car at the tail-end to maintain brake pipe pressure and to hold an Independent brake application (on the compressor car) at a predetermined setting in order to manage the recurrent slack action as a pusher arm indexes the consist through the dumper. It was intended to have PMAC communicate directly with the radio-based ECP system to provide real-time braking requirements as the consist was moved through the car dumper, thus affording accurate control over its movement and eliminating the high-maintenance compressor cars.

The EPx Direct Braking system - Description

EPx is designed to operate independently of other railcar systems:

The level and number of interfaces with other locomotive and car systems balances the need for integration of these car system functions with the need for autonomy of the safety-critical braking resources. As a result, less items requiring to be coupled between locomotive and other car system components and the braking system components helps to ensure the safety-critical braking functions are less-affected by failures in these other systems. It effectively isolates the safety-critical braking resources from others less critical.[78]

EPx operates in two modes: 'Emulated ABDX' and 'ECP'. In Emulated ABDX mode, the system imitates an ABDX braking system, sending and receiving brake commands through the brake pipe. In ECP mode, EPx sends and receives brake commands via a radio network that links each car. The amount of brake effort to apply is specified in brake commands sent every second through the radio network to each car. Extensive use of fault-tolerant design techniques ensures high availability of the electronically-controlled braking services.

The system is composed of two major components, the Car Control Devices (CCD) and the Head End Units (HEU). EPx employs a decentralised design that distributes the braking services—including the detection of system-wide failures—to each CCD on each car. The effects of the failure of a single CCD or even a small group of CCDs is limited in scope and often results in leaving the overall braking service minimally affected. The design of communications pathing between CCDs is such that the complete failure of one communications path to a CCD would not cause loss-of-control of the braking services on that car. Also, the failure of one CCD would not cause loss-of-control of the braking services on the other cars.

The basic principle adopted in defining the architecture to best meet the system's functional goal is fault tolerance through replication. All brake control commands are duplicated within each brake command message and each message transmitted more than once. The primary principle in defining the architecture used to meet the system's safety goals is to 'fail-safe' by applying emergency braking until the train comes to a stop. In the event of loss-of-communications with the HEU, the failure of the CCD at the EOT, or a rapid

77. Process Measurement and Control.

78. Knorr-Bremse.

loss of brake pipe pressure, each CCD will make an Emergency brake application. All Emergency brake applications are propagated through replicated "Train Stop" messages transmitted over the radio network and through the reduction of brake pipe pressure.

The system architecture extensively follows the applicable FRA and AAR brake rules as required for interchange service, thus providing EPx with a safe and solid foundation. The handling of brake control services is separated from the processing of information exchanged over the radio network. All ECP brake commands exchanged within the system use the radio network. All processing of incoming or outgoing radio messages is allocated to the Intra-Train Communications Unit (ITCU). The ITCU implements the TrainTalk communications protocol, transforming incoming messages to the required internal format and outgoing messages to the required external format. By this means, the impact of interface changes or new interfaces is localised and minimises the effect on the application software. This approach is intended to enhance the level of safety and cost effectiveness associated with the introduction of new interfaces into EPx.

Testing EPx

In entering into a contract for the supply of an ECP brake system, BHPBIO recognised the level of potential risk associated with a product that had not been fully developed. The company also recognised that it was very likely to be the first railroad to have a radio-based ECP brake system. As a consequence, after an intense series of FMEA workshops, a contract and specification were prepared that adequately addressed all issues to provide a level of safety acceptable to the company and that called for the fit-out of one train consist of 240 ore cars and four locomotives. The intent of the agreement was for a six-month trial period after the successful testing and commissioning of the system. The trial period was to allow BHPBIO to evaluate and determine the effectiveness of the ECP system and its suitability to company operations.

Trials were conducted to prove the system met contract requirements and the company's expectations (including the realisation of a 50 percent reduction in stopping distance and the almost total elimination of slack action).

Key parameters—including operational mode, train loaded condition, track conditions, and track gradient—were tested and an immense amount of technical and operational data acquired.

The following three figures show selected results from these tests. Figure 48 compares the stopping distance for a Minimum Service application at various speeds for an empty conventionally-braked train, an empty ECP-braked train,

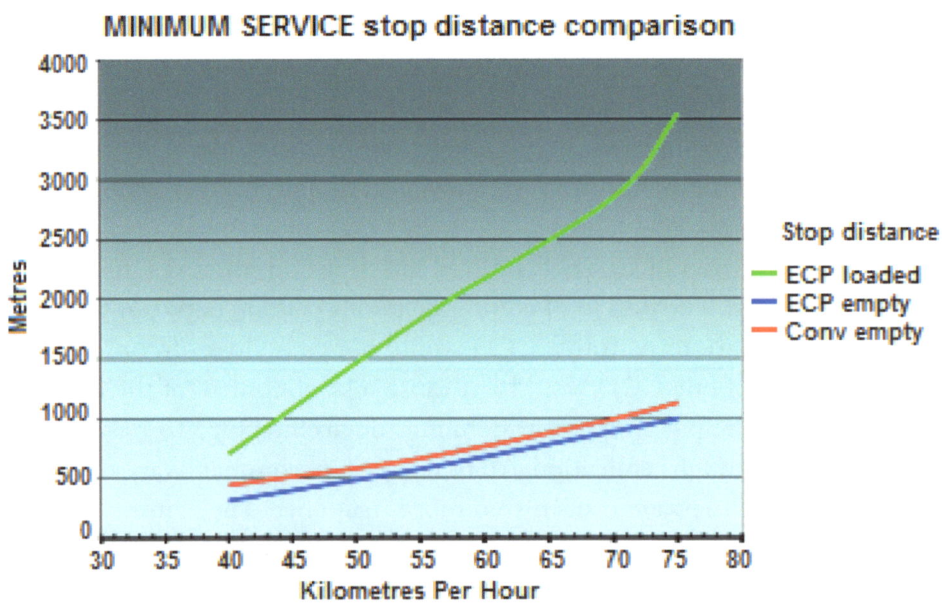

FIGURE 48: EPX STOPPING DISTANCE COMPARISON - *MINIMUM REDUCTION* SERVICE BRAKE APPLICATION

and a loaded ECP-braked train. As it happened, too few stopping distance tests with loaded conventionally-braked trains were able to be performed that would provide statistically relevant results to publish; these tests were to be conducted during the evaluation period. It should be noted that locomotive crew comments served to indicate the exceptionally better stopping performance with a loaded ECP train over a conventionally-braked loaded train.

Note that the difference between the conventionally-braked and ECP-braked trains is less pronounced for light air brake applications. The reason is that the effect of slow conventional pneumatic command propagation is not a significant factor given the slow deceleration of the train. Still, the ECP stopping distance is about 18 percent better than the conventionally-braked train. What cannot be obvious from this graph is the elimination of slack action on the ECP-braked train.

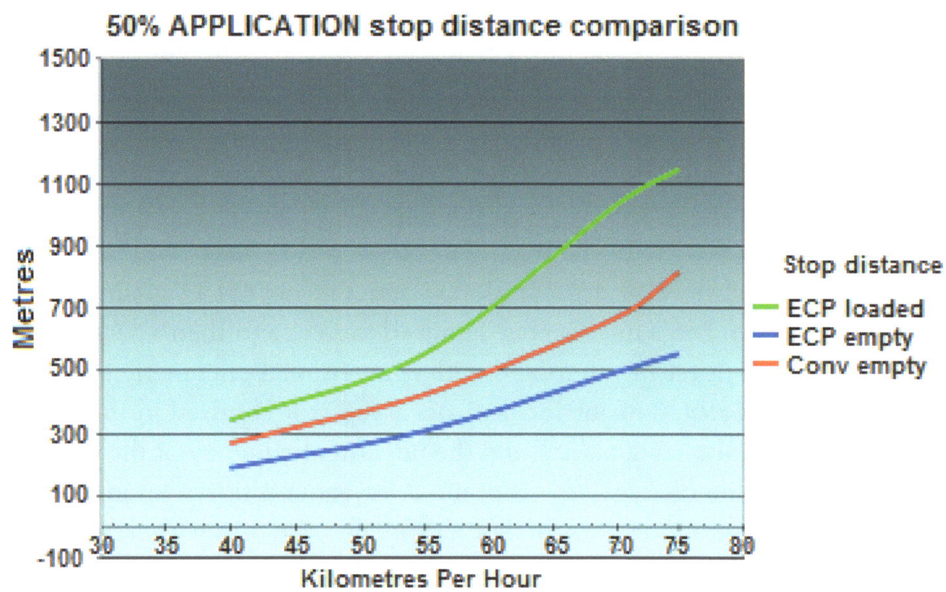

FIGURE 49: EPx STOPPING DISTANCE COMPARISON - *50%* SERVICE BRAKE APPLICATION

Figure 49 shows the average results of the 50 percent application stopping distance tests. Notice the (approximate) 28 percent improvement in stopping distance for the ECP-braked train.

The advantages of ECP braking in terms of stopping distance become obvious with a Full Service application as seen in Figure 50. Note the improvement—exceeding 50 percent—in stopping distance with the ECP-braked train. In fact, up to about 63 km/h,

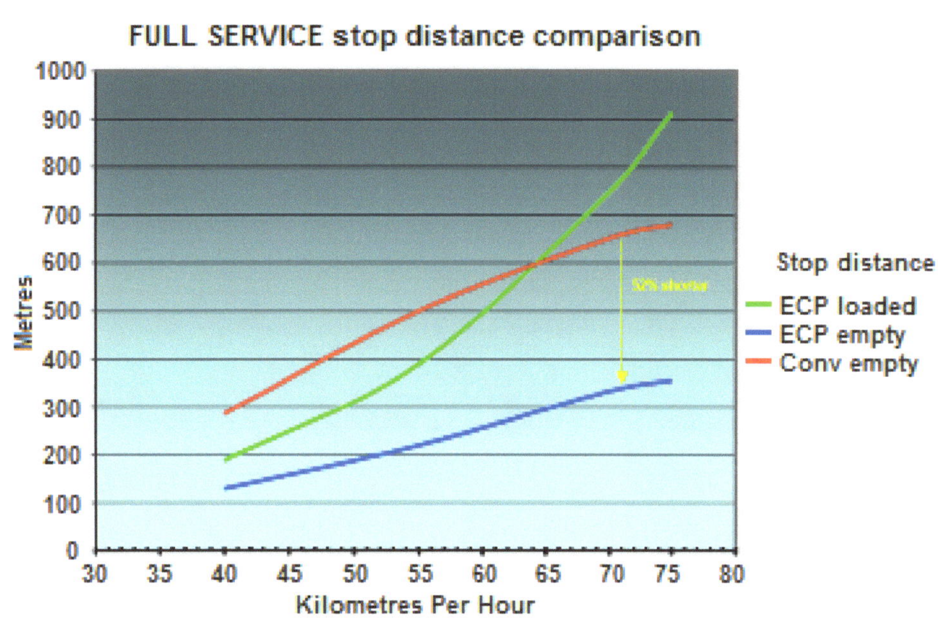

FIGURE 50: EPX STOPPING DISTANCE COMPARISON - *FULL SERVICE* BRAKE APPLICATION

the ECP stopping distance for a fully loaded train is better than that for an empty conventionally-braked train. Again, this marked improvement in performance is accompanied by the elimination of dynamic slack action.

Perhaps as important as the gathered data are the reactions of the train drivers involved in the tests. The ECP technology is truly revolutionary. The locomotive drivers' reaction to the enormous improvement in braking control was enthusiastically positive. Not only is the train-braking response virtually instantaneous, but it is incredibly smooth as well. It is intriguing to watch a driver make his or her first ECP Emergency application from road speed and instinctively brace for the run-in of slack that never occurs. Recovery from an ECP Emergency application is noticeably shorter as well since the brake pipe is not dumped for an operator-initiated Emergency stop.

The graduated release feature allows unheard-of ease of operation and precision in control of the train. Crew members instantly recognised the benefits of being able to effortlessly make whatever degree of release was desired. One of the many ECP tests was to make an intentional over-application with a fully-loaded train on a 1 percent downgrade to come up short at a signal stop. A graduated release was made to permit the train to run a further four hundred meters and then a re-application to stop safely in rear of the signal. This could never be attempted with conventional pneumatic brakes.

[In October 1999, a major incident occurred that resulted in the runaway of a loaded train down a 2 percent grade on the Yandi branch and the uncontrolled entry of the train onto the main line. The runaway train—with no-one on board[79]—continued for around thirty kilometres until it encountered an adverse track gradient whereupon it came to a stand. The train was withdrawn from service and it was discovered that the test version of the software that had been installed on the locomotive in the Hedland rail yard to facilitate faster static testing of the EPx system had not been removed and replaced with operational software prior to the train's departure. Engineers hastened to write a software modification that would prevent the movement of a locomotive with test software installed, and specific operational practices and procedures were initiated to further prevent recurrence. Revenue trials were recommenced in January 2000 and were completed by July. The incident, though, unnerved BHPBIO, and they did not proceed any further with EPx. To the best of the author's knowledge, EPx is not in operation on any railroad. BHPBIO now utilises the NYAB EP-60 wire-line ECP system.]

79. The train had stopped on the downgrade with an unknown technical fault. The driver made a brake application to hold the train, had left his cab, and was moving along the train applying car hand brakes when the air brake application "kicked off" and the train moved away before he could take any action to prevent it. Engineering test personnel who were accompanying the train had also vacated the cab at the time.

3.

The Operating Valve in Australasia

Westinghouse Down Under

THE TECHNOLOGY OF air braking came early to the colonies, albeit not without controversy. While many railway engineers of the time were visionary, some were less so. Thomas Higinbotham, engineer-in-chief of the Victorian Railways[80] throughout the mid-1880s, for example, originally held a cautious and comfortable view—apparently quite prevalent in Mother England at the time—that separate car braking was eminently preferable to the artless notion of continuous brakes then being propounded in its various forms throughout the industrialised world. Other influential people—including some of those fortunate few who populated the hallowed corridors of power—also held obstructionist and ill-informed opinions and, frustratingly, also the purse strings. Yet, still others—both within and beyond the railway administrations—acknowledged the self-evident need to have trains equipped with a reliable and effective driver-controlled "stopping system." The embryonic issue of the application of power brakes to the nation's railways generated nothing if not passionate debate.

Within Australia, the state of New South Wales—with the emergent nations' first public railway—clearly led the way. Within the mechanical engineering fraternity in colonial NSW, there was considerable curiosity about railway air brake systems and, as in the United States, much difference of professional opinion. A certain Mr. List had introduced a patented braking system to the Engineering Association of NSW in 1891, and an association member, Mr. Norman Selfe MICE, MIME[81] had presented a paper to the association—also in that year—describing the train braking system invented in the United States by an acquaintance of his, a Mr. William Hanscom[82] "one of the best-known engineers of the Pacific Slope." This latter individual had succeeded in having his air brake arrangement installed upon a Californian railway[83] and was apparently prominent in the cable tramway industry in the United States. He had adapted the Westinghouse brake in an attempt to overcome three perceived drawbacks:

80. Refer to chapter 7 for the biographies of some of these personalities.

81. Ibid.

82. William Wallace Hanscom. (San Francisco, California, USA). US Patent 326,646, 22 September 1885, Hanscom Straight-Air Automatic Brake Company, 1886.

83. The North Pacific Coast Railroad (NPC), a common carrier narrow gauge steam railroad begun in 1874.

...the entire severance of the driver's control from the brake cylinders during application of the brakes, and the impossibility of graduating off. When the brakes are on he can keep them on and keep up the supply of air from his pump to the cylinders for any period.[84]

Since the paper was delivered in the shadow of the recent Peats Ferry (or Peats Ridge) train runaway accident in NSW[85] (attributed by the authorities, perhaps a little complacently, to the driver's inexperience with the Westinghouse brake), Selfe—perhaps with a self-serving eye on the possibility of future equipment purchases by the NSW railways—took advantage of his presentation to suggest that the accident resulted from *"certain inherent defects in the Westinghouse brake...."* Nonetheless, Selfe did appear to passionately believe that *"the brakes then in use could be made in the colony for far less than they cost the Government (apart from royalties)"* and, furthermore, that there was plenty of room for the Westinghouse product to be improved, and there was no reason to believe this could not be achieved with local expertise.

Selfe was not directly connected with the railways, although he had apparently designed the gas lighting system then in railway use. He had, however, developed a mechanical and scientific interest in the action and efficiency of the railway brakes then in use and, indeed, had received from his US contacts the full particulars of the Burlington brake trials. He thought it possible to devise a brake that combined the good qualities of the Carpenter, Hanscom, and Westinghouse brake systems without what he considered to be their inherent design defects. As a result, Selfe prepared some drawings and, in 1889, eventually submitted a proposal to the Commissioners for Railways through their Special Brake Board (a group of experts appointed to inquire into the whole question of railway brakes within the colonies). This board consisted of the following: W.D. Cruickshank, Marine Superintendent, NSW; T.F. Rotheram, Locomotive Superintendent, New Zealand Government Railways (who would eventually move to superintend the railways in Western Australia and whose early advocacy for the adoption of air braking on the NZGR would come to nought until the New Zealand government released the purse strings following the Rakaia collision);[86] W. Thow MICE, MIME, (and a member of the American Railway Master Mechanics' Association), Locomotive Engineer, NSW Railways (and prior to that, South Australia); Professor Warren, Sydney University; and D.H. Neale, an expert railway engineer.[87]

Sometime during 1891, this Special Brake Board conducted some braking trials, although only two systems appear to have been assessed – the Westinghouse and a vacuum brake that is not further identified. We are left to wonder why the results of the much more comprehensive Burlington trials in the United States did not provide sufficient information for the worthy members of the board. Suffice to say that when

84. Minutes of Proceedings, The Engineering Association of NSW, 1891, vol. VI, 13–132.

85. On June 21, 1887 (two years prior to the opening of the Hawkesbury River bridge), as Australians celebrated Queen Victoria's Golden Jubilee public holiday, an excursion train ran out of control on the 1-in-40 downgrade of the Cowan Bank from Peats Ridge and crashed into the Hawkesbury River at Brooklyn (originally Peats Ferry) 65 km north of Sydney, killing six people and injuring seventy. The result could have been worse but for the brave act of a railway pointsman, Patrick Proctor, who—realising the approaching train was a runaway and was likely to collide with another passenger train on an adjacent track at the Brooklyn station—sat on a points lever within extremely close proximity to the speeding train to divert it onto a stub-end siding, whereupon it collided with some freight wagons and plunged into the river alongside the rail embankment. Driver Thomas Wilson was killed, as was an engineer employed by railway contractor Amos Bros. who was riding in the cab. Fireman Pye fell into the river and survived. Driver Wilson was found to have manipulated the brake injudiciously and to have exhausted the car auxiliary reservoirs, effectively running out of air pressure with which to control the train. Moreover, the brakes were not tested properly as required, either before departing Sydney or after remarshaling the train at Hornsby; the locomotive was not sufficiently powerful for this train; and Railways Department supervision was deficient.

86. On the evening of Saturday, 11 March 1899, in gale-force driving rain, the second of two excursion trains returning from Ashburton to Christchurch ran into the rear of the first at the Rakaia Railway Station, killing four passengers and injuring twenty-two. Only the locomotives were equipped with air brakes, and the accident led to the fitting of air brakes to rollingstock and improved signaling (Rakaia station had no fixed signals), interlocking, and safeworking (Tyer's Tablet system). Driver Charles Carter was acquitted of manslaughter but was dismissed after a Commission of Inquiry found he was negligent in not observing the regulations governing an approach to a station. Carter was considered to have been the victim of a "railway system suffering from severe growing pains and using operational methods that had not kept pace with increased traffic."

87. Refer to biography in chapter 7.

it became obvious that his proposed system had not been as enthusiastically received as he had hoped for, Selfe was derisive of those who were more enamoured of Westinghouse and indignant at the lack of interest in Hanscom's design (which it appears that Selfe had strongly identified with). Selfe proclaims bitterly in his presentation to the Engineering Association of NSW:

The author [i.e. Selfe] *knows well that there is hardly any greater sin in New South Wales than for a person to try and serve his country. He had a desire, however, that he might in some small way assist in building up a New South Wales Railway Brake, and thus help keep in the country the hundreds of thousands of pounds which are sure in the near future to be spent on these appliances.*

Selfe's design did not prevail. However, as an invention introduced and championed so vigorously by an eminent Australian engineer—and for its technical distinctiveness—it is worthy of acknowledgement and discussion as part of the story of railway air braking in these colonies.

Describing it as a "Differential Brake," Selfe's claims for the system included that:

- it was a pressure brake that would utilise the air brake plant then existing on NSW locomotives and rollingstock;
- it was graduable in Release;
- it would maintain against reservoir and cylinder leakage (and therefore improve upon both the Westinghouse and vacuum systems);
- it eliminated guesswork on the part of the locomotive driver when making brake applications and releases as to what brake cylinder pressure he was carrying;
- it left the driver in control of his brake application at all times between the initial application and the final release and did not require a brake release in order to replenish the reservoirs;
- the position of the "train pipe" cock for each vehicle (the system did not use conventional headstock angle cocks) was made obvious to an observer, meaning that any defective/isolated brake was clearly indicated; and
- it could be coupled to and used with the Westinghouse system then in use.

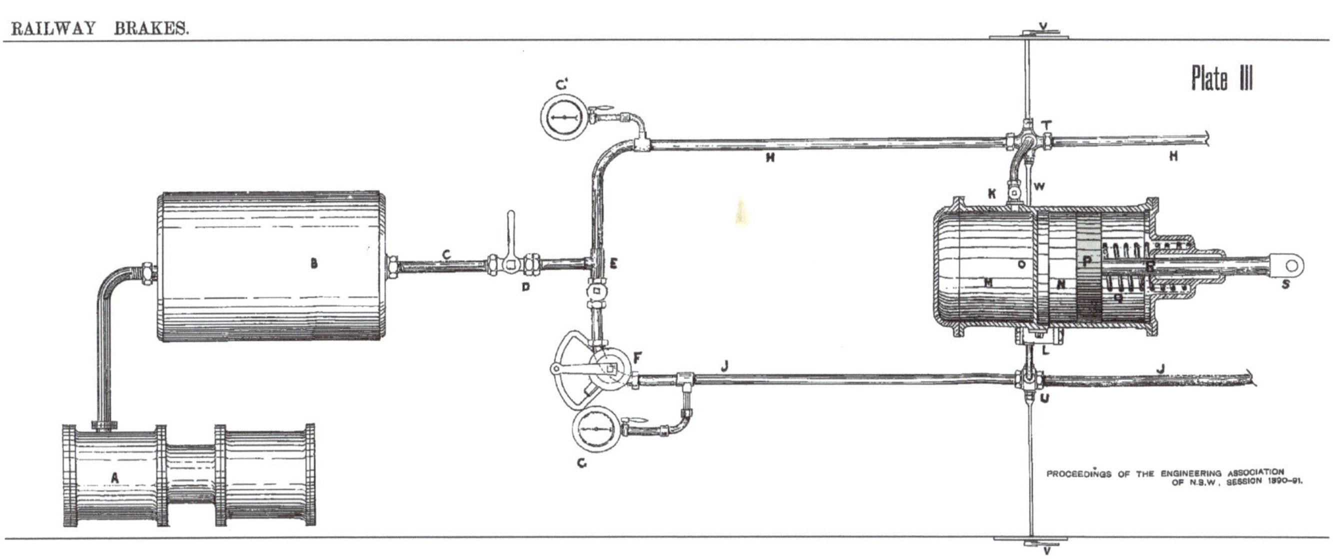

FIGURE 51: THE SELFE BRAKE (PLATE III)

Figure 51 shows the arrangement as applied to rollingstock. In this diagram **A** represents the locomotive air compressor, **M** the "brake reservoir" (today we might call it the auxiliary reservoir) supplied from the main reservoir **B** by train pipe **H** via the train pipe cock and retention valve **K**. The brake reservoir was thus kept continuously charged at a slightly lower pressure than the locomotive main reservoir according to the value of the spring in valve **K**. Brake cylinder **N** was divided from **M** by division **O**. **L** is the operating valve itself. A second train pipe **J**, variously referred to as either the "controlling" or "governing" train pipe was managed by the driver's brake valve **F**. Each train pipe had a cab pressure gauge.

It appeared that Hanscom eschewed conventional nomenclature. Selfe referred to the proposed operating valve variously as a *Differential-* or a *Controlling Valve*. The device—Figure 52—consisted of two pistons (or diaphragms if desired; refer also to Figure 55, Plate VII) connected so as to move together and thus occupying two separate cylinders and creating four separate chambers.

The right-hand cylinder cover of the differential valve connected that chamber of the right-hand piston to atmosphere at **Y**, thus pressure in chamber $\mathbf{J^{II}}$ had to equal the sum of that in chambers $\mathbf{J^{I}}$ and $\mathbf{L^{I}}$ to produce equilibrium. Since $\mathbf{L^{I}}$ was open to brake cylinder **N**, $\mathbf{J^{II}}$ to the brake reservoir **M**, and $\mathbf{J^{I}}$ to the controlling train pipe, the pressure developed in the brake cylinder would be whatever was the difference between the reservoir and train pipe pressures. This pressure relationship provided for brake cylinder pressure maintenance as well as graduated release when the controlling train pipe was partially recharged from the drivers' valve. Springs (Figure 54, Plate X) were provided to inhibit any tendency of the valves to "dither" under equilibrium and to keep them centred. Since brake cylinder pressure was entirely dependent upon variation in the relative pressures of the two train pipes, the pressure gauges provided the driver at all times with real-time brake cylinder pressure information.

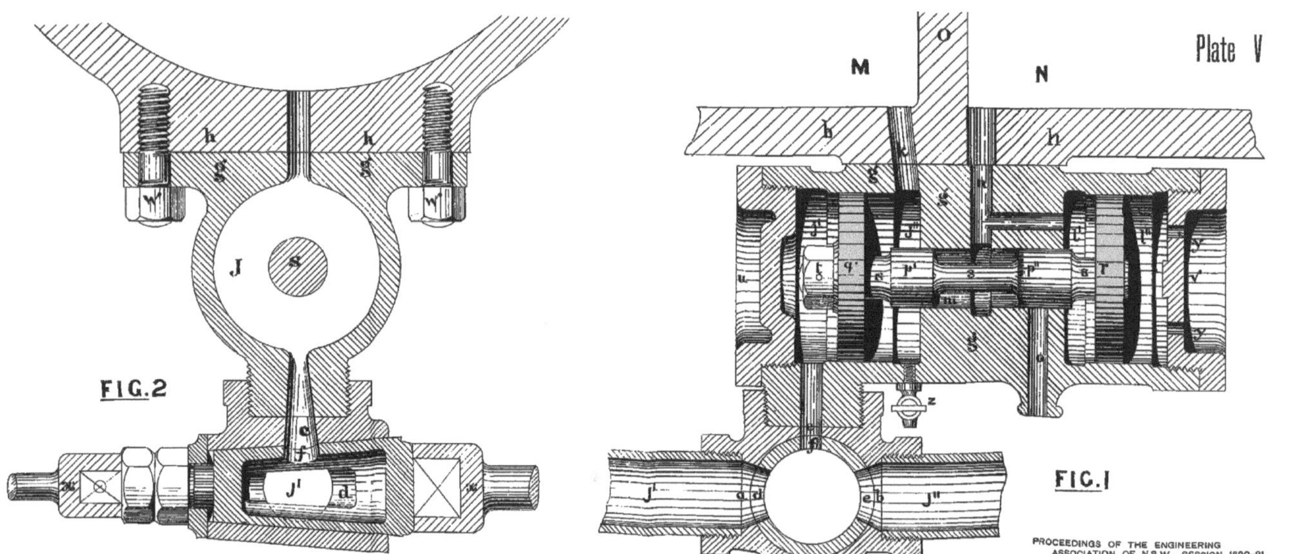

FIGURE 52: THE SELFE BRAKE (PLATE V)

Train pipe cocks were connected to a handle on each side of the vehicle (**V** on Plate III; see also Plate VI) with an indicator to visually convey their settings (as described by dot-point on the previous page), being effectively OPERATING, ISOLATED, or PLUGGED (this last applying to the last vehicle, whereby its brake was operating but the train pipe beyond that point was isolated). While this arrangement would trap a considerable amount of air in the train pipe to be released during uncoupling, a small port in the cock plug would provide the necessary "bleed" or pressure relief function. Selfe presented three versions of a unique method for coupling dual brake pipes via a single gladhand or "palm" (Figure 57, Plate IX), and—while acknowledging the added complication of an extra train pipe—argued that there was a Westinghouse precedent for this,[88] invoking an old colloquialism by suggesting that "*....it is for the members to discuss whether the mackerel of simplicity, security and certainty aimed at in this brake is worth the sprat of an extra train pipe.*"

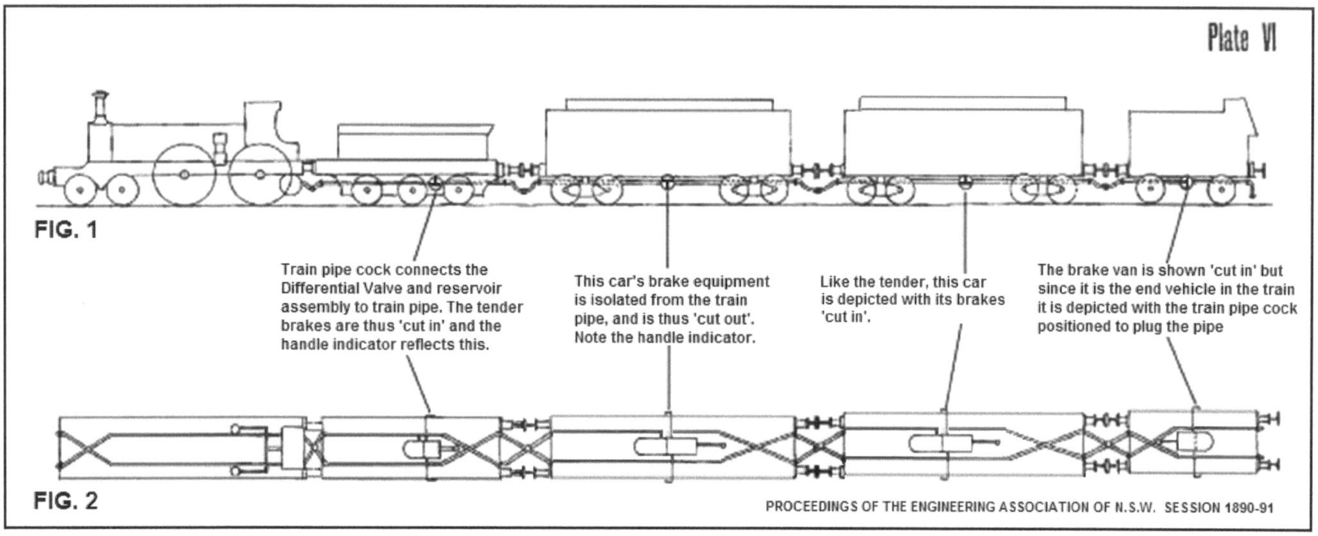

FIGURE 53: THE SELFE BRAKE (PLATE VI)

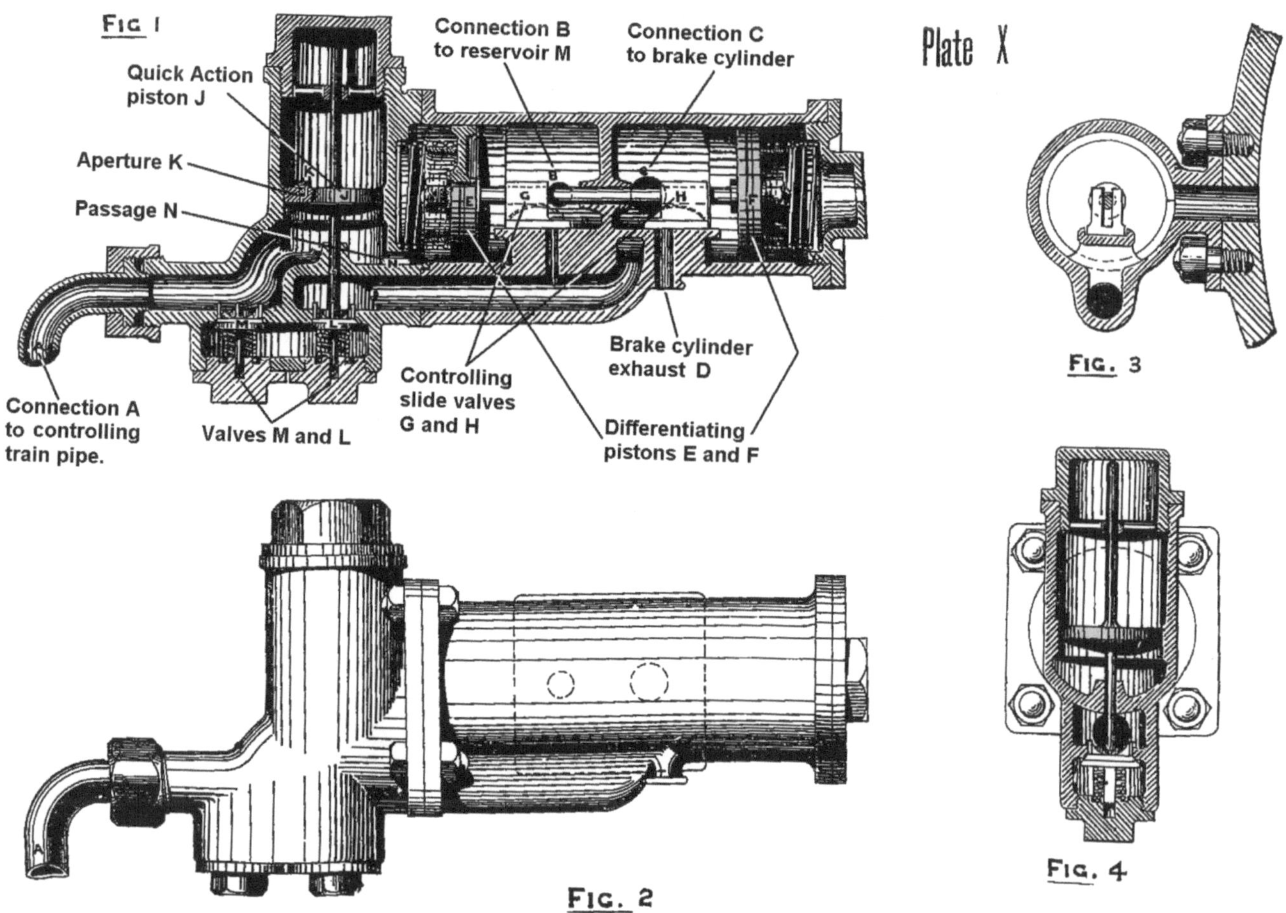

FIGURE 54: THE SELFE BRAKE (PLATE X)

88. The early design intention of George Westinghouse's air brake system was, indeed, that it included two train pipes. See footnote 87.

Finally, Selfe contends that the design pursues Westinghouse's precedent in two further ways: (1) in providing for electric operation (Figure 56, Plate VIII) that *"would only be required for emergent* [sic] *stoppages on very long trains"* and (2) in devising a supplementary "quick application" valve (Fig. 1, Plate X) that would supply into the brake cylinder air that would otherwise be exhausted at the driver's brake valve during a heavy application; effectively a version of Westinghouse's proven Quick-Action. This latter feature was proposed to operate as follows (Figure 54):

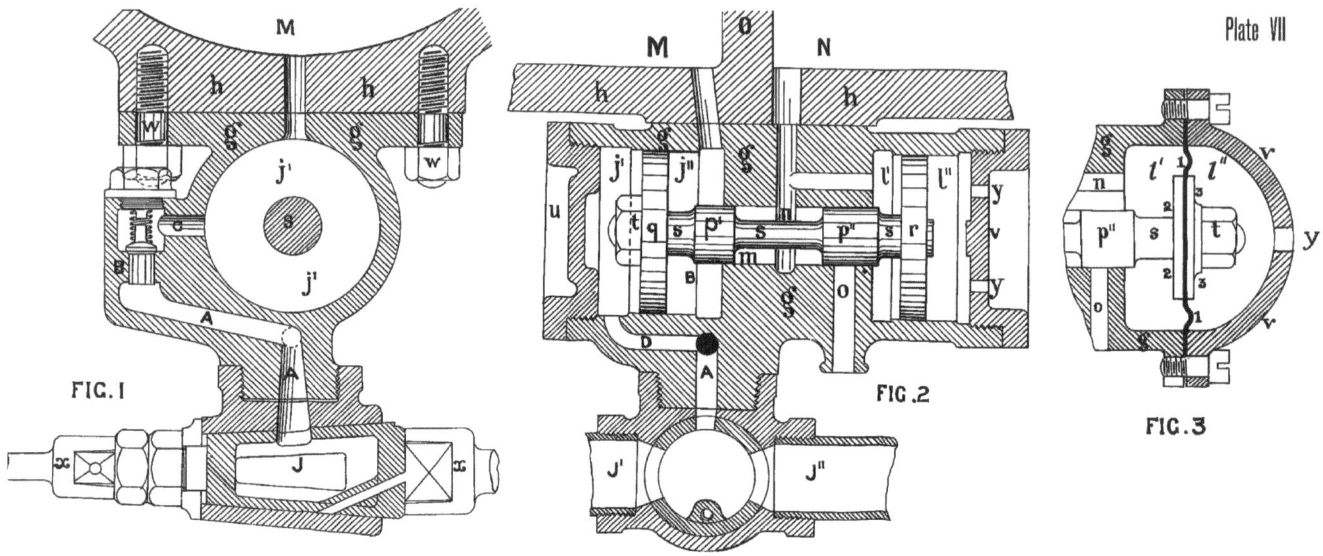

FIGURE 55: THE SELFE BRAKE (PLATE VII)

Valves **M** and **L** permitted direct communication between the controlling train pipe and the brake cylinder. The opening **K** was so adjusted (by a set screw) that with any ordinary and gradual change of pressure in train pipe connection **A** the brake system operation was conventional as already described. However, a sudden, heavy reduction in **A** would cause piston **J** to descend—before pressures could equalise via aperture **K**—forcing valve **L** open and supplying controlling (pipe J) train pipe pressure into the brake cylinder. Should the train pipe be completely exhausted, brake cylinder pressure would be retained by valve **M**, and

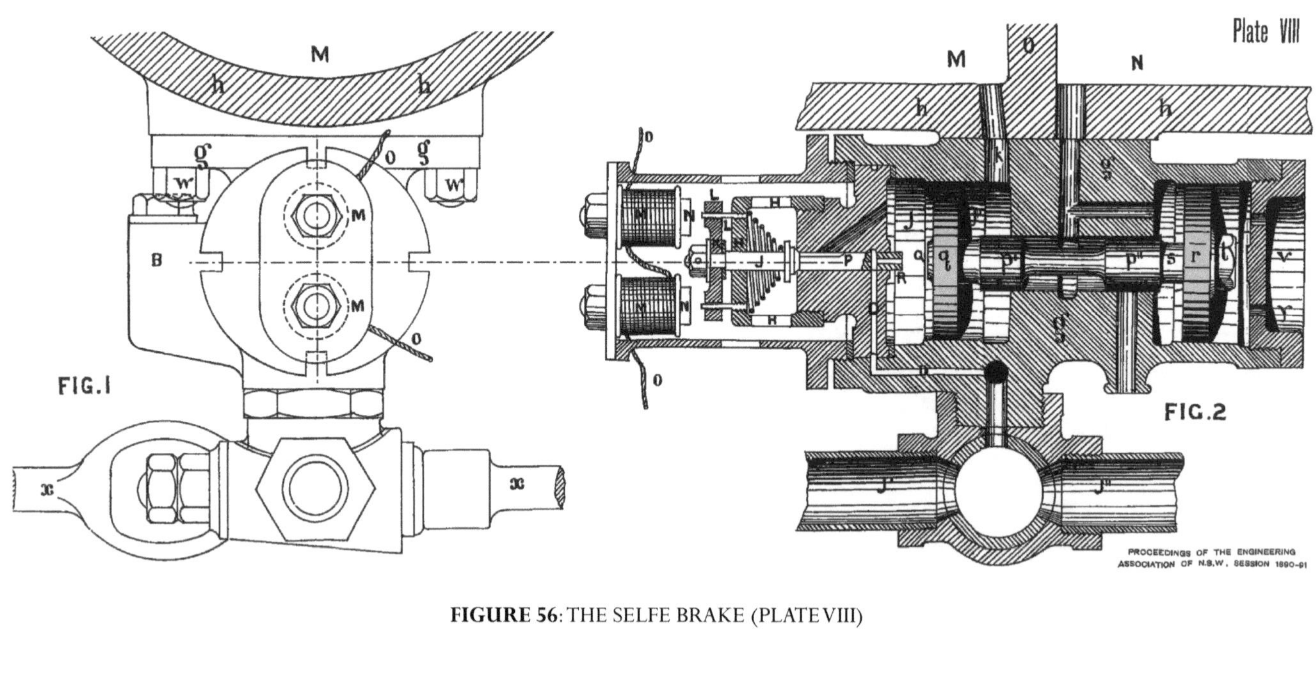

FIGURE 56: THE SELFE BRAKE (PLATE VIII)

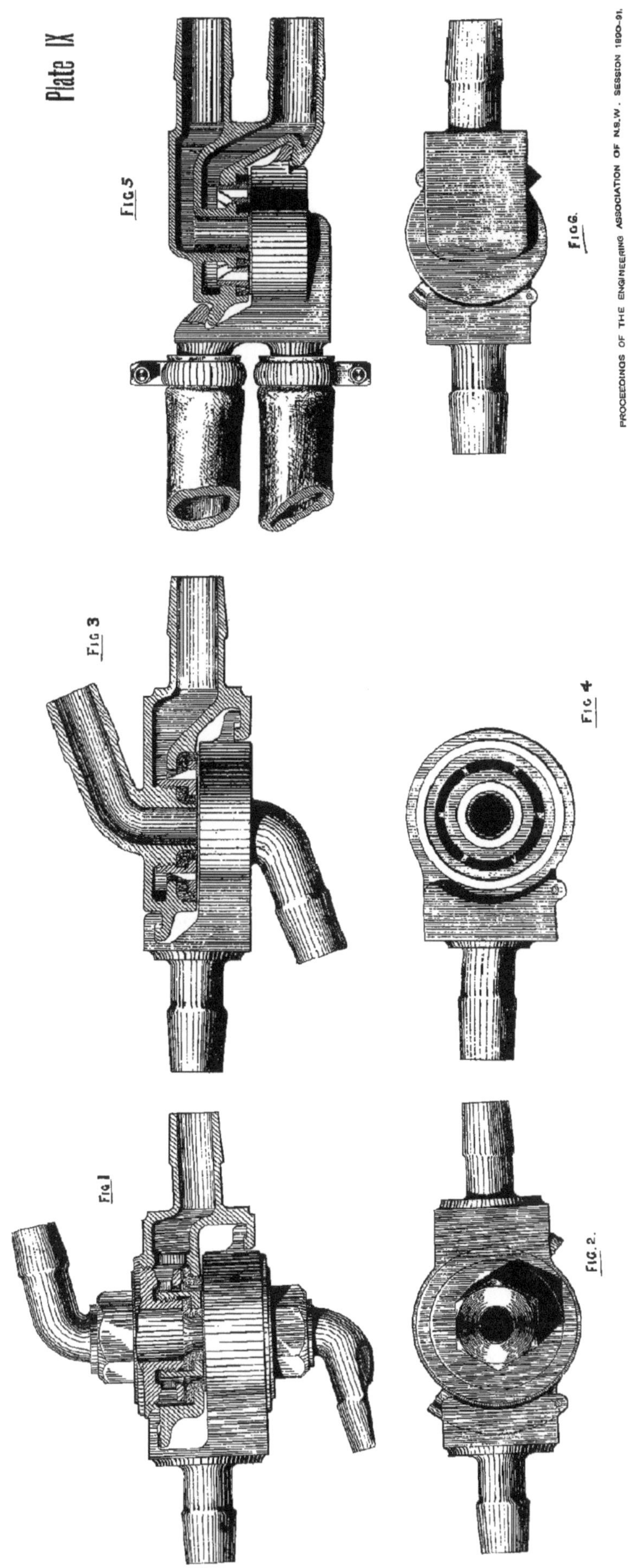

FIGURE 57: THE SELFE BRAKE (PLATE IX)

the differential piston assembly would remain directly controllable by train pipe pressure because passage **N** is not affected by vertical piston **J**. The reservoir above **J** would recharge via aperture **K** as pressure in **A** was restored and the brakes gradually released.

In concluding his presentation, Selfe—seeking from his peers the validation that his design had not received from the railway establishment—asked them to provide opinions freely on the following points:

- Did they think this proposed brake held sufficient promise to warrant asking the railway authorities to look into it?
- Did they think any harm would have been done if the NSW Railways and the Special Brake Board had extended the courtesy of asking for him to personally explain it?
- Was it desirable for the colony to manufacture its own brake equipment?
- Did they think his proposal was worth the time and effort bestowed upon it at their meetings?

Some notes from the record of discussions following Selfe's presentation provide illuminating insight into these times and the perceptions and opinions among the engineering sorority that so influenced the outcomes for train braking in Australasia:

Mr. W. D. Cruickshank stated that he attended the meeting with the intention of listening rather than taking part in the discussion. As a member of the late Board appointed by the government to inquire into the relative merits of the Westinghouse and Vacuum brakes, it would be very questionable taste on his part to deal with anything beyond the constructional details of these appliances. His opinion of the question was that it was one that required a considerable amount of study before it would be possible to express an opinion of the merits of any particular type with authority, and with regard to the invention under discussion he had had but very little time to consider its various points.

There was not the least doubt that the author, in pointing out that the Westinghouse brake could be only graduated one way, had called their attention to the weakest feature in this appliance, and one that was of especial importance on the New South Wales railways. For example, on the mountains from Katoomba to Emu Plains the line had gradients that he did not think were equalled in any part of the world [a comment that revealed a significant lack of international awareness], *hence the necessity for trains being fitted with brakes that could be graduated both on and off. As far as the author's paper was concerned, he considered that it concentrated itself in the invention of the controlling valve, which appeared as far as he* [Cruickshank] *could judge to be a most ingenious, sensible and practical appliance. He was under the impression when first glancing over the paper that nearly the whole of the advantages of this brake were represented by the two train pipes, and if such had been the case it would not be difficult for the Westinghouse Company to attain the same object by the addition of another pipe.*[89]

The Westinghouse brake was the outcome of some of the highest intelligence both in England and America, and, although it was a complicated appliance, it had been practically demonstrated in hundreds of thousands of instances, that it was reliable. With regard to the new quick-acting brake, the quick-acting gear only came into play in an emergency, and the valve worked in a very satisfactory manner.

On some points there was a resemblance between the author's and the Westinghouse brake. He had examined Mr List's brake, which in theory was much better than the Westinghouse. It had many good points but the exception that he took to it was that with the constant pressure behind the pistons there was a possibility of the air leaking through while the train was running, and if the brakes were required at full power in an emergency, the necessary pressure might be wanting. In other respects it was a very good invention, and would prove far more economical than the Westinghouse.

89. The first Westinghouse design configuration (patent no. 124,404 [1872]) consisted of a two-pipe system. Refer to the Westinghouse biography, chapter 7..

At the late [Australian] *trials in testing the Westinghouse and Vacuum brakes for leakage, each train was put on a grade of 1-in-30. The brakes were put on with full power, and the locomotive then cut off. From memory he believed the Westinghouse had from 62 to 64 lbs pressure. In 30 minutes one of the brakes leaked off, and in 83 minutes the train began to move. With the vacuum brake the results were very similar. He was prepared to admit that when the trials commenced he had an instinctive feeling that the Westinghouse was the better brake of the two, but as the trials progressed and the results obtained were compared he was compelled to change his opinion. They were both good brakes, but for graduation, simplicity and general usefulness the results of the trials had forced the conviction on his mind that the Vacuum brake was the best of the two, and he was glad of this opportunity of giving the members his opinion on this matter. He considered Mr Selfe's brake a much better appliance than the present Westinghouse system, and was aware of the difficulties involved in introducing such an invention; but possessing real merit it must eventually assert itself.*

Mr Middleton,[90] *a visitor, stated that he had studied the* [presented] *design and wished to compliment him on the ingenuity exhibited. At the same time he thought, although not the cheapest appliance obtainable, it certainly deserved to be tested, and before spending £300,000 in fitting the goods rollingstock with brakes, a few more trials should be made to prove beyond doubt which was the best brake in existence.*

Mr W. Shellshear,[91] *after complimenting the author on his invention, stated that on an examination of the Board of Trade returns it would be found that nearly eighty per cent of the failures of the Westinghouse brake were due to broken hosepipes. To get over the exhaustion difficulty there must be two train pipes. With one pipe, the difference between the author's and the Westinghouse brake was not great.*

The president then called the attention of the members to the questions asked at the conclusion of the paper. There was an idea existing among a certain section of the community that whatever was required of us must be imported, but surely this was not necessary when we had the skill and appliances for manufacturing here. He considered that the author should have received the courtesy due to him from the gentleman who received the drawings of his brake and have accepted his offer to explain the features of his invention. If it proved better than either the Westinghouse or Vacuum brake, why should it not be adopted?

Mr Cruickshank considered the Railway Commissioners the most abused men in New South Wales and they had been so ever since they had taken office. They were constantly having all sorts of schemes laid before them, and they were not justified in making tests of all of them, but he was confident that they wished to obtain the best brake for their rollingstock. They were not personally acquainted with the author and did not know that he was one of the leading engineers in the colony. If they had known, there was no doubt that they would have given this invention greater consideration.

Mr N. Selfe wished it to be understood that the remarks made in his paper regarding the Railway Commissioners were not the result of any personal feeling, but he thought that the official who had to deal with the matter might have treated him more courteously.

The Westinghouse air brake was eventually adopted in NSW in 1877 and originally applied to rollingstock operating over the Sydney suburban, southern, and western main lines. One historical outcome of the formative NSW connection with colonial air braking was that, for the next ninety years, many of the senior managers and technicians of Westinghouse Australia would come from the ranks of the NSW railway engineering divisions.

Victoria and South Australia were not far behind in the 'air brake stakes'. In Victoria, John Woods,[92] a professional engineer with a background in mining and public service (including—propitiously for him—a couple of periods as Minister of Railways) had invented a hydro-pneumatic power brake that

90. Possibly Thomas Middleton, Locomotive Engineer NSWGR, around 1882.

91. Walter Shellshear (1856–1939). http://adb.anu.edu.au/biography/shellshear-walter-4569 (retrieved 2012).

92. Woods is described in some biographical commentary as also having been a 'railways commissioner' and 'Commissioner of Railways and Roads'. Although the Act of Parliament establishing the VR Commission was not passed until 1883, Woods is identified on the 1879 VR Service Timetable as 'Commissioner of Railways'. Refer to chapter 7 for a well-known biography of Woods.

used water at boiler feed pump pressure and air to distribute the required force along the train. Being professionally both well-positioned and well-connected, Woods had his appliance fitted to some locomotives and passenger cars and tested in service. The equipment and its performance were favourably regarded by many 'knowledgeable' observers; however, despite its head-start in the race, it may not be surprising that it did not eventually prevail. Although more than thirty locomotives and a significant amount of rollingstock were equipped, Woods' eventual successor as Victorian Railways Commissioner was more disposed toward the Westinghouse brake, and tests of it were commenced in late 1882. Two years later, a Board of Inquiry resolved to direct a series of comparative tests, and—although the results were inconclusive—large quantities of Westinghouse equipment were nonetheless ordered toward the end of 1884, as well as some Smith's vacuum systems. The Westinghouse patent appeared to have been the most satisfactory and rapidly replaced the other types to become the VR standard.

In South Australia, the first air brake trials were conducted in November 1875. Some vacuum brake equipment was placed into service in 1884 on two broad-gauge train consists; these being converted to Westinghouse air brake operation in 1889. In February 1890 the SAR appointed a Board of Inquiry to investigate the most satisfactory form of braking for South Australian trains and to make recommendations accordingly. For various reasons, and despite some field testing (further successful air brake trials having been conducted during September 1890), this inquiry was also indecisive although it did recommend the Westinghouse patent be introduced to the narrow-gauge system. This proposal engendered considerable dissent within the SAR, not least because there existed at the time a tentative proposal to link the South Australian narrow-gauge with Queensland—which was at the time fitting some vacuum equipment—via a route through Hawker (Leigh Creek) and Innaminka. In 1891 the South Australian parliament declared that the Westinghouse Automatic air brake was the only system to be used on the SAR, but the situation languished until some progress with fitting air brake equipment was made in 1908. By the end of the Second World War, only a few freight wagons remained unfitted with the Westinghouse air brake.

The Queensland government, meanwhile, had been considering the question of train braking since the 1875 Newark trials in England. In their Annual Report for 1876, the commissioners noted that the Westinghouse brake was in service in New Zealand,[93] South Australia, and New South Wales and was apparently very successful. Taking an each-way bet, they imported both Smith vacuum and Westinghouse air-equipped locomotives about 1877.[94] Westinghouse air braking of trailing rollingstock commenced in a modest way in 1884 on one passenger train consist—running in comparative evaluation against a vacuum consist—on Brisbane's Sandgate Branch. After an early decision to adopt vacuum braking, the commissioners appear to have had a change of heart, and the decision was taken to progressively equip the Queensland Government Railways with the Westinghouse system beginning with passenger cars, since it was thought that fitting goods wagons would be prohibitively expensive. By 1909 most passenger stock had been equipped, and the QGR embarked upon a large-scale program to fit the equipment to freight rollingstock. This was apparently largely completed by 1912–13, although it took at least another decade to equip some of the more far-flung sections of the system.

93. This is not strictly correct. It seems that Westinghouse air brake equipment had indeed been acquired by the New Zealand Government Railways and tested to some degree or other during 1876, but could not be described as being 'in service', which did not occur until much later. Both the NZGR and the W&MRC did not apply air braking across their trailing rollingstock fleets until the 1900-1902 period.

94. Australian Railway Historical Society publications state that the air-braked units were two E-class Fairlie locomotives imported from the Avonside Engine Company in 1877; however, I cannot substantiate this and am doubtful. Queensland Government Railways' experience with Fairlie locomotives—of both the Single and Double type—was problematic, and I can find no reference to a QGR E-class Fairlie. In New Zealand, Double Fairlie locomotives were received from Vulcan in 1872 and Avonside in 1875, and eventually classed "E"; the last of them being finally withdrawn in 1920. It is unlikely that they arrived equipped for air braking.

Meanwhile, the New Zealand Government Railways were nothing if not aware of current overseas practice. They had experimented with Westinghouse brakes on locomotives in the North Island as early as 1876—this equipment being fitted under the supervision of NSW engineer Mr. G. Ashcroft[95]—and later ordered six locomotives from US manufacturer Baldwin that were equipped with Westinghouse brakes and were delivered in 1885. However, despite warnings from well-informed insiders, the NZGR did not embrace air braking for trailing rollingstock until the turn of the century. In March 1899, following the Rakaia collision in the South Island, public and political pressure resulted in the wholesale adoption by both the NZGR and the private Wellington & Manawatu Railway Company of the Westinghouse brake. Commencing in 1900, both organisations began to acquire their Westinghouse equipment; the NZGR paying almost £400 000 (approximately AUD$50 m today) to furnish 224 locomotives and 11 222 other vehicles while the W&MRC outlayed £20 000 for their more modest task. It is ironic that air brakes—that would prove so critical to the development of train operations throughout this mountainous country—were only hastened into service following a tragedy that occurred on one of its most level regions, the Canterbury Plains.

Perhaps the swiftness with which air braking was instituted on the NZGR left little time for much early operational assessment of the technology. In 1909, H.H. Jackson[96]—at the time Chief Mechanical Engineer—carried out some air brake testing in the South Island. The following result accrued from one of these runs; with a train consisting of a WF-class 2-6-4T tank engine (44 tonnes), 30 iron LA high-side wagons, and a bogie brake van (total trailing weight approximately 212 tons, speed 33 mph on level track), the train stopped within 16 seconds and 330 feet. One assumes the hierarchy were satisfied.[97]

In vacuum-braked Tasmania,[98] the Australian National Railways (ANR) introduced air braking after its acquisition of the Tasmanian Government Railways in the early 1980s, this conversion not being completed until 1991. The Western Australian Government Railways continued as a vacuum-braked system until the "air" arrived with the establishment of the standard-gauge link between Kalgoorlie and Perth/Fremantle/Kwinana in the mid-1960s. The advent of standard-gauge air braking in Western Australia initiated the conversion of the narrow-gauge rollingstock, although even in 2009 there was still some vacuum-braked 'work train' rollingstock in operation.

---

95. Later to be Works Manager at the NZGR workshops at Petone in Wellington's Hutt Valley.

96. As did most NZR CMEs of the early 20th century, Harry Hughlings Jackson made his mark. He was born in Auckland in 1860—into an eminent legal family—and educated locally. He was not a professional engineer but after leaving school did pass both the junior and senior Civil Service examinations. He was a nephew of Dr John Hughlings Jackson, the English physician who pioneered the development of neurology as a medical specialty during the reign of Queen Victoria. H.H. Jackson joined the Locomotive Branch of NZ Railways in October, 1876, and spent a period of apprenticeship firing and driving locomotives (as mechanical engineering staff did almost up until the end of steam). After completing his indentures he was transferred to Addington Workshops, Christchurch, where he worked as a fitter before moving to Wanganui Workshops to fill a similar position, which he occupied for nearly three years before being promoted to locomotive foreman. A year later, Jackson took charge of the isolated Nelson section as acting district manager, and after four months became locomotive foreman at Wellington. Subsequently he filled the position of relieving officer in the Locomotive Department on the Hurunui-Bluff section, performing various duties including those of foreman fitter at Hillside Works for six months, and locomotive foreman fitter at Dunedin and Invercargill. He was then successively acting district manager at Picton, locomotive manager and acting district manager at Napier, and again relieving officer on the Hurunui-Bluff section. While on this section, Jackson was entrusted with the special project of testing the steaming qualities of all New Zealand coals. In 1888 he was promoted to the position of locomotive engineer in Auckland which he held until June 1894, when he was promoted to Locomotive Engineer in Head Office. In 1913 he was appointed to the position of Chief Mechanical Engineer which he held until his retirement in 1919. Jackson and his Chief Draughtsman, Sidney Hartley Jenkinson inherited the classic NZR AB class 4-6-2 locomotive at an intermediate design stage, and brought the project to finality in 1915. Many AB-class locomotives remained in operational service until 1969, and several are still in Tourist & Heritage operation today. H.H. Jackson died in October 1937.

97. R.E. Robertson, Westinghouse Brake Engineer, NZR, *The Evolution Of The Westinghouse Automatic Air Brake,* THE NEW ZEALAND RAILWAYS MAGAZINE, Volume 1, Issue 3 (24 July, 1926)

98. The isolated 14-mile Bellerive-to-Sorell line (1892-1926) of the Tasmanian Government Railways near Hobart utilised a brake system that was apparently operated manually by a chain coupled alongside the rollingstock but of which I know little else.

The Westinghouse Brake Company in Australia

The Westinghouse Company came to Australia from its UK subsidiary when, in 1887, Imray & Company[99] of Kent St., Sydney, were appointed agents for the Westinghouse Brake Co. Ltd. In July 1894, the company registered its trademark in Australia. Afterward, arrangements for the company's representation in Australia changed, and William H. Nisbet MIE, became the Westinghouse agent, located in the Mutual Life Buildings in what is now Martin Place in Sydney. In 1899 he accepted a position with the Queensland Government Railways as a consulting engineer and was replaced from London by his brother John Nisbet. When the latter died unexpectedly in 1900[100], William Nisbet—now Chief Mechanical Engineer of the QGR—returned to Sydney and resumed his old job.

William Nisbet—echoing Selfe's earlier views—was soon stressing to the parent British company the essentiality of local manufacture. Whether or not this opinion was influenced by George Westinghouse's loss of control of his companies as a result of the 1907 financial panic is unknown, but, in any event, Nisbet recommended that an Australian Westinghouse Company be formed to take over and continue the British company's business within the Commonwealths of Australia and New Zealand. This resulted in the incorporation of the Westinghouse Brake Company of Australasia Ltd in NSW in July 1907 and the establishment of manufacturing operations—as the Railway Brake Division—at Concord West in 1911. In March 1937, the firm was registered as a proprietary company, and its name changed to Westinghouse Brakes (Aust.) Pty Ltd.

In England the interests of the Consolidated Signal Company and the old Westinghouse Brake Company (separated since 1916 from the US Westinghouse Company) had been merged in 1920, and this influenced the next phase of the evolution of the Australian company when Donald F. Brown (Managing Director, Westinghouse Brake & Signal Co. Ltd) visited Australia in 1944. He and J. B. Jacobson (Managing Director of McKenzie and Holland Pty Ltd, Newport, Victoria) agreed upon a closer relationship between the brake and signal interests in Australia, resulting—in 1951—in the creation of the Westinghouse Brake & Signal Co. (Australia) Pty. Ltd. as a holding company to report the trading results of the separate Australian subsidiaries. Under the early direction and management of W.R.C. Foster, the company prospered and became a well-established and successful enterprise.

Its story was soon, however, to become somewhat complicated. In 1957, the Westinghouse Brake & Signal Company Ltd was purchased by British aerospace and industrial conglomerate Hawker Siddeley Pty Ltd who formed Hawker Siddeley WESTCODE and continued to develop and manufacture air brake equipment in the UK and Europe. This organization was—in turn—acquired by BTR Pty Ltd (previously

99. A company thought to have been established in Sydney by John Imray (1820-1902). Imray was born at Peterhead, the son of a Scottish Minister from Longside, Aberdeenshire. After graduating as a Master of Arts from the Marischal College of Aberdeen, he was employed as an apprentice engineer and later in the 'Ventilation Office' at the Houses of Parliament. From 1850 to 1867 he ran an engineering business in London, during which time he was granted numerous patents for inventions in the field of railway engineering and steam engines; later becoming a consulting engineer and patent agent. Among the early clients with whom he collaborated was the English philosopher-scientist Matthew Piers Watt Boulton (named after his grandfather, Matthew Boulton and his grandfather's business partner, James Watt). In 1868 Boulton and Imray jointly patented inventions related to 'Propelling vessels'. Some of Imray's works were also featured in the 1862 International Exhibition in London. In 1871 Imray joined Charles Denton Abel, brother of the late Sir Frederick Abel, in partnership as Consulting Engineers and Patent Agents, under the name Abel & Imray. He was frequently employed as an expert in cases of patent litigation, including notable cases such as the "Telephone actions", the "Otto gas-engine actions", the "Westinghouse air-brake actions", and the "Welsbach gas-lighting actions". Abel & Imray, with offices in the Holborn area of London as well as Bath, Cardiff and Munich, still operates under that name today. Imray assisted in founding the Chartered Institute of Patent Agents in 1882, and served as its President from 1884 to 1886. He was also a Member of the Institution of Mechanical Engineers and of the Royal Institution. In 1900, at the age of 80, he became an elected councilor in the Metropolitan Borough of Holborn, representing the South East St. Andrews ward.

100. John Nisbet, who had practiced as a barrister both in England and Australia, had only recently returned from a trip to New Zealand to complete contracts for the supply of air brake equipment for the rollingstock in the North Island. While in New Zealand he contracted the cold which ultimately resulted in his death, aged 34.

the British Tyre & Rubber Co., originally derived from the B.F. Goodrich Co., and from 1995 called BTR Nylex in Australia). In 1999, BTR merged with Siebe Pty Ltd (derived from Siebe Gorman) to form Invensys Pty Ltd[101]. Invensys split the company into Westinghouse Signals Ltd—which became Westinghouse Rail Systems (UK) and Invensys Rail Systems Australia (both part of Invensys Rail)—and Westinghouse Brakes Ltd. In April 2000 Invensys announced the sale of Westinghouse Brakes to the Knorr-Bremse Group ("Bremse" meaning *brake* in the German language).

The German company had originated in Berlin in 1893 when Georg Knorr acquired his late employers' company, Carpenter & Schulze. Founded—by American expatriate Jesse Carpenter—on the early commercial success of a ten-year contract with the Prussian State Railways to supply braking systems for passenger trains, the company fell on hard times when the contract ended. Forced to accept financial support via a partnership agreement with a Berlin machine tool manufacturer, Knorr established a limited company, Knorr-Bremse AG. This company went on to develop air brake systems for road trucks, followed—between 1913 and 1927—by the successful Kunze-Knorr railway air brake (a system combining the direct release rapid-action brake with a dual chamber brake cylinder, forming a *"gradual-release, single-chamber combined brake with low exhaustibility."* [102]) This was followed by the pre-eminent Hildebrand-Knorr (HiK) equipment for express trains (1931–39) and in 1955 by the elegant KE distributor valve.

By the early 1990s, K-B had picked up the challenge from now-defunct Hawker Siddeley Westcode to compete against Wabco in the US market and had developed their DB-60 control valve—incorporating some design characteristics reminiscent of the short-lived Hawker Siddeley Westcode Z1AW valve—and had introduced it successfully into the United States with their purchase of New York Air Brake in 1993. At the beginning of 2003, Westinghouse Brakes (Australia) Pty. Ltd. became Knorr-Bremse Australia Pty. Ltd. and in July of the same year relocated from their historic Concord West premises to Granville. Through the Knorr-Bremse ownership of New York Air Brake (founded in 1876 by Frederick Eames of the Eames Vacuum Brake Co.[103]), the Australian subsidiary now officially represents that US company and its well-known products; however, it continues to provide product and maintenance support for the Wabco products previously sold in Australasia.

In 1990, the US assets of the Westinghouse Air Brake Company and the Wabco name were purchased in a management buyout, and a new WABCO entity was created that listed publicly in 1995. In November 1999, the Westinghouse Air Brake Company merged with MotivePower Industries to form the Wabtec Corporation which has a corporate presence in Sydney and an agency in Melbourne. Nowadays, the WABCO brand is associated with a direct successor company, WABCO Vehicle Control Systems, which manufactures commercial road vehicle air brake, stability, suspension, and transmission systems. The iconic 'Westinghouse' name no longer exists in the context of the railway brake and signal industry.

---

101. Augustus Siebe, a former Austrian cavalry officer, had arrived in London in 1819 and is credited with a number of innovative products, including the world's first diving suit. The company—later becoming Siebe Gorman & Company Ltd, after Siebe passed his business to his son Henry and his partner William Augustus O'Gorman, who had married Siebe's daughter—developed into a marine engineering company specialising in safety and rescue systems and became a supplier to military naval forces, developing breathing apparatus and submarine emergency escape equipment. This technology was later adapted for use in the aviation high-altitude environment and ultimately to safety and rescue equipment for mining operations.

102. Prof. Manfred Pohl, *Safety First by Road and Rail: The History of Knorr-Bremse AG* (Munich: Piper Verlag GmbH, 2005).

103. See biography in chapter 7.

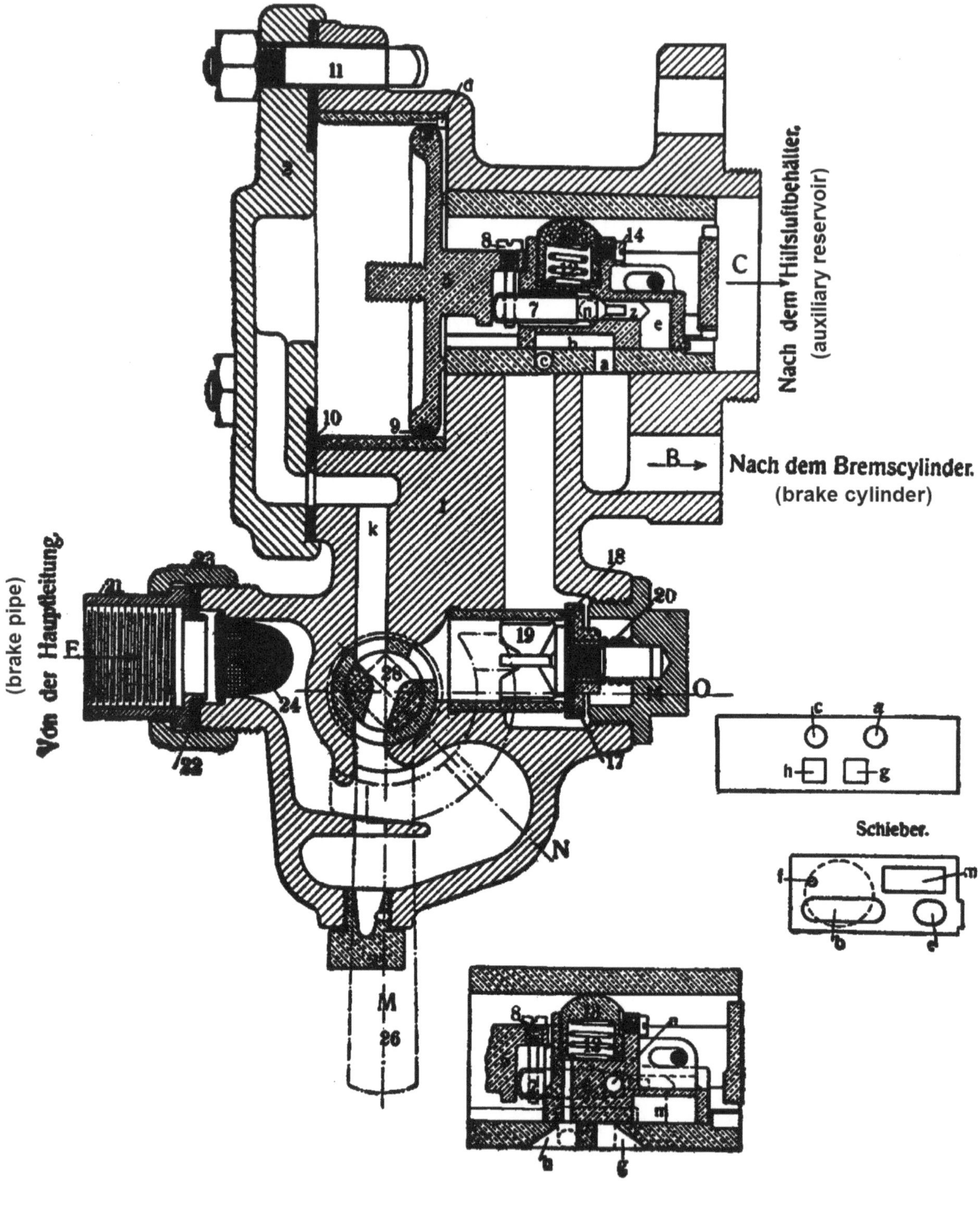

Steuerventil Bauart Knorr mit Füllnut

(Knorr control valve with Filling Slot)

FIGURE 58: GEORG KNORR'S DIRECT-RELEASE "RAPID-ACTION" CONTROL VALVE OF 1899 EVOLVED FROM AN EARLIER (1892) PROJECT TO SIMPLIFY THE DESIGN OF THE WESTINGHOUSE VALVE OF THE TIME AND COULD BE INTERMIXED WITH THE WESTINGHOUSE VALVE IN FREIGHT TRAIN CONSISTS - DIAGRAM *(© Knorr-Bremse, used with permission)*

The Ordinary Triple Valve

The original triple valves used in Australia and New Zealand were versions of Westinghouse's modified Plain Triple (piston-and-slide valve action instead of the original poppet style). This valve had a three-inch piston and was soon superseded by the Quick-Action triple for the same operating reasons as had occurred in the United States.

The Quick-Acting Triple Valve

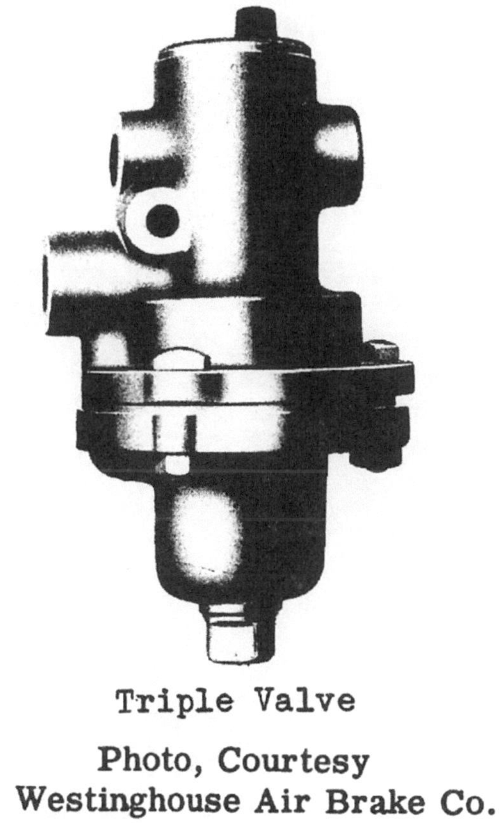

FIGURE 59: THE WESTINGHOUSE (AUSTRALIA) ORDINARY TRIPLE VALVE – c1910

The limitations of the Ordinary 3-inch Triple were soon evident, resulting in the development of the Standard 3½-inch Quick-Acting Triple Valve as shown in Figure 61. This triple valve incorporates two distinct sets of mechanism. One is the triple valve piston and stem assembly that includes the slide valve and graduating valve with the graduating stem and graduating spring. Also referred to as the *Service part* of the triple valve, it operates when making Service brake applications and when releasing the brake. The other mechanism, called the *Emergency* or *Quick-Action* part, comprises the emergency piston, emergency valve, and brake pipe check valve. It comes into play upon an Emergency application, rapidly admitting brake pipe pressure directly into the brake cylinder and augmenting the auxiliary reservoir supply to provide increased brake cylinder pressure. At some point, the Quick-Action feature was deemed either inappropriate or not required for New Zealand Government Railways requirements, and in the NZGR diagram depicted in Figure 62, the operating part of this feature—the emergency valve—has been removed or the passage to it blanked. I have been unable to establish why this was done; however, it is probable that axle loads and braking ratios of the time—especially considering the multitude of lightweight, four-wheel wagon types then in use (many without empty/load equipment)—were considered to be too modest for the application of Emergency brake cylinder pressures (which occurred rapidly to a value of around 70 kPa more than for a Full Service application) that might cause wheels to slide.

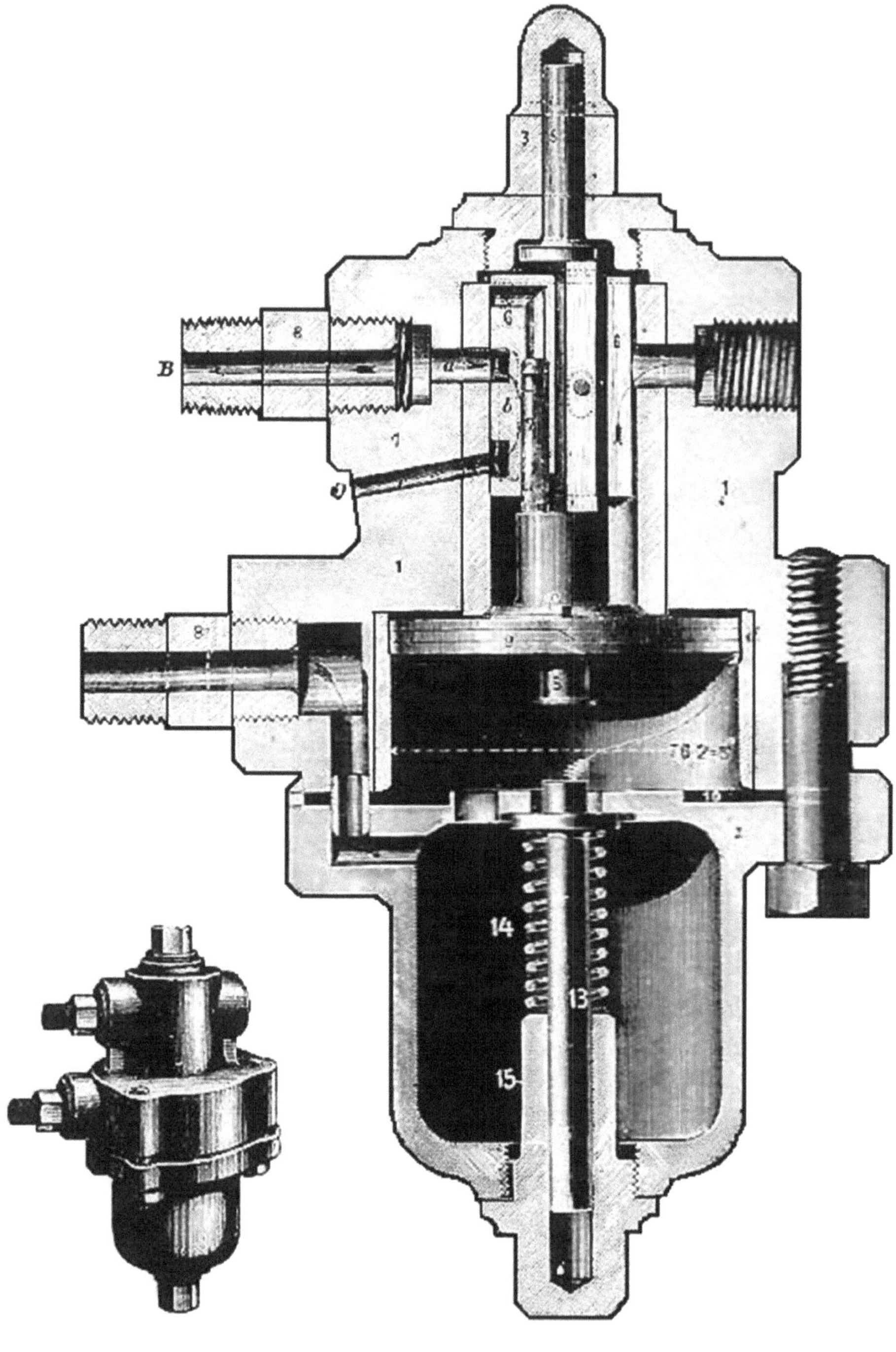

FIGURE 60: THE WESTINGHOUSE (AUSTRALIA) ORDINARY TRIPLE VALVE – SECTIONED

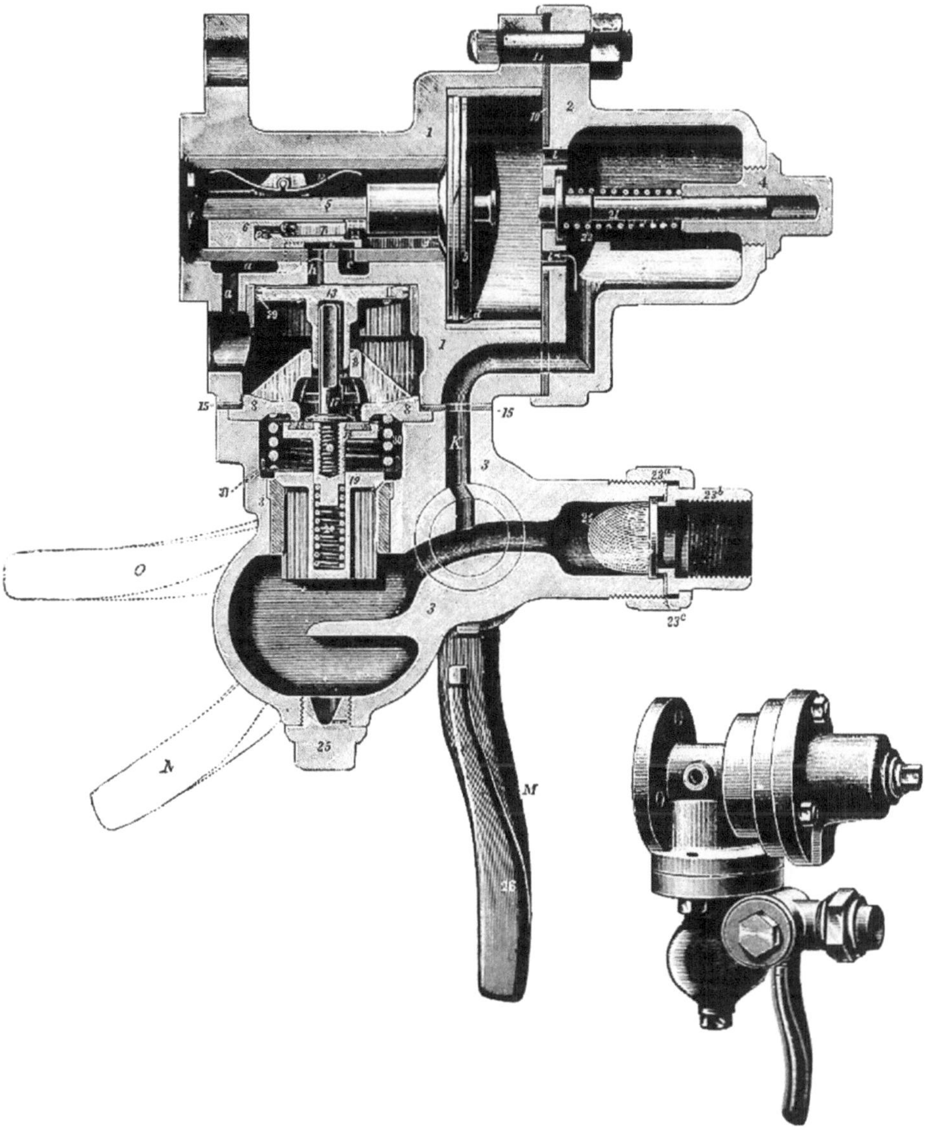

FIGURE 61: THE WESTINGHOUSE (AUSTRALIA) STANDARD 3½-INCH QUICK-ACTING TRIPLE VALVE - SECTIONED

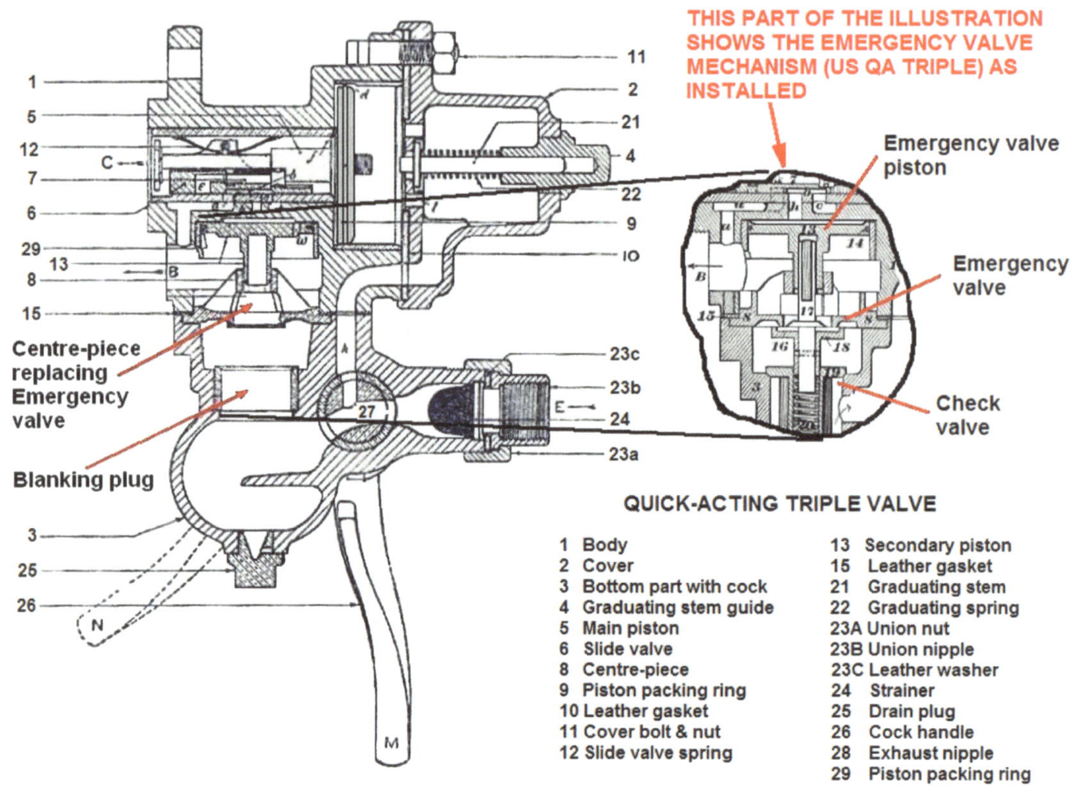

FIGURE 62: THE WESTINGHOUSE (AUSTRALIA) QUICK-ACTING TRIPLE VALVE - DIAGRAM

The Improved Triple Valve

The Quick-Acting triples were eventually replaced with the 3½-inch Improved Triple Valve, providing Quick-Service that was designed about 1906 and was also fitted to shunting locomotives. The major feature of the ITV was the bulb fastened to the underside of the triple valve body that provided an empty volume into which brake pipe pressure was directed during the early stages of a brake pipe reduction. The result was a rapid, local brake pipe pressure reduction of around 25 kPa at each vehicle, thus producing the required serial *Quick-Service* activity that rapidly put the blocks lightly against the wheels thus tending to minimise slack action during a brake application. The siphon tube seen extending into the Quick-Service bulb enabled any accumulated moisture to be removed during brake cylinder exhaust.

Being designed in England just after the turn of the century, this valve was intended to cope with conditions then extant on air-braked railways outside North America; trains were comparatively short, and this valve was quite adequate. As train lengths increased though, the ITV was found to be less satisfactory and the same problem that had earlier become apparent in America of the head-end triple valves "robbing" all the brake pipe recharge air from—and thus starving—the rear-end valves now afflicted some air-braked train operations in Australia and New Zealand. In addition, since brake pipe volume within a train actually increased by the total volume of the number of triple valve Quick-Service bulbs, a further problem appeared. With a brake application in effect, each bulb was connected and open to the brake pipe, and the pressure in the bulbs had to also increase—along with that in the brake pipe—before each vehicle could go to Release. Add the inevitable piston ring leakage, and it is obvious that before an adequate pressure differential across the triple valve piston could be established, brake pipe pressure had to rise faster than the rate at which ring leakage raised auxiliary reservoir pressure. This problem of slow BP pressure build-up in ITVs that were subject to indifferent maintenance often resulted in brakes failing to release, and this engendered throughout Australia a widespread and somewhat timid view among operations personnel of the capabilities to be reasonably expected of a good air brake system when competently-handled. The author was never convinced that there was clarity around the subject of the critical necessity for proficient system maintenance. It is noteworthy that the then-National Rail Corporation—heavily Victorian-influenced at the time of its inception—was still using ITVs (that were not particularly well-maintained) late in the 1990s[104].

Two further weaknesses existed in the integral Regulating valve arrangement (the Regulating valve being the equivalent of the Emergency valve in the US Quick-Action triple) that, while better than nothing and quite suitable for short passenger trains, did not provide very good control of brake cylinder pressure. The first flaw was that a heavy brake pipe reduction would result in rapid head-end brake cylinder pressure development and thus compressive slack action ("run-in") that could derail lightly loaded wagons further back in the train. A second inherent defect was a Quick-Service bulb design deficiency. If a single bulb were made large enough to provide adequate Quick-Service with no other assistance, it would be liable—during

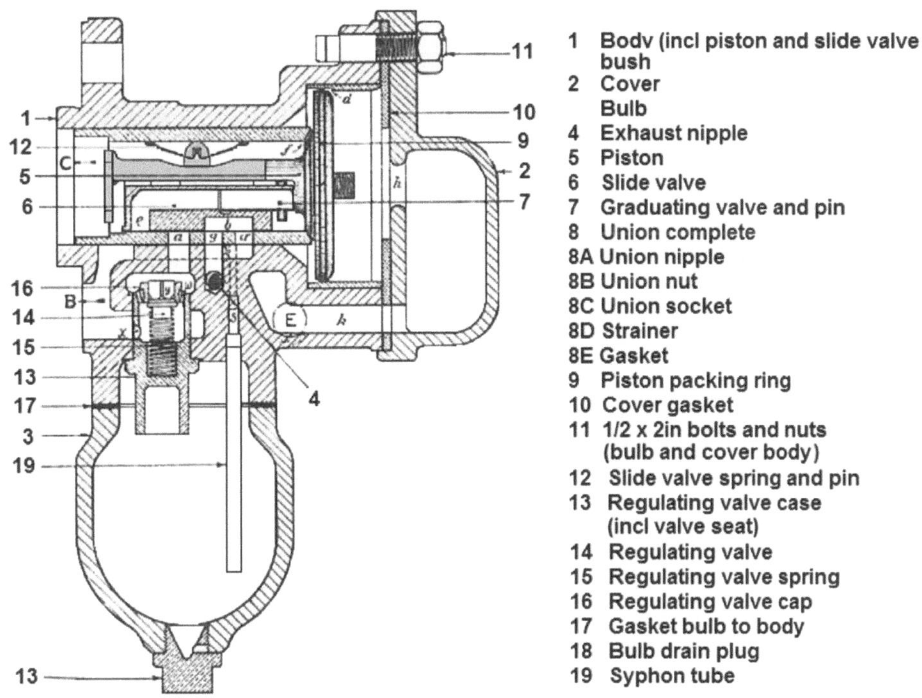

FIGURE 63: THE WESTINGHOUSE (AUSTRALIA) IMPROVED TRIPLE VALVE – DIAGRAM

104. It may be that the single-most defining aspect by which air braking on the Victorian Railways developed differently to other states was the practice throughout the decades leading up to the 1970s of appointing senior locomotive drivers—of inspector or instructor rank—to the functional position of corporate air brake management. This was a ready indication of the degree to which railway air braking was an incomprehensible science to many Australian railway senior managers, and for this reason (although these managers invariably knew enough to be aware that air braking was vital to the safety of train operations) the responsibility for managing this technical function did not always reside with a professional engineer but was often ceded to the locomotive running or workshops departments. On the Victorian Railways, supervisory drivers were elevated to Westinghouse (meaning "air brake") Inspector. These worthy individuals were given advanced air brake instruction by Westinghouse Australia and no doubt maintained good and safe practices across their railway system. But they could hardly be expected to take the initiative in pursuing and embracing newer air brake equipment that might—were it to be introduced—render them impotent in terms of the positional power conveyed by their expertise in whatever was the equipment of the day and with which they were comfortably familiar. Nor were they equipped to make sophisticated engineering-based judgments about the suitability of new air brake technology. It is my view that this reason contributed most to the longevity of obsolete air brake equipment that has characterised Victoria's railway air braking culture; indeed, it should not be overlooked that when VicRail progressively scrapped four-wheel wagons during the 1990's, the parsimonious decision was taken even then to salvage the ITVs and recycle them onto new bogie rollingstock. The VR did not appoint a professional engineer to the air brake management function until 1978, when Gerry DeBont became their first air brake engineer. Refer also to chapter 6, the Air Flow Indication section for additional comment.

filling—to cause a "surge-and-stop" effect within the brake pipe, thus producing a pressure wave that could result in an undesired release. Bulb sizes were matched to the length of the brake pipe on the wagon, and long vehicles required a larger bulb than shorter vehicles.

Of course, the above weaknesses applied only when the triple valve was operating on longer trains, and the device was generally suitable for shorter freight trains and was indeed quite suitable for passenger trains and shunting locomotives. Both New South Wales and Victorian state rail authorities appear to have been satisfied with the performance offered by their ITV-equipped freight trains—characteristically of modest length—and for this reason, neither would rush to embrace the next operating valve development to appear in Australia.

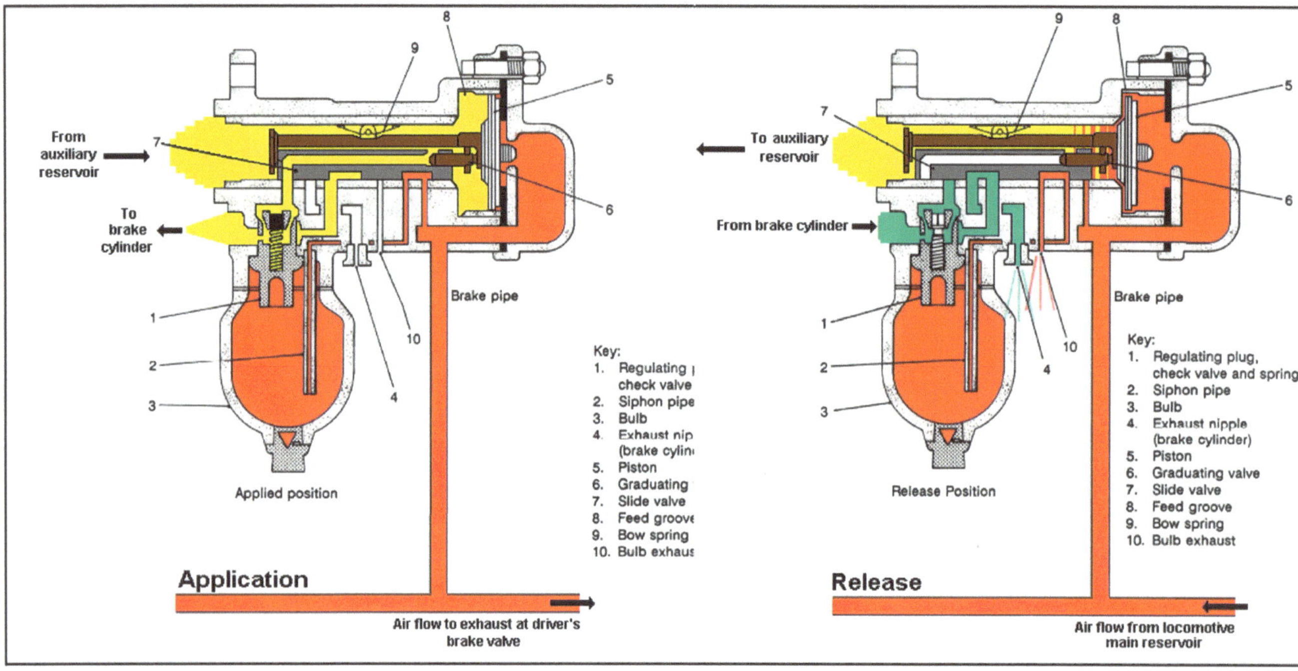

FIGURE 64: THE IMPROVED TRIPLE VALVE – SIMPLIFIED OPERATING DIAGRAM

FIGURE 65: IMPROVED TRIPLE VALVE ATTACHED TO A 10-INCH BRAKE CYLINDER

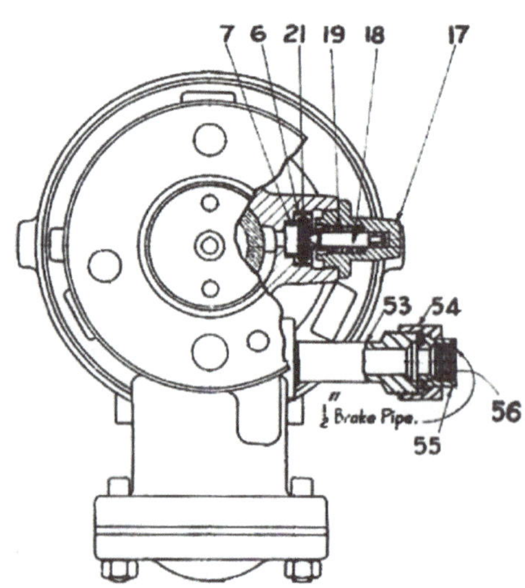

FIGURE 66: THE WESTINGHOUSE (AUSTRALIA) AF TRIPLE VALVE - END VIEW

The AF Triple Valve

There were some aspects of the Improved Triple Valve that made it expensive to overhaul, and it was rendered obsolescent as soon as a better device became available. This improvement appeared during the 1930s when the Australian Westinghouse Company developed a new valve that incorporated features of modern European and American valves. Piston diameter was increased to four inches, providing greater piston force for any pressure differential across it. This greater force was required to move a slightly larger slide valve but also ensured more positive action at lower differentials.

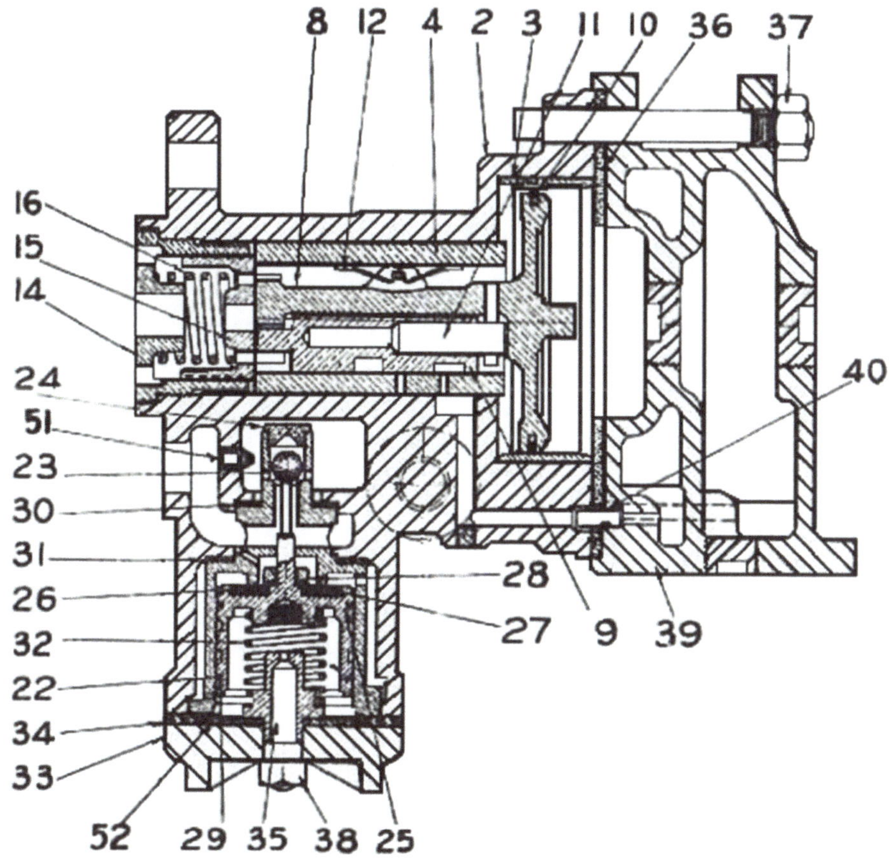

3 Piston bush
4 Slide valve bush
6 Prake pipe check valve bush
7 Brake pipe check valve seat
8 Piston
9 Slide valve
10 No. 1 ring
11 Graduating valve
12 Slide valve spring
14 Retarded recharge piston bush
15 Retarded recharge piston
16 Retarded recharge piston spring
17 Brake pipe check valve cap
18 Brake pipe check valve stem
19 Brake pipe check valve spring
21 Brake pipe check valve
22 Regulating valve bush
23 1/2in ball vale
24 Regulating valve bush cap
25 Inshot piston with stem
26 Inshot piston seat
27 Inshot piston washer
28 Inshot piston lock nut
29 Inshot piston ring
30 Regulating valve bush top seal
31 Regulating valve bush bottom seal
32 Inshot piston spring
33 Regulating valve cover
34 Regulating valve cover gasket
35 Regulating spring guide
36 Cover gasket

FIGURE 67: THE WESTINGHOUSE (AUSTRALIA) AF TRIPLE VALVE - DIAGRAM

The most important new feature of the ***A****ustralian* ***F****reight* triple valve, however, was its introduction of *retarded-recharge*, referred-to earlier in this book with the description of the American Type-K triple valve. Operation of the AF, though, differed from the K triple in that compression of the retarded-recharge spring during the release action permitted the feature to operate via a special position of the piston—that was able to move to a secondary (narrower) part of the feed groove—rather than by a special position of the slide valve as with the American version. The result, though, was the same; brake pipe recharge pressure was conserved at the head-end of the train and more was able to flow down the brake pipe and thus be available to the rearward triple valves, which would not be in their Retarded-Release position. Note that to move a triple valve to its Retarded-Release position, brake pipe pressure must rise about 20 kPa above auxiliary reservoir. Since on longer trains it is impossible to obtain this pressure difference beyond about the first thirty wagons, the triple valves beyond that point do not go to Retarded-Release. This characteristic provided a more rapid rise in brake pipe pressure toward the rear of the train, resulting in an earlier and more certain release of brakes thereat.

To provide more positive control of rapid brake cylinder pressure build-up, the older US concept of a spring-loaded valve opened by the rush of air was abandoned. With the American Quick-Action triple valve, a large auxiliary reservoir flow was directed to the triple valve's emergency valve, which was forced open to provide a large-volume brake pipe supply to the brake cylinder. Instead, for the local triple valve, Westinghouse Australia added a new *inshot* portion, being the part of the valve below the main portion. As with the Quick-Action triple, there is a rapid build-up in brake cylinder pressure initially but a slower rate thereafter.

Because of the possibility of brake pipe surge problems with a large bulb, the Quick-Service volume was redesigned. Instead of being a *bulb* (arranged below the triple valve), this volume was separated into two chambers and integrated as part of the cover. The large chamber is fed by a large passage, while the smaller chamber is fed via a choke in a smaller passage. This configuration ensured that Quick-Service activity was started by brake pipe pressure flowing into the large chamber and then completed—without the danger of surging—by the final flow into the small chamber. The following describes the process:

When brakes are applied, brake pipe air flows freely into the large chamber and gives very rapid Quick-Service activity, thus applying the brakes on the whole of the train even more rapidly than would be possible with Improved Triple Valves. Owing to the choke in the passage to the smaller chamber, the pressure therein will be low at the time that any surge tends to take place in the brake pipe, and the slight rise in pressure caused by the surge will pass into the smaller chamber and there be dissipated, instead of acting against the brake pipe face of the triple valve piston and forcing it to the Release position.[105]

Because the Improved Triple Valve had a direct connection between brake pipe and bulb through the slide valve when the brake was applied, the Quick-Service bulb became part of the brake pipe and added to its volume when the piston and slide valve were in the Service and Service Lap positions. While being satisfactory as far as Quick-Service activity was concerned, this was a disadvantage during the remainder of the brake pipe reduction when brake pipe pressure was exhausting at the driver's brake valve. During this time the air flowed out of the bulb and back into the brake pipe again, thus slowing down the actual rate of brake pipe reduction. Extra time was likewise consumed during Release as recharge pressure flowed into the bulb at each vehicle (as well as the brake pipe) before assuming a value sufficient to move the triple valve to Release.

In the AF triple, a check (or non-return) valve was placed in the passage leading from the brake pipe to the Quick-Service ports of the slide valve and its seat (and thus the QS chambers). When the slide valve

105. *Handbook of Instructions for the Guidance of Locomotive Operating Staff*, New Zealand Government Railways, (1960).

moved during a Service application, the connections made resulted in a flow of brake pipe air into the empty Quick-Service chambers, opening this check valve against spring pressure. Once the two chambers were charged, the check valve closed under the influence of its spring, and the chambers were isolated from the brake pipe. Thus, at any further brake pipe reduction, only air from the brake pipe was discharged and not also from the charged Quick-Service chambers as was the case with the ITV.

At brake release, no brake pipe pressure could flow into the already-charged chambers. As a result, the brake pipe recharge was more rapid, and the triple valves moved to Release earlier than they otherwise would have. This, plus the Retarded-Recharge feature, provided a timelier and more uniform rate of recharge throughout the train.

Meanwhile, as the release progressed, Quick-Service chamber pressure would exhaust from the large chamber via a larger passage than the choked exhaust from the small chamber. With the Quick-Service chamber exhaust split in this manner, Quick-Service activity in the case of a re-application would be reduced should grade control (retainer) valves be used in the High Pressure position.

The AF triple valve was purchased extensively by the New Zealand and Queensland railways but not by their NSW or Victorian counterparts; they apparently being happy with the performance of the ITV—given the unexceptional train sizes of the day—and preferring to stay with the particular freight timings (i.e. brake cylinder filling times) provided by this equipment.

The QSH Triple Valve

Dating from the early 1900s, the QSH Triple Valve was fitted to older Sydney (1927) and Melbourne Electric Multiple Unit passenger trains and, for this reason, was optimised for the brake cylinder capacity installed on such rollingstock. Although differing somewhat in its arrangement by comparison to the Improved Triple Valve, it was very similar in operation. Probably the major difference in layout is that the Quick-Service bulb is incorporated on the end of the valve body instead of below it, possibly to reduce the profile of the valve for more practical installation under EMU stock. In addition, there is no regulating valve and spring. The NSW State Rail Authority also used this valve—designated QSE—on 1953-era passenger rollingstock.

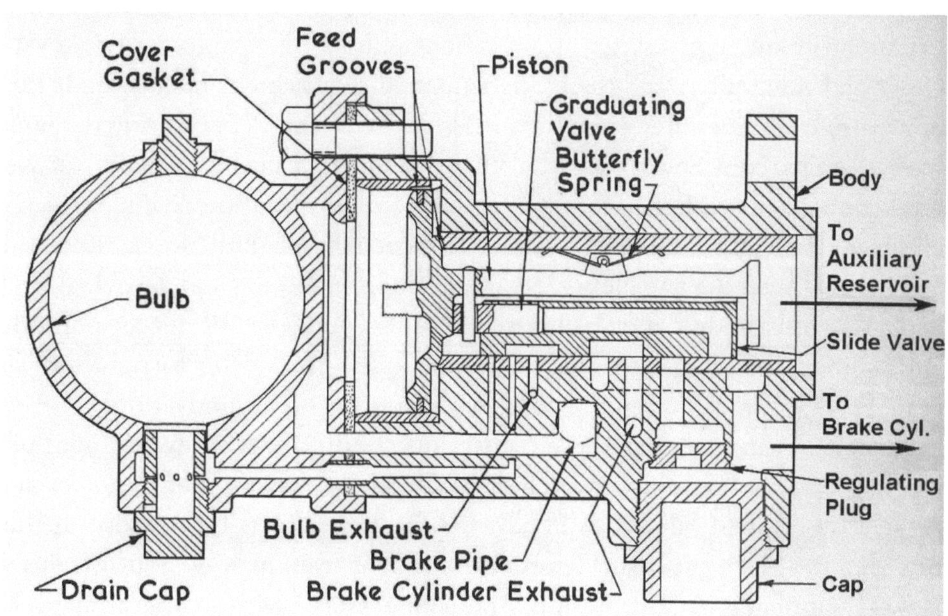

FIGURE 68: THE WESTINGHOUSE QSH TRIPLE VALVE

Another feature of this valve was that the brake pipe was flange-connected at the pipe bracket, rather than directly to the triple valve, as was the case with the ITV. This meant that removal of the valve was a simpler process due to not having to disturb the brake pipe connection.

The W Triple Valve

Prior to the late 1960s, when the realities of the potential coal-haulage task facing the Queensland Government Railways began to materialise, there was little about railway operations throughout either Australia or New Zealand that could be construed as "large scale." Indeed, train sizes on the Australasian rail systems—for the most part state-owned general freight and passenger common-carrier operations—were decidedly unpretentious. In New Zealand at the time, wagons of modest size and mass, the use of chopper couplers and drawgear of modest capacity, as well as a rugged topography and a network bifurcated by a 93-km sea crossing and the necessary Roll-On/Roll-Off vehicle-and-rail ferry connection, limited the potential for larger train sizes. Throughout Australia, with the exception of the geographical confrontation posed in several strategic places by the Great Dividing Range and other demanding coastal ranges, railway networks were less topographically-challenged than were their Kiwi cousins and persevered into the modern era with their nineteenth-century infrastructure. Continental Australia unarguably offered the railway builder and operator a more benign geographical environment than did its trans-Tasman neighbour. This is not to say that Australian railways did not have some serious grades; indeed they did and still do—in fact, if one further characteristic differentiates Australian rail topography from that of New Zealand, it is the many 1-in-40 grades of country and regional lines laid across the sprawling vastness of the rolling and otherwise benign-looking Aussie interior—and these grades, as well as those of the coastal ranges, were a major factor in early decisions to adopt continuous train brake systems.

In Western Australia and Tasmania, the predominant limiting factor to train size was vacuum braking (not to mention the infrastructure constraints that had evolved as a result, e.g. short crossing loop and siding lengths, tight track curvature, etc.), while in South Australia it was the sparse population and relatively less significant traffic requirements of an underutilised transcontinental link. On the eastern seaboard, state government rail operations were limited by infrastructure and operating practices characteristic of the 400-metre, 35- to 40-wagon maximum train lengths suited to operations utilising the Improved Triple Valve. English-style screw-link couplings and side buffers and the aforementioned requirement to surmount the continental divide or to ascend otherwise from tidewater to hinterland also contributed to preserving the unpretentious form of these humble "goods" trains. New South Wales and Victorian railways, unable even in later years to completely sever their cultural connections with the Mother Country, continued to use side buffers *and* AAR automatic couplers, a rather indulgent slack attenuation combination (for coping with "buff" forces at least) that lasted into the 1980s.

In both Australia and New Zealand, political bickering and parochialism between states and regions also had a detrimental effect on rail development. The legacy of decades of indifferent government ownership and political meddling and—in many cases—the average performance of sinecurist public service bureaucrats at all levels was often the persistent suppression of innovation.

Especially in Australia, these constraints resulted in the development of railway corporate philosophies that impacted unhelpfully in such disparate areas as equipment choices and operating practices. Some of these attitudes and values would have far-reaching effects throughout the decades to come. Indeed, even in the mid-1990s, the standard length for a freight train in New South Wales was still 400 metres - a fact not lost on National Rail Corporation[106] personnel (some with an operating background in other than Australian state-owned rail systems) who later struggled valiantly in western parts to operate interstate trains up to 1 200 metres in length over a network not designed to accommodate them. On this subject—operating at

106. Now Pacific National.

the time out of the Parkes depot—the author well recalls the delays and legwork involved in the infamous and unlamented "long cross" [107].

The impetus for the development of the W-type equipment originated with New Zealand Railways who, by the mid-1970s, were experiencing maximum capacity on their major spine, the 680-km North Island Main Trunk between Auckland and Wellington. Approximately 50 percent of this route consists of rugged, mountainous country typically on extensive grades of 2 percent (1:50) and greater, and sharply curved, and much of the rest of the nation's rail network was similar. At this time, NZR still operated a considerable fleet of four-wheel rollingstock and continued to utilise the colonial "chopper" hook-and-central-buffer coupling system[108]—albeit by then of greatly-enlarged dimension and capacity—on most freight trains. These hooks were possibly the heaviest chopper couplers on any railway system; indeed, the author recalls them being as heavy to carry as an automatic coupler knuckle. Running—for most of the period since the inception of air braking—with a 550 kPa (80 psi) brake pipe, NZR, by the 1970s was operating freight trains of the type described above up to almost 1 000 metres in length. Modest perhaps by North American standards, these were, however, "flexible" trains (to say the least!) in a topographical operating environment that was anything but benign. These environmental factors, plus the crucial challenge at all times to preserve train integrity with the slight NZR couplers, slender drawgear, and lack of any cushioning from side buffers, set the NZR operation apart from most Australian railway operations of the period; in particular Australian narrow-gauge. These influences also served to engender a local train-handling culture that was grounded, at the time, in competent technical instruction and that emphasised rigorous attention to the control of slack action.

In the years following WWII, several air brake equipment manufacturers carried out research into the use of newly developed materials and new approaches to braking problems. In Europe, this produced innovative brake valves and a decision to use only graduated release distributor valves on all new rollingstock. These new valves incorporated synthetic rubber valves, seats, seals, and diaphragms, and in the United States, the preeminent, contemporaneous development was the AB (and then the ABD) control valve.

By 1961 the new 26-L locomotive air brake schedule was in use in New Zealand, and it was soon considered by NZR mechanical department engineers that to make the most of this equipment, a more advanced triple valve was needed. European valves, however, being of the graduated release type were not suitable, and the American ABD equipment—designed to cope with trains up to 2 km or more in length—was considered more sophisticated and expensive than was necessary for Australian and New Zealand railways. NZR managers had been impressed by a new, lighter control valve assembly (the author suggests possibly a recent model of the KE distributing valve) manufactured by Knorr-Bremse in Europe wherein the integral slide valve componentry had been discarded in favour of more sensitive diaphragm and poppet valve operation. NZR urged the Australian Westinghouse Railway Brake Division to investigate the development of a new operating valve incorporating these latest engineering design features to suit Australasian brake systems.

107. The complicated process by means of which a crossing of opposing trains was effected on the distinctive, offset, crossing-loop-and-siding arrangements unique to the single-line, electric staff-controlled Main West line between Parkes and Broken Hill, when one or both trains was too long for either the loop or siding. Much legwork by train crews was involved as was the uniquely intricate Annett key operation of interlocked point levers.

108. The earliest rollingstock coupling device used in New Zealand was the British centre-chain-and-side-buffer system of the time. In terms of the safety and health of operating staff, it is to the eternal good fortune of local staff that the NZGR never adopted the infernal American link-and-pin equipment that preceded the Janney automatic coupler. The chopper-type coupler is, in fact, a much-improved adaptation of the link-and-pin system that also used a central buffer. However, instead of the connecting link that had to be hand-guided by a switchman (who, in the process, was placed greatly in harm's way) into a horizontal slot in the buffer and then manually locating a vertical pin to complete the connection, the chopper hook was deployed in the vertical plane and designed such that it would self-engage with the buffer to which it was coupling.

FIGURE 69: TYPICAL OF FREIGHT TRAIN WORKING OVER THE NEW ZEALAND NORTH ISLAND MAIN TRUNK IN THE 1970'S—AND ONE OF ABOUT 24 TRAIN MOVEMENTS EVERY DAY OVER THIS SECTION AT THE TIME—THIS IS TRAIN № 651, RUNNING SOUTHBOUND THROUGH KAKAHI AND ABOUT TO TACKLE THE LONG 2 PERCENT GRADE TO NATIONAL PARK BEHIND 26-L-EQUIPPED EMD AND GE POWER. NOTE THE MIXTURE OF 4-WHEEL AND BOGIE WAGONS. IN THE CONTEXT OF TRAIN HANDLING, THIS IS A LONG AND FLEXIBLE TRAIN. NZR INFLUENCE IN THE CREATION OF THE 'W' TRIPLE VALVE STEMMED FROM THIS OPERATING ENVIRONMENT. *Photo: G.J. McClare, courtesy NZ Railway & Locomotive Society.*

A gifted air brake engineer at Westinghouse (UK), Jack Washbourn, was despatched to Australia to commence design work on the New Zealand request. Washbourn had a solid background in his field, having been a key participant in designing the seminal P4 Distributing Valve and its descendent, the P4a. In addition, he had originally set up the Westcode operation in the United States and was the principal designer of Hawker Siddeley's innovative Z1A control valve and its successor, the Z1AW. Indeed, whether the "W" designation given to the new Australasian equipment represented 'Wilson' (acknowledging the contribution made by Wabco engineer Jim Wilson in designing the "*W" accelerated application* feature into the ABD valve in the United States), whether it stood for 'Westinghouse', or whether it was chosen by Jack to immortalise his contribution to this project remains—I think—unclear. The collaboration of Australian Westinghouse personnel resulted in the WP/WF (***W***estinghouse [or ***W***ashbourn or ***W***ilson?] ***P***assenger/***W***estinghouse ***F***reight) family of triple valves introduced into service in New Zealand in 1968 and subsequently in Australia.

The basic W triple valve—WP or WF—was conceived as the core device in a new modular automatic air brake system designed to incorporate the latest refinements in air brake equipment practice and to perform reliably in the great variety of Australian and New Zealand operational and climatic conditions. From the outset, the intention at Westinghouse Brakes Australia was for a basic valve assembly, to which optional features could be added separately or in combination suitable for application to locomotives as well as passenger and freight wagons, to provide both direct and graduated release capability as required. Additionally, the equipment was designed to fit conveniently to existing ITV and AF triple valve brackets and include the unique option of a multi-compartment reservoir with four separate chambers of varying capacity, all individually connected to a common mounting face that, in turn, connects to the pipe bracket.

Mr. Jack Washbourn

FIGURE 70: JACK WASHBOURN, WESTINGHOUSE BRAKE & SIGNAL CO., UK

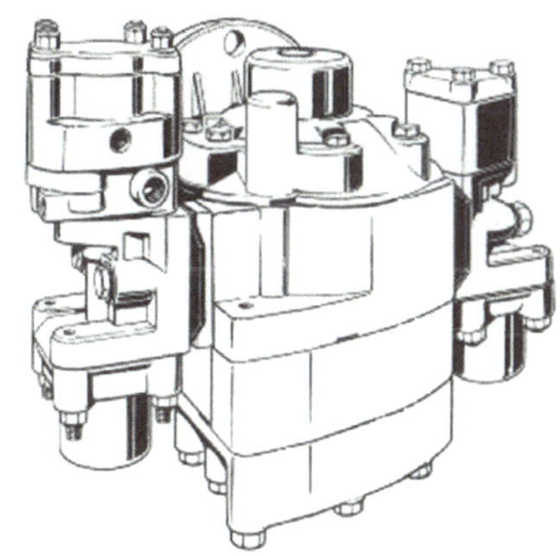

FIGURE 71: THE WF2 TRIPLE VALVE. THE DRAWING ON THE RIGHT DEPICTS WF2 WITH *IMPROVED ACCELERATED RELEASE* (THE PORTION ATTACHED TO THE LEFT OF THE VALVE AS VIEWED).

The W triple valve introduced the option of brake cylinder relaying to Australasian state-owned rail operations and, despite both New South Wales and Victorian Railways persevering into the 1980s with obsolete 3½-inch ITVs[109], all freight wagons manufactured in Australia since then (except some recent additions to Pilbara iron ore fleets) have used relayed brake systems. Relayed car brake systems have been used in the United States since the development of the AB control valve, although—with the exception of the bulk-commodity unit train and intermodal operations—not in widespread application. Often employed where multiple brake cylinders are fitted (e.g. bogie-mounted direct-acting brake systems or multi-unit/semi-permanently coupled/articulated rollingstock), relaying is an effective means of augmenting the braking on the wagon by using the auxiliary reservoir air during a brake application simply as a relatively low-volume 'control' pressure to pilot a brake cylinder relay valve rather than as an immediate brake cylinder supply.

During a brake application, the triple valve feeds auxiliary reservoir pressure to a dummy brake cylinder volume instead of straight to the brake cylinder(s). This dummy volume is protected against leakage and—in conjunction with the total volume of the control piping—provides a reference pressure to the relay valve for the desired brake application (brake cylinder pressure). Thus controlled, the relay valve

109. Victorian Railways acquired its first W triple valves with the so-called 'Hitachi' EMUs in 1972 and half of a fleet of container flat wagons around that time (the other 25 wagons of this acquisition being equipped with the D&M ESR500 triple valve).

then provides the higher-volume brake cylinder supply from a *supply* or *supplementary* reservoir fitted to the wagon or car; this reservoir being recharged from the brake pipe, along with the auxiliary and emergency reservoirs, during brake release. For its *Empty or Loaded* function, the relay valve responds either to a pressure signal or the lack of one from a load detection valve that is mounted on a bogie and is activated by spring deflection. When conditioned in this manner, the relay valve provides a fixed proportion of the desired braking effort when the wagon is empty and full braking when it is loaded.

The system provides the primary dual benefits of pressure maintenance (from the brake pipe so long as it is being maintained by the locomotive) against leakage from the brake cylinder and associated piping during applications, plus the capability of a higher-pressure supply to the brake cylinders for an Emergency application, should the latter feature be required. Secondary benefits include the capacity for finer control over the Empty/Load braking differential as well as the inherent ability to compensate for variations in brake shoe wear and thus brake cylinder piston travel.

Relaying does result in higher air consumption; although, since the WF equipment relays from triple valve and brake cylinder volumes equivalent to six inches as distinct from the AAR ten-inch volumes, this is not a defining issue. Although it should not be required (due to the relay feature), the *accelerated-release reservoir* pressure nevertheless exists as a reserve supply in the event that there is insufficient pressure stored (in other words, an incomplete recharge) to stop the train with a Service application.

Modern US control valves incorporate an *accelerated-release* feature that makes use of the pressure stored in the emergency side of the combined reservoir. With a relay system providing brake cylinder pressure maintenance (in other words, compensation for reasonable leakage), and in the context of the smaller train mass and size when compared to Class 1 North American operations, Australasian state-owned rail operators did not consider full Emergency to be an operational necessity and—due to its cost premium, higher air consumption, and operational non-compatibility with older-style brake equipment—did not desire it. For this reason, the W triple valve brake system was not designed to provide for the full Emergency function; an accelerated-release reservoir being provided instead on WF valves for this extra brake pipe supply.

The design intent of Emergency braking, of course, is to reduce the stopping distance (provided the accelerated release reservoir has not been depleted due to a recent brake release) and regarding this attribute, Australasian practice diverges significantly from North America. The idea of 'Emergency' as part of air brake functionality on Australasian freight and passenger trains[110] is something of a misnomer since Emergency brake pipe exhaust occurs either only on the locomotive (if the Emergency has been driver-activated) or only at the point of BP separation (if the Emergency has resulted from a burst hose or broken pipe) and not also on each trailing vehicle, and this cannot be over-stated. The resultant functionality of our triple valves with respect to brake cylinder filling in response to an 'Emergency' application plays a very minor role in hastening the time-of-development of brake cylinder pressure. In addition, the following should be understood: <u>*Car brake cylinder pressure derived from an Emergency application is the same as for Full Service*</u>. The singular meaningful difference between a Full Service and an Emergency brake application with our equipment is that the wagon or car triple valves are activated marginally earlier under Emergency application conditions than for a Service application. In fact though, under these circumstances, the time-to-activation of the triple valves during a driver-initiated or brake pipe separation Emergency—as distinct from a Service application—will be only minimally faster, and such an Emergency application can reduce the stopping distance over that obtaining from a Full Service application by only a modest, perhaps even an insignificant, value. In the context of our trains the notion

110. This lack of "true" Emergency does not apply to Pilbara iron ore railroads and much of the contemporary intermodal rollingstock operating in interstate service.

that there exists an enhanced or 'reserve' stopping force that—under emergency conditions—will serve to halt progress by some superior or dramatic degree is both a technical fiction and a reprehensible safety deception. Unless the rollingstock is equipped with control valves that include an Emergency portion (and the trailing rollingstock with a fully-charged emergency reservoir) no train has true Emergency air braking functionality.

The operating sensitivity of the W-type triple valves varies depending upon the operating (i.e. brake pipe) pressure. At the Australian standard 500 kPa (70 psi), it is more sensitive than the equivalent US control valves. At 620 kPa (90 psi), as used on the US Class 1 and Pilbara iron ore railroads, it would be about the same as the American valves, and at brake pipe pressures above this value it becomes progressively less sensitive. The W-type triple valve sends its Service brake application signal down a 32-mm brake pipe at around 280 m/sec (almost 850 ft/sec). This is fast by any measure, and it is noteworthy that if its AAR counterpart—the ABD control valve—were to operate at this propagation rate during Service brake operation, it would risk activation of its Emergency feature. For this reason it is not, theoretically, the most advantageous practice to mix W-type triple valves with Emergency-equipped AAR control valves on Australian freight trains, although it must be acknowledged that some railways have done so with no adverse result that the author is aware of.

Some industry observers believe that the later versions of the W triple valve exceed the ABD in performance, and in this author's view, it could well be argued that this device—the winner of an Australian Design Award for design and technical achievement—might more deservingly be referred to also as a *control valve* rather than its current "colonial" and somewhat archaic designation.

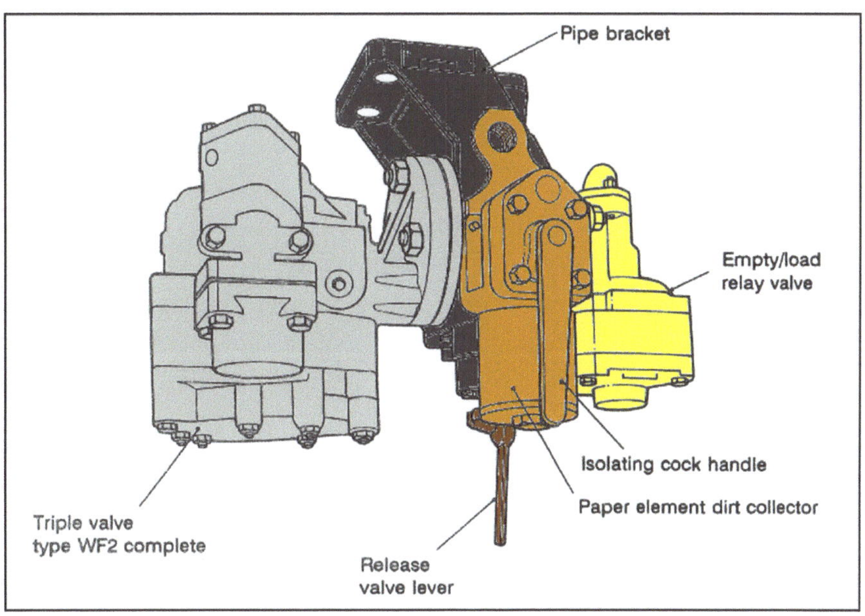

FIGURE 72: THE WF2 TRIPLE VALVE AND PIPE BRACKET MOUNT

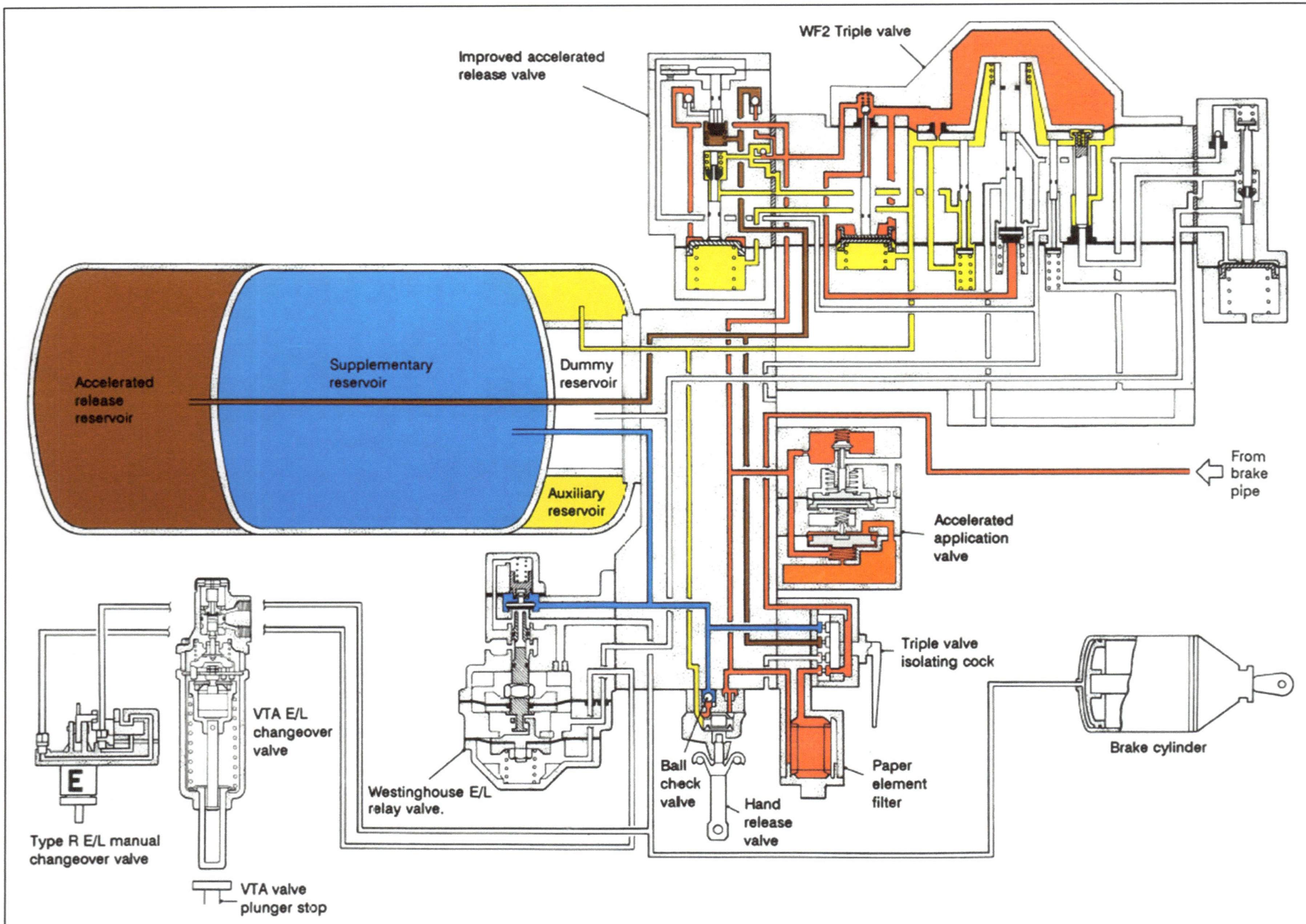

FIGURE 73: THE WESTINGHOUSE AUSTRALIA (NOW KNORR-BREMSE) 'W' TWO-STAGE FREIGHT BRAKE EQUIPMENT – SCHEMATIC

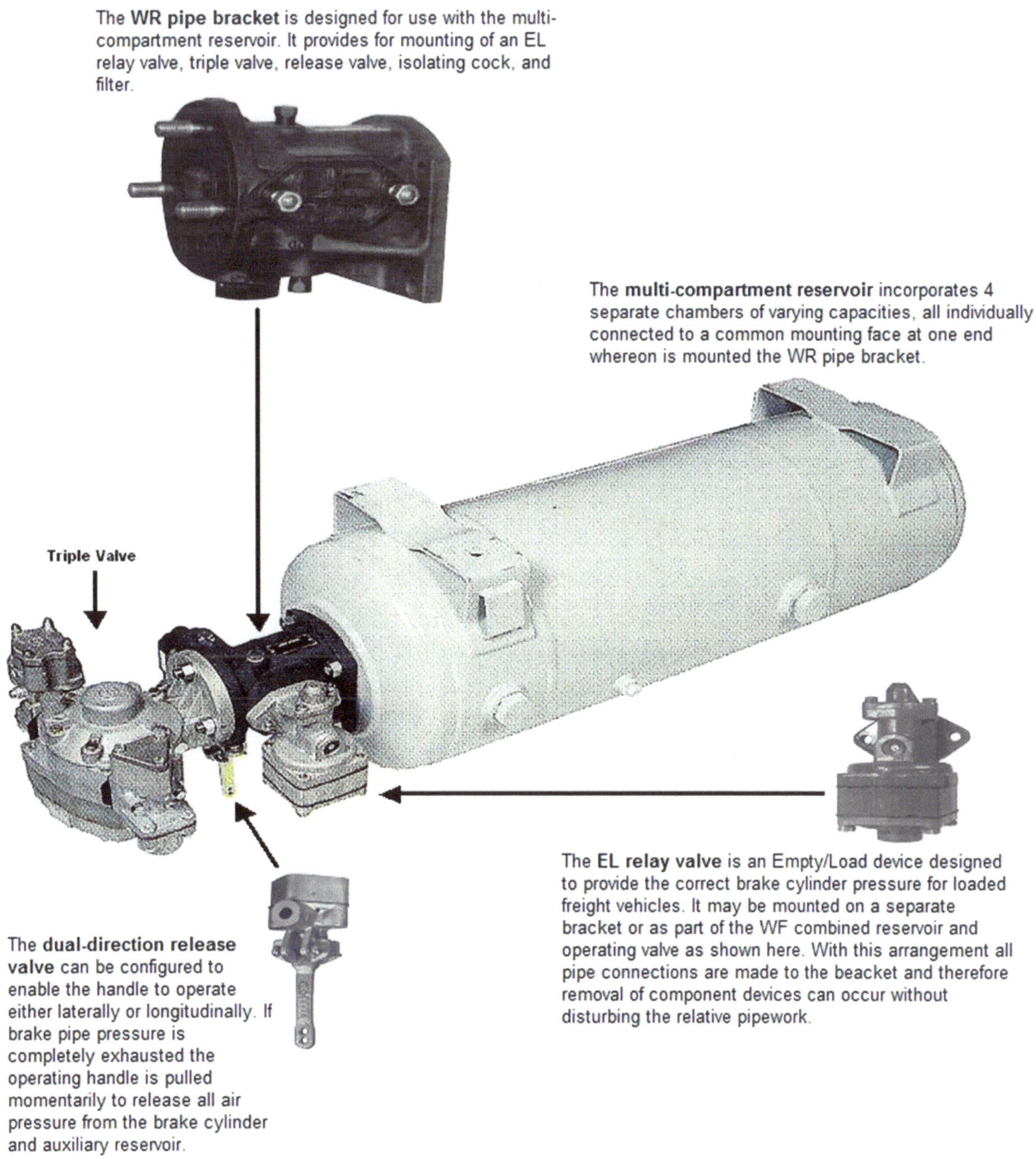

FIGURE 74: THE WF COMBINED ARRANGEMENT

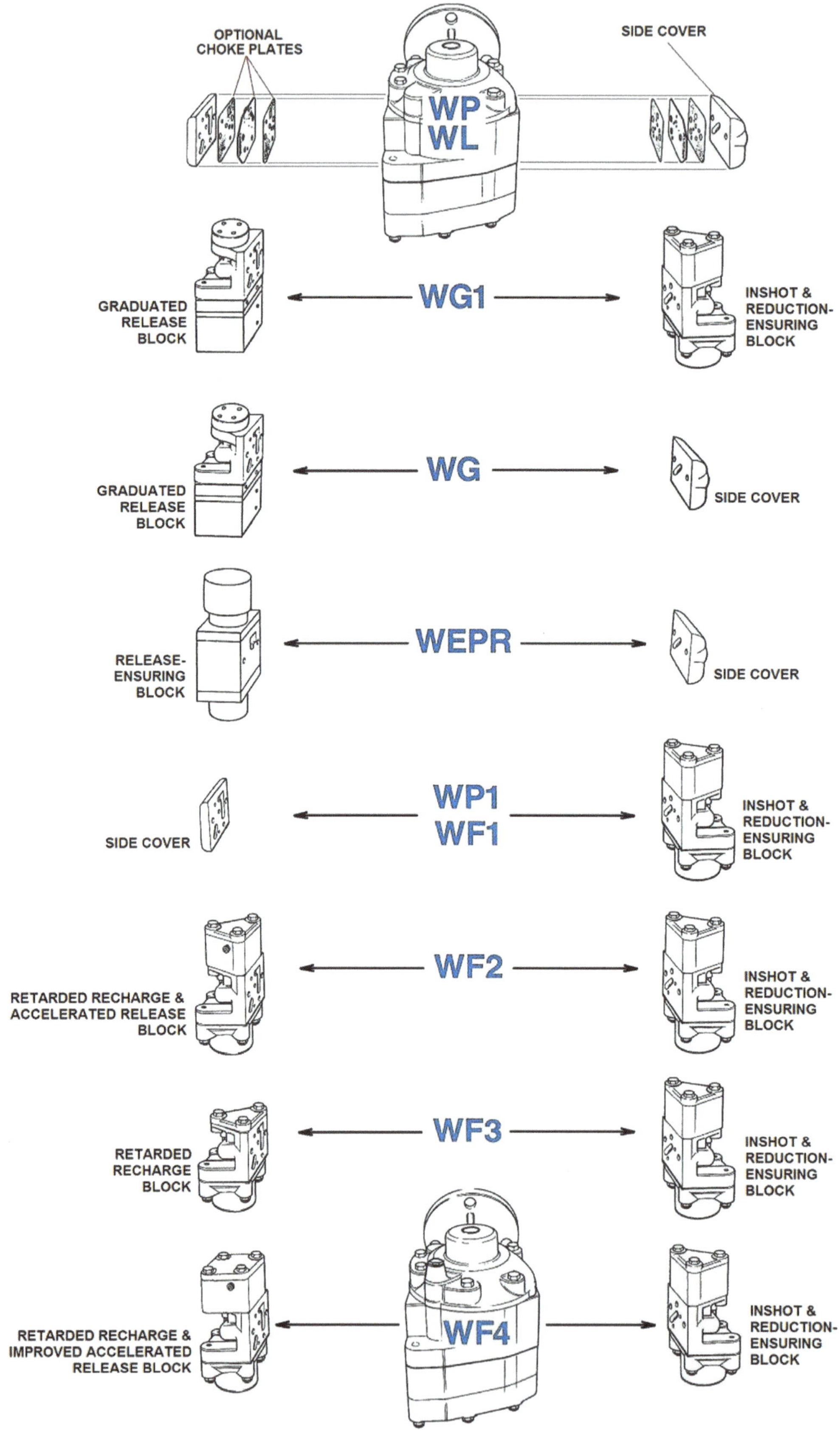

FIGURE 75: THE W TRIPLE VALVE TYPES AND MODULAR OPTIONS

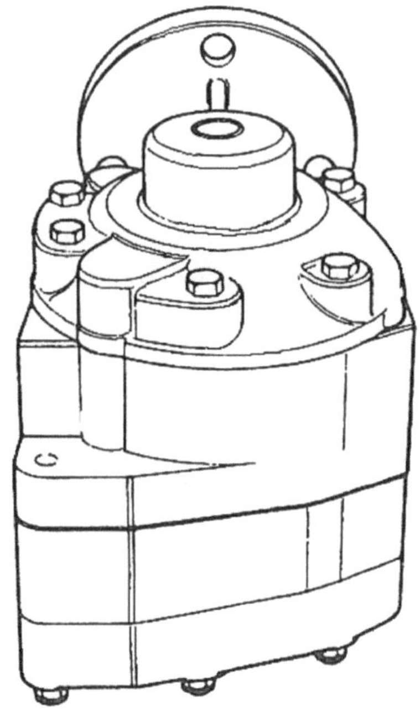

FIGURE 76: THE BASIC W TRIPLE VALVE

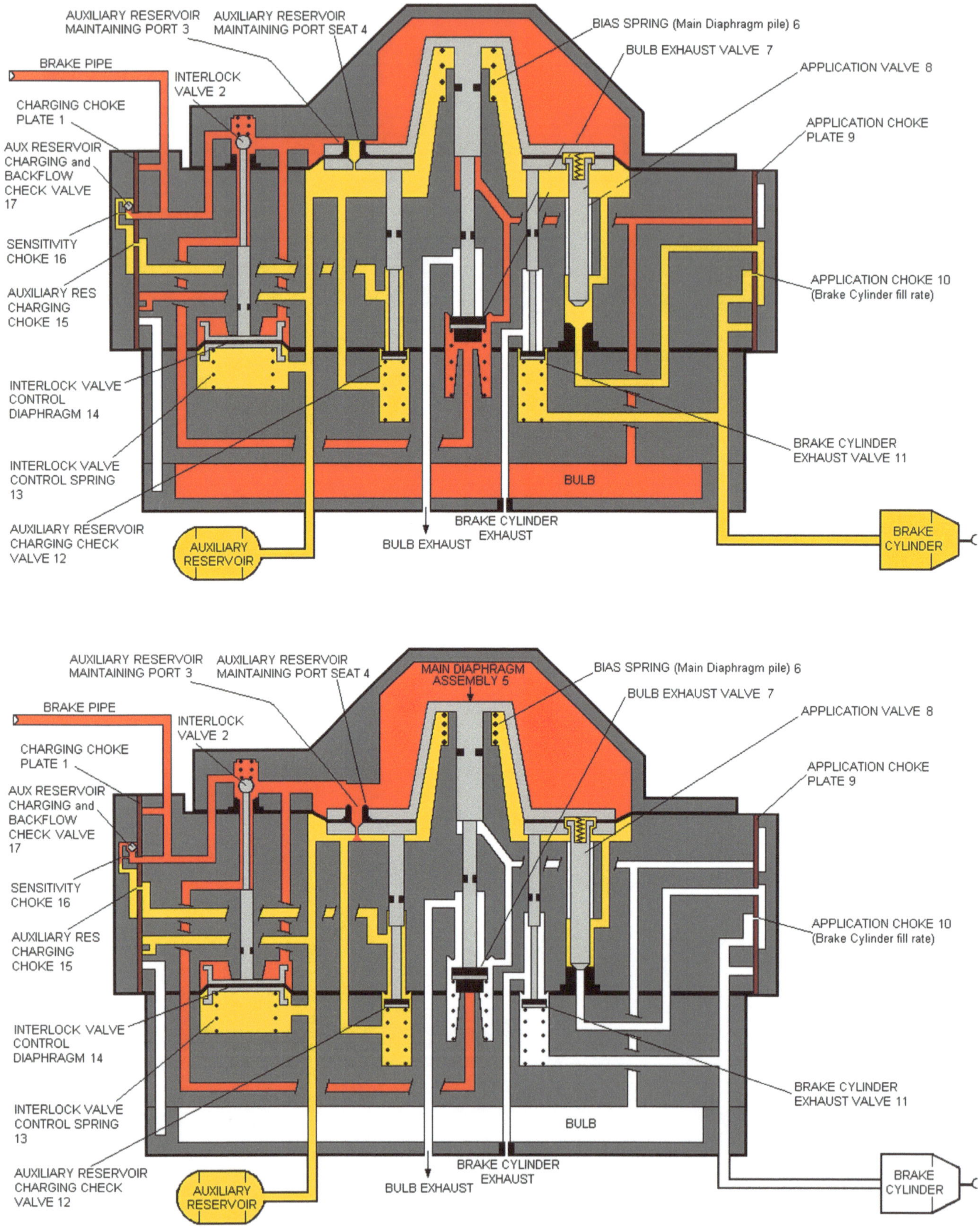

FIGURE 77: THE BASIC W TRIPLE VALVE - DIAGRAM: APPLICATION (UPPER) AND RELEASE

The WP Triple Valve

A basic valve (Figures 77 and 78) was initially designed for use on vehicles that did not demand any special features to cope with long trains. This is the WP triple valve for use on passenger trains and shunting locomotives; diaphragm-operated and with a synthetic rubber coating on a nylon sheeting being used for the diaphragm material. This amounts to a modern-day version of George Westinghouse's original concept. He was forced by the limitations of the materials then available to change to a piston-and-slide valve operation, so it is interesting to see his initial ideas validated by this latter-day development.

A main diaphragm assembly **5**, having brake pipe pressure on its upper face and auxiliary reservoir pressure below, controls an auxiliary reservoir charging check valve **12**, a brake cylinder exhaust check valve **11**, a brake cylinder inlet valve **8**, and a combined QS chamber inlet and exhaust valve **7**. The diaphragm followers of this assembly contain an auxiliary reservoir maintaining port **3** and associated choke. Contained in the top cover is the interlock valve **2** controlled by a diaphragm **14** working between brake pipe and auxiliary reservoir pressures.

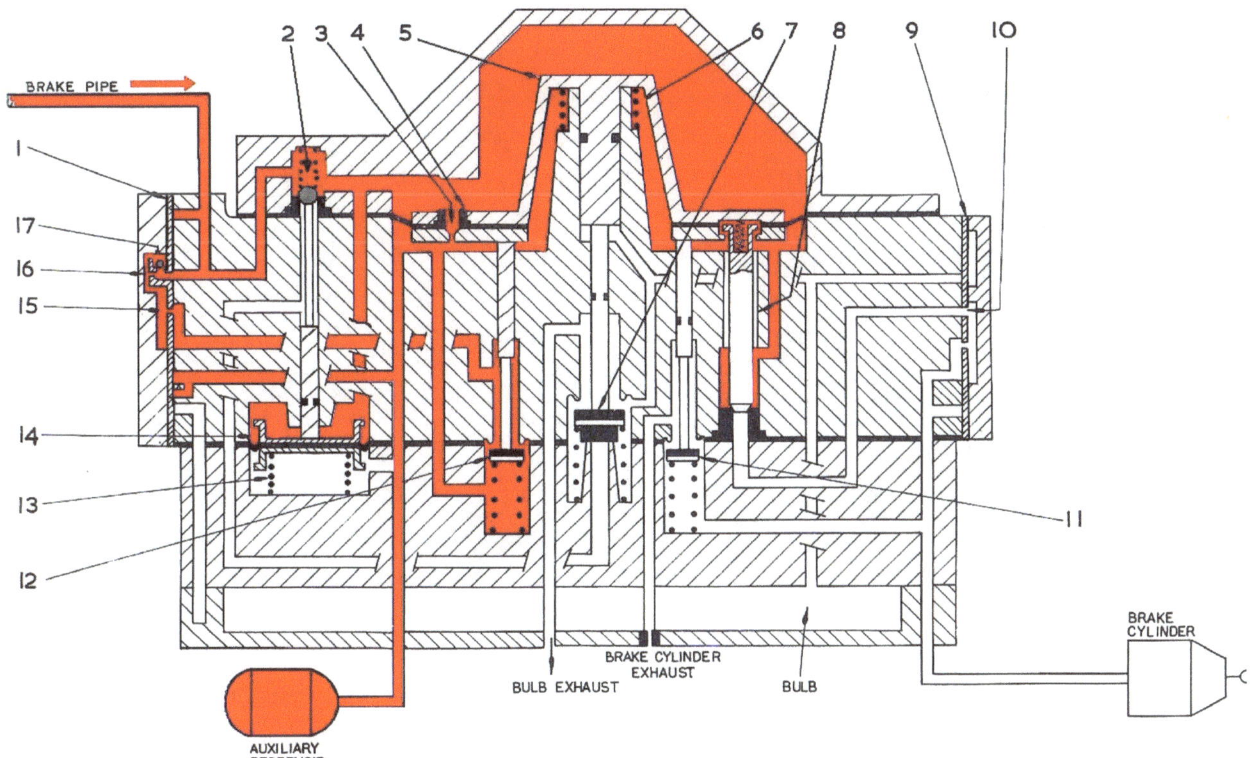

FIGURE 78: THE WP TRIPLE VALVE - DIAGRAM

1. Charging Choke plate
2. Interlock valve
3. Auxiliary Reservoir maintaining port
4. Auxiliary Reservoir maintaining port seat
5. Main diaphragm assembly
6. Main diaphragm pile bias spring
7. QS chamber exhaust valve
8. Application valve
9. Application Choke plate
10. Final filling choke (Primary Application)
11. Brake cylinder exhaust valve
12. Auxiliary Reservoir charging check valve
13. Interlock valve control spring
14. Interlock valve control diaphragm
15. Charging Choke
16. Sensitivity Choke
17. Backflow check valve

Application:

During a brake application, a reduction in brake pipe pressure above the main diaphragm assembly of as little as 7 kPa will permit it to move upward under the influence of auxiliary reservoir pressure. This reduction in brake pipe pressure also induces backflow check valve **17** to close, thus backflow from the auxiliary reservoir can only occur at a limited rate via sensitivity choke **16**. The combined Quick-Service exhaust and bulb inlet check valve **7** rises to close the bulb exhaust and open the QS inlet port. Brake pipe pressure flows from open interlock valve **2** to the QS bulb producing a sudden, localised reduction in brake pipe pressure that propagates the reduction and thus the brake application down the length of the brake pipe.

The main diaphragm assembly is moved positively upward aided by QS bulb pressure newly developed against the underside of the main pile stem, closing auxiliary reservoir charging check valve **12** and brake cylinder exhaust valve **11**. Inlet or application valve **8** is opened, allowing auxiliary reservoir pressure to flow to brake cylinder via application choke **10**. Brake cylinder pressure now develops until it slightly exceeds auxiliary reservoir, whereupon the main diaphragm assembly reverts slightly to assume its LAP position, closing brake cylinder inlet valve **8**. With auxiliary reservoir and brake pipe pressures now substantially equal, any tendency for the diaphragm assembly to continue its downward movement toward RELEASE position is resisted by QS bulb pressure holding up exhaust valve **7**.

Any further brake pipe pressure reduction results in another upward movement of the diaphragm assembly that re-opens brake cylinder inlet valve **8**, increasing the application. This graduation of the application can be continued as desired until auxiliary reservoir pressure equalises with that in the brake cylinder; this being *Full Service equalisation*. Any brake pipe pressure reduction beyond this value will not increase the application but causes the main diaphragm assembly to lift to its "top stop," sealing auxiliary reservoir maintaining port **4** and preventing backflow of auxiliary reservoir and brake cylinder pressures to the brake pipe.

Release:

When brake pipe pressure is increased, the main diaphragm tends to move downward, but this is resisted by QS bulb pressure (originally from the brake pipe) acting against exhaust valve **7**. This resistance is overcome once brake pipe pressure exceeds the auxiliary reservoir by approximately 10 kPa, whereupon QS bulb pressure is released to atmosphere. Since the latter pressure is also beneath the main pile stem, its rapid exhaustion leaves a sudden overbalance of force, causing the main diaphragm to snap to RELEASE position, opening the QS bulb exhaust valve plus brake cylinder exhaust valve **11** and the auxiliary reservoir charging check valve **12**.

The brake cylinder now vents to atmosphere via a choked exhaust port, and the auxiliary reservoir is recharged.

WP1 and WF1 Triple Valve

The WP1 and WF1 Triple Valves are identical apart from different brake cylinder application choke sizes; larger for the passenger and smaller for the freight version. To the right-hand side of the basic W triple valve depicted on the previous page is added—in place of the cover that clamps the choke plate containing the brake cylinder application rate choke—an Inshot and Reduction Ensuring (or *brake cylinder maintaining*) block. The Inshot Control Valve **24**—supplying auxiliary reservoir pressure—and the Quick-Service chamber exhaust valve **7** plus check valve **19** and Reduction Ensuring Valve **26** (supplying brake pipe air) operate in parallel to provide a rapid brake cylinder fill (the Quick-Service feature) until a pressure of 70–80 kPa (10–12 psi) is reached, at which point both are closed by the air loading above operating diaphragm **23**. Flow into the brake cylinders now continues at a reduced rate via the primary application choke.

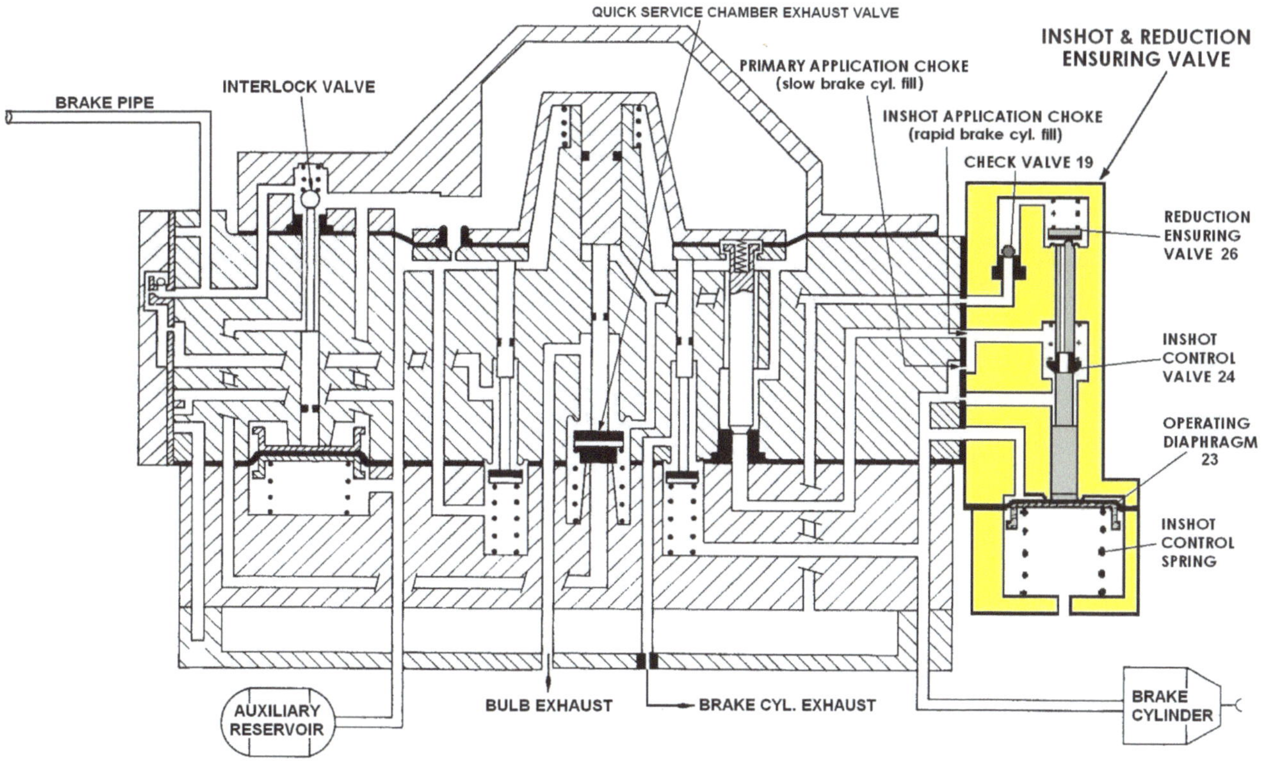

FIGURE 79: THE WP1 TRIPLE VALVE – DIAGRAM. NOTE *INSHOT & REDUCTION ENSURING VALVE* ATTACHED TO THE RIGHT-HAND SIDE (APPLICATION) CHOKE PLATE

WF2 Triple Valve

1. Charging Choke plate
2. Interlock valve
3. Auxiliary Reservoir maintaining port
4. Auxiliary Reservoir maintaining port seat
5. Main diaphragm assembly
6. Main diaphragm pile bias spring
7. QS chamber exhaust valve
8. Application valve
9. Application choke plate
11. Brake cylinder exhaust valve
12. Aux res charging check valve
13. Interlock valve control spring
14. Interlock valve control diaphragm

To the WF1 triple is added a block on the left-hand side of the valve body in place of the cover clamping the choke plate containing the auxiliary reservoir charging choke. This is the Retarded-Recharge and Accelerated-Release block.

A diaphragm-driven stem controls the *retarded-recharge* valve such that there is either one charging feed (via just the Inshot Application choke) or two (via both the Inshot Application *and* Primary Application chokes) to the auxiliary reservoir, depending on whether the vehicle is toward the front of the train (accepting less air from the brake pipe) or toward the rear (accepting more air). As with all operating valves so-equipped, this feature ensures the auxiliary reservoirs at the front of the train do not absorb all the recharge pressure, and therefore charging at the rear of the train is assisted. This action equates to that of the AF triple valve Retarded-Recharge feature but is under more precise control, as the chokes are drilled to an accurate size instead of being cut into the piston bush as they must be in the AF triple.

Accelerated service release is the latter-day solution to the problem whereby brake pipe gradient (or "taper") is responsible for triple valves toward the rear of the train releasing later than those toward the front. The WF2

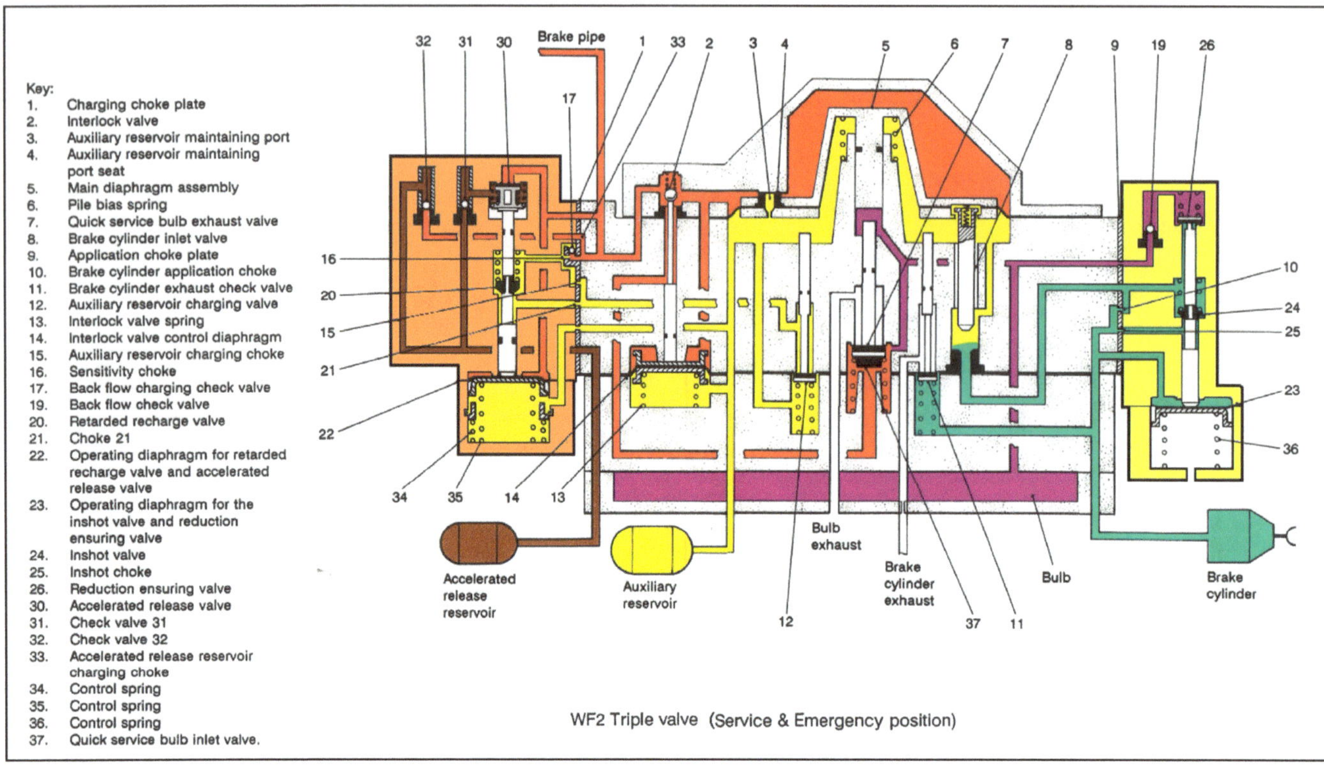

FIGURE 80: THE WF2 TRIPLE VALVE – DIAGRAM. NOTE THE COMBINED *RETARDED RECHARGE/ACCELERATED RELEASE* BLOCK ATTACHED TO THE LEFT AND THE *INSHOT & REDUCTION ENSURING* BLOCK ON THE RIGHT

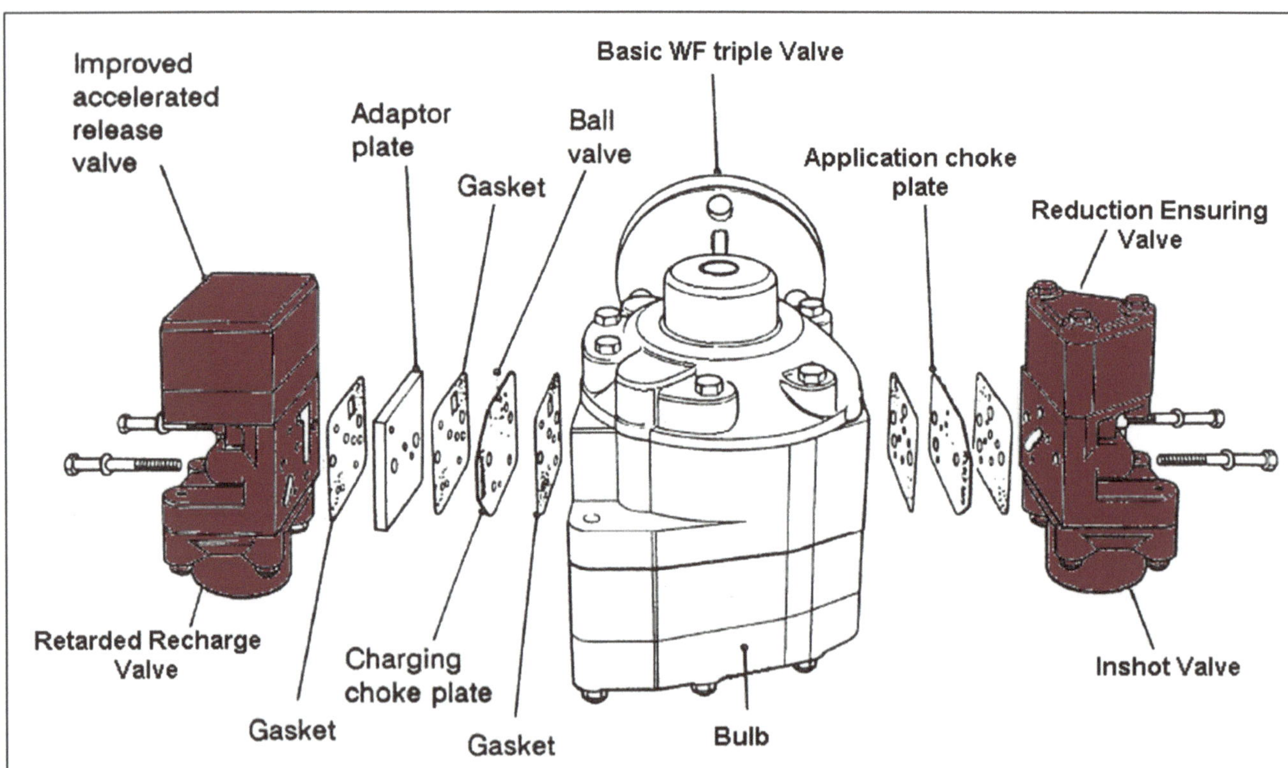

FIGURE 81: THE WF TRIPLE VALVE – ILLUSTRATION. AN EXPLODED VIEW SHOWING ADAPTION TO WF2-TYPE

overcomes this problem by providing a store of compressed air that is released into the brake pipe to provide a local increase in pressure on each individual wagon during the early stages of a brake release. Whereas with the ABD control valve, this supply of air pressure comes from the emergency reservoir, on the WF2 it comes from the accelerated release reservoir, which is an extra one fitted to the vehicle and which is sized to match the length of the brake pipe (24 litres for the obsolete four-wheel wagons and 37 litres for bogie wagons). The positive operating effect of these valves on a train is in direct proportion to the number of wagons fitted with them.

Accelerated Release increases the initial rate of brake pipe charge and assists the retarded-recharge valve in providing a more even and quicker response in brake release.

WF3 Triple Valve

The WF3 triple valve had all the features of the WF2 except for Accelerated Release. The the left-hand block referred-to in the WF2 description—incorporating only one valve on the diaphragm stem—provides for Retarded-Recharge only and is thus smaller.

This triple valve was used as a replacement for the Improved and AF triple valves on older wagons, which would have required considerable modifications to accept the WF2 valve (e.g. the extra reservoir mentioned above and its piping would be required and, generally, WF2 valves are best fitted to the special pipe bracket that incorporates all the porting for the various connections; this bracket being best fitted when the wagon is new or during a complete rebuild).

FIGURE 82: THE WF2 TRIPLE VALVE – PART-SECTIONED

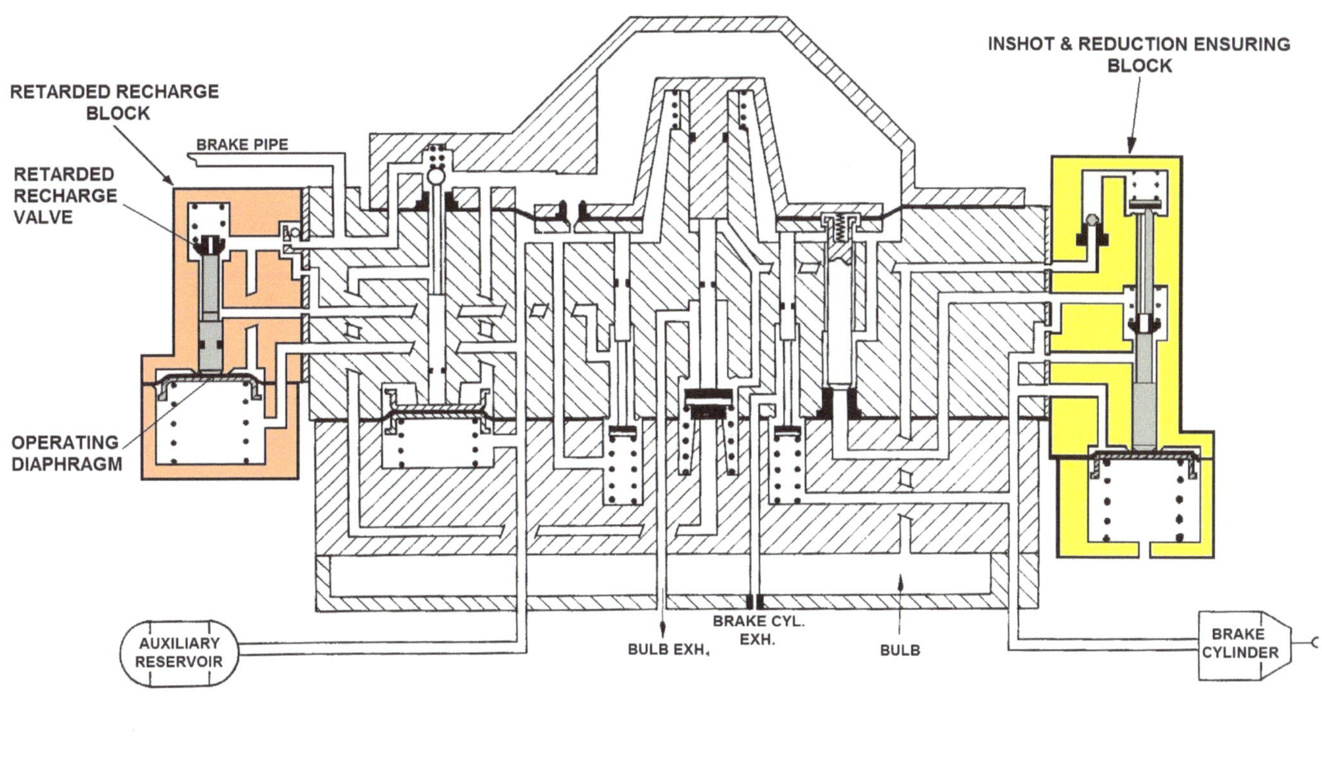

FIGURE 83: THE WF3 TRIPLE VALVE, *RELEASE* POSITION – DIAGRAM

WF4 Triple Valve

Introduced in 1987, the WF4 was designed specifically for larger freight trains; 120-plus bogie wagons then being introduced into operations on Leigh Creek coal trains in South Australia. The WF4 incorporated the following improvements:

Split bulb volume

When air flowing rapidly within the brake pipe is suddenly checked, a surge wave will develop on the flow, and this is what occurs with Quick-Service as brake pipe pressure flows rapidly into the bulb during a brake pipe reduction. At the rear of a long train, once Quick-Service activity ceases, the pressure drop in the brake pipe is not rapid; thus, any large surge is more noticeable and may cause a momentary rise in brake pipe pressure sufficient to cause the triple valve to go to Release. (Note: this potential exists where brake pipe pressure during QS activity flows into a single bulb volume.) In this case, a re-application will occur as brake pipe pressure continues to fall (but to a lesser brake cylinder pressure than desired), and—depending on the rate of brake pipe pressure reduction—the triple valve may "hunt" in this manner several times.

To avoid this phenomenon, the WF4 bulb chamber has a **surge suppression choke** drilled through the wall separating the two chambers. The action of the quick-service inlet valve allows brake pipe air to flow rapidly into the first section of the bulb and then, via the choke, into the second section. This action suppresses the magnitude of the surge to a value sufficient to ensure the most sensitive triple valve will not "kick" to Release. The propagation rate is not adversely affected; the combination of brake pipe reduction from flow into the QS volumes plus into the brake cylinder, in fact, results in a very high propagation rate approaching 300 m/sec.

Improved accelerated release

The failure of brakes to release following light brake applications has been a prevalent problem since train lengths began to increase above about 400 metres. US air brake manufacturer pamphlets and other instructions have always recommended that train brake releases should only be initiated from *effective applications*[111] of not less than 70 kPa (10 psi) brake pipe reduction. However, this instruction is difficult for footplate inspectors (or "workplace assessors" as they are known these days) to monitor; indeed, from this author's observations over many years, a large proportion of drivers and their supervisors are unaware of this basic requirement! Thus, due to the sometimes unwelcome prospect that brake applications—when correctly made—might result in the train coming unintentionally to a stand,[112] this wisdom is not always followed by drivers. In other words, locomotive drivers tend too often to make brake pipe reductions that are too light and then to release too early. As originally designed, the accelerated release valve did not operate unless the total brake pipe reduction was around 90 kPa (13 psi) and, therefore, the device could not assist in the release of light brake applications.

Occurrences of stuck or dragging brakes arising from brakes failing to release following an exceptionally light application on long trains, then, would be greatly reduced or even eliminated if the accelerated release feature were to operate under these circumstances. This outcome was duly achieved by designing

111. A brake pipe reduction becomes "effective" only when sufficient time has been allowed for *Service Equalisation* to occur (i.e. the brake pipe exhaust has ceased).

112. Locomotive drivers, when making a brake application to control train speed (as distinct from intending to stop), will often release before the application has equalised (i.e. become effective) in order to prevent an unintended stop. Such hasty action is not a desirable practice.

the **improved accelerated release valve** to use the pressure of the air exhausting from the brake cylinder/s to trigger the function.

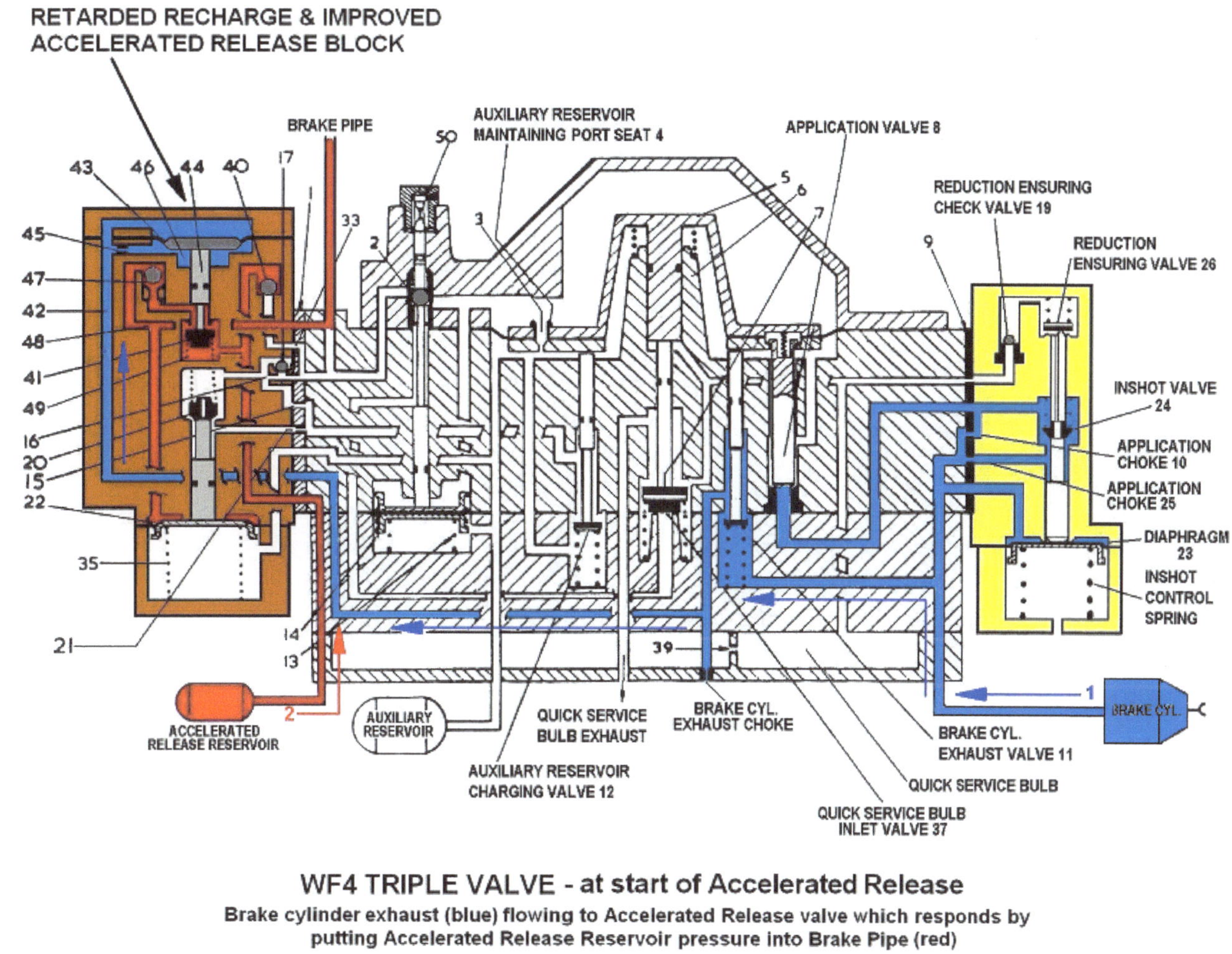

FIGURE 84: THE WF4 TRIPLE VALVE - DIAGRAM

1. Charging choke plate
2. Interlock valve
3. Aux Res maintaining port
5. Main diaphragm pile assembly
6. Pile bias spring
7. Quick-Service bulb exhaust valve seat
9. Application choke plate
10. Interlock valve control spring
11. Interlock valve control diaphragm
12. Primary Aux Res charging choke
17. Back flow ball check valve
18. Retarded recharge valve
19. Secondary Aux Res charging choke
20. Retarded recharge valve operating diaphragm
21. Accelerated Release Res charging choke
39. Surge suppression choke
40. Check valve
41. Accelerated Release valve
42. Passage – brake cyl exhaust to Accelerated. Release valve
43 Accel. Rel. valve operating diaphragm
44. Accel.Rel. valve operating piston
45. Accel.Rel. equalisation choke
46. Accel. Rel. valve diaph lower chamber
47. Ball check valve
48. Passage – Accel. Rel. valve to brake pipe
49. Accel.Rel. valve balancing spring
50 Initial charge lock

Initial Charge Lock

This feature is designed to prevent the loss of brake pipe pressure from the Quick-Service bulb exhaust port during the initial stages of brake pipe charging on the long trains for which it was designed, when the rate of brake pipe pressure increase might be extremely slow. In this event, the rate of pressure increase in the brake pipe might be retarded to the extent that insufficient pressure differential exists between the brake pipe and auxiliary reservoir to keep the **interlock valve** closed. If this should occur, brake pipe pressure will flow to the **quick-service inlet valve**, with the potential to cause the triple valve to momentarily assume the Applied position. In turn, the triple valve will cycle or "hunt" during initial charging with subsequent loss of pressure from the Quick-Service and brake cylinder exhaust ports, and this action can preclude the complete charging of the brake pipe. The **initial charge lock** prevents the interlock valve from opening in this event, ensuring that brake pipe air does not flow to the quick-service inlet valve until there is sufficient pressure above the main diaphragm assembly to create the downward force required to hold the triple valve in its Release position.

WF5 Triple Valve

Originally called the WF4A, this development—introduced in 2001—improved upon the WF4, incorporating its features and adding the Piloted Reduction-Ensuring feature.

Piloted Reduction Ensuring

The basic **reduction-ensuring valve** augments the Quick-Service feature and ensures that a sufficient initial brake pipe pressure reduction occurs throughout the train to initiate braking action. With the slow rate of brake pipe reduction experienced at the rear of long trains, pressure surges generated during the initial brake application can create sufficient pressure differential across the main diaphragm to sometimes cause triple valves to assume the Release position. Airflow from the bulb cavities through the reduction-ensuring valve decreases pressure below the main diaphragm pile stem, reducing operational stability and assisting the inherent tendency of the triple valve to assume the Release position. The triple valve will re-apply as brake pipe pressure continues to reduce but can continue to apply and release until brake pipe pressure stabilises. This condition will occur only at or near the rear of long freight trains subject to light brake applications (this, as previously mentioned, not being a desirable train-handling practice).

The **piloted reduction-ensuring valve** is "piloted" in its operation during the initial brake application from brake pipe pressure entering the Quick-Service bulb cavities. The flow of air through the reduction-ensuring valve is ported from the brake pipe cavity above the diaphragm, and therefore the service stability of the triple valve is not affected under the operating conditions previously described.

Whereas the WF3 triple valve—which includes the **Inshot**, **Reduction-Ensuring**, and **Retarded-Recharge** features—is suitable for freight trains *up to* 900 metres in length, the WF5 is designed for trains *exceeding* 900 metres in length and replaces the WF2 and WF4 types. Knorr-Bremse Australia has now standardised the manufacture of W triple valves on WF3 and WF5 types; and all existing WF2, 3, and 4 types can be upgraded to WF5 during refurbishment.

The successive modifications to the WF triple valves herein described—designed to introduce and enhance the desirable operating features of both stability and sensitivity—overtook the original necessity for the suppression of Quick-Service activity. These modifications have proven to be all the more necessary since suppressing QS activity is detrimental to the propagation of light brake applications on long trains. For this reason, some Australian rail operators have requested the removal of QS suppression, and this has been achieved by modifying the bulb gasket that removes an obstruction to the free flow of air pressure between the cavities and by increasing the diameter of the orifice through the QS inlet valve seat. Such modifications will become standard for new and refurbished W-type triple valves.

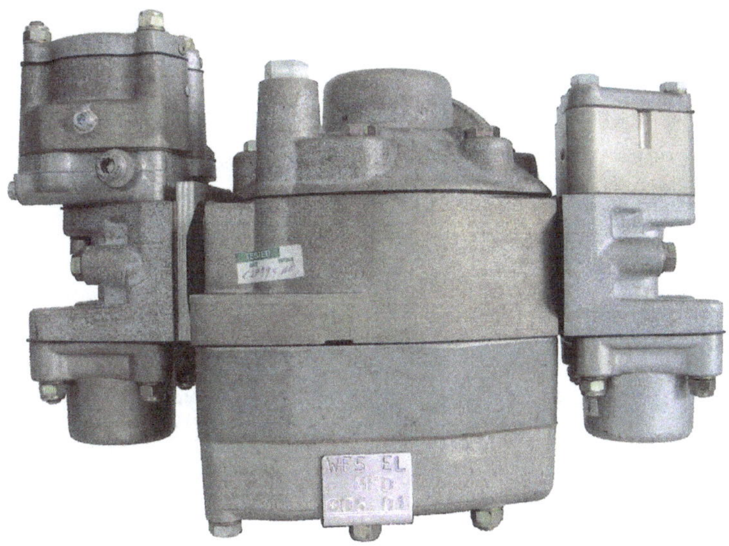

FIGURE 85: THE WF5 TRIPLE VALVE – ILLUSTRATION

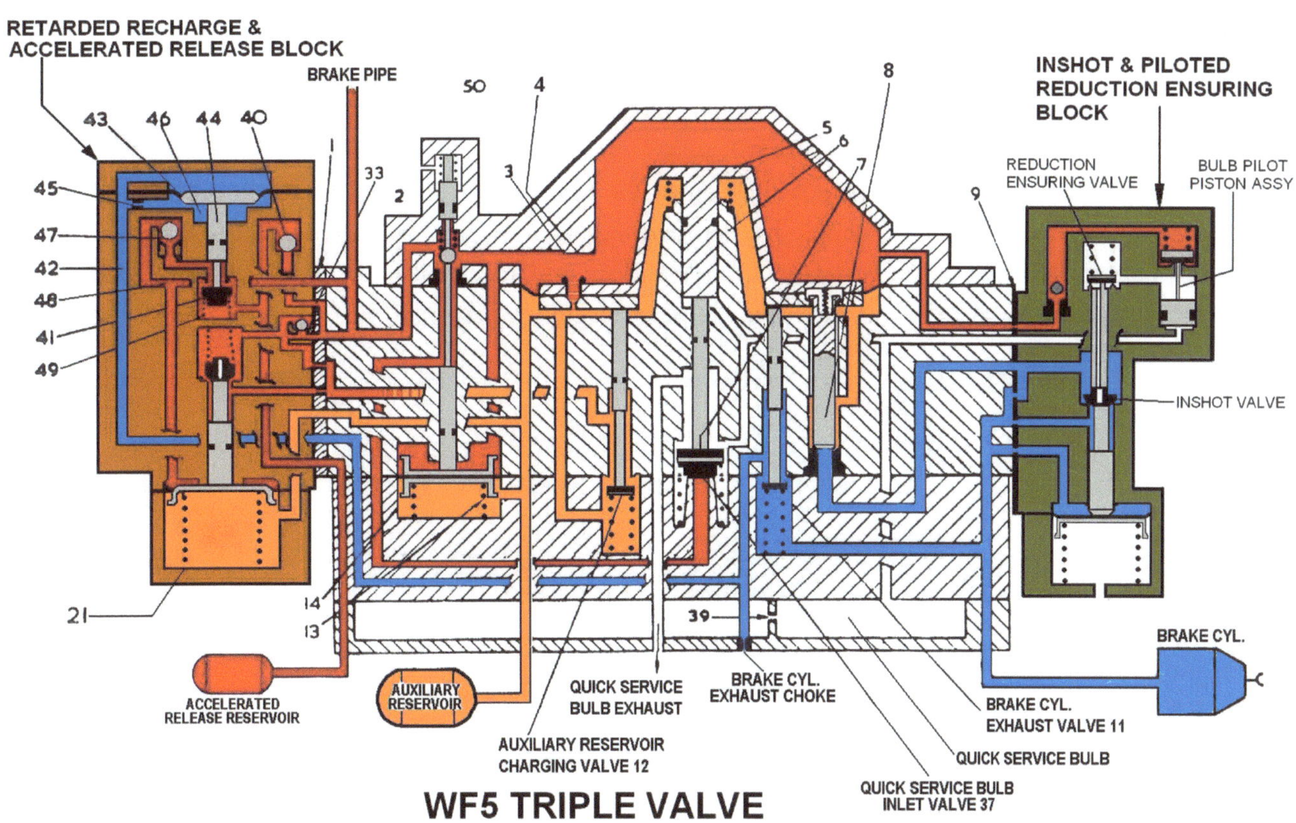

FIGURE 86: THE WF5 TRIPLE VALVE – DIAGRAM

The WFE-W Control Valve

This was developed by Westinghouse Brakes Australia as a result of Mount Newman Mining Pty Ltd (now BHP Billiton Iron Ore Ltd), giving concession for local manufacture on two small contracts of wagons, and an expression of interest by other iron ore mining companies in Western Australia for an air brake control device of poppet valve design. With its designation denoting the full ***E***mergency and "***W***" (accelerated application) features, this equipment was based on the WF4 triple valve, which was used as the *Service portion,* and included a newly designed *Emergency portion*. To address the Mount Newman brief, the new control valve was designed for freight wagons regularly used in trains of extreme length and thus requiring the capability of full Emergency brake operation per North American practice. But it was also intended to be suitable for trains of lesser length running at higher speeds over sections where signal spacings demanded a higher-than-normal brake cylinder pressure in Emergency. It was designed for Australian conditions and to operate compatibly with other types of AAR-compliant equipment. Since there was no intention of marketing the WFE-W in North America, AAR approval was not sought.

The function of the Emergency Portion was to monitor the rate of brake pipe pressure reduction. Should this have been at the normal Service rate, only the Service portion acted to feed auxiliary reservoir pressure to the brake cylinder/s (or to the relay valve if used with a relayed system); however, if the rate of brake pipe pressure reduction was *greater* than the Service rate (from whatever cause), then the Emergency portion would vent the brake pipe rapidly and feed emergency reservoir pressure to the brake cylinder/s to augment the auxiliary reservoir pressure, thus providing a 20 percent increase in brake cylinder pressure.

The WFE-W control valve included the following:

- An **accelerated-release valve** (part of the WF4 Service portion) that—when triggered during brake release—fed emergency reservoir pressure into the brake pipe to assist recharge and thus ensured a prompt release of brakes to the rear of the train. This valve was mounted on the side of the Service portion; and,

- An **accelerated-application valve** (part of the Emergency portion) that vented brake pipe pressure locally at each vehicle by a controlled amount during brake applications, thus propagating the brake pipe reduction along the train and accelerating brake cylinder pressure buildup. This valve mounted to the side of the Emergency portion.

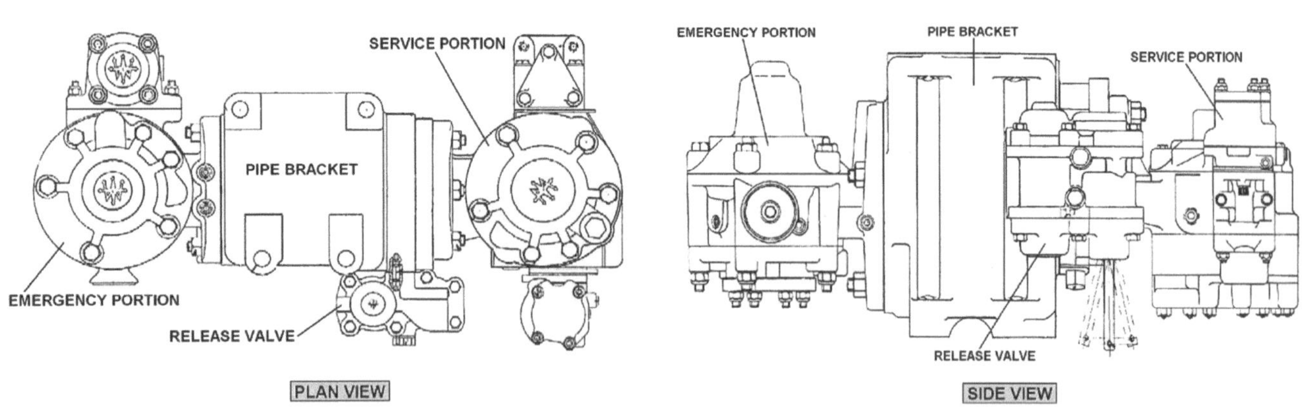

FIGURE 87: THE WFE-W CONTROL VALVE – GENERAL ARRANGEMENT

Several issues, however, led to the early demise of the WFE-W. It was required to be fully compatible with the AAR equipment then in use on the iron ore railroads—with respect to interchangeability and release-valve operation—and for this reason a special adapter plate was required to suit the AB-type pipe bracket; the other constituent parts (Emergency portion, Release Valve, and Accelerated-Application Valve) were too expensive to produce in the quantities required; and since Westinghouse Brakes Australia already provided a similar (US) product through their agreement with Wabco, the Australian-designed product amounted to a potential infringement of that.

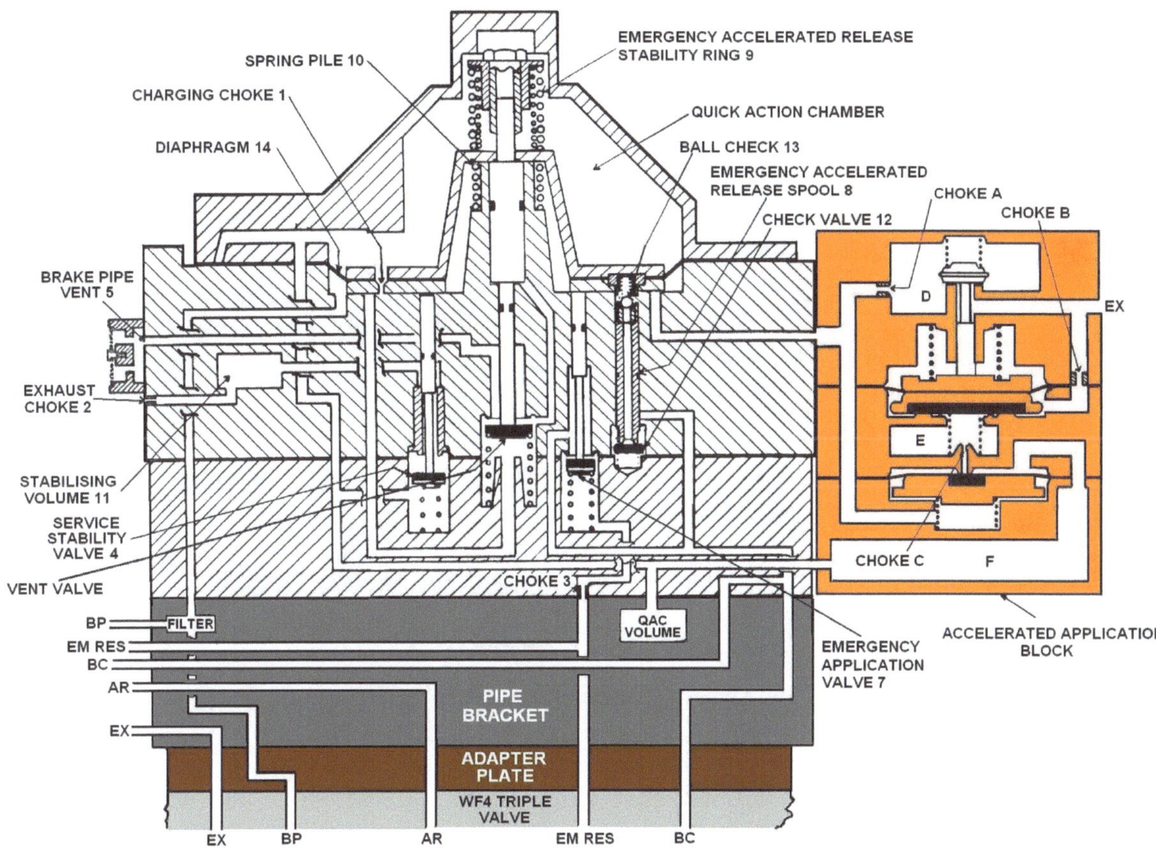

FIGURE 88: THE WFE-W CONTROL VALVE EMERGENCY PORTION - DIAGRAM

The WG Triple Valve

This is the ***G***raduated-release version of the W-triple valve. The WG (Figure 89) is created from the basic WP triple valve to which is added—in place of the Choke Plate 1 cover plate (left-hand side of valve in diagrammatic view)—a Graduated-Release portion. A control reservoir is included to provide the necessary reference pressure, and this, as well as the Graduated-Release portion, is charged from the brake pipe simultaneously with the auxiliary reservoir.

The functioning of the valve during brake application is identical to the basic WP, except that the Graduated-Release valve assembly holds the control reservoir fully charged. During a brake release, the WG valve functions identically to the WP valve during the preliminary stages of the event; however, the brake cylinder exhaust is piped to the Graduated-Release valve rather than to atmosphere as is the case with the WP and WF versions. At the Graduated-Release valve, the brake cylinder exhaust pressure is either vented to atmosphere or trapped, depending upon the balance of forces of the brake cylinder pressure (assisted by a bias spring) versus the control reservoir pressure acting on opposing diaphragms.

If the brake pipe pressure increase is only partial (in association with a graduated release-capable driver's automatic brake valve), then the amount of brake cylinder pressure permitted to vent to atmosphere at the Graduated-Release valve will also be partial by a comparative amount. By manipulating the automatic brake valve, the driver is able to graduate brake pipe pressure both down and up in small steps with car brake cylinder pressure being increased or reduced accordingly. A complete brake release will occur when brake pipe pressure is restored to within approximately 15 kPa of its maximum value (i.e. to 485 kPa in Australia or 555 kPa in New Zealand).

Due to the intrinsic requirement of locomotive dual-brake (air *and* vacuum) systems for graduated release capability on the vacuum brake operations, the WG triple valve is used as part of the WL schedule supplied to the Western Australian narrow gauge P-class (GE Dash 8, re-classified 2500 Class) built by A. Goninan & Co. in 1989. In this case, the triple valve pilots the operation of a proportional valve to provide vacuum brake control.

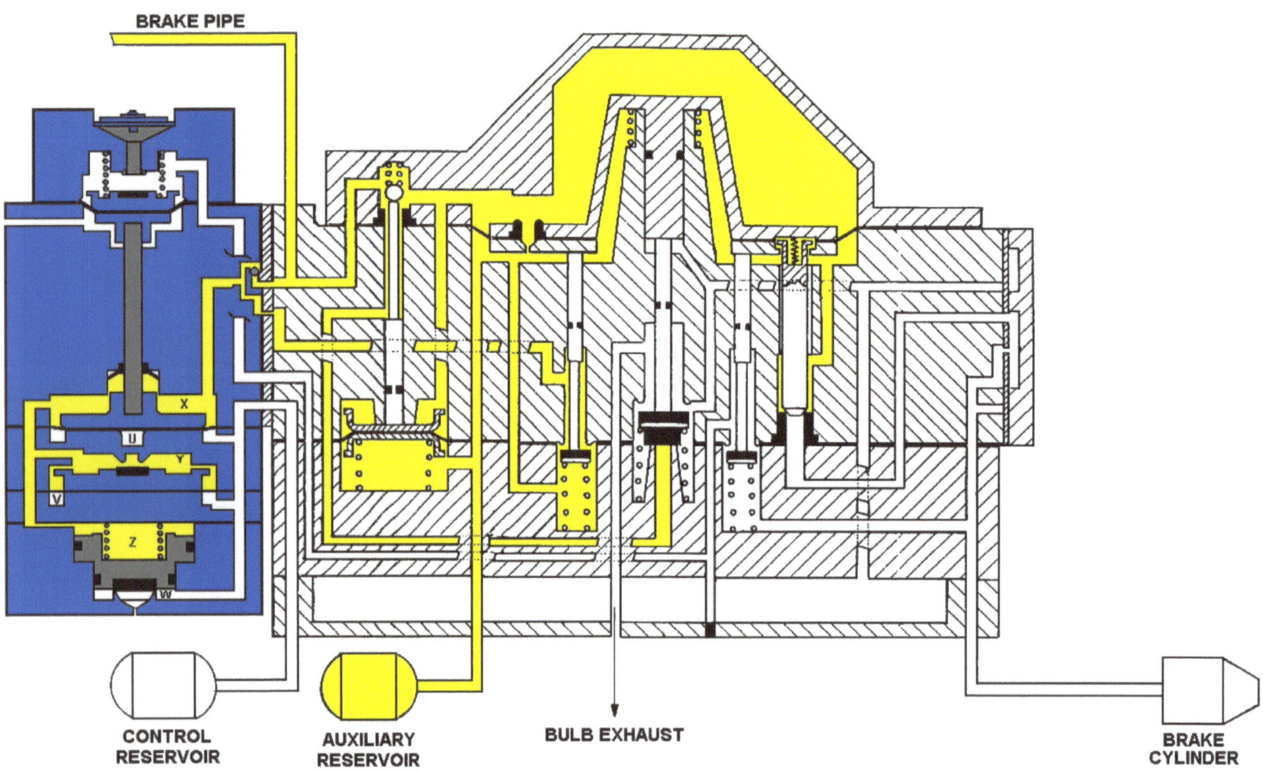

FIGURE 89: THE WG TRIPLE VALVE, RELEASE/CHARGED POSITION – DIAGRAM. NOTE *GRADUATED RELEASE BLOCK* ATTACHED AT CHARGING CHOKE PLATE

The WG1 Triple Valve

As with the WP1 and WF1 triple valves, the WG1 version is created by adding an Inshot & Reduction-Ensuring portion in place of the choke plate (9) cover plate (right-hand side of valve in diagrammatic view). The function and operation of this portion is identical to that of the WP1 and WF1 versions.

The WG1 is used as the backup automatic brake for the *Tangara* EMU trains in Sydney. This system is "relayed" and provides for automatic load compensation.

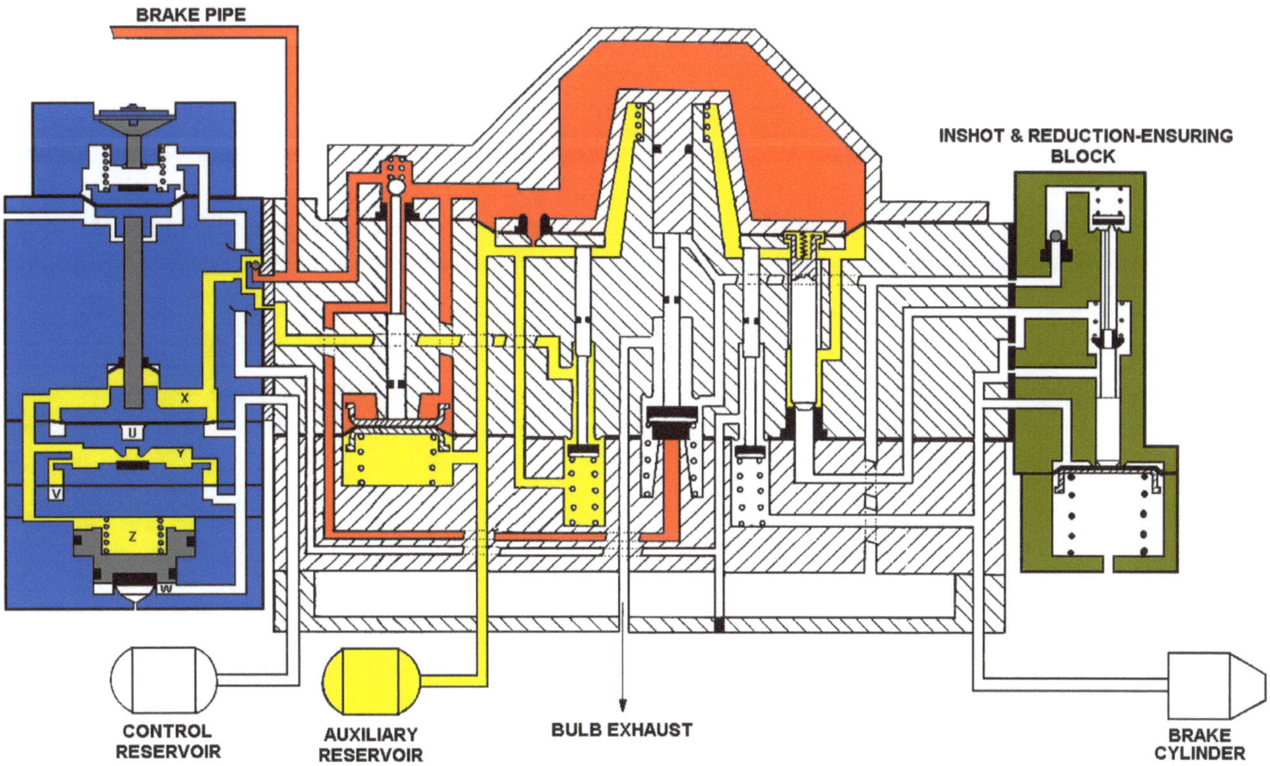

FIGURE 90: THE WG1 TRIPLE VALVE – DIAGRAM

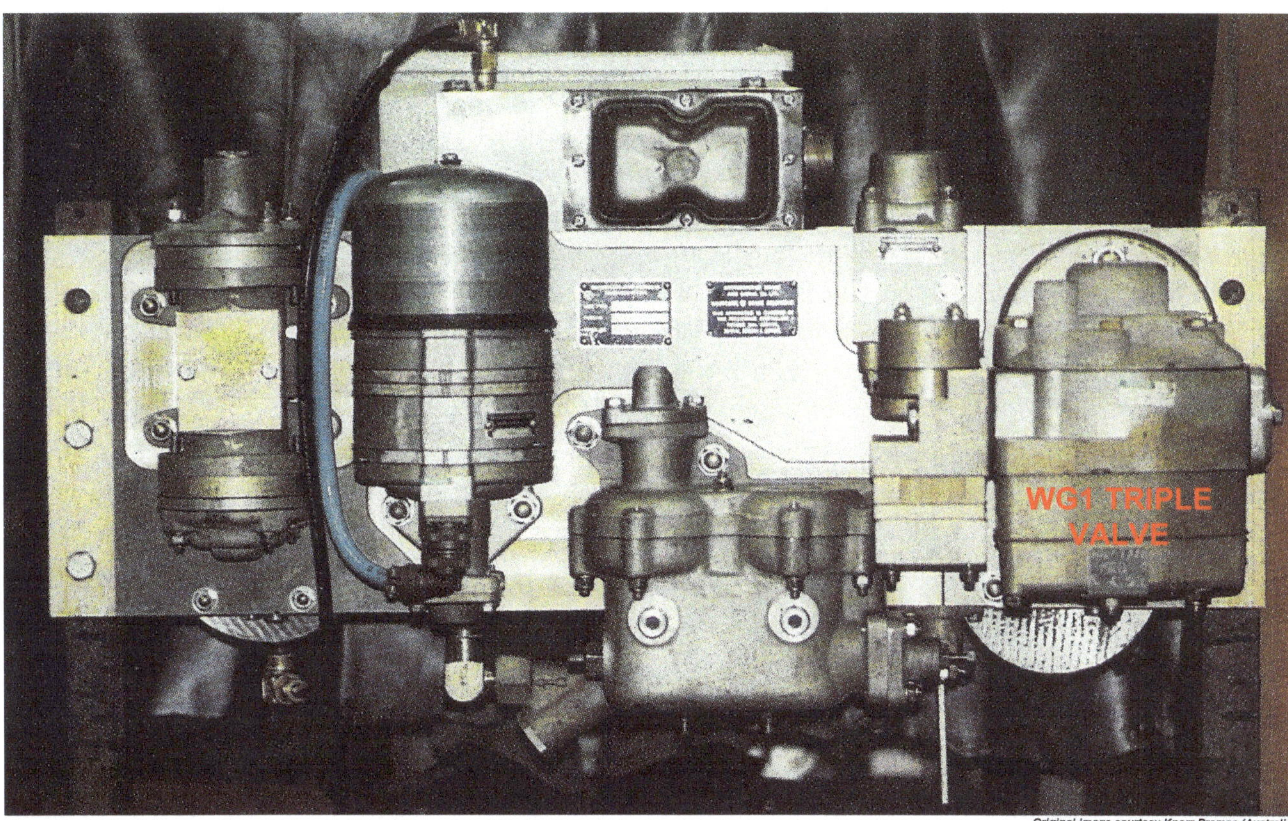

FIGURE 91: THE WG1 TRIPLE VALVE ON *TANGARA* EMU BRAKE EQUIPMENT MANIFOLD

The WEP Triple Valve

Yet another variant of the "W" triple valve family, the WEP forms part of the Westcode ***E***lectro-***P***neumatic brake unit for systems utilising E-P operation for Service braking and full-pneumatics for standby and Emergency braking. This direct-release triple valve is the main operating valve of the fully pneumatic automatic air brake element and functions similarly to other W-type direct release triple valves. As with all W-type variants, the basic W triple valve is used with internal and/or add-on modifications to provide the required customised functions. In the case of the WEP, the Application choke plate (9 in Figure 78) is rendered redundant by placing the **brake cylinder application rate choke** (normally marked 10 in the choke plate) in the mounting flange connection to the brake cylinder. This can be seen in Figure 92.

Should there be a failure of the electrical supply (and consequent loss of Westcode control) during a brake application, the automatic brake operates to provide a similar braking effort without any selective action by the driver. Thus, the automatic brake serves as a standby service brake using the same controller handle positions.

The "Westcode" System

The Westcode system of brake control for multiple unit trains is a self-lapping electro-pneumatic brake giving simultaneous and rapid operation of all cars in a train with seven equal steps of braking in both application and release. The brake can be graduated on or off at will under the driver's control and brake cylinder pressure is maintained against leakage. The system is also used to control electric braking.

A special feature of the equipment is that while the driver has overall control of braking throughout the train the brake cylinder pressure on each car is controlled individually on that car. The brake force may, therefore, be regulated on each car according to the load so that the braking percentages are maintained at the correct values throughout the train whatever the variation of load on the different cars. With the brake fully applied, a closer approach to the maximum possible brake retarding force can be achieved on each car with a consequent improvement in stopping distance.

The Westcode system is based on a simple code of electrical supply for three train wires. The driver's brake controller operates switches to energise or de-energise the three wires in coded sequence to give seven steps of braking. Alternatively, with four wires, fifteen steps can be obtained, but seven steps generally give sufficient range of control. The system can be arranged to apply the brakes either by energising the three train wires according to the code, or by de-energising them.

Electric braking equipment, which can be arranged to provide variable brake force in response to a varying electrical signal, may be integrated with the Westcode brake in such a way that both the air brake and the electric brake are controlled from the same positions of the brake controller handle. Maximum use is made of the electric brake by giving it precedence over the air brake for normal service, and the electric brake as well as the air brake can be regulated by the variable load control.[113]

The WEP triple valve was supplied by Westinghouse Australia to New Zealand Railways as part of the Westcode E-P brake system for application to the Hitachi-built *Silver Fern* diesel-electric railcars in 1972 and to the Hungarian Ganz-Mavag 1 500 Vdc EMU train sets supplied for the Wellington city suburban system from 1981. In Australia the WEP was applied to EMU rollingstock in both Sydney and Melbourne, and from this came the rather interesting reason for the subsequent provision by Westinghouse Australia of yet another sub-variant, the WEPR.

The WEPR Triple Valve

In both Sydney and Melbourne, rail operations management noted that EMU drivers had developed the habit of routinely manipulating their brake controllers such that when making a normal electro-pneumatic

113. H. R. Broadbent, *An Introduction to Railway Braking* (Chapman & Hall, 1969), chapter 3.

Service brake application, they were moving the brake handle past the Service zone into the Emergency position, momentarily opening the emergency valve and thus partially applying the automatic air brake over the E-P brake. The reason given by these EMU drivers for this irregular practice was that *"they liked to hear the reassuring sound of air exhausting from the brake valve, just like old times."* It is possible that, as well as preferring the comforting audible cue of a brake pipe exhaust, the EMU drivers were uneasy with the small initial brake cylinder pressure obtained under electric braking due to the **restricted application feature**.

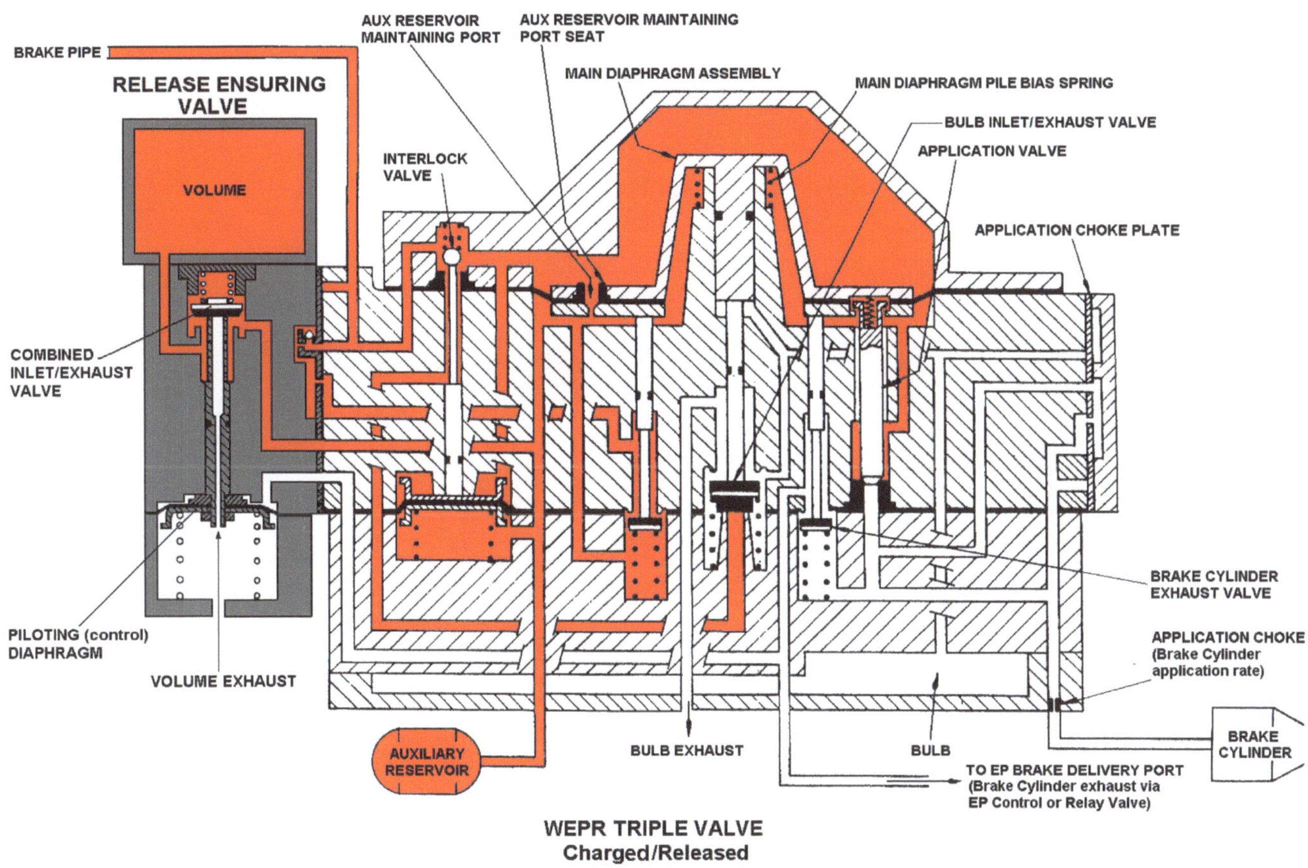

FIGURE 92: THE WEPR TRIPLE VALVE – DIAGRAM. RELEASE ENSURING VALVE ATTACHED TO CHARGING CHOKE PLATE

This practice often resulted in the automatic backup brake failing to release due to the manner of its application. Presumably, the rail authorities concerned lacked confidence that their drivers would respond adequately to retraining in the correct use of the brake valve, and so Westinghouse Australia was asked to develop a mechanism that would effectively prevent the drivers' actions from causing mechanical harm. The result was a WEP valve with a ***R***elease-ensuring feature added.

The WEPR valve is distinguished by a **Release-Ensuring Valve** attached to the left-hand side (referring to the basic W-Type diagram) in place of the choke plate cover. This feature ensures that the automatic air brake element of this Westcode system will assume the RELEASE position when an E-P brake application is released. In addition to the matter of inappropriate manipulation of the brake controller, it was also acknowledged that when changing ends, a situation can occur where the automatic brake is applied but auxiliary reservoir pressure is not reduced sufficiently below the pilot valve (regulating valve) setting, creating a differential of sufficient magnitude to initiate a brake release when brake pipe pressure is restored. This can occur when brake pipe pressure is reduced by an amount less than the Minimum Reduction value, or when the pressure settings of the brake cylinder safety valve, the auxiliary reservoir dump valve, or the brake controller pilot (regulating) valve have not been adjusted to the correct values. The Release-Ensuring Valve provides the required differential mentioned above by locally reducing the auxiliary reservoir pressure.

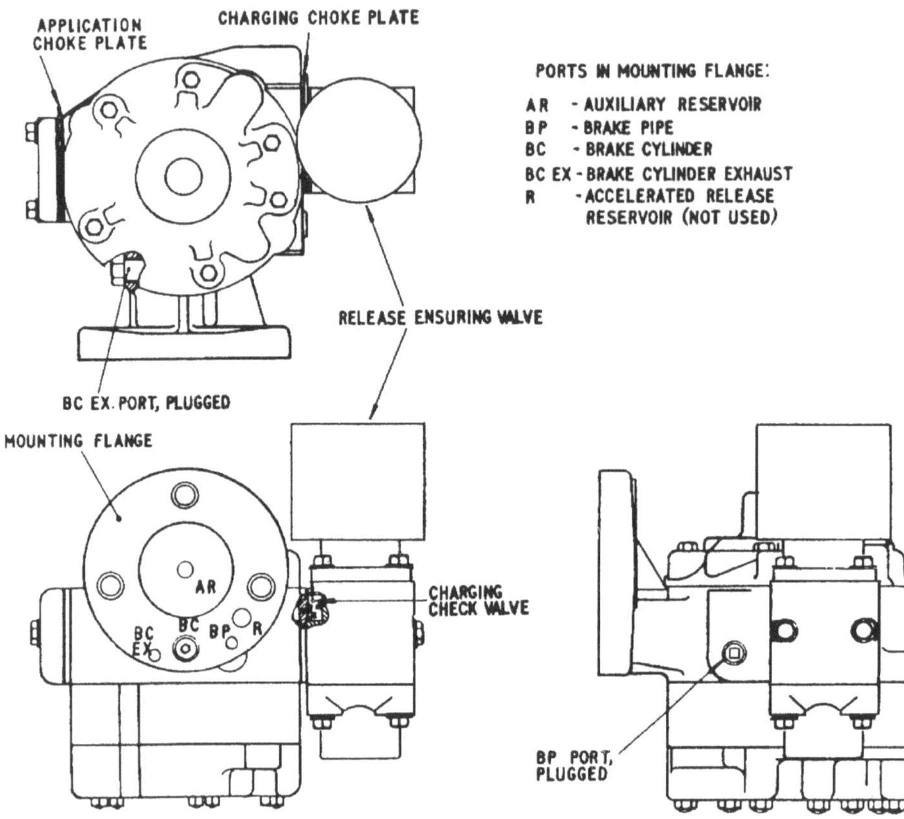

FIGURE 93: THE WEPR TRIPLE VALVE – GENERAL ARRANGEMENT

The device consists of a combined inlet and exhaust valve that connect a volume to the auxiliary reservoir chamber directly below the main diaphragm pile of the triple valve. This volume creates a pressure differential to Release of approximately 14 kPa (2 psi) when the controller handle is moved to the RELEASE position following an E-P brake application, thus ensuring the triple valve also assumes its Release position. Upon application of the E-P brake, the volume is disconnected and exhausted. The WEPR triple valve is used as the backup automatic brake for the Westcode E-P brake systems fitted to other EMU coaching stock in both Sydney and Melbourne.

It is worthwhile noting one last fact about the Australian W-type triple valve: with the advent of long, multi-unit intermodal and bulk cargo wagons (that have a greater length of brake pipe), operating valves with a more vigorous response activity are often nowadays required, and this has been addressed by such US products previously discussed in this book as ABXL and DB60L. The W-triple can also be configured—by providing it with a larger Quick-Service volume (bulb)—to operate satisfactorily on the above wagon types. Such installations are currently used on rollingstock operated in transcontinental intermodal service as well as some Queensland and New South Wales coal fleets. In such an installation, an extra W triple (a 'basic' valve) is used—mounted with a small auxiliary reservoir—as a reduction relay or ***P**ropagation-**E**nsuring **V**alve*. Through its local Quick-Service activity, the PEV assists the pressure reduction on a wagon with a long or extremely long brake pipe, and during a brake pipe reduction what would normally be the brake cylinder feed supplied from the auxiliary reservoir is exhausted, unused, to atmosphere.

While some multi-unit wagons in Australia are fitted with a pneumatic park brake that will apply in response to the loss of brake pipe pressure, one must assume that the rail operators—not to mention the designers—of those other classes of wagon not so-equipped are somehow satisfied that the individual portions that comprise the wagons are unlikely to separate from each other in service. Having learned, however, from bitter experience, of the immutability of Murphy's Law (and having some awareness of British safety expert Professor James Reason's "Swiss Cheese" theoretical model of accident causation), this author is one observer who cannot but wonder what might happen if a solid drawbar were to fracture (or the drawgear package or a yoke draft pin fail) on the rear-most wagon while the train was ascending a grade. Consider a typical such installation:

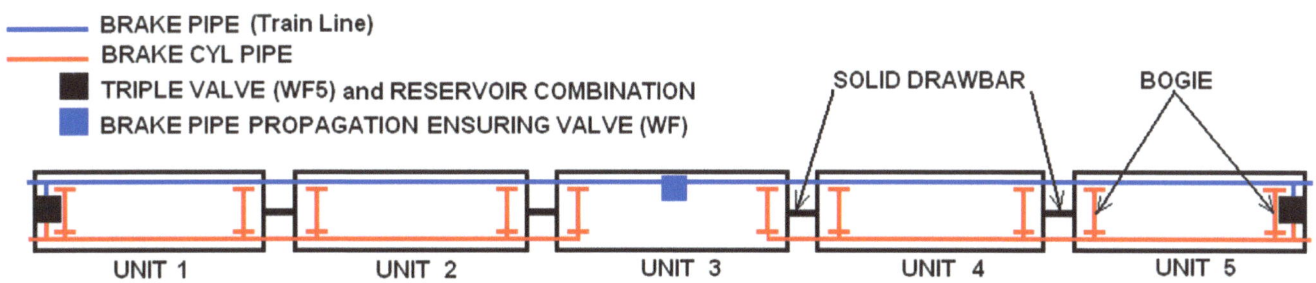

FIGURE 94: 5-UNIT WAGON. TYPICAL AIR BRAKE ARRANGEMENT – SIMPLIFIED PLAN VIEW

In the diagram above, the train is traveling from right to left, and this multi-unit wagon is the rear-most vehicle in the train. If the brake cylinder pipe (red line) supplying the rearward units of the wagon were to part or rupture—for whatever reason—then braking to all wheel sets supplied by the triple valve controlling that portion of the wagon (in this case, units 4 and 5 complete, plus one bogie of unit 3) would be lost due to brake cylinder pressure being lost to atmosphere. If the rupture were due to a unit separation (e.g. between units 3 and 4), then that separated portion of the wagon (in this case, units 4 and 5, being the trailing portion) would be uncontrolled if no pneumatic park brake were fitted. The author's view is that the annals of transport accident investigation are littered with examples of incidents caused by factors that were foreseeable—and for which the risk was considered acceptable based on statistical data—but that in the event they were to surprise us with their occurrence, could be defended with muttered excuses about the wonder of hindsight.

The Metcalf-Oerlikon ESR/500B Distributor Valve

Of European heritage, this is a relay-type valve, but it is also available without the relay feature (ES/500B). The ESR is visually differentiated from the ES by the inclusion of a two-stage relay portion (making it suitable for use with either manual or automatic empty/load systems) and a dummy cylinder volume mounted below. Internally, the major distinguishing feature is the brake cylinder Inshot element, that—on the ESR—is controlled by chokes and on the ES by an additional valve. The valves incorporate **accelerated release** (optional - requiring the provision of an accelerated release reservoir), **retarded-recharge**, and the **brake cylinder Inshot**.

Construction is described as 'typically Oerlikon', employing rubber-seated valves operated by the patented self-sealing diaphragms. Valve seats, diaphragms, and seals are comprised of oil-resistant synthetic rubber characteristically tolerant of significant temperature variation. Tension-free when assembled, the diaphragms are fully supported across their working surfaces and are self-sealing under the influence of compressed air. Other design features include corrosion resistance (both internally and externally), ease of assembly, and maintainability through the use of interchangeable component parts.

Operating characteristics include excellent sensitivity to Service brake pipe reductions and brake pipe recharge, yet good values of resistance to lesser rates of reduction (e.g. leakage) as well as brake pipe pressure surges, plus a Service-propagation rate exceeding 275 m/sec (900 ft/sec), and an effective retarded-recharge system (called "regulated recharge").

Features:

- *Accelerator Valve* – Vents brake pipe pressure to atmosphere via the expansion chamber during a brake application. This is the Quick-Service feature and it operates only during the initial brake application and is thereafter disengaged automatically by the Locking Member. It is reset by the reduction in brake cylinder control pressure upon brake release;
- *Locking Member* – Operated initially by air pressure from the accelerator valve expansion chamber and then by brake cylinder pressure. This ensures the accelerator valve remains disengaged when expansion chamber pressure has dissipated to atmosphere after the accelerator valve has closed. It is reset by the reduction in brake cylinder control pressure upon brake release;
- *Inshot* – When a brake application is initiated, the rate-of-flow of the control pressure being supplied from the auxiliary reservoir to the relay valve secondary inshot chamber is greater than its outflow into the timing chamber, thus providing a rapid increase in pressure on the relay valve control diaphragm, which then moves to open the relay inlet valve to supply auxiliary reservoir pressure to the brake cylinder. Some of this pressure flows via a choke to a balancing diaphragm (to eventually "lap" the valve), and the controlled development of pressure thereupon permits a sufficient volume of air to be supplied to the brake cylinder to cause initial brake piston movement. This helps to cushion coupling slack action as the brake application propagates along the train;
- *Sealing Valve* – Closes communication between the brake pipe and pilot reservoir immediately a brake pipe reduction is made, and restores communication upon brake pipe restoration (Release);
- *Pilot Reservoir Charging Choke* – Also referred to as the "Insensitivity choke," this regulates the speed at which the Pilot Reservoir is charged;
- *Supplementary Reservoir Check Valve* – Ensures that supplementary reservoir air does not backflow into the brake pipe during a brake pipe reduction. It also enables this reservoir to be maintained directly from the brake pipe during braking, thus compensating for brake cylinder leakage (that may otherwise cause the supplementary reservoir pressure to fall below that of brake pipe);
- *Retarded-Recharge Valve* – The supplementary reservoir (auxiliary reservoir on ES/500B) is charged from the brake pipe via this valve as well as through choke 3. If the pressure differential across the Retarded-Recharge diaphragm has resulted in the supplementary reservoir being 20 kPa (3 psi) or more below brake pipe pressure, this valve closes, restricting further charging of the reservoir to the rate permitted by the choke. This action permits more brake pipe charging pressure to pass toward the rear of the brake pipe;
- *Main Control Valve* – Comprises two concentric valves. The outer (inlet) valve controls air flow from the pilot reservoir to the relay valve during a brake application (i.e. the relay valve control air supply), while the inner (exhaust) valve controls the venting of this control pressure via the release choke (or grade control valve if fitted), during brake release. An override spring

operates in conjunction with the controlled relative lift between these two valves to ensure that the Main Control Valve and Direct Releasing valve operation is synchronised;

- *Main Control Assembly* – Comprises a hollow valve stem. The lower end of the valve stem is attached to the upper and lower pressure plates between which the assembly diaphragm is sandwiched, thus providing support to the diaphragm when pressure differentials are acting across it. The upper pressure plate also incorporates the carrier for the rocking arm that controls accelerator operation during brake application and actuates the Direct Releasing Valve;
- *Accelerated Release Assembly* – Located directly below the Main Control Valve stem. When the distributor valve goes to Release (or in the event of leakage in the pilot reservoir), the brake cylinder is vented to atmosphere, and the accelerated release valve is opened by a downward movement of the Main Control Valve assembly, permitting accelerated release pressure to flow into the brake pipe to assist the brake pipe recharge toward the rear of the train. The accelerated-release reservoir is charged from the supplementary reservoir (auxiliary reservoir on the ES/500B);
- *Direct Releasing Valve* – During brake pipe recharge, the Main Control Valve assembly moves down, which permits the Direct Releasing Valve to also move down and allow the brake cylinder to exhaust via the Main Control Valve spool. The Direct Releasing Valve also controls operation of the Sealing Valve during brake applications;
- *Two-stage Empty/Load Relay* – Attached to the underside of the distributor valve, the relay portion provides alternative brake cylinder pressures to cater for variations in the vehicle's weight. It also enables the distributorvalve to accommodate multiple brake cylinders, should they be fitted. The relay inlet valve controls airflow from the supplementary reservoir to the brake cylinder in proportion to the degree of brake application in effect. A control diaphragm and two equalising diaphragms (having different effective diameters) are mounted on the relay's hollow valve stem. In the arrangement illustrated, a three-way cock mounted on the side of the relay enables the appropriate equalising diaphragm to be selected manually (the larger, low-pressure one for Empty braking and the smaller, high-pressure one for Loaded braking).

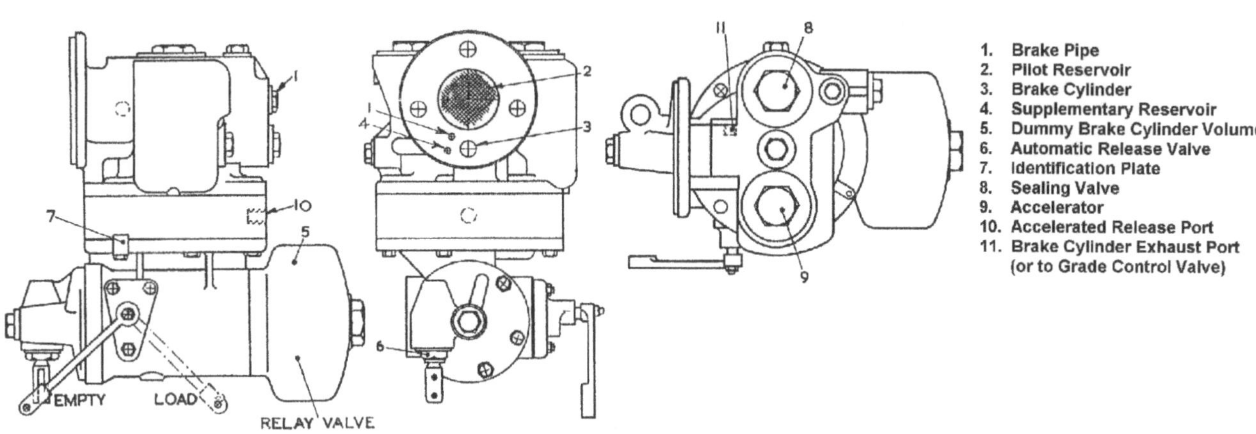

FIGURE 95: METCALF-OERLIKON ESR/500B DISTRIBUTOR VALVE – GENERAL ARRANGEMENT

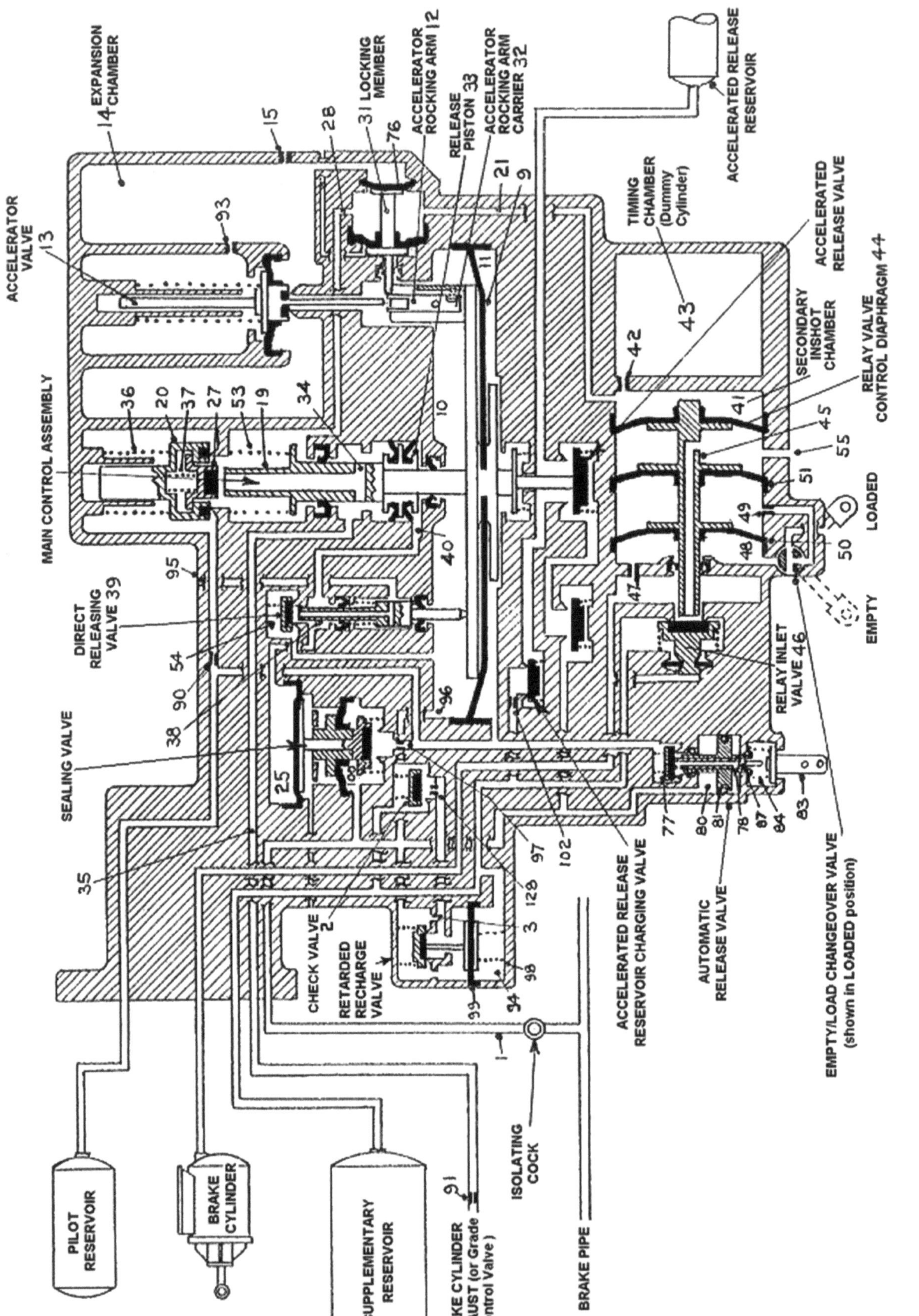

FIGURE 96: METCALF-OERLIKON ESR/500B DISTRIBUTOR VALVE – DIAGRAM

The Davies & Metcalf ES500W Distributor Valve

Released onto the Australian market in 1989, this valve is a development of the ES/ESR500B valves and—able to be mounted directly to the Westinghouse pipe bracket—was conceived as a competitor to the W-triple. Interchangeability with the W-triple was not possible with the previous Metcalf-Oerlikon distributor valve (see previous pages), and this new feature is a distinctive selling point. Although well engineered (and perhaps this is part of the problem), the ES500W has not found wide acceptance, being more expensive to buy and maintain than the W-triple. This valve, and the ESR500, are in service in Queensland and Victoria.

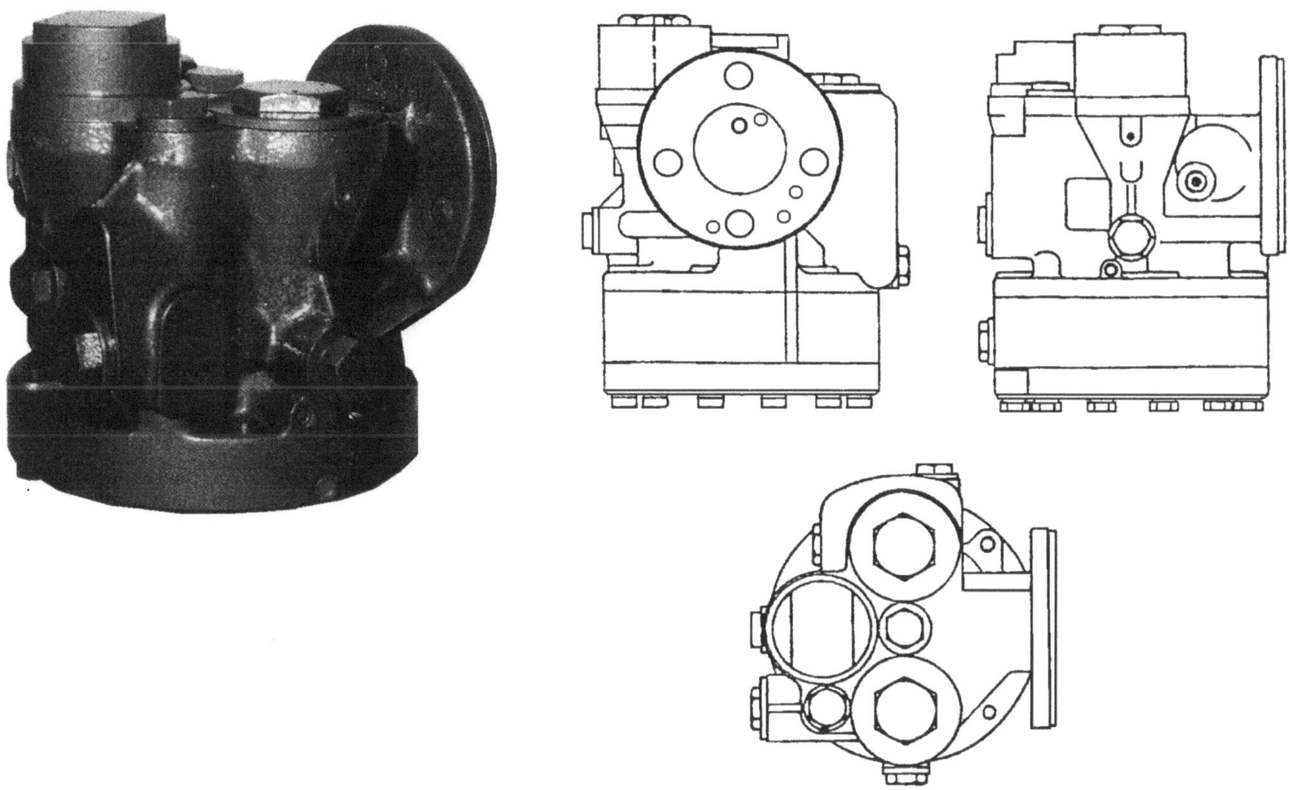

FIGURE 97: METCALF-OERLIKON ES500W DISTRIBUTOR VALVE – GENERAL ARRANGEMENT

The Metcalf Oerlikon EST4/di Distributor Valve

This graduable-release distributor is designed to UIC specifications and can be used with either single or two-pipe brake systems. It embodies the familiar Oerlikon principles of design simplicity, employing durable, oil-resistant, synthetic rubber-seated valves; self-sealing diaphragms; and corrosion-resistant metal parts. Separate supplementary relays that are interchangeable with its lower cover may be attached to the valve body to provide Empty/Load, Two-Stage Speed-Controlled, or Load Proportional braking. Electrical control may also be superimposed in a simple manner.

Operating characteristics include the following:

- Responsive within 0.5 seconds to a brake pipe reduction rate of 55 kPa (8 psi) in 6 seconds;
- Insensitive to a brake pipe reduction rate of 51 kPa (7½ psi) in 60 seconds;
- Insensitive to “normal” brake pipe surges;
- Service propagation rate exceeding 274 m/second (900 ft/sec);

- Brake pipe exhaust is through the *accelerator valve* via an expansion chamber;
- In Release condition, the auxiliary and control reservoirs are in direct communication with the brake pipe, this communication being interrupted immediately the Distributor operates;
- Highly resistant to overcharge; thus—during brake release—a high-pressure brake pipe charge may be utilised without the risk of a reservoir overcharge. (NB: It should be understood that since *graduated-release* operating valves were inherently slower in Release than were *direct-release* types, the manufacturers of these UIC systems strove to improve this operating characteristic, typically by improving drivers' brake valve capacities to provide a momentary increase in the brake pipe pressure at the head of the train. An important feature of some later European locomotive brake valves—for example, as described in the section on the Metcalf-Oerlikon FV4—was an ability to provide a controlled "brake filling surge" or Full Release. This is a main reservoir charge supplied unreduced into the brake pipe for a limited time. This charge was considered to reduce by about one half the time to release the brakes on longer trains by comparison to having the brake valve handle placed initially in the normal Release or RUNNING position. The duration of this brake-filling surge had to be controlled in order to avoid an objectionable overpressure within the operating valve control chamber or auxiliary reservoir. This difficult judgment was generally left to the locomotive driver, with typically inconsistent and imprecise results; thus, modern operating valves and drivers' brake valves incorporate design protection against this excess pressure. This may be by limiting the maximum pressure or duration of the brake-filling surge or by enabling a slow and progressive diminution of the overpressure by taking account of the sensitivity limits of the operating valves used (an example being the FV4 driver's brake valve, which limits the filling surge to between 70 and 75 kPa);
- Provides an "inexhaustible" brake. (Author's note: This term is commonly used by European brake equipment manufacturers to promote the *graduated-release* philosophy. As implied, "graduated-release" permits the brake to be released in steps; the principle being that brake cylinder pressure is reduced only to the extent that reservoir pressure is restored on each car. Thus, the brake cannot be completely released until the system is completely recharged. While UIC advocates are resolute in their conviction that this "inexhaustibility" is of paramount consequence, supporters of *direct-release* systems might consider it to be but a theoretical advantage of lesser value since, in practice, the *direct-release* brake does not itself become unsafely "exhausted," and there is the backup of the separate emergency feature—where it is provided—supplying an on-call reserve of braking power.);
- Provides easy and effective regulation of brake cylinder pressure down to low values, due to the *accelerator valve* resetting only when brake cylinder pressure falls below 4 psi (27.5 kPa) in other words, approximately when the brake shoes commence to leave the wheel treads;
- When used with a two-pipe brake system, an HB-type pressure-limiting valve that pre-controls the maximum brake cylinder pressure replaces the Distributor inlet valve assembly. This is necessary, particularly when the EST4/di distributor valve is used for railcar and multiple unit installations where the auxiliary reservoir is fed directly from the main reservoir.

Features:

- *Locking Member*: This device is operated initially by air pressure from the accelerator valve expansion chamber 14 acting on diaphragm 76. As the brake application develops, this is superseded by the brake cylinder pressure acting on diaphragm 30, ensuring the accelerator remains disengaged until brake cylinder pressure is less than 4 psi (27.5 kPa);

- *Inshot*: Inshot chamber 91 is provided to ensure a rapid flow of air to the brake cylinder during the initial development of the brake application to provide sufficient volume to displace the piston and bring the brake shoes quickly against the wheel. This activity reduces the development of slack action during the early stage of the brake application;
- *Sealing Valve*: Interrupts the normal communication between brake pipe and auxiliary reservoir immediately the brake application is initiated, and automatically restores it upon brake release;
- *Main Control Valve*: This simple assembly—actuated by the well-known "three-pressure principle"—operates the inlet valve to control the flow of air from auxiliary reservoir to brake cylinder during Application and exhausts brake cylinder to atmosphere during Release. The rate of brake cylinder pressure buildup is controlled by choke 90, while the rate of brake cylinder exhaust is controlled by choke 92. The main components of the assembly are the hollow valve stem 19 onto which equalising diaphragm 33 and control diaphragm 9 are mounted. The open tip of the hollow stem seats on the inlet valve, thus controlling the exhaust of brake cylinder pressure during Release;
- *Equalising and Filling Valve*: The equalising valve provides the chokes controlling the air flow from brake pipe to control reservoir. During braking, this sensitive assembly isolates the control reservoir, preventing any loss of air pressure by causing equalising valve 44 to seat against filling valve 36 while simultaneously opening the filling valve to permit the brake pipe to recharge auxiliary reservoir via choke 3. When the auxiliary reservoir attains a pressure slightly less than its normal working pressure, the equalising valve operates to allow the filling valve to close, and the auxiliary reservoir recharge is thereafter completed by the sealing valve. Note that since the control of brake cylinder pressure is dependent upon the difference between control reservoir and brake pipe pressures, brake cylinder pressure during Release in no way depends upon auxiliary reservoir pressure;
- *Release Valve*: Provides manual release of brake cylinder pressure.
- *Check Valve*: Preventing backflow of auxiliary reservoir pressure to brake pipe

Other variants of this Distributor Valve include the earlier EST4/d used on the Queensland SX-class locomotive-hauled suburban cars and the EST4f/g, which is included as part of the P85 locomotive brake equipment described in section 4.

OPERATION OF THE TYPE EST4/di DISTRIBUTOR

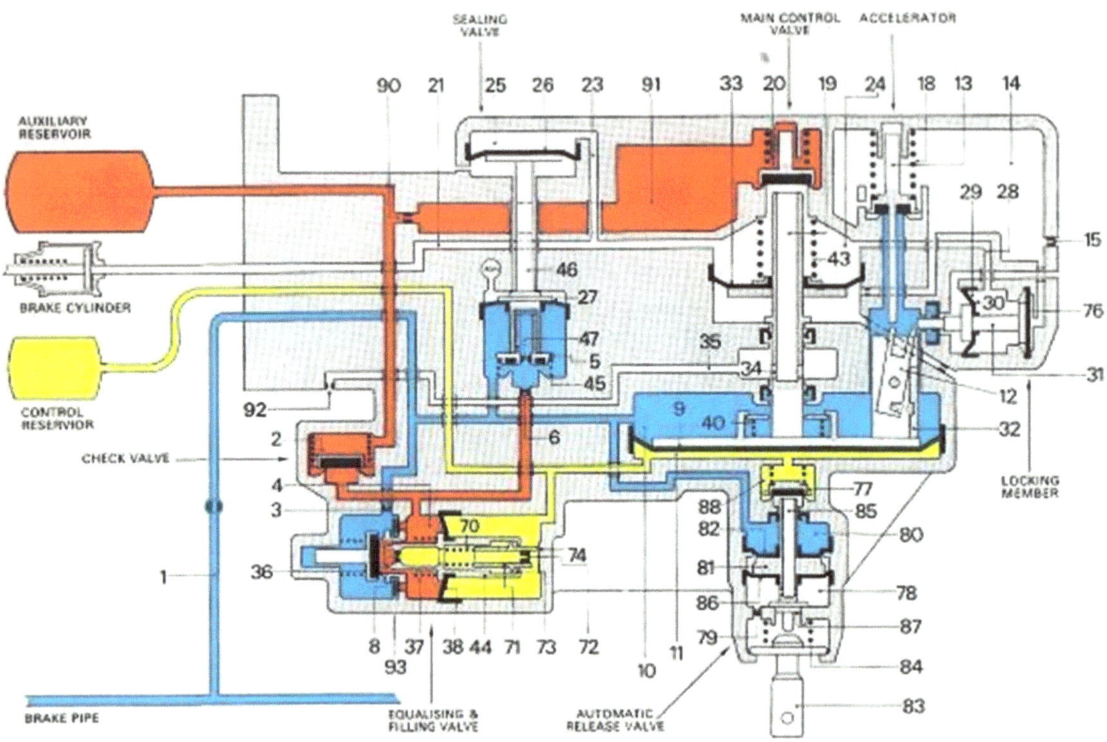

CHARGING THE RESERVOIRS

The system for the intial charging of the control and auxiliary reservoirs is simple and reliable, and charging is achieved in the shortest time. Brake pipe (1) pressure flows to the auxiliary reservoir via choke (6) and check valve (2). The auxiliary reservoir is charged to brake pipe pressure; the rate of charge being controlled by the bore of choke (6), the choke size depending upon the size of the auxiliary reservoir. The control reservoir is charged via choke (8) in the equalising valve. Air pressure in chamber (4) acting on diaphragm (38) moves the equalising valve to the right so that telescopic valve (71) is held open by contacting stop (74) on the body of the distributor; thus air flows into the control reservoir via the annular space formed between the telescopic valve and the body of equalising valve (44). This annular space plus the bore of choke (72) exceed the area of choke (8), thus the rate-of-charge of the control reservoir is controlled by choke (8).

In this manner, both the auxiliary and control reservoirs are charged to the same pressure as the brake pipe under normal running conditions. It will be seen that the Inshot chamber (91) is also charged to brake pipe pressure via choke (90).

FIGURE 98: METCALF-OERLIKON EST4/Di DISTRIBUTOR VALVE – DIAGRAM (CHARGING)

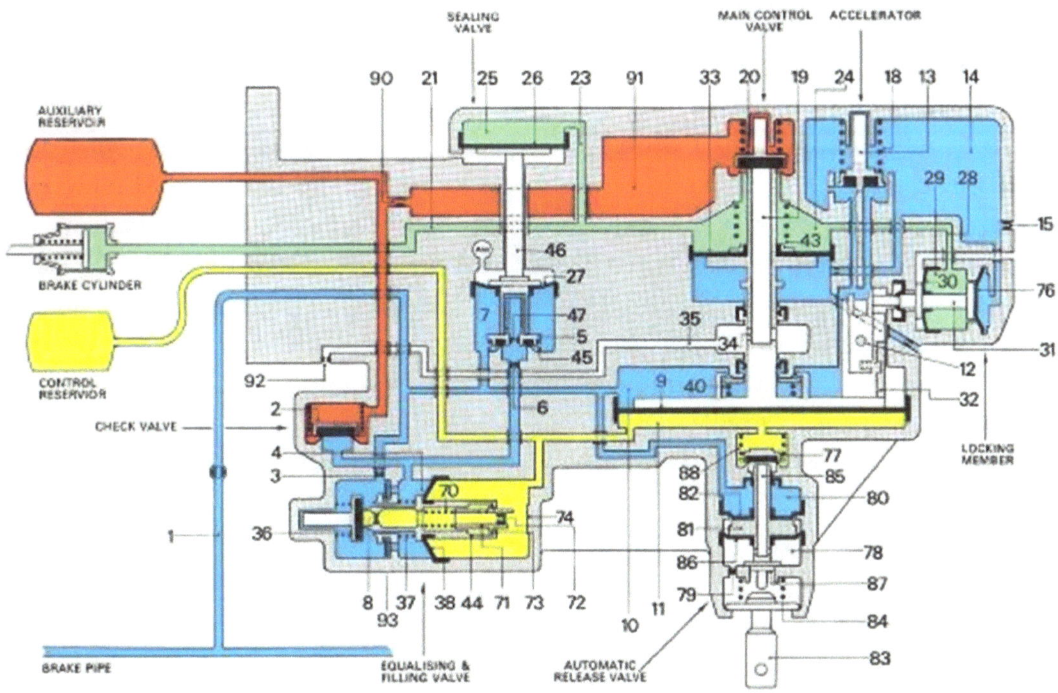

BRAKING

When brake pipe pressure is reduced, there is a corresponding reduction in pressure in chamber (4) of the equalising valve due to the flow of air past check valve (93) and via choke (3) into the brake pipe. Thus the sensitive equalising valve assembly moves to the left under the influence of control reservoir pressure acting on diaphragm (38), and the tip of the equalising valve body (44) seats on the face of filling valve (36), completely isolating the control reservoir from the brake pipe. The efficient design of this mechanism provides protection against loss of control reservoir pressure during braking.

Simultaneously, check valve (2) is seated, preventing loss of auxiliary reservoir pressure. The reduced brake pipe pressure acting in chamber (10) allows the control diaphragm assembly to rise under the influence of control reservoir pressure in chamber (11). As the control diaphragm assembly rises, it takes with it the rocking arm carrier (32). The tip of rocking arm (12) contacts the lower stem of accelerator valve (13) and lifts this valve from its seat. Brake pipe pressure from chamber (10) thus flows past the accelerator valve and into expansion chamber (14), thence to atmosphere via choke (15). The accelerator valve is held open by the flow of air from chamber (10) to chamber (14) acting on the underside of the head of the valve and remains open until pressures in the two chambers equalise. Simultaneously, the pressure in chamber (14) acts on diaphagm (76), pushing the locking member (31) to the left and moving rocking arm (12) so that the accelerator valve may re-close under the action of spring (18). The action of the accelerator valve --- as described above --- causes a rapid, controlled flow of air from the brake pipe to atmosphere on each vehicle and is thus instrumental in promoting a rapid rate of propagation.

FIGURE 99: METCALF-OERLIKON EST4/Di DISTRIBUTOR VALVE – DIAGRAM (APPLICATION)

Westinghouse (UK) P4a Distributor Valve

This well-known (although perhaps not in Australasia) British graduated-release operating valve has evolved from the P4 Distributor, designed by British Westinghouse with principal involvement from Jack Washbourn, who was later closely involved with the Australian W triple. The P4a entered New Zealand during the 1990s as part of the equipment on imported second-hand British Rail Mk 2 passenger cars. These cars have now been extensively rebuilt as the S class (in Wellington) and the SA/SD classes (in Auckland). The P4a has not operated in revenue service in New Zealand - the rebuild having resulted in the removal of this equipment in favour of WP triples. Some second-hand diesel locomotives imported from the UK by Silverton Rail[114] carried the P4a distributor. One can expect that Silverton might desire to retain this equipment in Australian service, and it will be interesting to see whether—despite its design incompatibility with Australian equipment—the regulator will permit this (assuming they are aware).

The Qiqihar 120K Control Valve

The railway network of The People's Republic of China—rapidly expanding in concert with the country's burgeoning economy—is comparable in its vastness to that of India, and rail is easily the foremost freight and passenger transport modality. Given the realities, both logistical and others, of this dynamic and rapidly expanding rail environment—such as the renowned reputation of the Chinese for ingenious engineering self-sufficiency and the fact that they have long utilised copies of the venerable American K triple—it is hardly surprising that they would eventually begin manufacturing modern railway rollingstock and contemporary air brake equipment for export. Perhaps we unintentionally assisted their efforts when—some years ago, as the result of an abortive and perhaps clumsy attempt at creating a Korean agency—Westinghouse Australia allowed all of their technical drawings and data for the production of the W triple to fall into the hands of an enterprising local individual who promptly "disappeared". However, although industry observers and participants within Australasia may have been expecting to see China enter the global supply industry, they could not have known exactly when it would happen or how it might be accomplished.

Prior to becoming a republic in 1949, most Chinese railway rollingstock and parts were imported from North America and Europe. No doubt this included their triple valves (H-types converted to K, plus K-1 and K-2 types) mounted in combination with the brake cylinder and reservoirs. Indeed, to cope with their braking requirements over rugged, mountainous country, the Chinese fitted two K triples to their high-capacity freight cars - one for each bogie! In the early 1960s, due to unique local requirements pertinent to the operation of even larger freight wagons as well as a need to address wheelslide problems arising from the 590 kPa (85 psi) brake pipe pressure used in mountain-grade territory (and the resultant braking ratio problems), Chinese railway engineers adapted the K triple valve to work with a 14-inch brake cylinder and a simplified Empty/Load system. This became their GK equipment and was applied to most freight wagons for many years thereafter. The characteristics of this setup, however, produced problems that resulted in the Chinese commencing development of their own freight control valve, with design features owing much to both U.S. and European operating valves. This is not altogether surprising considering that Chinese railway engineers have studied and worked overseas, while American and European air brake engineers have frequently visited and instructed in China. Meanwhile, the preeminent system used for Chinese passenger cars was LN-style equipment (described in section 2), eventually to be replaced by a development of the new control valve.

114. Southern & Silverton Rail was created when South Spur Rail Services (WA) purchased NSW company Silverton Rail in 2006. The company became a wholly owned part of the Coote Industrial Group in 2007. In 2010, Coote sold the SSRS business to P&O Trans Australia which became Qube Logistics in 2011

Over the next thirty years, this new equipment, as well as the Chinese railway equipment manufacturing industry, evolved and matured. With the arrival in Australia during 1999 of a fleet of skeletal container wagons for stevedoring and transport logistics company Patrick Corporation Ltd, the Chinese railway supply industry dipped a toe into the Australian railway pool; but when, in 2000, Chicago Freight Car Leasing Australia took delivery of their new fleet of skeletal container flat-top wagons, the Chinese manufacturers entered with a splash. National Rail Corporation would soon sign up for some articulated intermodal flats and ATN Access[115] likewise for a fleet of grain hoppers. All of these wagons would mount the 120A version of the Qiqihar [*CHI-CHI*-ha] control valve (the differences—such as internal choke sizes—being relatively minor), while the 120K would eventually make its appearance on Pacific National 5-Pack articulated container flats and 5-Unit double-stack well wagons.

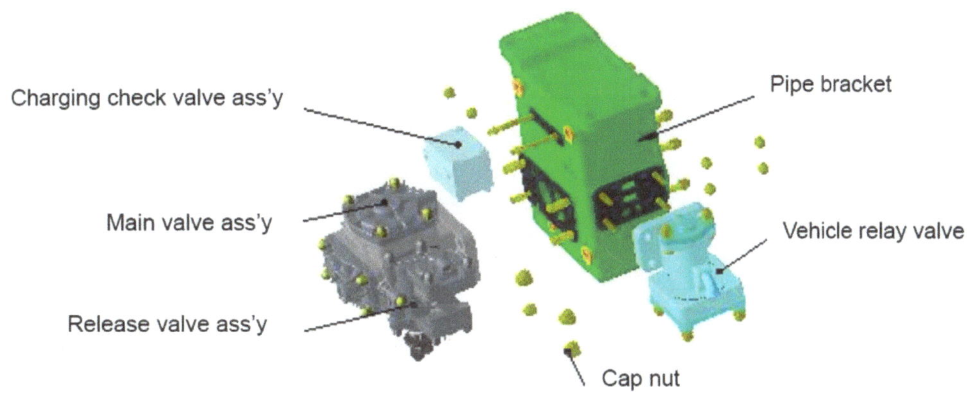

FIGURE 100: QIQIHAR 120K CONTROL VALVE – COMPOSITION (EXPLODED VIEW). *CIMC, courtesy Les Smith*

The 120K is functionally equivalent to an ABSD control valve (see following section under 'AB Freight Car Brake Equipment') and its design clearly owes more than a passing nod of acquaintance to US practice of the ABD era. Described by the Qiqihar Railway Rolling Stock (Group) Co. Ltd. as being of the "duo-pressure mechanism type", which they describe as referring to activation by two pressure sources (a reference, perhaps, to its brake cylinder relaying functionality), the 120K is a conventional direct-release freight control valve utilising a rubber diaphragm and metal slide valve structure. It provides accelerated-release and retarded-recharge, automatic Empty/Load changeover, Quick-Service and brake cylinder inshot, brake cylinder pressure maintaining, and built-in stability. The major elements of the 120K are:

- a *pipe bracket* (with integral emergency and local pressure-reduction chambers, plus strainers and filter). The charging check valve assembly is externally attached;
- a *main valve assembly* (i.e. Service portion) composed of an application portion (diaphragm and slide valve assembly) called the "action part," a retarding (retarded-recharge) portion called the "decelerating part," a quick-service limiting (brake cylinder inshot) valve, an accelerated-release valve (brake cylinder exhaust pressure being used to activate the accelerated-release valve and supply accelerated-release cylinder pressure to the brake pipe), a two-step emergency valve (rapid/slow development of brake cylinder pressure during Emergency brake application), and an attached semi-automatic release valve ("one-pull" brake cylinder exhaust, "pull-and-hold" to exhaust entire car system);

115. An Australian joint venture between US railroad operator Wisconsin Central and (then) New Zealand rail operator Tranz Rail who were part-owned by Wisconson Central Transportation Corp. In February 2004, Australian Transport Network was purchased by Pacific National.

- a *vehicle relay valve*, reacting conventionally to the presence (or lack thereof) of a control pressure delivered from the auxiliary reservoir via the control valve to supply (from a supplementary reservoir), maintain, and exhaust brake cylinder pressure in response to brake pipe pressure reduction or restoration; and
- a *charging check valve assembly*, controlling charging of the supplementary reservoir from the brake pipe.

The Qiqihar Railway Rolling Stock (Group) Co. Ltd. now offers—should the customer opt for conversion at a later date—a pipe bracket adapter that accommodates the Australian WF control valve; an interesting option indeed (and one that Pacific National is already reported to have taken advantage of due to problems with the Chinese product on new wagons). This option might be a marketing strategy designed to get Australasian customers to at least try the Chinese valve, which might be expected to improve over time; and time will indeed tell whether the Type-120 control valve has the durability of the American and Australian products.

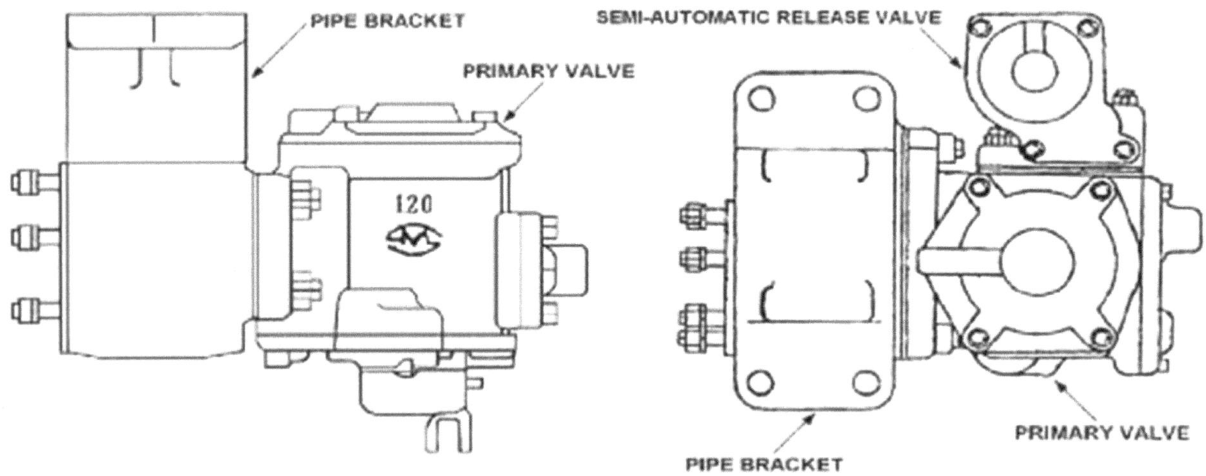

FIGURE 101: QIQIHAR 120K CONTROL VALVE – GENERAL ARRANGEMENT

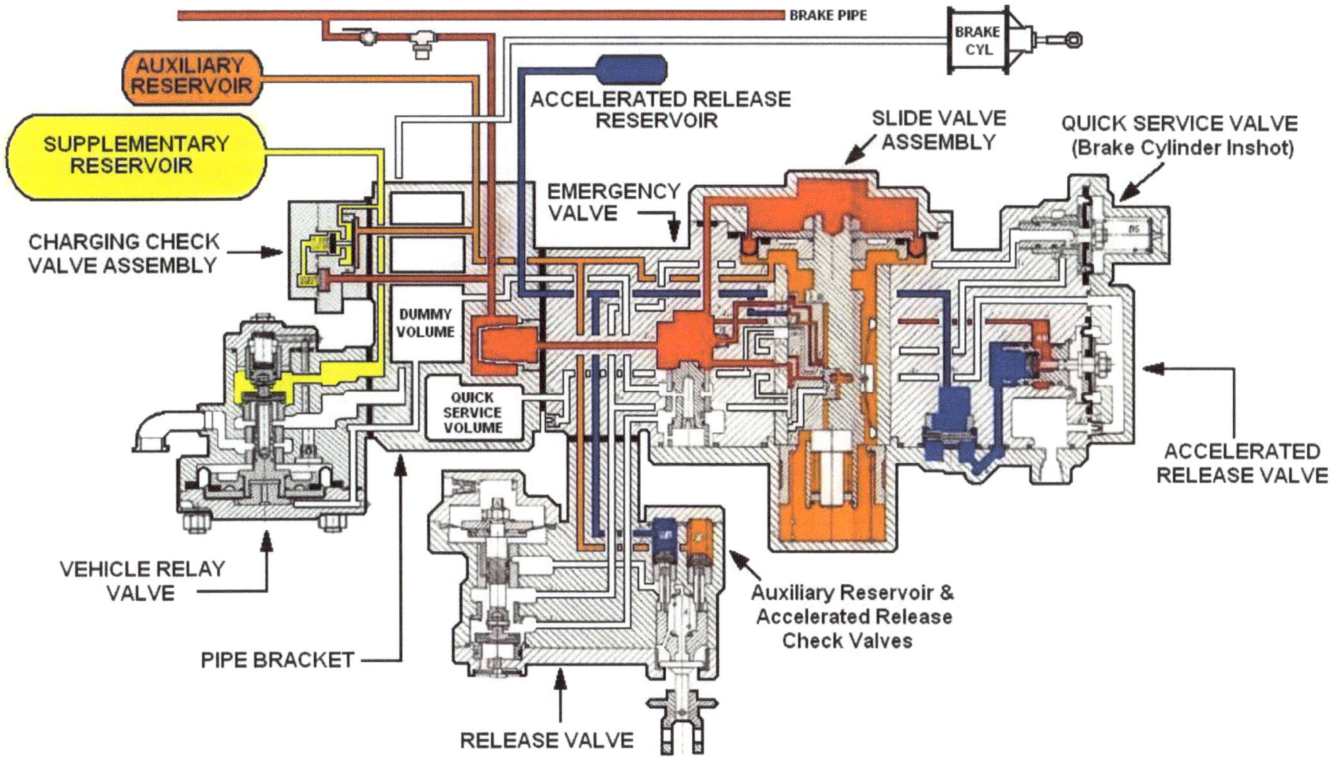

FIGURE 102: QIQIHAR 120K CONTROL VALVE – DIAGRAM, *RELEASE/CHARGING* POSITION. *Courtesy Les Smith*

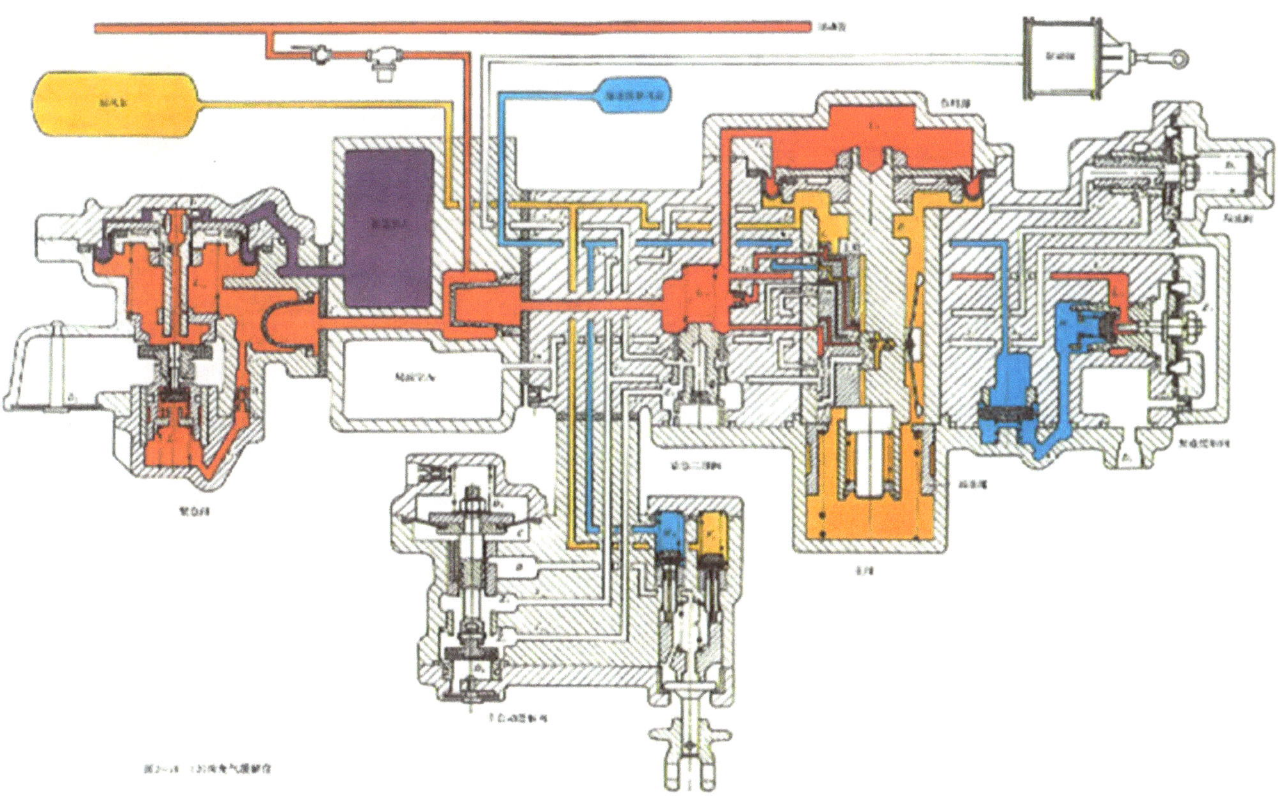

FIGURE 103: QIQIHAR 120 CONTROL VALVE – DIAGRAM, *RELEASE* POSITION. *Courtesy John Scott*

The Knorr-Bremse KE-Oc Distributor Valve

While the very effective KE distributor valve in its various forms is well-established throughout Europe and Southeast Asia, its inclusion on a group of EMD diesel-electric locomotives imported second-hand from Denmark is its first application within Australasia. Designed and approved for air brake systems conforming to UIC specifications, this equipment is modular and can be employed in a variety of brake systems, including switchable Direct/Graduated-Release, two-stage or continuous automatic load-controlling, speed-controlling, or combinations of such. An EPZ electro-pneumatic control unit can be attached to achieve simultaneous braking of all cars in a train.

The KE distributor valve provides Inshot, Quick-Service, Accelerated-Application, and brake cylinder and control reservoir pressure-maintenance.

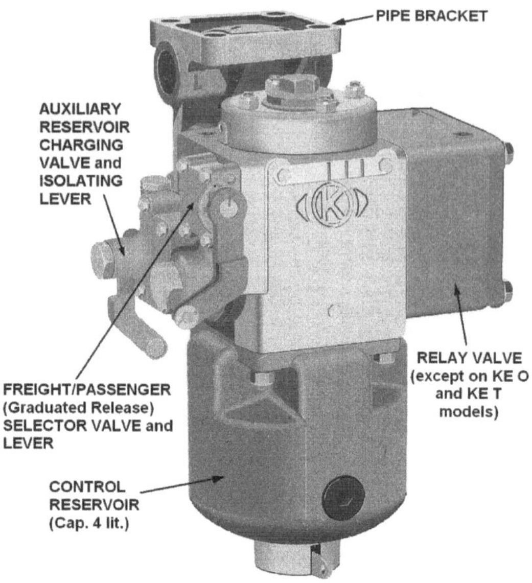

FIGURE 104: KNORR-BREMSE KE DISTRIBUTOR VALVE

Other features include:

- *Generational interchangeability* (existing systems are upgradeable as required);
- *Fast response* (brake application resulting in less than 1 second (UIC specification: 1.2 seconds) from a brake pipe pressure reduction of 60 kPa (8.5 psi) within 6 seconds;
- *Propagation rate* of 285 m/second (935 ft/second);
- *Stability* against uncontrolled brake pipe pressure changes and Full Release surges (i.e. overcharge protection);
- *Maximum brake cylinder pressure control* irrespective of auxiliary reservoir size or pressure or brake cylinder stroke;
- *Inexhaustible supply* (auxiliary reservoir replenished from brake pipe via the check valve in the auxiliary reservoir charging valve or cover).[Author's note: This is "inexhaustibility" per UIC definition, which means it must always be possible when stationary—and after operation of the driver's brake valve—to obtain a brake cylinder pressure on any vehicle in the train of at least 85 percent of that obtaining from an Emergency application under "normal" (i.e. fully charged) conditions. An additional requirement is that a minimum brake cylinder pressure of 0.3 kg/cm^2 (30 kPa) must be always available. It is not, in fact, inexhaustibility in its strictest sense, since 15 percent exhaustion of the system is permitted; but it is a compromise between brake release times and the system's ability to maintain auxiliary reservoir at the pressure considered necessary by the UIC for the descent of long, steep gradients.]

Operation (Figure 105)

Initial Charging:

1. Brake pipe pressure is admitted to the chamber above the control reservoir diaphragm plate and choke switch **H**, which is forced back against its spring to permit charging of the control reservoir. Brake pipe pressure on the diaphragm plate moves the piston and

slide assembly down to cut off control reservoir exhaust, and the control reservoir is now charged via choke switch **H**, a control valve and a restricting choke.

2. Brake pipe pressure also flows via a check valve within the auxiliary reservoir charging valve to both the auxiliary reservoir and the relay valve inlet valve (not applicable to the Oc version), which is being held closed by its compression spring. When normal brake pipe pressure has developed within the distributor valve (i.e. both auxiliary and control reservoirs), then the valve is in its Charging Lap position.

3. The control reservoir diaphragm plate is restrained by balanced spring and air pressures, and the quick-release valve slide is held at its Cut-Off position, preventing the exhaust of control reservoir pressure at the bottom of the distributor valve. The distributor valve is now ready to brake at its normal response rate and will maintain normal working brake pipe pressure relative to control and auxiliary reservoir pressures. Any auxiliary reservoir leakage is maintained via an auxiliary reservoir charging valve choke port in the charging valve (190) and via check valve V190.4.

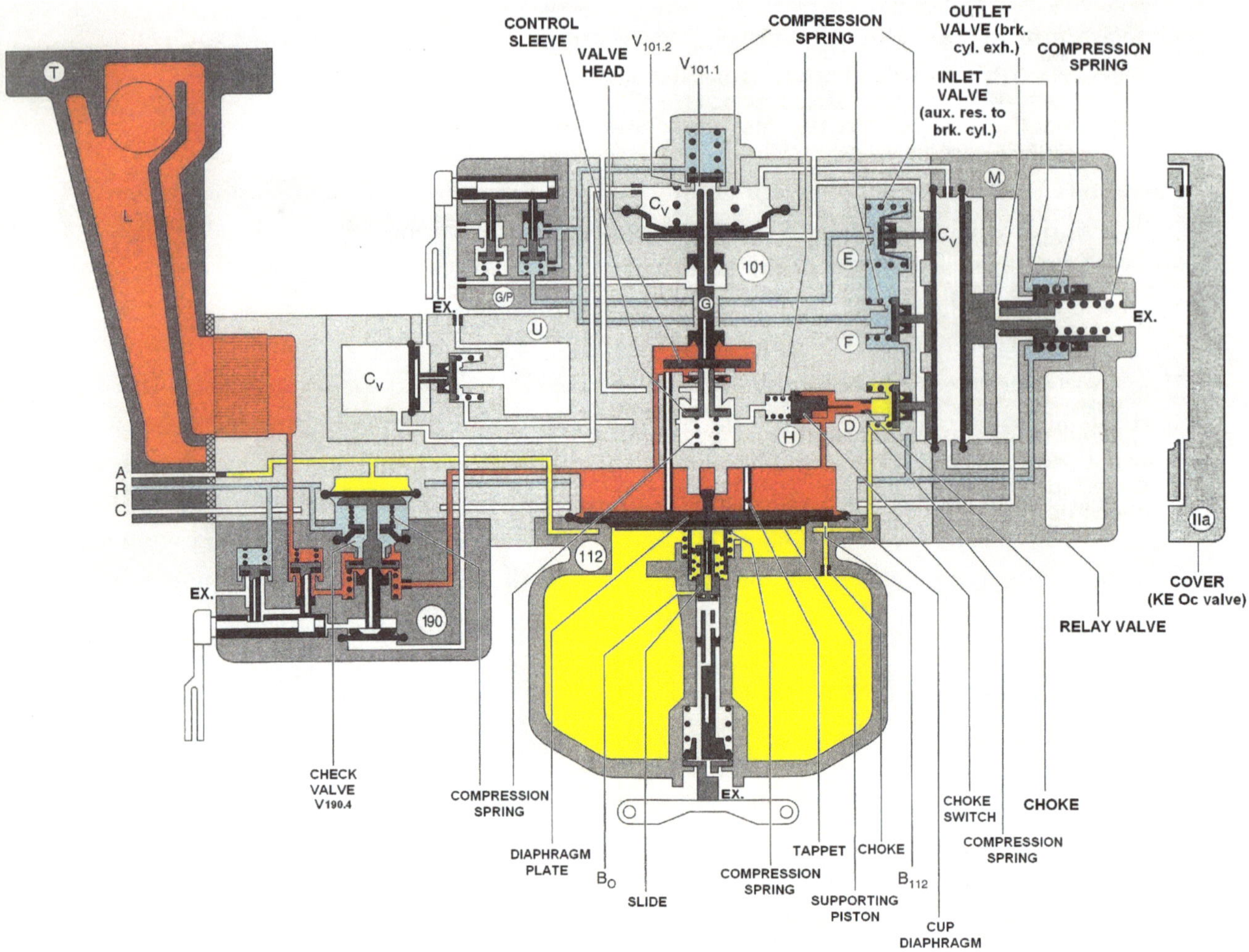

Pressures

A Control reservoir pressure
C Brake cylinder pressure
C_V Precontrol pressure
L Brake pipe pressure
R Auxiliary reservoir pressure

B... Port
V... Valves

112 **Control reservoir A** with quick-release valve
190 **R charging valve** with cut-off valve
G/P **G/P selector** (Goods/Passenger)
M **Relay valve**

Functional assemblies

D Control reservoir (A) control valve
E Maximum pressure limiter
F Minimum pressure limiter
G Triple-pressure valve
H Choke switch
T Valve bracket
U Quick-service (Ü) control valve

FIGURE 105: KNORR-BREMSE KE DISTRIBUTOR VALVE – DIAGRAM (*INITIAL CHARGING*)

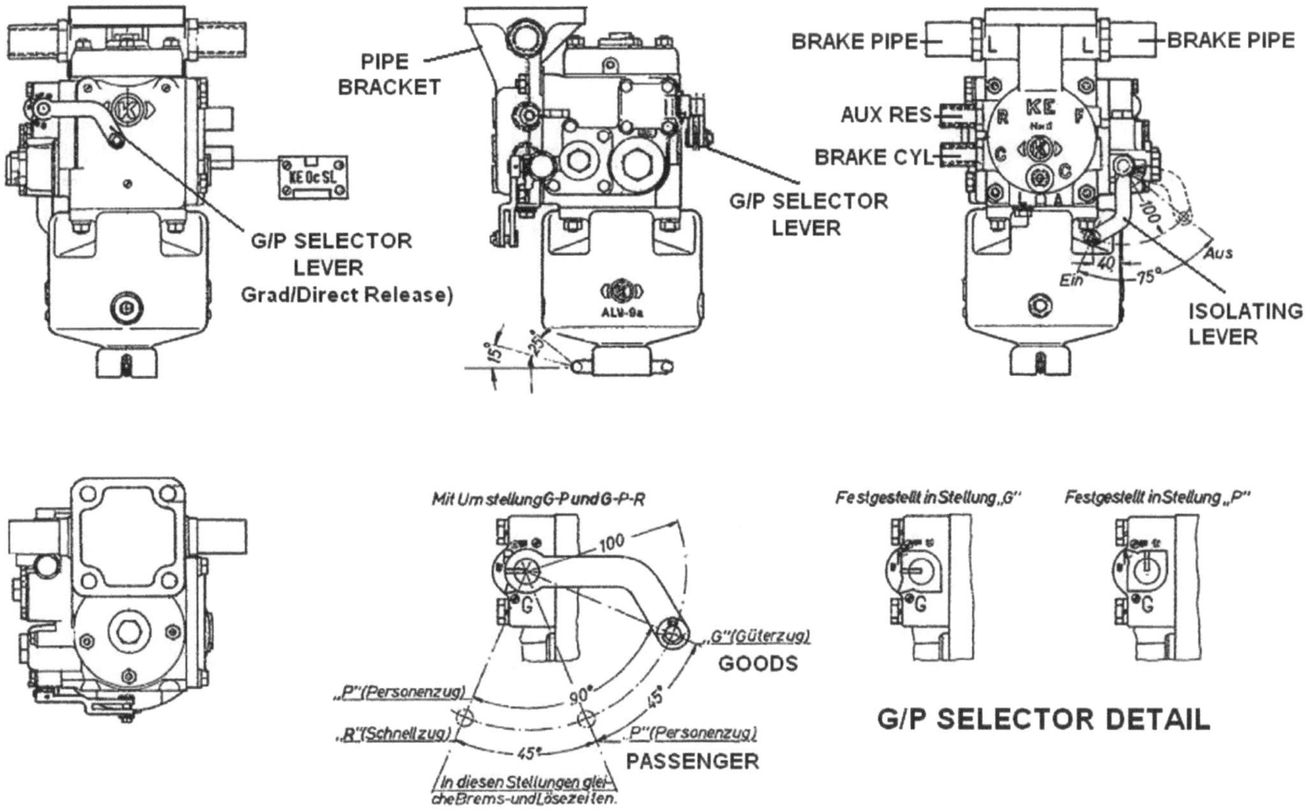

FIGURE 106: KNORR-BREMSE KE DISTRIBUTOR VALVE – GENERAL ARRANGEMENT

The AB Freight Car Brake Equipment

This equipment resulted from the development of the AB Control Valve in the United States. Its evolution has resulted in its improved descendants finding their way onto Australian railways either because they preceded suitable Australian equipments or were cheaper to buy.

The ABD and ABDW Control Valve

Despite the introduction of the capable W triple valve, the American ABD control valve gained a foothold in Australia following its introduction on pre-owned iron ore cars imported from the United States for the Pilbara railroads under construction by American company Bechtel Pacific during the 1960s. This equipment complemented the US-sized nature of these resource projects and provided the necessary operational capabilities prior to development of the W triple. Once the service performance of the ABD valves was proven to the Australian iron ore railroads, it is understandable that these companies—for whom capital funding was not a major issue—would continue to purchase this equipment. More recently, it is likely that this equipment is cheaper than the Australian-made product.

The first ABD installations were **non-relayed** systems in which, during brake applications, the control valve supplied auxiliary reservoir air direct to the car brake cylinder in the orthodox manner, applying the brakes via conventional foundation brake rigging. In this case, a basic Empty/Load feature was provided by which bogie spring deflection—as the car was being loaded—resulted in the brake rigging fulcrum point being altered, in turn changing the leverage of the brake rigging. As well as in the Pilbara, this equipment is in use in South Australia (by Genessee & Wyoming Australia) and far north Queensland (by Comalco).

Later installations have utilised a relay system whereby—during the brake application—the control valve supplies auxiliary reservoir air as a control pressure to a 'dummy' brake cylinder volume *and* a relay valve which in turn functions to supply air from a separate source (the supply or "supplementary" reservoir) to the car brake cylinders. This system was optimised to manage the large volume presented by the four brake cylinders of Truck-Mounted Brake equipment and also provided for finer control over the Empty/Load braking differential. In the latter case, a changeover valve, providing a second auxiliary reservoir pressure feed to the relay valve, is automatically activated as the car is loaded. This enables the relay valve to supply brake cylinder air pressure at two different values - one for when the car is empty, and an increased pressure for when it is loaded. Relay brake systems were discussed in the W triple valve section.

It has been mentioned previously in this book that the AB equipment tends to be a relatively heavy user of air. This may be one reason that one Pilbara iron ore railroad has reverted to non-relay equipment for its more recent ore car fleets.

ABD equipment is designed to properly control the brake systems on long, heavy trains and so must—under AAR regulation—possess specific characteristics that produce particular and identifiable performance outcomes. These results, required of all AAR-endorsed systems, include the following:

- All brakes must be capable of applying on a 35 kPa (5 psi) Service brake pipe reduction;
- The 150th car (where this applies) must have developed 70 kPa (10 psi) brake cylinder pressure within twenty seconds of the commencement of the brake application;
- The 150th car (where this applies) must start to release in no more than forty seconds from a 105 kPa (15 psi) brake pipe reduction;
- A 140 kPa (20 psi) brake pipe reduction must develop approximately 350 kPa (50 psi) brake cylinder pressure [the 2:5:7 rule];
- The brake must be of the *direct-release* type;
- The brake must be stable against undesired emergency applications (UDEs) during a Service brake application. However, Emergency applications must be available at any stage of Service brake application or release. [Note: This requires sufficient brake pipe pressure to reliably transmit the Emergency reduction, generally at least 280 kPa of brake pipe pressure.]

<u>ABD Control Valve Operation</u> (assumes 630 kPa/90 psi brake pipe pressure. For lower brake pipe settings, some operating pressures as quoted will be less).

- *Service Application:*
  As with any operating valve, a brake pipe pressure reduction triggers the control valve to supply auxiliary reservoir pressure either directly to the car brake cylinder/s (non-relayed system) or into the control pipe to the brake cylinder relay valve (relayed system). A brake pipe reduction of as little as 15 kPa (2 psi) will cause **Preliminary Quick-Service**. The driver has no control over the amount of brake pipe reduction caused by this QS function other than the initial reduction made. Quick-Service occurs only during the first part of a Service application (and not thereafter) and results in the following two actions:

1. The control valve ports brake pipe pressure into the small QS volume (integral to the Service portion of the control valve);

2. With relayed systems, some brake pipe pressure also flows to the control pipe (and thus to the dummy reservoir and brake cylinder relay valve).

The effect of this QS activity is a local reduction of brake pipe pressure at each car that promotes the transmission of the reduction to the next car. Its purpose is the early activation of the control valve to ensure the rapid development of a minimum brake cylinder pressure throughout the train in order to diminish the development of compressive slack action. When output from the auxiliary reservoir reaches approximately 50 kPa (7 psi), QS activity is stopped by action of the **Quick-Service Limiting Valve**.

An **Accelerated-Application Valve** (AAV) is integral to the ABDW control valve. Its purpose is to further hasten a Service brake pipe reduction and consequently reduce brake cylinder pressure buildup time. It functions during all Service brake applications by venting the brake pipe in a large number of very small steps or "bites." The AAV function will operate so long as the rate of brake pipe reduction is greater than 48 kPa (7 psi) per minute and will automatically cut itself out when not required during a heavy Service or Emergency reduction. The AAV will not add to apparent brake pipe leakage unless the leakage rate is 50 kPa/min or greater. Note that while both ABD and ABDW control valves share the same Service portion, the ABDW Emergency portion includes the AAV.

- *Emergency Application*:
  Any brake pipe pressure reduction of 140 kPa (20 psi) within 1.2 seconds or less is at an Emergency rate and can be initiated from the driver's automatic brake valve or from a burst or separated air brake hose bag or a broken or separated brake pipe. This may be compared to the Service rate of reduction where, during a Full Service application, the first 140 kPa of brake pipe reduction will take some 7–9 seconds.

  The rapid reduction of brake pipe pressure allows the **emergency piston** to move and connect Quick-Action chamber pressure to (1) the **Vent Valve**, to open it wide, and (2) the **High Pressure Valve,** which moves to exhaust QA chamber pressure at a rate sufficiently slow to ensure the vent valve will remain open for approximately one minute. This ensures rapid continuation of the Quick-Action and prevents any attempt by the driver at releasing the brakes before the train comes to a stand. (Note: The QA chamber is integral to the pipe bracket.)

  The emergency reservoir is connected to the control pipe (relay system) or brake cylinder (non-relay system) and augments the flow of auxiliary reservoir pressure, providing a 20 percent increase in maximum brake cylinder pressure. The first stage of emergency reservoir pressure buildup (to around 105 kPa/15 psi) occurs very rapidly to provide a quick, positive brake retarding force throughout the train to snub the tendency for harsh slack run-in. At approximately 105 kPa (15 psi) control pipe/brake cylinder pressure, the inshot valve operates to slow the rate of pressure buildup. This pressure now continues to increase toward its maximum value (approximately 540 kPa/77psi), taking about thirteen seconds from the time the control valve went to Emergency. Thus, Emergency brake cylinder pressure development of the ABD/ABDW control valve occurs in two distinct stages.

- *Release*:
  The control valve responds conventionally to the restoration of brake pipe pressure by recharging the auxiliary and supply (or supplementary) reservoirs and exhausting control pipe/dummy reservoir pressure (relayed system) or brake cylinder/s (non-relayed system). With a relayed system, removing control pressure from the brake cylinder relay valve allows it to exhaust brake cylinder pressure.

  <u>Accelerated Service Release:</u> When brakes are released following a Service application, the action of the control valve is somewhat different to that in initial charging, since the emergency reservoir will have been held to its original (brake pipe) pressure during the application. If brake pipe pressure increases above auxiliary reservoir by approximately 10 kPa (less than 2 psi), the control valve functions to feed emergency reservoir pressure to the brake pipe and thus also into the auxiliary reservoir. When the differential between the brake pipe and auxiliary reservoir diminishes to less than 10 kPa, the **accelerated release valve** closes and Release-and-Recharge proceed normally.

  Emergency reservoir pressure will assist in rapidly raising the pressure in the brake pipe of the next car to the value necessary to trigger that car's accelerated release valve, and so the action will propagate along the train. This shortens the overall recharge time for auxiliary reservoirs on a long train and allows the rear end of long trains to release in one-half to one-third the time of earlier control valves, particularly if existing leakage is concentrated toward the rear. The major benefit is the option to make running releases at a much lower speed (saving considerable time and distance should, for example, a restrictive signal—that may have been the reason for the brake application—clear).

  The accelerated service release feature will operate on all ABD and ABDW control valves during release after an Emergency application as soon as brake pipe pressure exceeds auxiliary reservoir by 10 kPa. The Accelerated Release feature will operate from as little as a minimum brake pipe reduction.

  <u>Accelerated Emergency Release:</u> This feature hastens release from an Emergency application. Without it, release and complete recharge would take more than twice as long as from a Full Service release due to having to recharge the brake pipe and *both* reservoirs. The first increase in brake pipe pressure will cause the control valve to allow both auxiliary reservoir and control pipe (and dummy reservoir) pressure to "back-dump" into the brake pipe. This provides a substantial local rise in brake pipe pressure at each car, thus greatly reducing the amount of recharge air required to be supplied from the locomotive.

This backflow continues until brake pipe pressure has increased to within approximately 70 kPa (10 psi) of equalising with the combined brake pipe and auxiliary reservoir pressures. Concurrently, auxiliary reservoir pressure is being partially reduced while brake pipe pressure is being increased, thus the supply of recharge air via the automatic brake valve only has to increase brake pipe pressure by about a further 70 kPa (10 psi) to actuate the brake cylinder relay valve (relayed system) and initiate release.

As brake pipe pressure increases, the Quick-Action chamber is being recharged. When brake pipe and QA chamber pressures equalise, the backflow from the auxiliary reservoir and control pipe (relayed systems) into the brake pipe ceases, and brake pipe pressure will continue to increase from the locomotive only. When brake pipe exceeds auxiliary reservoir pressure, the control valve goes to Release, the remaining control pipe pressure is exhausted, and reservoirs will be recharged.

Retarded-Recharge: Since cars located closer to the locomotive will be subject to a higher brake pipe pressure during recharge than those further back in the train, this feature (in the Service portion of the control valve) is provided to prevent those head-end cars from absorbing most of this restoration pressure. When brake pipe exceeds auxiliary reservoir pressure by at least 20 kPa (3 psi), the control valve connects brake pipe air—entering through the charging choke—to a restricted charging port. The emergency reservoir connection to the auxiliary reservoir is similarly restricted in this condition. Since the head-end control valves can now only use the available brake pipe air at a restricted rate, more of it can flow toward the rear of the train.

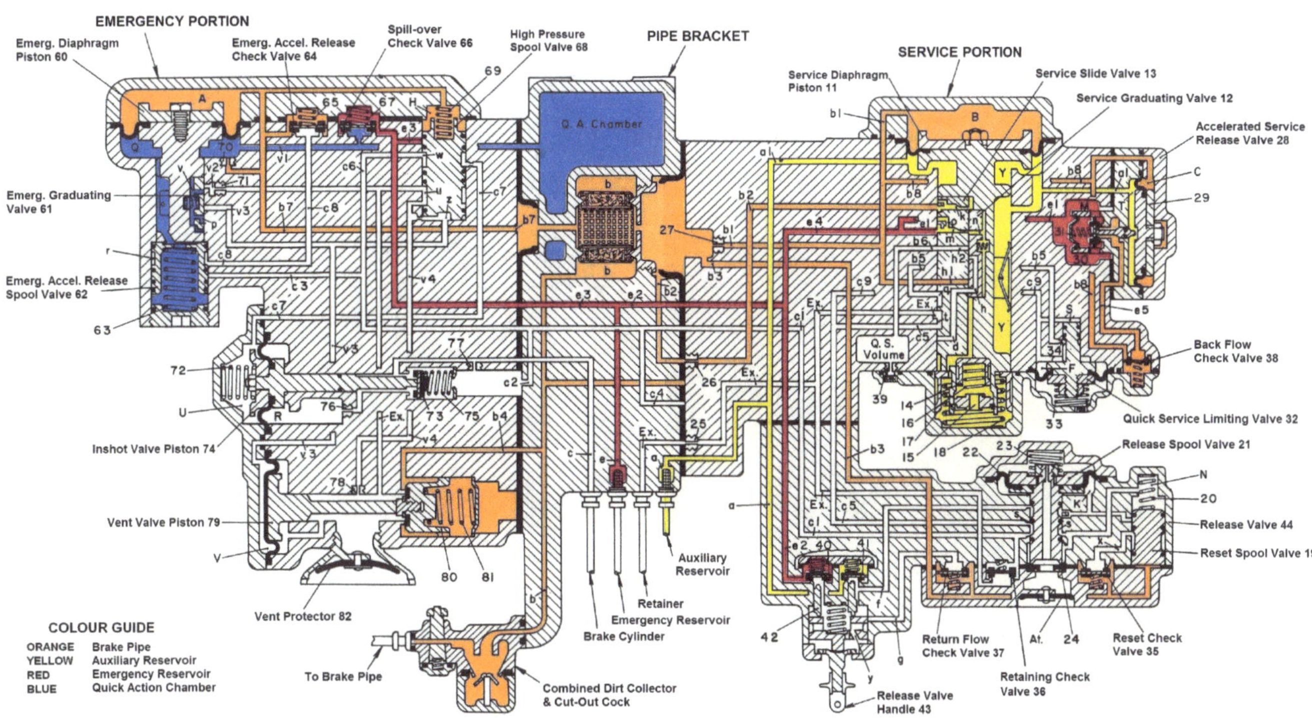

FIGURE 107: ABD CONTROL VALVE – DIAGRAM, *RELEASE/CHARGING* POSITION

The ABSD Control Valve

This is an ABD control valve without an Emergency portion (i.e. only the Service portion is attached to the pipe bracket) and with different application and recharge timing chokes. Lacking the Emergency portion, this control valve does not provide the Accelerated Application function; however, the Accelerated Service Release function prevails. This equipment was acquired by Westrail and is used by ARG (now QR National) on some mineral and grain rollingstock.

The ABDX Control Valve

This control valve evolved from the ABDW to meet the railroad industry's demand to control trains of longer and heavier cars at higher speeds. One operating challenge quite unique to North American railroading is the UDE (***U***n***D***esired ***E***mergency) brake application. These "mystery" brake applications occur infrequently but nonetheless regularly in their train operations but are not—as far as the author is aware—a common occurrence in Australia, probably because we do not run long ABD/ABDW-equipped trains (Pilbara excepted) and probably because we do not operate through the extreme climate range experienced in North America. After a total of thirteen years railroading in the ABD/ABDW-equipped Pilbara between 1981 and 1996, I am not personally aware of any major UDE problem there. Although UDEs are believed to result from several causes—some being cold-weather related—one cause is believed to be the small but sharp fluctuations of brake pipe pressure, possibly due to slack action, that may coincide with normal Quick-Service control valve activity at a given location within the train. A unique ABDX feature absorbs these pressure spikes.

Other design improvements result in significantly faster brake cylinder pressure development (which assists in controlling slack action), and reduced possibility of leakage into the brake cylinder during sustained grade braking.

ABDX is used on new-generation, non-relayed ore cars in service with Pilbara Iron (Rio Tinto) and BHP Billiton Iron Ore, and ABDX Service portions are also used as control valves by ARG/QR National (presumably because ABSD valves are no longer available), despite the fact that the major benefits of ABDX cannot be realised without including the Emergency portion.

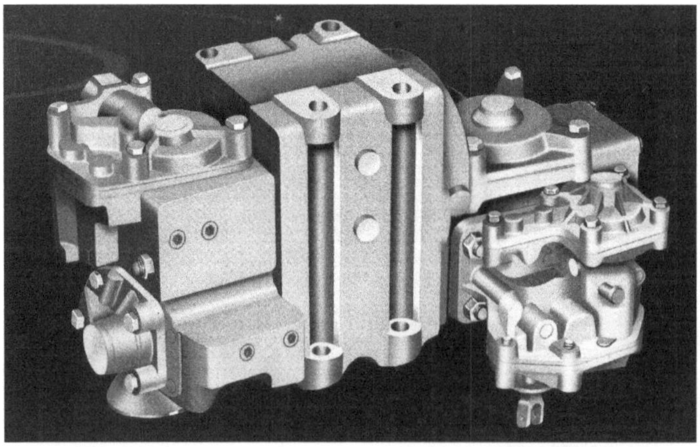

FIGURE 108: ABDX CONTROL VALVE - ILLUSTRATION

The Knorr/New York Air Brake DB-60 Control Valve

The DB-60 designation derives from the sum of the numerical segments of the principal components of the assembly, namely the DB-10 Service portion, DB-20 Emergency portion, and the DB-30 pipe bracket. The DB-60L version—with different choke rates accounting for a greater brake pipe length—is provided for installation to long cars.

Since this control valve possesses the same generic functional attributes as other American control valve equipments (e.g. Quick-Service, Accelerated-Application, Service Accelerated-Release etc.), these will not be re-examined here. DB-60 is widely used within North America and is well-established in Australia, both with state-based rail systems as well as within the Pilbara iron ore industry. These unique features afford this equipment its competitive profile:

- Soft-seated poppet valves: Functioning as check valves, these lack the detrimental properties common to slide and spool valves, namely friction (often a vital property) and **stiction** (an alternative term for the phenomenon of *cold-welding*, probably better describing the most objectionable aspect of friction). This is claimed to eliminate UDEs and extend service life;

- K-Rings: Previously described in chapter 2, these unique seals provide a smaller contact area, that, due to their distinctive design, is a significant benefit;

- Super diaphragm clamping: A further contribution toward extended life expectancy is the design objective of using a special thickened bead for clamping diaphragms rather than using the membrane, which is—technically—the diaphragms' weakest part.

Major selling points for the DB-60 are its interchangeability with other AAR-approved control valves, reliability and stability of performance, freedom from leakage, superior signal transmission times, and quality construction providing extended *between-service* periods and ease of maintenance. Another distinctive option is the single-sided pipe bracket for a more accessible mounting of the control portions. (Note: All versions are now manufactured in lightweight aluminium.)

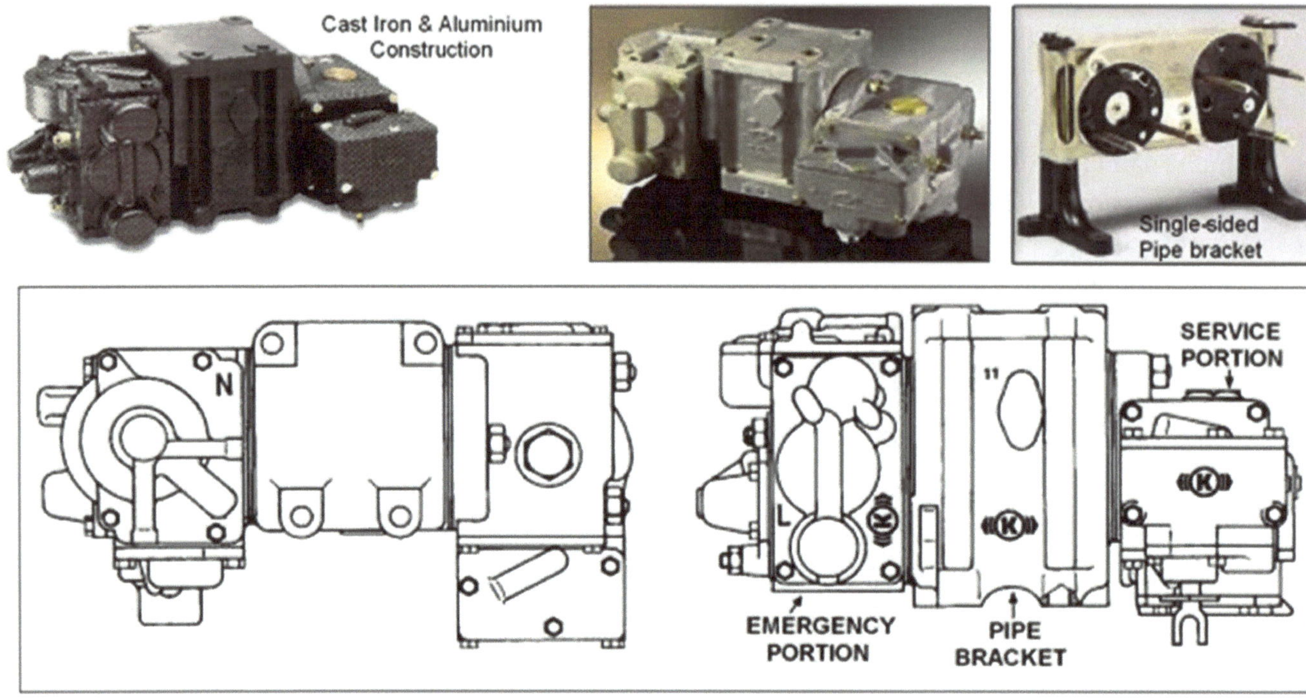

FIGURE 109: NYAB DB-60 (AND -60L) CONTROL VALVE

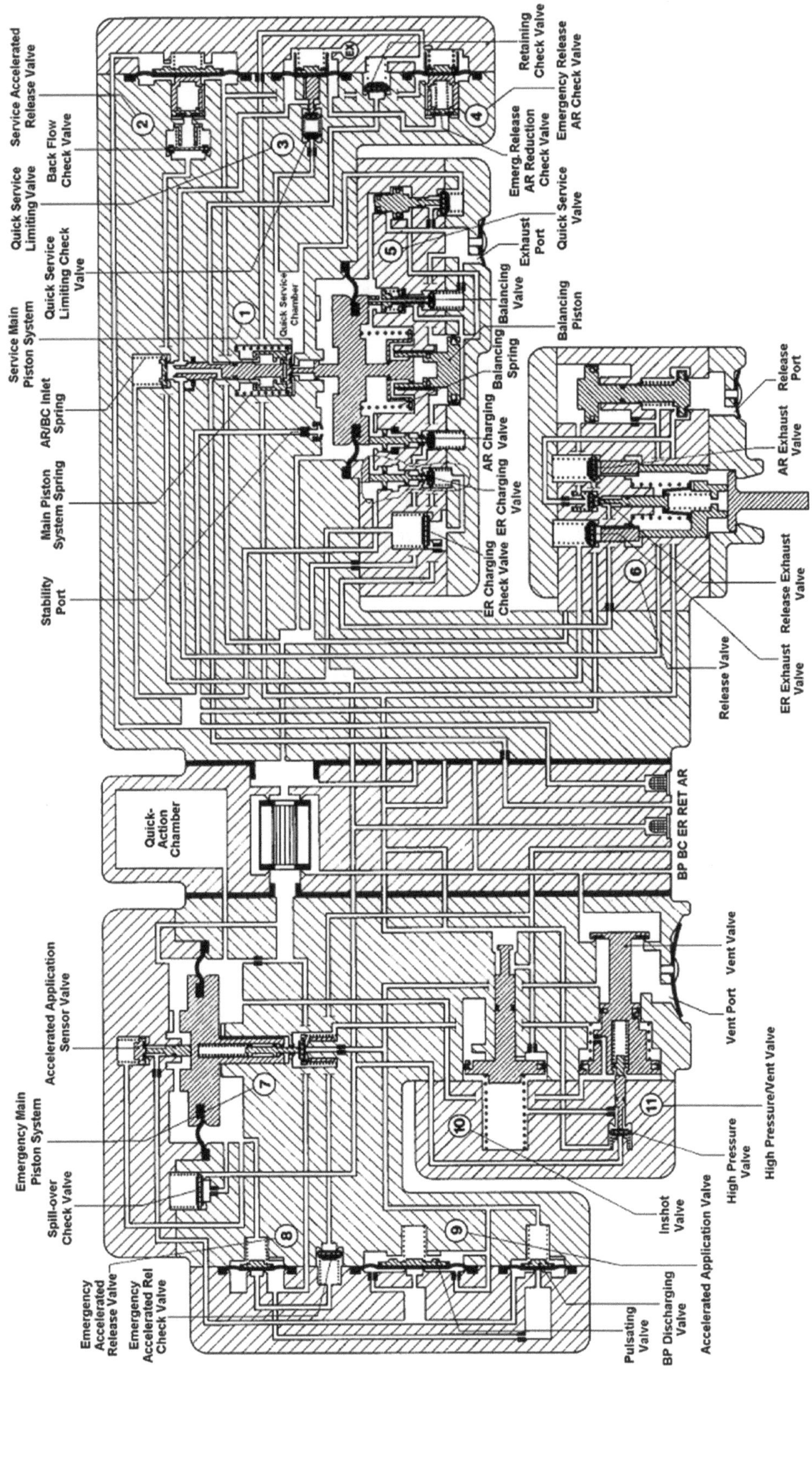

FIGURE 110: DB-60L CONTROL VALVE - DIAGRAM

Why Is It a "Triple Valve"?

Those who seek to become familiar with the specialty subject of railway air braking very soon learn that in order to speak meaningfully on their subject, they must master a bewildering series of component names and model numbers. In some cases the industry has even managed to provide completely different devices with identical designations. For example, a listener hearing reference to a '26C' might be perceiving a comment on either a locomotive driver's automatic brake valve or a passenger car control valve. Similarly, reference to a 'J2B' might be either about a freight car brake cylinder relay valve or a slack adjuster. In listening to air brake technical personnel speaking about air brake system design and operation, one might be forgiven for thinking that they exist on a diet of alphabet soup! Every device has a name and some sort of model or type designation (e.g. the A-19 flow adapter). Some may have letters that have direct relevance as a descriptor (e.g. the A7-EL locomotive airbrake schedule, where 'A' refers to it being an Australian adaptation, the '7' is an incremental value indicating the schedule's position in the continuum (from the № 4 equipment, through the 5-ET and the 6-ET) and 'EL' refers to the electric locomotives for which this schedule was originally designed. Or the J-1.6-16 locomotive brake cylinder relay valve, where '16' denotes the valves' capacity to supply a brake cylinder pressure at either 100 or 160 percent of control pressure (depending upon whether the car is empty or loaded).

However, sometimes there is neither an obvious reason for the designation given to a piece of equipment nor any obvious pattern to the continuum of model development, and the reasons (if there ever were logical ones) for a device having a particular designation are often lost in time. From time-to-time though, an evolutionary pattern is discernible (as with the W triple valves) as devices were developed and modified, gaining extra contiguous letters and/or numbers as engineering and design designators. This study of the triple valve would be incomplete without an examination of its designatory origins.

The device is the quintessential manifestation of Westinghouse's brilliance. Generically, it is an air brake *operating valve*. It began its developmental life as the 'triple valve' and evolved in North America to become the 'control valve' and, in Europe, the 'distributor' or 'distributing valve' (the latter usually controlled in its operation by two pressures rather than one). In Australia and New Zealand, the quaint 'triple valve' nomenclature persists within the rail industry despite considerable use of US control valves and despite the local manufacture of a complex operating valve (the W-type), which is optimised for Australian and New Zealand conditions and, in many respects, is the technical equal of the ABD/ABDW family.

The triple valve was (and is) the central component of the Westinghouse atmospheric air brake system and the key to its successful operation. It is almost universally held that the name "triple valve" refers to three often-stated 'principal functions' of the device; to apply and release the brakes and to recharge the reservoirs. Many Westinghouse technical publications refer to this as the reason for the name, and such US air brake luminaries as the late David G. Blaine (see biography, chapter 7) and C. M. Drennan also explain its name in these terms. However, while these men may have been air brake experts, they may not necessarily know any more of the early history of how the device came to receive its name than does anyone else. In fact, despite what is repeated dogmatically in industry literature and thus air brake training classes everywhere, the triple valve really only exists to achieve two outcomes; quite simply, it is designed to apply and release the brakes on a railway vehicle. It responds to variations in brake pipe pressure to either permit the flow of stored air pressure from the auxiliary reservoir to the brake cylinders (in which sense it depletes the auxiliary reservoir) or from the brake pipe to the auxiliary reservoir (in which sense it replenishes the auxiliary reservoir). In carrying out this latter function it must simultaneously permit the brake cylinders to exhaust.

Another expert source describes the three functions as *"to apply or release,* ***or hold a partial application,"***[116] the latter (emphasised by this author) referring to Lap. There are yet other explanations that attribute the name 'triple valve' to the three pipe connections: brake pipe, auxiliary reservoir, and brake cylinder. So, the origin of the term is open to conjecture even among the experts. On the basis of fundamental technical logic, I believe all of this reasoning to be mythical despite what may *appear* to be an overwhelming body of evidence to the contrary. Nothing that I have ever read about George Westinghouse actually quotes him explaining his rationale for the name of the device, although we can perhaps obtain a clue[117] from which readers may draw whatever conclusions seem reasonable.

In fact, the inventor was notorious for not writing notes or minutes and for leaving little in writing (apart from many detailed patent specifications) about his thoughts or rationale in conceiving and designing. He may have assumed that it would be obvious that the device was tri-valvular in composition; certainly, there is a record of his referring to the appliance as his *"triple valve device"* and a *"triple-valve arrangement"* (see next page). In the specification for a patent application for an improvement in October 1873, Westinghouse describes the operation of his device and the functioning of the three valves, A, B, and C, mounted on one stem and operating in three chambers. Refer to the drawing reproduced at Figure 111 and to Figure 8 in chapter 2 for the original diagram of his early steam- and air-operated triple valves (yes, Westinghouse dabbled with steam as the medium before he realised its limitations and settled on compressed air!).

In his specification, Westinghouse states:

Thus, by means of this triple-valve arrangement, when the air pipes communicating with E are charged with compressed air the auxiliary reservoirs are charged and the brakes are taken off...

Note *"triple-valve arrangement"*! Of the doubters, one might ask, how else can this phrase be interpreted?

Many air brake valvular devices are themselves comprised of parts that are also valves, and of course Westinghouse's original triple valve consisted of three integral valvular elements – separate but functioning as part of the whole. I contend that it is entirely plausible that the name "triple valve" refers to *three valves* rather than *three functions*. Even if Westinghouse's own words had not been heard, it is far more logical, in my view, that the inclusion of these three central components—(1) the piston, which provides the function of a 'valve' with respect to its control of the feed groove and is sometimes referred to as the "piston-valve"; (2) the slide valve; and (3) the graduating valve—are the reason for the device's designation.

Other comments support this view. One such observation appears in the New Zealand Government Railways Air Brake Handbook as issued to locomotive staff up to the 1970s. In a section describing the operation of the W triple valve someone from the NZR Chief Mechanical Engineer's office (Mr Noel Mather, perhaps?) appears to have some insight. The reference reads as follows:

These three valves [referring to the auxiliary reservoir charging check valve, the brake cylinder exhaust valve, and the brake cylinder inlet valve], *it is of interest to note, correspond to the three valves of George Westinghouse's*

116. *Engineering and Design of Railway Brake Systems*, The Air Brake Association, Wilmerding, PA, 1975.

117. Mr W.W. Nichols (see footnote 47, p 22), presenting a paper at the *GEORGE WESTINGHOUSE COMMEMORATION: A Forum Presenting the Career and Achievements of George Westinghouse on the 90th Anniversary of His Birth*. Conducted by the ASME, December 1, 1936, (The American Society of Mechanical Engineers, New York, 1937). Nichols states: "In the 1887 competitive air-brake tests, under the auspices of the Master Car Builders Association and following the defeat of Westinghouse by the Carpenter brake of Germany, "GW" suddenly appeared on the scene to the relief of the distraught Westinghouse Air Brake Company officials, whose agitation resolved itself into a sweet tranquility. Indeed, one high executive observed to me, 'Now that Mr. Westinghouse is here, everything will come right.' The trials had been at the West Burlington hill on the main line, and, as an engineer of tests I had charge of the dynamometer car. Following the formal trials, Mr. Westinghouse instituted a private series of tests under MCB conditions, and he spent much time during experimental runs studying the records and discussing, with me, their implications. Finally, one morning, he left impetuously before noon, inviting me to join him at lunch in his private car on the West Burlington Shops' siding. Later, as I passed through the corridor to the dining room, he emerged from his stateroom and waved a bit of brown paper, exclaiming with characteristic enthusiasm, 'Here's the triple that will revolutionize railroading!' How true that proved."

original diaphragm-operated triple valve of 1870. Because that valve did not have a Quick-Service bulb only three valves were needed, hence the name "triple valve."

A well-meaning friend

In his comprehensive biography of Westinghouse (see chapter 7), H.G. Prout[118] adds some extremely powerful words; a personal assertion—not referenced—that has never been attributed specifically to George Westinghouse, and yet, because it is stated so confidently in such an early and respected personal history[119], has almost universally become the authoritative explanation for the name of the device. Prout states:

It is interesting to note that the term 'triple valve' is not found in [the specification for patent 124,405] *or in that of the succeeding patent; but in patent 138,827* [filed February 1 and issued May 13, 1873] *in which it does appear for the first time, a reference is made to these earlier devices as such. The term 'triple valve' has become fundamental in air-brake language.*

He then expresses—as if it were a fact—an opinion on the rationale for its name:

It was coined to express the threefold function of the device, viz., to apply the brakes, to release them, and to charge the auxiliary reservoirs. Since the devices described in the previous patents were not designed automatically to release the brakes when automatically applied, the structures they cover can scarcely be called "triple valves" in the strict use of the term.

So, we appear to have a conundrum. The inventor refers to the device in descriptive terms. It is a *"triple-valve arrangement"*; in other words, its composition is *tri-valvular*. Yet, his friend and biographer states that the name was coined to express the device's "threefold functions." How does Prout come to this interpretation and how can we resolve this?

In my view, it is not so mysterious. Prout may well be quoting—but misunderstanding the context of—the words stated by Westinghouse that appear near the top of page 148, or he is simply repeating what he may have read elsewhere[120]. Despite Prout's earnest comment, it is but either an opinion or nothing more than a recurring account arising from his research. Once it was recorded within his biographical tome (not to mention the previously-published work of Sinclair referred-to in a footnote below), it seems natural that the comment would have been repeated. And once it began to appear in industry publications, it would have gained currency and soon become gospel within the industry and throughout the air brake world. Westinghouse company technical writers who compiled the early equipment pamphlets may have simply picked up on this—as well as written comment, perhaps even from George Westinghouse himself—regarding the intended purpose of the device and possibly explanations of its three main functions and descriptions of its three main physical connections within the brake system.

If we can assume that Sinclair was the first—and Prout the second—to speak of "three functions" as *being the reason for the name of the device*, then we might consider them jointly responsible for a universal myth.

A legal view

Perhaps, though, the most compelling source for elucidation is the specification Westinghouse himself prepared to support the patent application for his original valve and which is reported in the judges'

118. Colonel Henry Goslee Prout (1845–1927), a personal friend of George Westinghouse, hailed from Nutley (later renamed Franklin), New Jersey. He served in the Massachusetts Volunteers during the American Civil War and later with the American Military Mission to Egypt, led by discredited former Union army General Charles P. Stone. During that period, a total of fifty ex-Civil War Union and Confederate officers served in the Egyptian army. Prout, a civil engineer, was involved in conducting topographical surveys and geological reconnaissance expeditions as far afield as what are now Sudan, Uganda, and the Ethiopian borderlands. He then served as governor of Equatoria (at one time a province of Egypt, located in the extreme south of present-day Sudan along the upper reaches of the White Nile) for a short period in 1876. Later, Prout lived in New York and was editor of the eminent industry journal, *Railroad Gazette*.

119. Prout, though, was not necessarily accurate with all that he wrote. See the reference to Westinghouse v Gardner & Ranson in the Westinghouse biography in chapter 7.

120. In his own authoritative work, *Development of the Locomotive Engine* (originally The Angus Sinclair Publishing Co., later The MIT Press, 1907), Angus Sinclair states, "The device…was called a 'triple valve' because of its threefold function of applying the brake, releasing it, and recharging the auxiliary reservoir." Sinclair is a powerfully credible commentator, but is he too laboring under a common misapprehension?

summary of a legal action taken in 1898 by him in a case of alleged patent infringement.[121] The inventor describes the object of his automatic brake mechanism:

To enable the application of brake shoes to car wheels by fluid pressure to be effected with greater rapidity and effectiveness than heretofore, more particularly in trains of considerable length, as well as to economize compressed air in the operation of braking, by utilizing in the brake cylinders the greater portion of the volume of air which in former practice was directly discharged into the atmosphere. To this end, my invention—generally stated—consists in a novel combination of a brake pipe, an auxiliary reservoir, a brake cylinder, and a triple-valve device governing—primarily—communication between the auxiliary reservoir and the brake cylinder, and secondarily, communication directly from the brake pipe to the brake cylinder, whether the valves governing these passages or connections are arranged in one or more cases, and are moved by a piston, or its equivalent, or by a series of pistons, or their equivalents…

In using the terms "triple valve" and "triple-valve device," I refer to a valve device, however specifically constructed, having a connection with the main air or brake pipe, another with an auxiliary reservoir or chamber for the storage of power, and another with a brake cylinder or its equivalent, for the utilization of the stored power, and with a release or discharge passage for releasing the operative power from the brake cylinder.

Here, Westinghouse simply describes the connections to his triple valve. There happen to be three, each of them associated with the functioning of one of the three valvular components of his "device." It is possible that both Sinclair and Prout—and many others to follow—have presumed that comments about three "functions" or "connections" explained the reason for its name, without recognising the significance of the three components that provide those three functions.

The judges' report referred-to above goes on to describe:

…a device which was obviously the foundation of the triple valve which first made its appearance in patent No. 141,685 (1873), in which the main valve, which admitted air from the auxiliary reservoir to the brake cylinder, was of the poppet form; and, as a poppet valve can govern only one port, separate valves had to be provided for feeding in the air from the train pipe to the auxiliary reservoir, and for discharging the air from the brake cylinder to release the brakes. [Note the references to three valves.]

In subsequent patents, No. 144, 006 (1873) and No. 163, 242 (issued in 1875 to C.H. Perkins, and assigned to Westinghouse), Mr Westinghouse improved upon his prior devices by substituting a sliding-piston valve for the poppet form of main valve previously used by him. This enabled the piston to perform the feed-valve function of admitting air from the train pipe to the auxiliary reservoir; the main-valve function of admitting air from the auxiliary reservoir to the brake cylinder to apply the brakes; and the release-valve function of discharging the air from the cylinder to release the brakes.

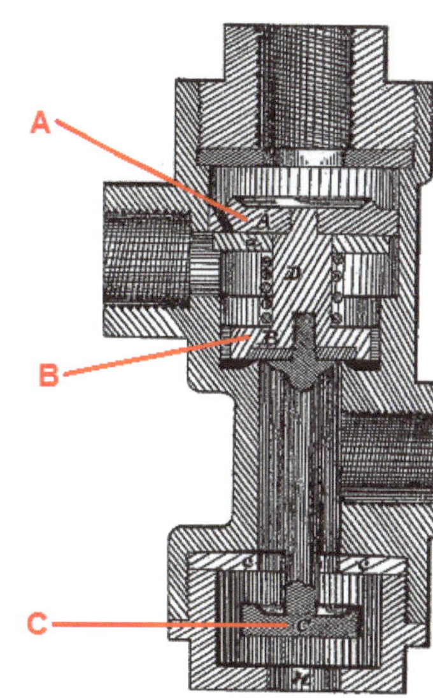

FIGURE 111: THE THREE INTEGRAL VALVES OF WESTINGHOUSE'S ORIGINAL 'TRIPLE-VALVE' CONCEPT

In this passage, the judges indeed discuss three functions. But this comment is in the context of what the newly-designed piston-valve was ultimately intended to achieve. Westinghouse has already referred to the device in the document as a "triple valve", and the judges do not seek, in this passage, to justify why it was so-called.

The judges continue:

In patent No. 168, 359 (1875) a piston actuating a slide valve was substituted for the piston valve, and, after a series of experiments, which did not seem to have been successful, he introduced into patent No. 217, 828 the idea of venting the train pipe not only at the locomotive but also under each car in order to quicken the application of the brakes…

121. http://laws.findlaw.com/us/170/537.html (retrieved June 2013).

Then comes an interesting part in the report, where an engineering drawing is obviously being described:

The main valve of the triple is black. Its office is to admit auxiliary reservoir air to brake cylinder. The quick-action valve is colored red. Its office is to admit train-pipe air to brake cylinder. The release port is colored green. Its office is to discharge air from brake cylinder in releasing the brakes. There is also shown, in yellow, what is known as the "graduating valve," the function of which will be hereafter explained. As at present used, the triple valve is in reality a ***quadruple valve****.* [Heaven forbid! But this is a comment from the learned benchers - not my opinion.]

One more comment from the judges' report appears to support the assertion that the Triple Valve is named for the number of its valvular parts rather than its number of functions:

In this device there is found a main air pipe, an auxiliary reservoir, a brake cylinder, a triple (or, rather, a quadruple) valve and piston 29, with three ports; First, for the admission of air from the train pipe to the brake cylinder through the feeding-in valve 26; Second, for the passage of air from the auxiliary reservoir to the brake cylinder through the apertures i, j, & k, in the stem slide valve 18; and, Third, for the release of air from the brake cylinder to the exhaust port by means of valve 17, colored green.

Whether this device has an auxiliary valve or not, is one of the main questions in the case – complainants' theory being that poppet valve 22, is an auxiliary valve, while defendants' claim is that it is in reality the main valve. In both the complainants' and defendants' devices there is (1) a "feeding-in valve" to charge the auxiliary reservoir; (2) a valve, which complainants call their "main valve," and which the defendants denominate a "graduating valve," which is opened by the preliminary traverse of the piston to admit reservoir air to the brake cylinder; (3) a "release valve" which discharges air from the brake cylinder to the atmosphere; and (4) a "quick-action valve"—41 in the complainants' patent, and 22 in the defendants'—which is opened by the further traverse of the piston to admit train-pipe air to the brake cylinder. In defendants' patent it may also be used to admit auxiliary reservoir air to the brake cylinder.

Note that at this point in its development, the triple valve is now comprised of four integral valves and might well deserve the designation "Quadruple Valve". The inference to be drawn is clear.

TWO LAST WORDS

A further compelling statement of historical significance cannot be overlooked:

I will particularly draw your attention to the triple valve, which has been made a regular bugbear by the opponents of the system, and has been called complicated, delicate, and liable to get out of order etc. The original triple valve used at the Newark experiments, a drawing of which I have now before me, has three distinct valves, and hence its name "triple valve."[122]

And a final comment on this matter is worthy of mention. A New South Wales Department of Railways (Railway Institute) publication detailing a fitters training course on the Elements of the Westinghouse Air Brake states, at the beginning of the section dealing with triple valves:

"The triple valve has three important parts: (a) the triple valve piston, (b) the slide valve, and (c) the graduating valve."

And with that, the author rests his case...

122. Michael Reynolds: *Continuous Railway Brakes - A Practical Treatise On The Several Systems In Use In The United Kingdom, Their Construction And Performance* (London: Crosby Lockwood & Co., 1882), p222. In this passage, Reynolds is quoting Mr. Thomas E. Harrison, Chief Engineer, North-Eastern Railway Co., at a meeting of company directors in April 1879. Reynolds was a member ofThe Society of Engineers and the author—also—of instructional texts of the time; *Locomotive Engine Driving, Stationary Engine Driving, Engine Driving Life*, etc.

4.

Locomotive and Train Brake Equipment in Australasia

The vast and unexampled use of the air brake makes a clear knowledge of it of vital importance to everyone actively interested in the movement of trains.

—M.M. Kirkman (*The Science Of Railways*)[123]

NO HISTORY OF the development of railway air braking would be complete without a study of the parallel development of locomotive brake equipment. When George Westinghouse placed an air brake system on freight and passenger cars, he had to provide the locomotive engineman with a means to control that system. This instrument has evolved from a simple three-way cock—allowing air pressure to flow directly to all cylinders on the locomotive and trailing vehicles—to its current form, the so-called electronic brake valve, which is less a 'valve' than an electric switch mechanism, and which is explained later in this section.

In the field of railway air braking, phraseology and nomenclature can become ambiguous with the passage of time. Some words, definitions, and descriptions have become corrupted by what at first is simply inappropriate or incorrect usage by industry participants such that eventually, through constant use, they acquire currency. The previous section of this book suggests one notable example. In a subtle way this phenomenon contributes to a diminution of the discipline and confounds many who might seek to develop an understanding of the subject. Therefore, this section is intended to be explanatory as well as instructional, and it is hoped by exploring the subject in a logical sequence that the "how and why" of the technical evolution of the locomotive driver's brake valve will be apparent and of interest to the reader.

Basic terminology is important; the late C. M. Drennan, US railroad air brake technician and car instructor, provided these generic definitions:

Air Brake System — All of the devices and parts included in making an air brake for controlling the speed and stopping a locomotive or train. It is made up of the operating devices, the pipes, fittings and foundation brake gear etc.

123. Marshall Monroe Kirkman, supplement to *The Science Of Railways - Book Two, New York Air Brake*, (New York and Chicago: The World Railway Publishing Company; 1900/02/03)

Air Brake Equipment – All of the air brake system except the foundation brake rigging.

Brake Rigging

The term 'foundation rigging', refers to the system of rods, levers, beams, hangers, and pins by which the force exerted in the brake cylinder or by the handbrake is increased and delivered by the brake shoes to the car wheels. There is usually one central brake cylinder attached to the car frame, some levers, perhaps a slack adjuster, and several long pull-rods as well as many pin connections. **Foundation rigging** was the preeminent system for freight rollingstock at the time that Mr Drennan was sharing his knowledge and experience with American carmen and mechanics, and although such setups are still widespread, the more recent (c1958) development of **truck-mounted brake systems**—"bogie-mounted" in Australasian parlance—that are now in their second generation, has rendered the former somewhat obsolete.

Although there are others, the New York Air Brake NYCOPAC equipment and the Wabtec WABCOPAC are probably the most common truck-mounted brake systems. Both manufacturers now have second-generation versions of these products, known as TMB (Truck-Mounted Brake) and TMX (because the abbreviation 'TMB' had been co-opted as a trademark by NYAB), respectively. Two more versions can be mentioned; AUSCOPAC, which is basically one of the above systems modified for local narrow gauge (1 067 mm) application, and ALPAC, which is a lighter-weight, fabricated version (Figure 115). Both are products of Westinghouse Australia, now Knorr-Bremse (Australia) P/L.

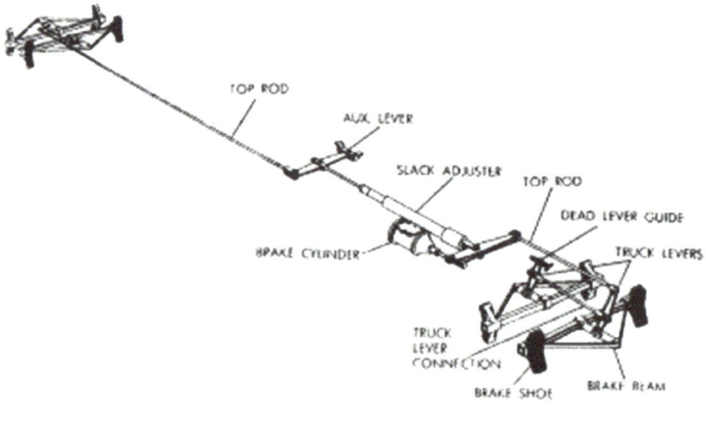

An example of foundation brake rigging
Courtesy of the Air Brake Association

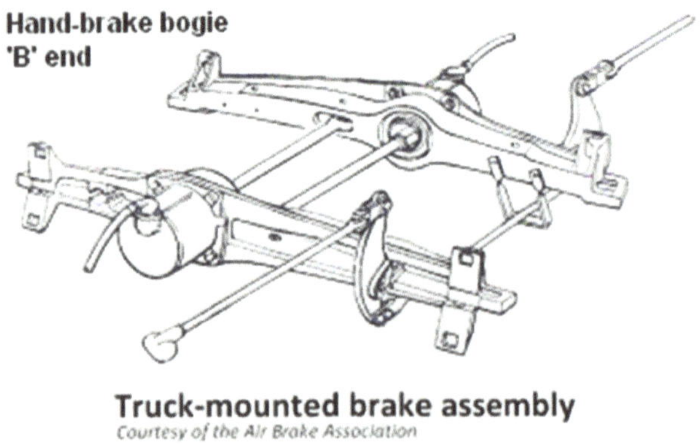

Truck-mounted brake assembly
Courtesy of the Air Brake Association

FIGURE 112: TRUCK-MOUNTED VERSUS THE CONVENTIONAL FOUNDATION BRAKE ARRANGEMENT

Truck-mounted equipment has traditionally utilised two **unit bodies** (shown to the left in Figure 112) that operate within guide lugs cast on the inside face of each bogie side-frame. These unit bodies perform similar to brake beams, each including an integral brake cylinder, that pushes directly on the opposite unit body. Each brake cylinder itself also pushes back on its own unit body – the resultant action and reaction directly applying force to the brake shoes on all four wheels. Thus, the equipment is said to be **direct-acting** since there are no intermediate levers to effect a multiplication of force between the brake cylinder and brake shoes. Some more recent versions of truck-mounted equipment have a somewhat different approach to providing this "direct-action" (e.g. see the lower illustration in Figure 115).

At this point, it is probably necessary to confuse the issue - and add a note of caution. The reader should be aware that there is also a brake equipment configuration known as **truck-confined**, which dates from the 1930s and should not be confused with **truck-*mounted***. Truck-confined systems also have their brake cylinders mounted on the bogie rather than the car body; however, instead of the unique features of the truck-mounted brake assembly, truck-*confined* systems utilise a conventional although much more compact arrangement of brake cylinders, rods, and rigging; a version of foundation rigging that is confined within the structure of a truck (a US version is depicted in Figure 114). Australian examples of this equipment include the ex-US Oroville Dam ore cars that were used for many years in the Pilbara by the Mount Newman Mining Co. (utilising two double-acting brake cylinders mounted low on each end of the bolster, with conventional transverse brake beams suspended from the bogie side frames) and one-piece, rigid, cast steel passenger car bogies (e.g. *Indian Pacific* and *Ghan* cars) incorporating locomotive-style brake cylinders and rigging. Other Australian railways—for example, V/Line in Victoria—also operate passenger cars utilising truck-confined brake equipment.

Note also that in referring to air brake equipment in the plural, the word "equipments" is generally used. The word "schedule," on the other hand (in the context of an arrangement or assembly of parts), is usually employed to designate a particular grouping of components that comprise a particular equipment set or fit-out and probably derives from manufacturers' original parts listings. Thus, for example, the **6-SL air brake schedule** of equipment might be expected to include (among other devices) a version of the H6 engineer's brake valve, the H6-A (or B) Relayair valve unit, and possibly the N1-A brake application valve. Similarly, and as a latter-day example, a **26-L schedule** will comprise the 26C driver's automatic and SA26 independent brake valves (or other compatible equipments such as 30A-CDW or W30D-G and their respective independent portions described in this section), and some or all of the following: 26-type Control Valve, J-type Relay Valve, MU2 or MU2A valve, F-1 Selector Valve, P-2 or P-2-A Brake Application Valve, A-1 Charging Cut-off Pilot Valve, etc. Still, the descriptors *equipment* and *schedule* are often used interchangeably these days. The distinction, and any reason for it, is probably fading.

Another inaccuracy often used in describing air brake equipment occurs when the "schedule" name is used as the designation for the driver's brake valve. Thus, we commonly hear of the "A6-ET brake valve" (from the steam era) when A6-ET is actually the **schedule** of equipment fitted to the locomotive, and the **driver's automatic brake valve** is in reality the A6 (Australasian designation for H6) or a derivative. Similarly, one might hear mention of the "26-L brake valve", when the standard driver's automatic brake valve utilised as part of the **26-L schedule** is in fact the 26C (although one of the compatible 30-series valves mentioned in the previous paragraph might also be used). Some might argue that these distinctions are inconsequential, and perhaps they are. But in the interest of technical—if not historical—accuracy, I feel compelled to make them.

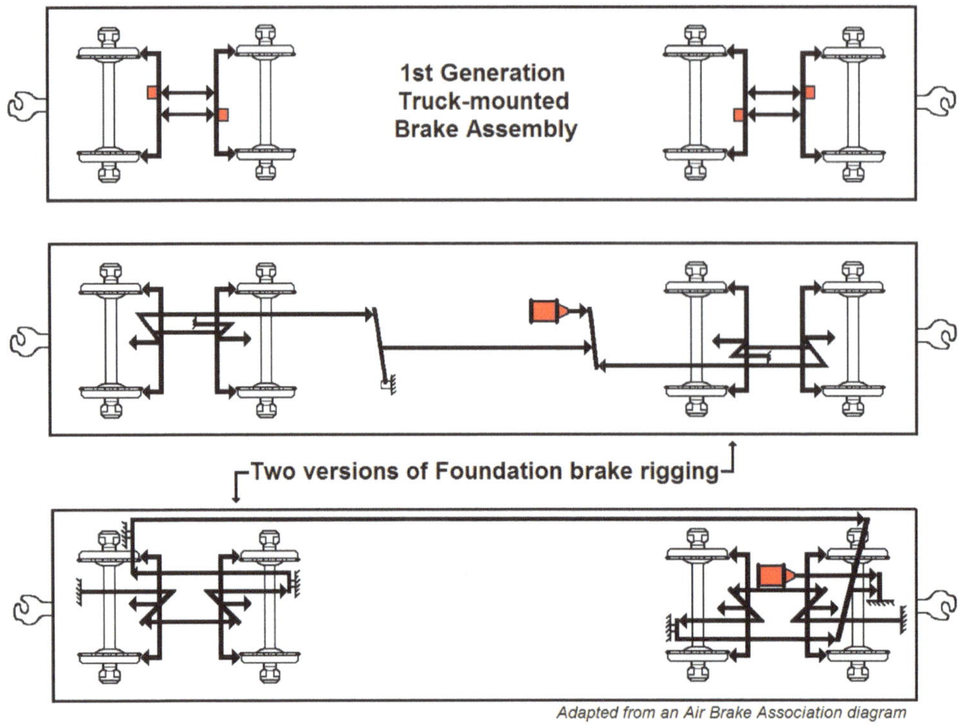

FIGURE 113: TRUCK-MOUNTED VERSUS FOUNDATION BRAKE RIGGING. THIS DIAGRAM SPECIFICALLY DEPICTS THE IMPORTANT FUNCTIONAL DIFFERENCE BETWEEN THESE ARRANGEMENTS

Truck-mounted brake systems, meanwhile, have moved to their next generation. Figure 115 displays the NYAB offering that includes an optional hydraulic handbrake. The design utilises a transversely-mounted pneumatic actuator with a self-contained, double-acting slack adjuster, two brake beams, two bell crank levers, and interconnecting push rods. The hydraulic handbrake features a mechanical lock that maintains the parking force with no fluid pressure assistance once applied. Because brake shoe loads are applied to the ends of the brake beam instead of the centre, wheel loading is uniform and bending loads are lighter; consequently, the brake beams can be lighter without sacrificing strength.

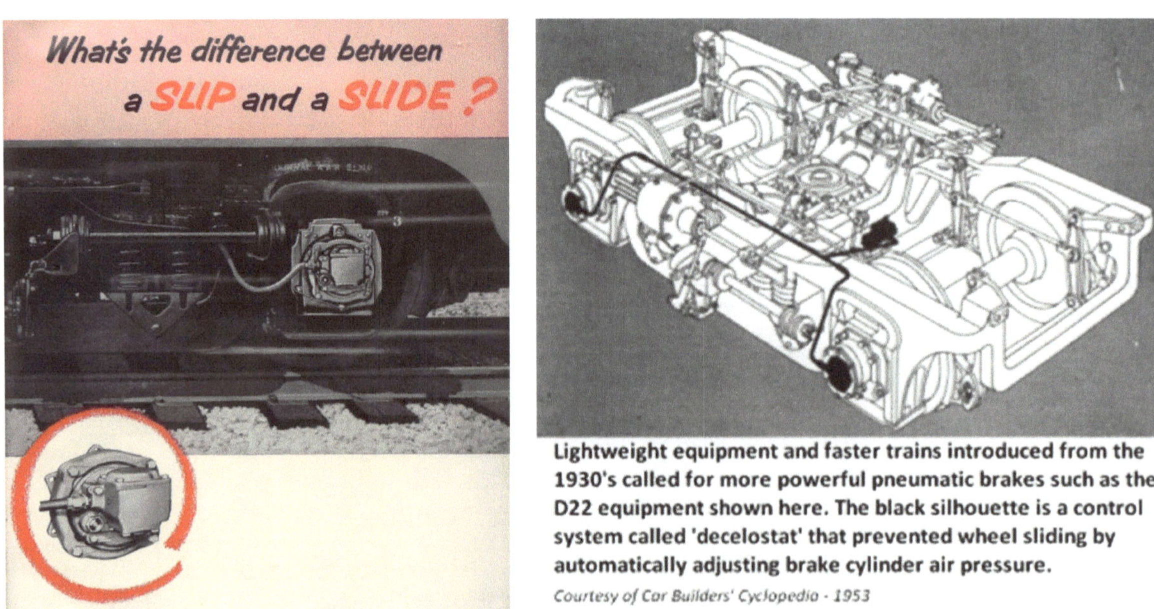

FIGURE 114: DECELOSTAT CONTROL

This equipment differs markedly from the earlier version (Figure 112) in that its layout forms an oblong rectangle with the counteractive forces being applied—via bell-cranks—along the ends to each corner. This rectangle, though, is flexible, and some operators have reported a tendency for the assembly to parallelogram or "tram," with the brake blocks being forced into the wheel tread/flange radius (corner), resulting in improper tread and flange wear. The manufacturer counters this assertion, claiming the problem arises not from any excessive suppleness of their assembly but from the flexibility of the standard three-piece freight bogie. They point out that bogies used in fast intermodal service in the United States, where TMB-60 equipment is extensively and successfully used, are braced to prevent their going "out-of-square".

ALPAC Truck-mounted brake equipment (1067mm)

The functional intent for ALPAC is identical to U.S. WABCOPAC and NYCOPAC, however the clever design features that shed weight while maintaining rigidity are apparent.

Photo courtesy K. Heslop, Knorr-Bremse (Australia) Pty Ltd

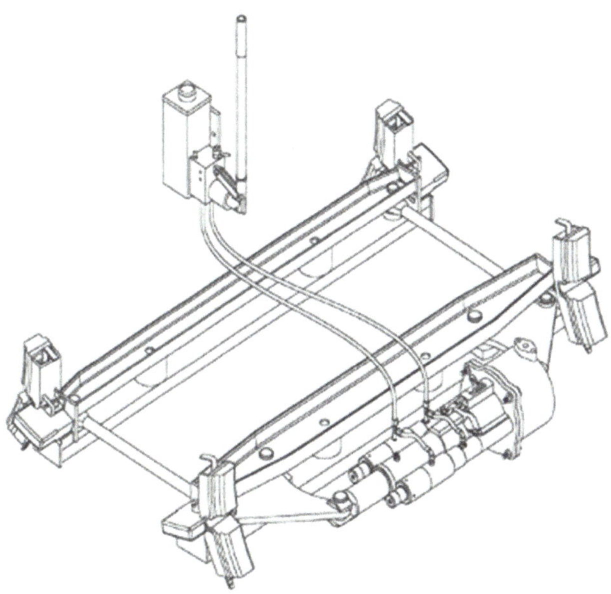

This illustration depicts a second-generation version of truck-mounted brake equipment. This is the NYAB TMB-60 arrangement utilising an hydraulically-operated handbrake.

Courtesy Knorr-Bremse (Australia) Pty Ltd

FIGURE 115: AUSTRALIAN ALPAC AND NYAB TMB-60 EQUIPMENT

The Locomotive Air Compressor

If both the *triple valve* and the driver's *brake valve* were vital components of the Westinghouse air brake system, then the very existence of an air supply to power that system was (and remains) no less vital. Part of this brilliant invention, then, was an on-board means to supply and store air at a useful pressure.

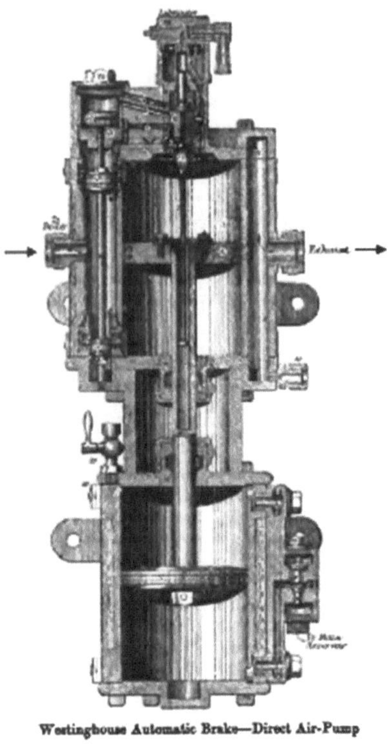

FIGURE 116: WESTINGHOUSE 7-INCH SINGLE-STAGE ('DIRECT') AIR COMPRESSOR. AN EARLY MODEL – ILLUSTRATION

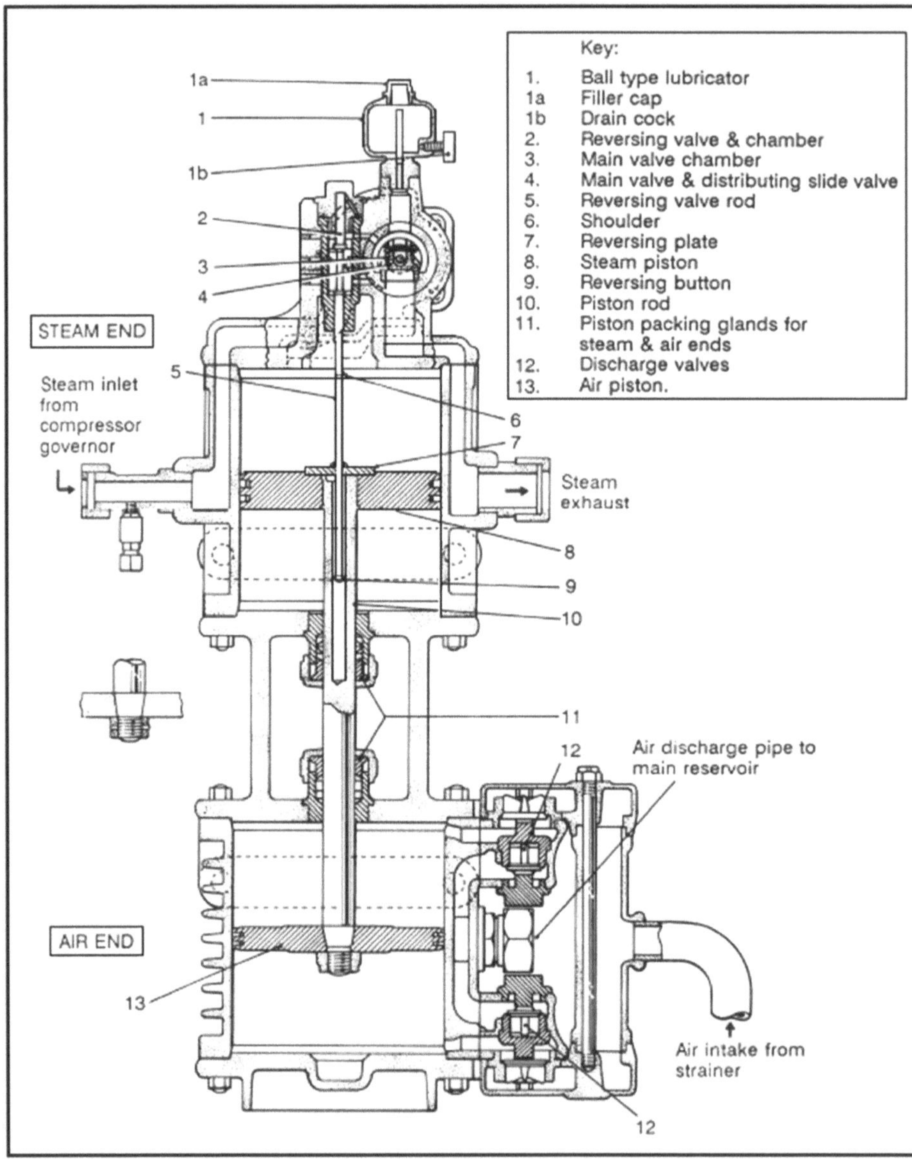

FIGURE 117: WESTINGHOUSE 7-INCH SINGLE-STAGE ('DIRECT') AIR COMPRESSOR. A LATER VERSION – DIAGRAM

The steam locomotive air compressor

For the historical record, the steam locomotive compressor will be studied in some detail. Westinghouse adapted the Cameron steam-driven, reciprocating industrial water pump, replacing the hydraulic cylinder with an air cylinder and adding a receiver system (eventually known as the *main reservoir*). Originally, the locomotive air compressor—which is an 'engine' in its own right—was a relatively humble machine having a single six-inch steam cylinder diameter, but by the end of the steam era in North America, compressor systems were no longer so unpretentious. As train lengths increased, so did the demand for huge volumes of compressed air, resulting in the compressor increasing incrementally in size to 8, then 9½, then 11 inches, until a machine that used its steam supply twice to compress its inlet air supply twice was required, and eventually (at least in the United States) two per locomotive were required! This machine was the **cross-compound** air compressor, and these large air pumps began to pose clearance problems until, in 1919, the Pennsylvania Railroad set a precedent by mounting them on the smokebox front (Figure 118, indicating the ultimate application of this practice). Cross-compound pumps comprised two steam cylinders and two air cylinders being arranged such that the high-pressure steam cylinder was placed above the low-pressure air cylinder, while the low-pressure steam cylinder was located above the high-pressure air cylinder (each vertical combination of pistons operating on a common rod) - hence, the *cross-compound* designation.

Steam passed first through the high-pressure cylinder and then through the low-pressure cylinder so that considerably less steam was required than in the case of a single-stage air compressor. In fact, while this compressor has a capacity of more than three times that of a single-stage unit, the steam consumption per 100 cubic feet of air compressed was about one third.

FIGURE 118: DUAL CROSS-COMPOUND AIR PUMPS MOUNTED ON THE FRONT OF A LOCOMOTIVE SMOKEBOX. *Henry B. Comstock, Galahad Books, NY*

The compressor assembly included a hydrostatic top-head steam cylinder lubricator and a steam-driven compressor governor to control the pressure delivered to the main reservoir(s). While some large North American steam locomotives may have required two cross-compound compressors to satisfy the air requirements of their lengthy trains, in Australia and New Zealand, air consumption requirements were slightly more modest, and compressor installations were either one or two single-stage machines (in the latter case, arranged in series) or one cross-compound unit mounted on the boiler running board or behind streamlined valances or headstock shields. Single-stage compressors were typically 9½ or 11 inches (High-Pressure steam cylinder diameter), while cross-compound pumps were either 7 or 8½ inches.

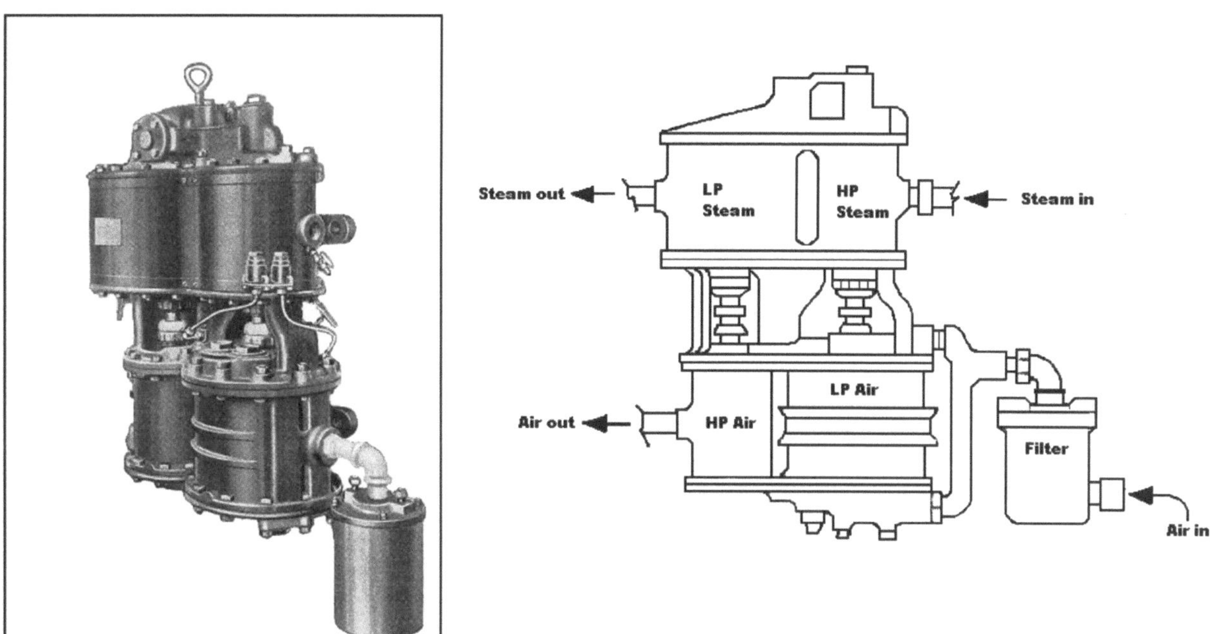

FIGURE 119: CROSS-COMPOUND AIR COMPRESSOR – ILLUSTRATION AND GENERAL ARRANGEMENT

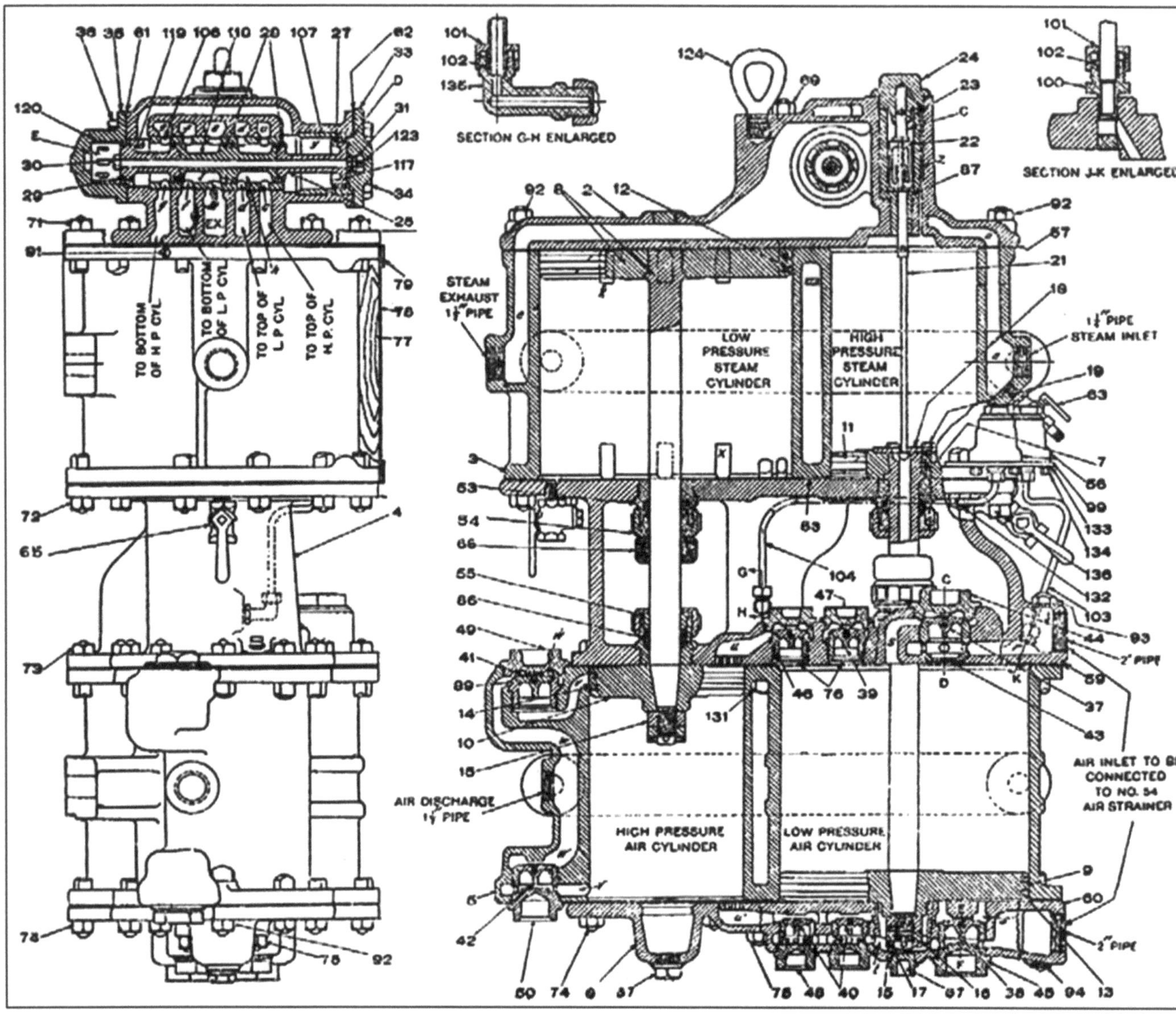

LEGEND

2. Top head.
3. Steam cylinders.
4. Centre piece.
5. Air cylinders.
6. Lower head.
7. High pressure steam piston.
8. Low pressure steam piston.
9. Low pressure air piston.
10. High pressure air piston.
11. High pressure steam piston ring.
12. Low pressure steam piston ring.
13. Low pressure air piston ring.
14. High pressure air piston ring.
15. Piston rod nuts.
16. Piston rod lock nut.
17. Piston rod cotter.
18. Reversing valve plate.
19. Reversing valve plate set screws.
21. Reversing valve rod.
22. Reversing valve.
23. Reversing valve chamber bush.
24. Reversing valve chamber cap.
25. Main piston valve.
33. Large piston cylinder cover.
35. Small piston cylinder cover.
37. Upper inlet valve.
38. Lower inlet valve.
39. Upper intermediate valve.
40. Lower intermediate valve.
41. Upper discharge valve.
42. Lower discharge valve.
53. Stuffing box.
54. Stuffing box nut.
55. Stuffing box gland.
56. Automatic air cylinder lubricator.
63. Drain cock.
65. Drain cock.
64. Drain cock.
67. Lower head plug.

FIGURE 120: 8½ INCH WESTINGHOUSE CROSS-COMPOUND AIR COMPRESSOR. DELIVERY: 150 CUBIC FEET/MIN – WORKING DIAGRAM

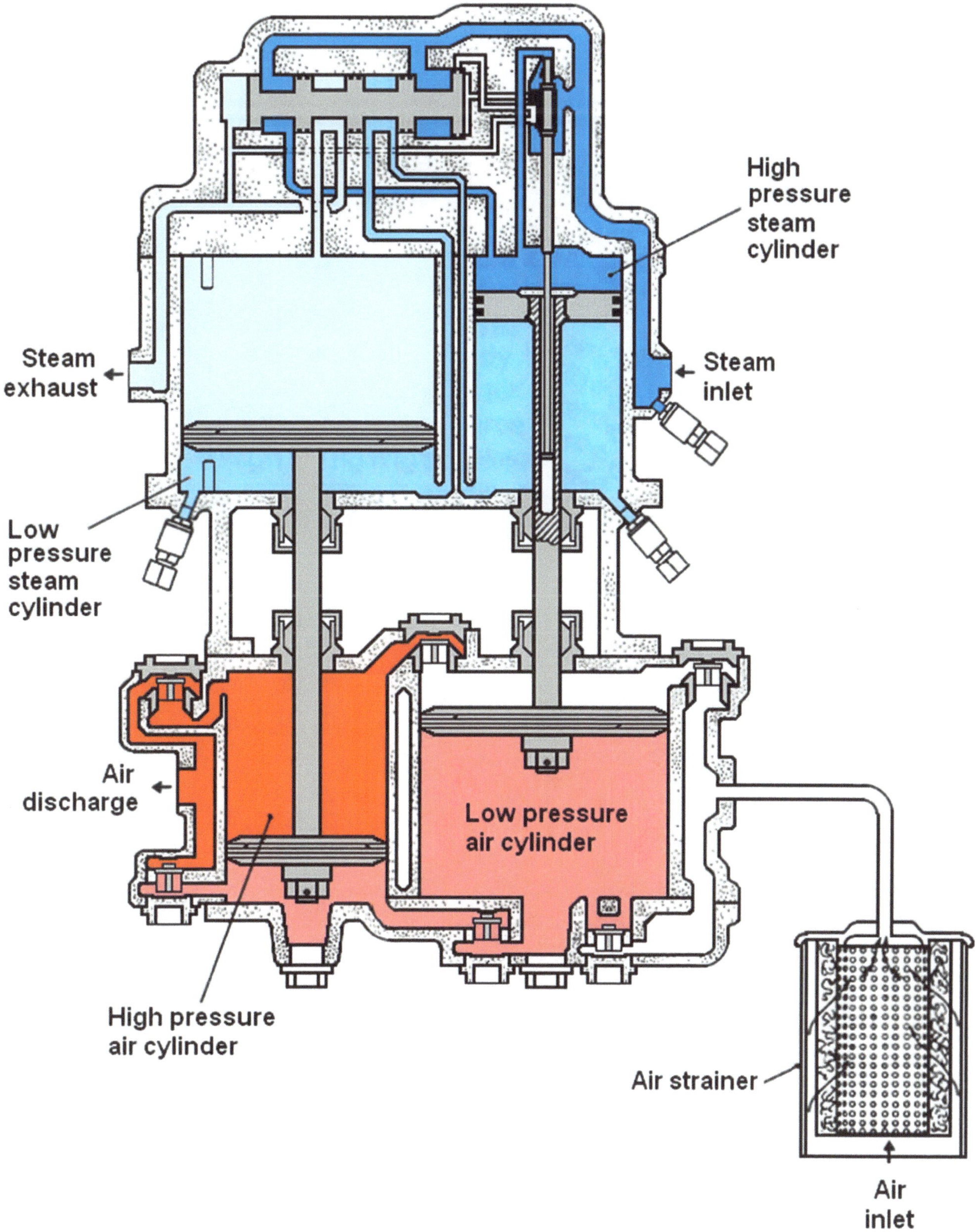

FIGURE 121: 7-INCH CROSS-COMPOUND AIR COMPRESSOR – SIMPLIFIED CUTAWAY DIAGRAM. SHOWING COMMENCEMENT OF DOWNWARD STROKE OF H-P STEAM (AND THEREFORE ALSO L-P AIR) AND UPWARD STROKE OF L-P STEAM (AND THEREFORE ALSO H-P AIR)

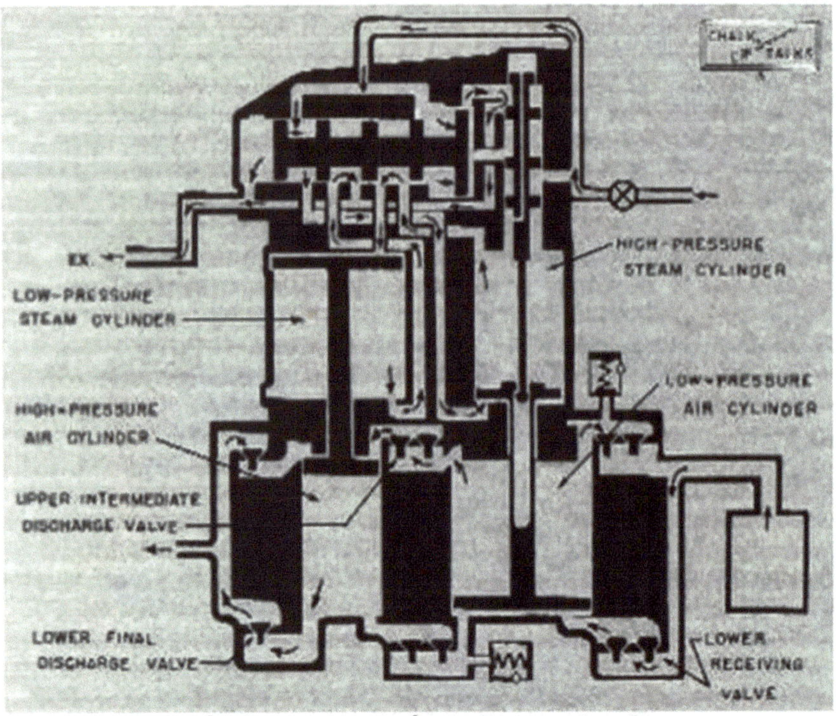

High-Pressure steam piston – Downward stroke

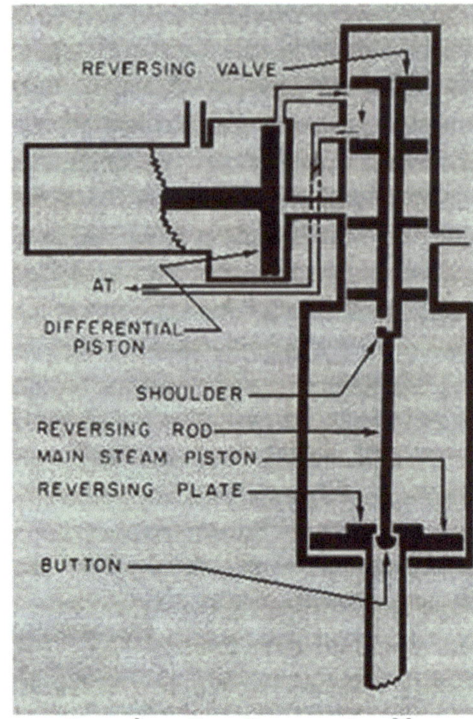

Reversing valve – Downward position

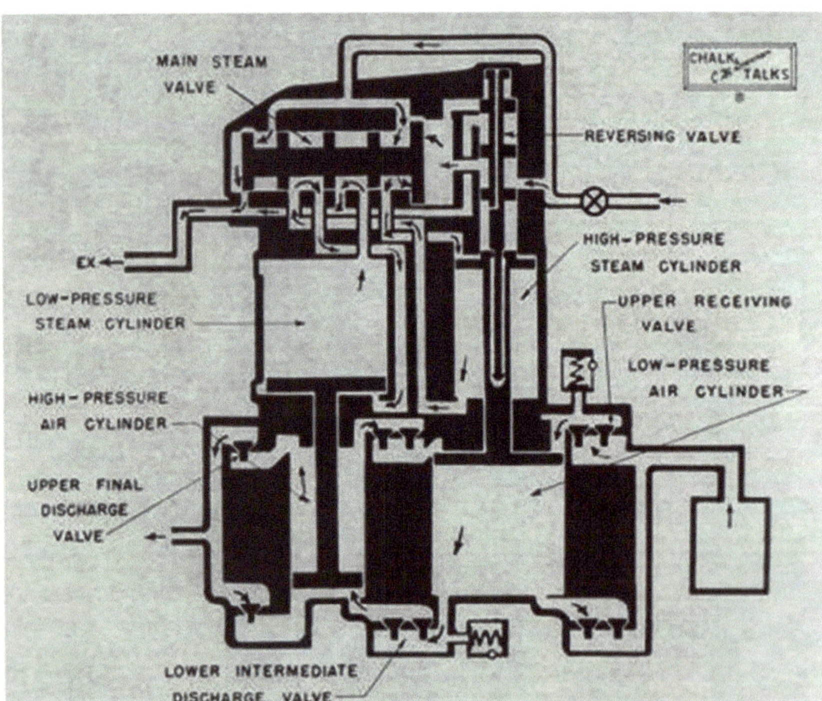

High-Pressure steam piston – Upward stroke

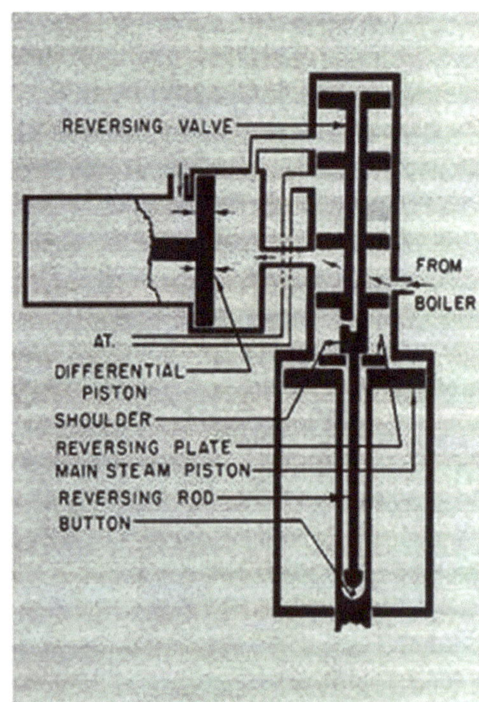

Reversing valve – Upward position

This set of diagrams, from Santa Fe Railway (and later Westinghouse) air brake instructor C.M. Drennan's historical "Chalk Talk" series of courses (© *Simmons-Boardman Publishing*), provide a readily understandable means of explaining the the operation of the Westinghouse cross-compound air compressor. They can be directly-related to the previous diagram depicting an 8½ compressor as fitted to many Australian steam locomotives. New Zealand and Queensland steam locomotives were equipped with the 7 inch version.

FIGURE 122: CROSS-COMPOUND AIR COMPRESSOR INSTRUCTION. HOW AMERICAN RAILROADERS WERE TAUGHT

Compressor Operation

The steam end: To effect a downward stroke of the H-P steam piston, steam entering the inlet port flowed to the top of the piston forcing it down. Steam remaining under this piston from the previous stroke was forced beneath the L-P steam piston forcing it up; steam remaining above this latter piston from the previous stroke was exhausted to atmosphere. When the L-P piston reached the end of its upward stroke, the steam driving it was bypassed through three grooves to its upper face, thereby preventing an accumulation of backpressure below the H-P piston due to possible ring leakage.

When the H-P piston reached the end of its downward stroke, the **reversing valve plate** (see Figures 120 and 122) struck the button on the end of the **reversing rod**, moving the rod and **reversing valve** to a position by which steam was then supplied to the **main valve**, which was positioned to provide steam pressure to the underside of the H-P piston to force it up. The upward movement of the H-P steam piston and resultant actions were accomplished similarly.

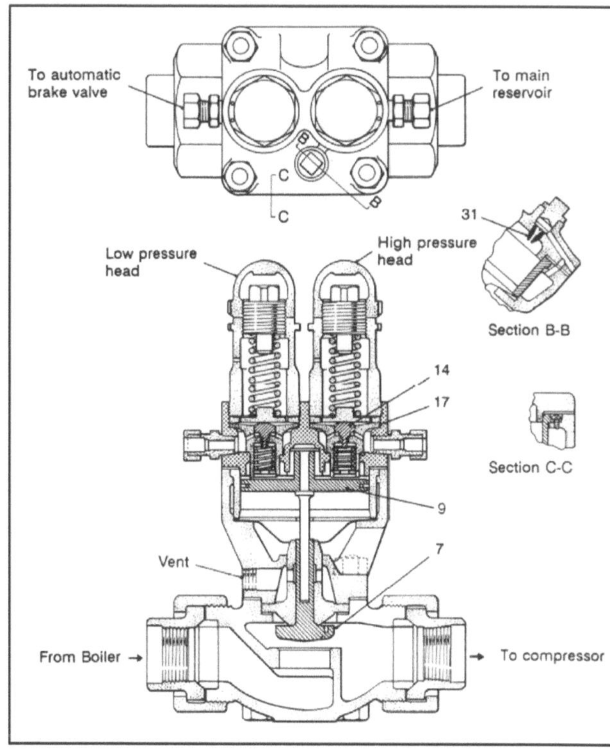

FIGURE 123: DUPLEX AIR COMPRESSOR GOVERNOR – DIAGRAM

The air end: During the foregoing operations, the L-P and H-P air pistons moved with the H-P and L-P steam pistons, respectively. Inlet air entered the L-P cylinder via the two **inlet valves** and was compressed, passing thence via the two **intermediate valves** to the H-P air cylinder where it was further compressed before being forced through the two **discharge valves** to the main reservoir.

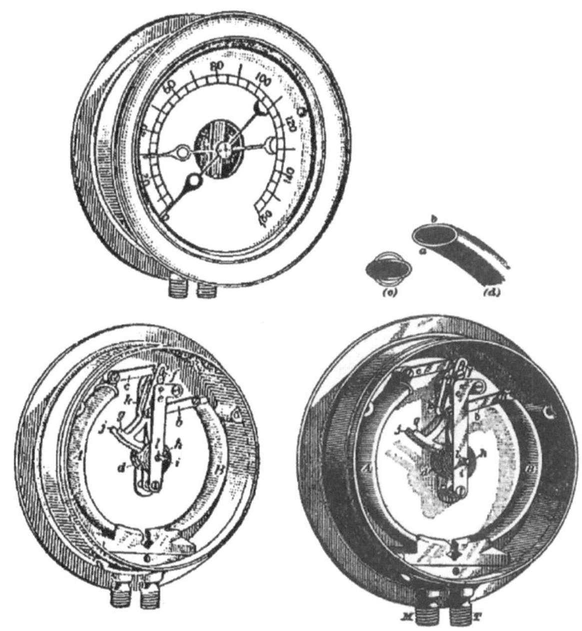

FIGURE 124: DUPLEX AIR BRAKE GAUGE

Once produced, information about the pressure and quantity (which, in this context, is much the same thing) of the compressed air stored on the locomotive needs to be available to the driver at all times, since most air braking decisions and actions will be taken on the basis of this knowledge. From the earliest days of air braking, this has been managed per medium of the 5-inch (later 3½-inch) duplex air pressure gauge, which indicates both brake pipe and main reservoir pressures. With a brass (later pressed steel) case and silvered dial, this instrument is still used today.

The gauge pointers are operated by two bent elliptical tubes having their lower ends held fast and their upper ends closed and free. When air pressure is supplied into these tubes, they tend to straighten, thus imparting movement to their free ends. Via a linkage, these free ends are attached to a toothed sector that engages with a spindle, thus rotating it. The pointer, attached to the spindle, is moved similar to the hands of a clock.

FIGURE 125: NSWGR 38-CLASS LOCOMOTIVE SHOWING 8½-INCH CROSS-COMPOUND AIR COMPRESSOR. *Leon Oberg*

FIGURE 126: NZGR JA-CLASS LOCOMOTIVE SHOWING 7-INCH CROSS-COMPOUND COMPRESSOR. *NZ Railways Publicity*

The diesel locomotive air compressor

Diesel locomotives have traditionally utilised two-stage, three-cylinder reciprocating air compressors—either water- or air-cooled—driven by an engine crankshaft extension or, often in smaller locomotives, a single-stage compressor belt-drive from the engine. In some cases, rotary screw compressors have supplanted the reciprocating machines and, more recently, locomotive microprocessor control schemes have permitted the introduction of AC electric-driven compressors that help reduce fuel consumption by switching off the machine when it doesn't need to pump, thereby reducing parasitic loads on the locomotive prime mover. Electric locomotives, of course, use an electric compressor drive that was simply switched on and off as required. This section offers some means of comparison with steam compressors; however, since information on diesel-electric locomotive compressors is commonly available, this equipment will not be discussed in great technical detail.

Whereas steam locomotives use just a pressure-controlled **governor** to control the pumping cycle of the compressor (which is accomplished by controlling the provision of the compressor's steam supply), diesel compressor systems require both a **governor** and an **unloader**. When engine-driven, a compressor will be rotating so long as the engine is running, even though air pressures may be within range and further production of compressed air is not required. In this case, the governor signals an electro-pneumatic switch that holds the inlet valves open so the compressor pumps to atmosphere instead of into the main reservoir. Even state-of-the-art diesel locomotive compressor control systems—that simply switch the electric drive on and off in response to a pressure control signal—may continue to use an unloader, as does General Electric. Here, the **unloader** functions only momentarily at compressor start-up to provide a pneumatic "soft-start", thus reducing the initial electrical load on the compressor drive motor.

The two-stage, cross-compound steam locomotive air compressors being used in Australia and New Zealand at the close of regular steam operations provided air at around 2 250–2 820 litres per minute (80–150 cu. feet/min) up to 840 kPa (120 psi) using approximately 120–130 single strokes per minute. Diesel locomotive air compressors today typically deliver net air quantities in the range of 4 500 l/m (160 cubic ft/minute) to 6 490 l/min (230 cfm) up to 1 000 kPa (140 psi) at 1 050 rpm when engine-driven.

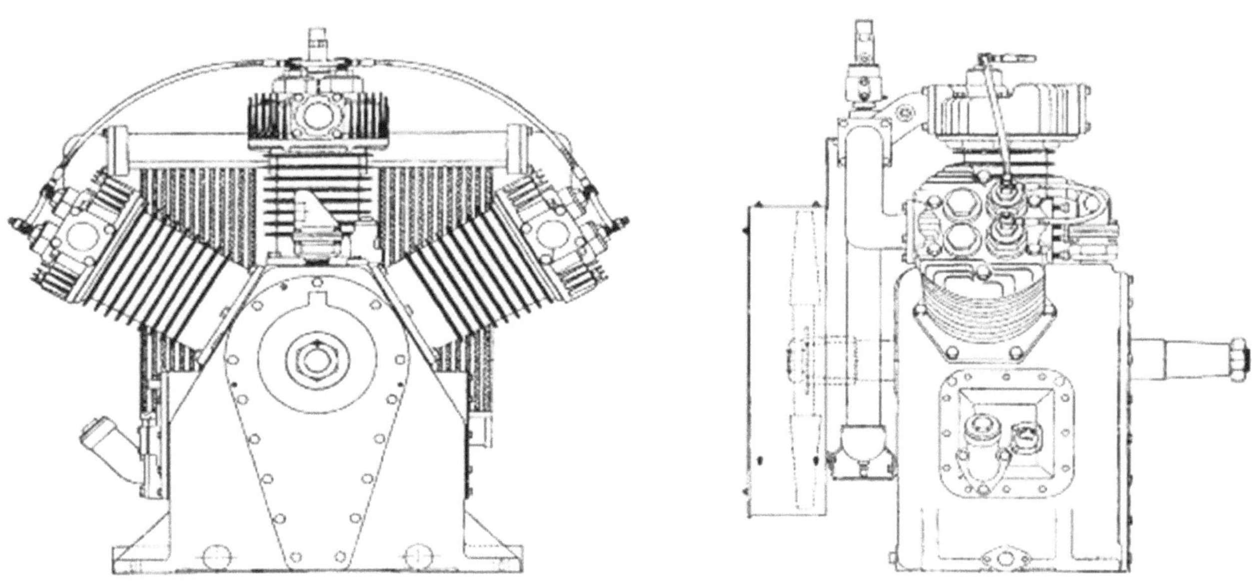

FIGURE 127: WABCO 3-CYLINDER, 2-STAGE, AIR-COOLED DIESEL LOCOMOTIVE AIR COMPRESSOR - GENERAL ARRANGEMENT. THIS DIAGRAM DEPICTS THE OPTIONAL LARGER CRANKCASE AND FAN-COOLED INTERCOOLER

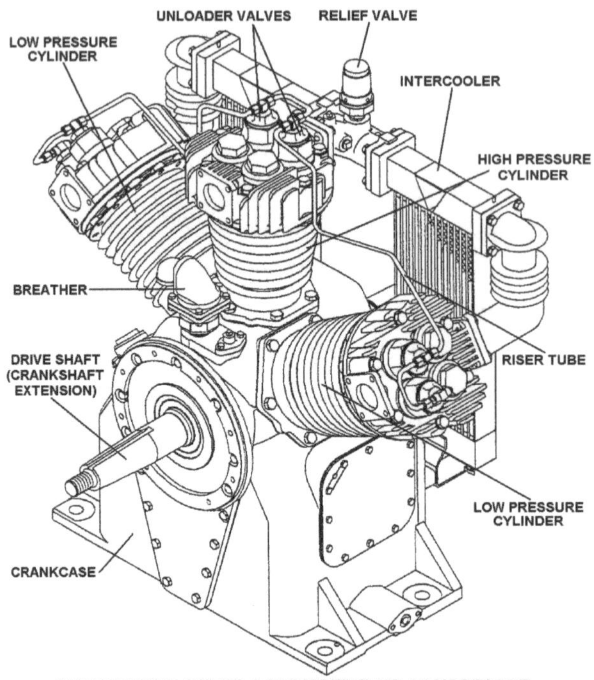

WABCO 3CDCL DIESEL LOCOMOTIVE AIR COMPRESSOR
(Note mounting flange and crankshaft extension to accommodate an ac electric drive motor)

FIGURE 128: WABCO 3-CYLINDER, 2-STAGE, AIR-COOLED DIESEL LOCOMOTIVE AIR COMPRESSOR. THREE-QUARTER VIEW

The Straight-Air Brake

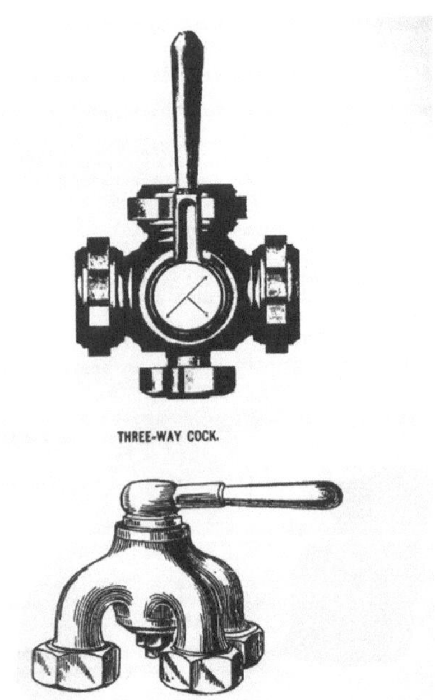

Photos Courtesy Westinghouse Air Brake Co.

FIGURE 129: ORIGINAL LOCOMOTIVE ENGINEER'S 3-WAY AIR BRAKE COCK FOR STRAIGHT-AIR BRAKE

The first locomotive power brakes were steam operated. With the locomotive boiler substituting for the main reservoir, this arrangement might well have been referred to as a *straight-steam* system. Once it became necessary to carry an on-board compressed air supply and storage system, train brakes became air-operated and were known as the **straight-air** type. This air brake, used from around 1868, was known in the United States as the Schedule A and has already been briefly described in an earlier section; however, a more detailed account is appropriate here.

The 'straight-air' designation derives from the fact that air is supplied—during a brake application—straight from the main reservoir (via the engineer's 3-way cock or brake valve) into the locomotive cylinder(s) and via flexible hose connections to the brake cylinders on the cars. In fact, when the straight-air brake first came into use, locomotives were not equipped since crews were used to controlling locomotives with the reverser and tender handbrake, and it was felt there was too great a risk of driving wheels being slid and flattened.

The *locomotive-only* component of this system, though, was soon provided as the need for maximum braking power increased. A braking system for the locomotive itself was developed and later greatly improved, even in some cases becoming a **relayed** system (while incorrectly retaining the *straight-air* nomenclature). Later, locomotive brake setups were significantly redesigned and became known as **independent** brake systems, chiefly because they could be controlled independently of the Automatic train brake, even during an Automatic brake application. Modern-day locomotive Independent brake systems no longer strictly qualify as straight-air brakes since the operating pressure is not admitted 'straight' to the cylinder but is supplied merely as a control pressure to a brake cylinder relay valve, the latter device supplying brake cylinder pressure.

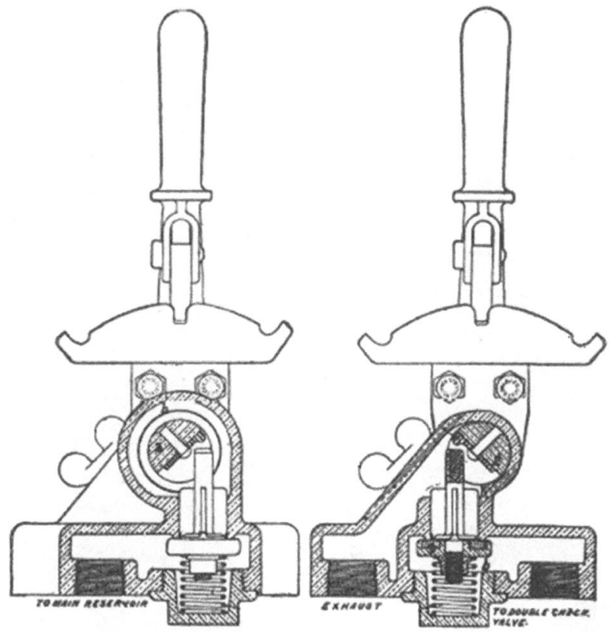

FIGURE 130: EARLY AMERICAN STRAIGHT-AIR BRAKE VALVE - SIDE VIEW

The earliest locomotive power brake systems had only one brake cylinder to supply. Figure 16 on page 20 depicts an early British version of a straight-air brake valve that is so basic, it is really little more than a redesign of the three-way cock mentioned earlier. Eventually, as ever-larger steam locomotives were built and fitted with Automatic air brake equipment, more than one engine brake cylinder was required, and brake cylinders were provided also on the engine truck(s) and tender (Schedules A-1, AD, and AG). These and later Independent brake arrangements are inevitably relay systems whereby the main reservoir feed supplied through the driver's valve is used as a "control" supply only. This feed—relatively low-volume and thus more controllable—was impressed upon a relay valve that, in turn, permitted a much higher volume to flow from the main reservoir to the locomotive brake cylinders. That said, as alluded to in the opening paragraph of this section, conventional nomenclature is sometimes no safer today than in earlier decades. The term 'straight-air brake' is still sometimes used in Australia to identify what is actually an Independent brake.

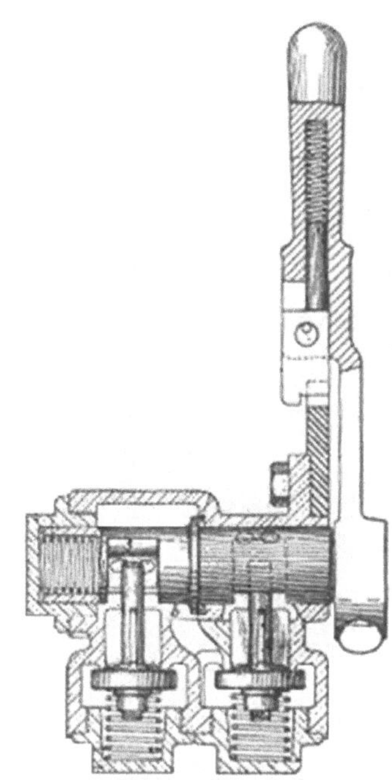

FIGURE 131: EARLY AMERICAN STRAIGHT-AIR BRAKE VALVE - END VIEW'

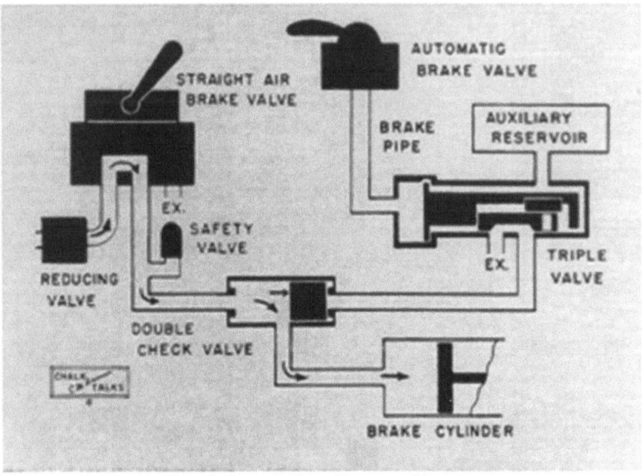

Straight air brake application

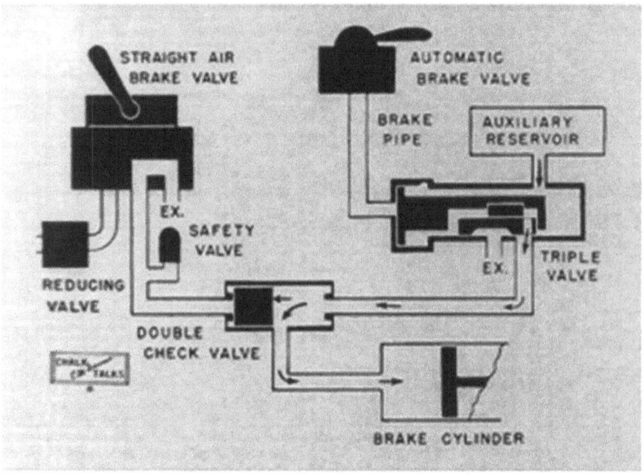

Automatic brake application

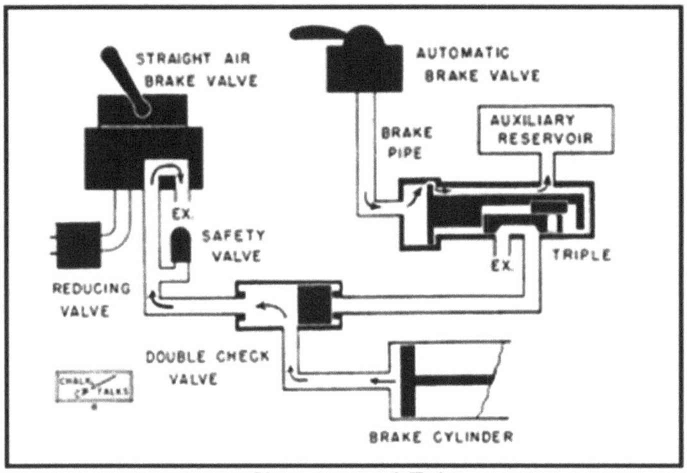

Charging and Release

Diagrams courtesy Simmons Boardman Publishing

FIGURE 132: COMBINATION STRAIGHT-AIR AND AUTOMATIC AIR BRAKE ARRANGEMENT – *CHALK TALKS*

On a locomotive engaged in shunting, the Automatic air brake system used air from the locomotive auxiliary reservoir as a brake cylinder supply. After a few applications and releases, the depleted auxiliary reservoir caused the locomotive to lose its braking power and required to be recharged. The straight-air brake was different in that it supplied main reservoir air directly to the brake cylinders on the engine and, where applicable, the tender. This provided the driver with greater control of the locomotive during manoeuvers where the Automatic brake, with its operational time lag, was either not required or not warranted; for example, in the aforementioned shunting operations. In due course, a combination Straight *and* Automatic air brake (Schedule SWA for locomotives and SWB for tenders, Figure 132) was developed so that—by means of a double-check valve in the pipe leading to the brake cylinders—the engine brake could be operated 'straight-air' independently of the Automatic. A reducing valve was used to supply main reservoir air to the straight-air brake at reduced pressure. In Australia and New Zealand, this was typically 45 psi (315 kPa).

The straight-air brake was a mature system by the time of the changeover from steam to diesel traction and is now operated—wherever it remains in use—only on the locomotive, as distinct from locomotive and train per its original application. When the brake was applied, main reservoir pressure flowed through a reducing valve, the driver's straight-air brake valve, and then through a double-check valve into the brake cylinders on the engine and tender. The reducing valve was usually identical to and acted as a feed valve for the locomotive brake. The double-check valve had three connections and permitted air pressure to flow to the locomotive brake cylinders from either the straight-air brake valve when applied or the auxiliary reservoir in the case of an Automatic brake application - but not both at the same time. In the United States, a straight-air brake valve was eventually developed to incorporate an extra double-check valve, thus permitting the locomotive brakes to be released without affecting the Automatic brake on the train. In Australia and New Zealand, this function was performed on this early generation of equipment by the Independent brake valve RELEASE position.

A development of the basic straight-air brake valve was the English Westinghouse No. 17 equipment that provided two extra operating positions: SLOW APPLICATION and SLOW RELEASE. This equipment was designed for operating either the straight-air brake on the locomotive only or on a double-braked train (i.e. fitted with the combined Straight-air and Automatic brake). By moving the handle either to one side or the other of the LAP position, gentle graduation of the train brakes—both on and off—was possible. This valve has been used extensively within Australia and can still be seen today in daily operation on locomotives of Tourist & Heritage operator Puffing Billy Railway in the Dandenong Ranges near Melbourne.

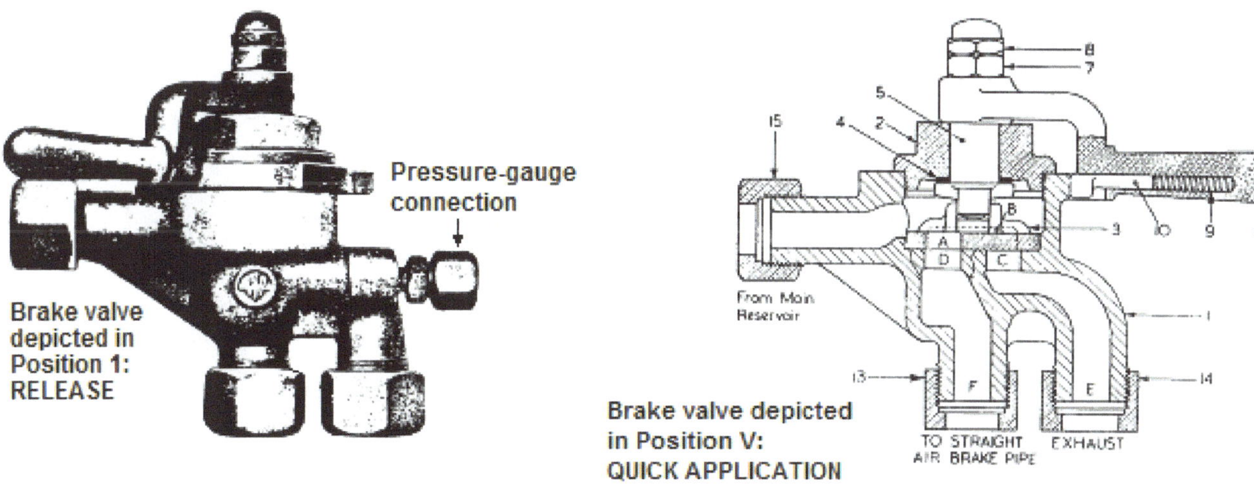

Numbers and Names of Parts

Ref. No.	Name of Part.	Ref. No.	Name of Part.
1.	Body.	9.	Handle Stop Spring.
2.	Main Valve Chamber Cap.	10.	Handle Stop.
3.	Main Valve.	11.	Union Nipple (pressure gauge) (not shown).
4.	Main Valve Spindle Gasket.	12.	Union Nut (not shown).
5.	Main Valve Spindle.	13.	Union Nut (straight air brake pipe).
6.	Handle.	14.	Union Nut (exhaust).
7.	Main Valve Spindle Nut.	15.	Union Nut (main reservoir).
8.	Main Valve Spindle Cap Nut.	16.	Stud and Nut, 2¼" × ⅝" (not shown)

FIGURE 133: THE NO. 17 STRAIGHT-AIR BRAKE VALVE

Well-developed straight-air brake versions, some of them electro-pneumatic, were also applied to railcars/railmotors (see the section describing the WEP brake valve).

Automatic and Independent Brake Valves

To illustrate the evolutionary process that has resulted in the valves used in Australia and New Zealand, this section will briefly examine the early American driver's air brake valves. Note that the terms 'train pipe' (British use) and 'brake pipe' (American use) have the same meaning.

The first Westinghouse driver's Automatic brake valve—type B-11 dating from 1876—is illustrated in Figures 134 and 135. Although it was listed in Westinghouse catalogues from the United Kingdom, I am not

FIGURE 134: THE WESTINGHOUSE B-11 AUTOMATIC BRAKE VALVE

aware that it was used in either Australia or New Zealand. The B-11 was the first application of a *rotary valve*; an attempt to make the brake valve 'automatic' in its termination of the discharge of brake pipe pressure. This brake valve utilised a *trainline discharge valve* (also referred-to in some illustrations as a 'trainline exhaust valve'; see **3** in Figure 135) held on its seat against train pipe pressure by a short, stiff, *compression spring* **4** located in the brake valve handle. This handle incorporated a quick-thread screw that, when the handle was turned to set the brake, relieved the spring so that train pipe pressure would lift the discharge valve and flow to exhaust. Simultaneous to varying the discharge valve spring tension, movement of the brake valve handle turned the *rotary valve* **6** to cut-off the main reservoir feed to the train pipe by closing *supply port* **a**. When train pipe pressure was sufficiently diminished, the compression spring was able to seat the discharge valve and close off the train pipe exhaust. Thus, the amount of reduction on the spring tension—commanded by the distance the handle was rotated or "unscrewed"—controlled the degree of train pipe reduction. Two drawbacks were; (1) if this spring was too weak, train pipe pressure could not escape before the supply ports were "lapped"; and (2) assisting engines engaged in double heading could not be equipped with this brake valve as it could not hold the trainline pressure when the driver on the leading engine released the brake.

For double-heading operations, either a cut-out cock was required in the trainline beneath the brake valve, or it was necessary to make a blind joint there. Referring again to the diagrams, the main reservoir *excess pressure valve* **7** was located within the rotary valve assembly.

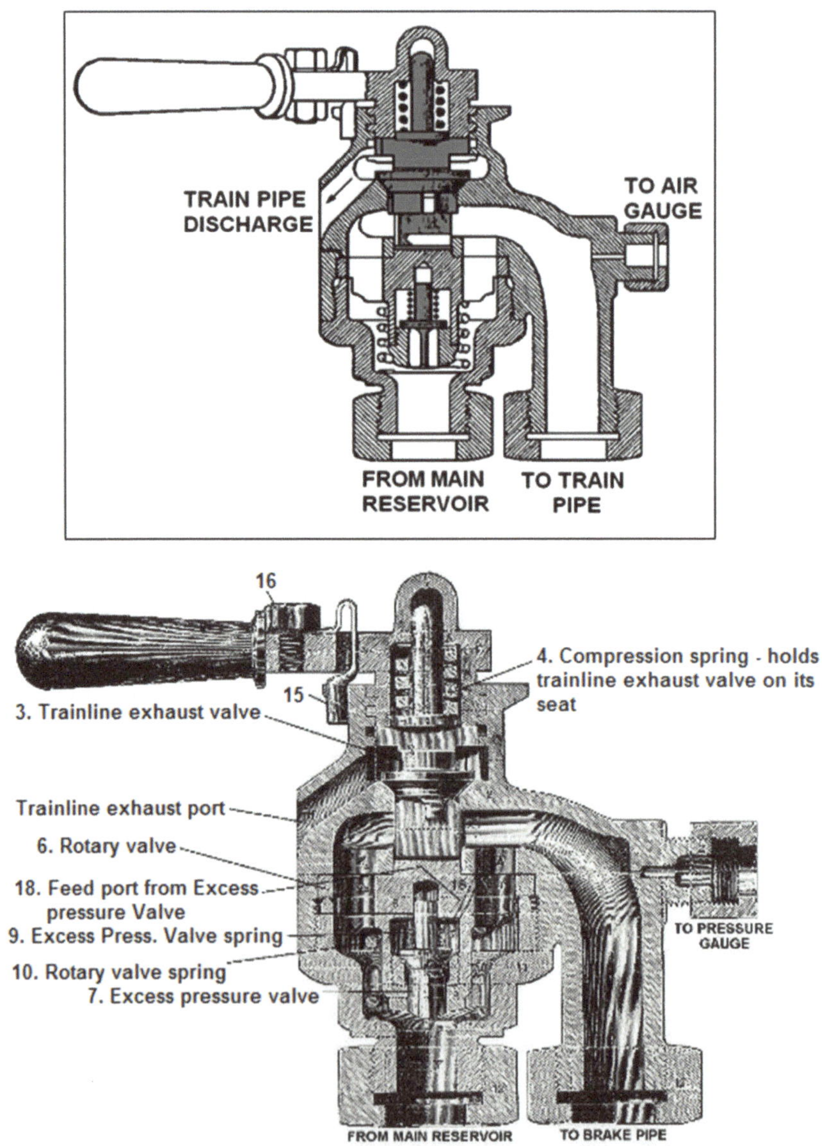

FIGURE 135: THE WESTINGHOUSE B-11 AUTOMATIC BRAKE VALVE – DIAGRAM AND SECTIONED VIEWS

With the brake valve in the Running position, the function of the excess pressure valve was to provide for the retention of an excess of pressure in the main reservoir over that in the train pipe, which was necessary to effect prompt release of all triple valves. *Feed port* **18** (dotted line) supplied the train pipe from the excess pressure valve spring cavity. A later version of the B-11 had an exhaust cavity in the rotary valve that bypassed the trainline discharge valve and connected the train pipe direct to atmosphere when the brake valve handle was at the Emergency position (was this the first "Big Hole"?). The B-11 brake valve possessed what would eventually become "conventional" handle operating positions: FULL RELEASE, RUNNING, LAP, SERVICE APPLICATION, and EMERGENCY. These positions were set and permanent for each use of the brake valve with the exception of LAP and SERVICE, which depended on compression spring tension.

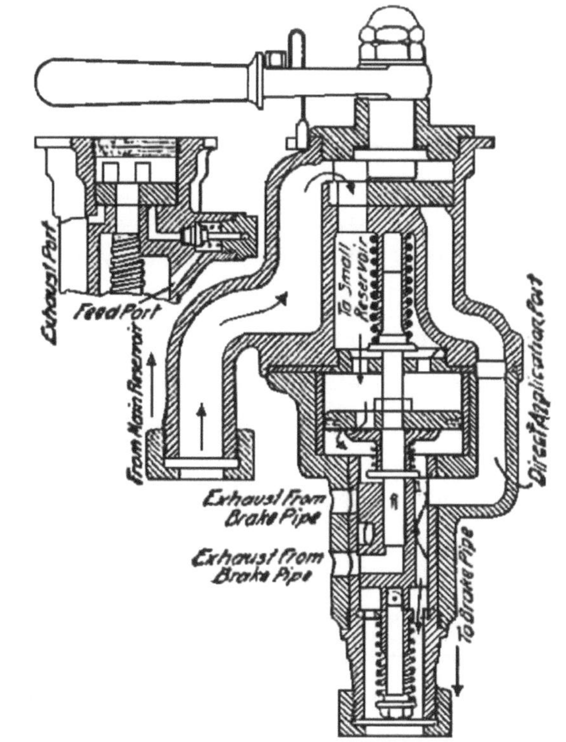

FIGURE 136: THE WESTINGHOUSE B-12 AUTOMATIC BRAKE VALVE - DIAGRAM

The B-11 could be used successfully only on very short trains; the air ports—in particular, the excess pressure valve port **18**—being too small. However, there is little doubt that the greatest drawback of the B-11 was its lack of the vital feature that would be the next (perhaps even the first) great incremental improvement in locomotive air brake equipment: **brake pipe equalising discharge**. Equalising discharge brake valves—early examples in the United States being the D-8 and the D-5 (these devices were also known as the E-6 and the F-6; in all cases the designations having derived from the relevant page number of the applicable Westinghouse parts catalogue)—were provided with a clever and crucial *relay* feature that "rendered independent of the train's length, the time for which the brake valve handle was required to be held in the SERVICE application position in order to obtain a given brake pipe reduction". In other words drivers only needed to know how much of an application they desired and reduce the equalising reservoir by that amount; regardless of the length of their train.

Instead, the driver was able to rapidly reduce (with reference to a pressure gauge) the pressure in a small relay reservoir to an easily measured value, and—since this action permitted movement of a pilot piston attached to a brake pipe exhaust valve—the brake pipe, with a vastly greater volume of air contained therein, had its pressure reduced by a like amount. This pressure egress continued until brake pipe pressure (indicated by a second needle on the same gauge) became equal with that in the small relay reservoir, at which point the brake pipe exhaust valve automatically cut off the exhaust of air. Thus, the small relay reservoir was named the **equalising reservoir**; the pilot piston became the **equalising piston**; and this style of driver's brake valve became known as an **equalising discharge brake valve**.

The 'equalising' function was important. The brake pipe discharge flow had to be gradual and be stopped gently when the brake pipe exhaust valve lapped. A rapid discharge of brake pipe pressure abruptly cut off would be liable to cause an artificial pressure increase in the brake pipe near the front of the train which, in turn, was likely to cause triple valves on vehicles near the front of the train to go to Release. To obviate this, the internal brake pipe exhaust (or discharge) valve assembly was arranged to ensure this latter valve always closed gradually—driven by the equalising piston—when equilibrium of pressure (i.e. brake pipe to equalising reservoir) was established across it.

The first such brake valve was the B-12, dating from 1886. It had an **equalising piston** with air pressure on both sides that drove a **slide valve** to provide a controlled brake pipe exhaust, and was also provided with a **direct-application port** that was used in an Emergency application only. The chamber above the equalising piston had a small reservoir attached to increase its volume. In the FULL RELEASE and RUNNING positions, main reservoir pressure passed through this chamber and then some small holes in the equalising piston en route to the brake pipe (note the line of arrows in the diagram).

There was an **excess pressure valve** in one side through which main reservoir pressure had to pass when the handle (and rotary valve) was in RUNNING position. In 1887 the C-7 automatic brake valve was introduced to improve upon the B-12. In this valve, main reservoir pressure passed through the rotary valve and its seat into the chamber connected to the brake valve reservoir and thence via openings in the equalising piston as in the B-12. The slide valve and springs of the B-12 were deleted and an **equalising discharge valve stem** incorporated instead, thus simplifying construction and making the brake valve more certain in its action. A shortcoming, however, was that its perforated equalising piston was troublesome as any dirt or foreign substance fouling its underside would allow air to pass either way, rendering the piston unresponsive to moderate pressure variations. To control leakage from its brake pipe exhaust elbow, a stopcock was provided that could be closed if the piston failed to seat the discharge valve. One inevitable result was that the stopcock was often kept closed and the brake valve manipulated by drivers into its EMERGENCY position in order to effect Service brake pipe reductions.

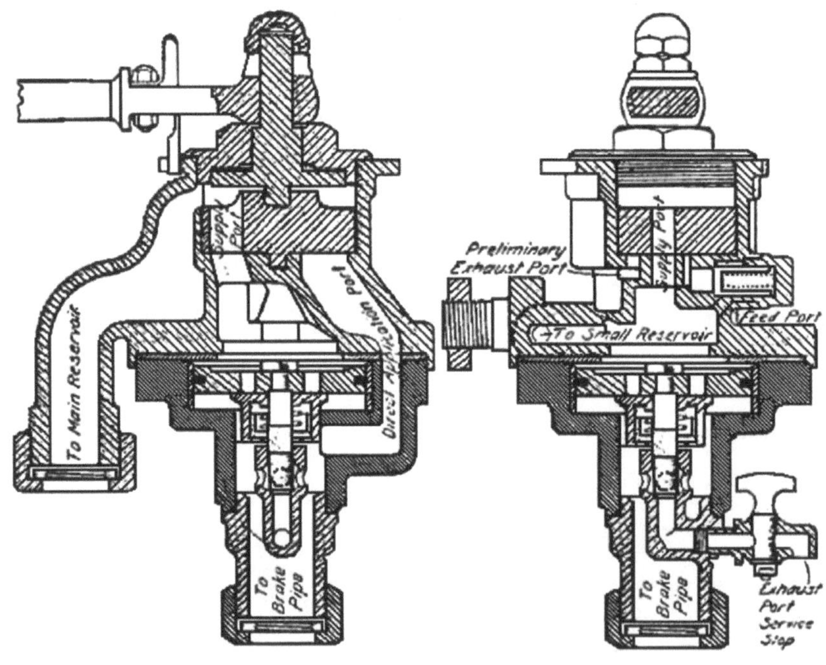

FIGURE 137: THE WESTINGHOUSE C-7 AUTOMATIC BRAKE VALVE – DIAGRAM

By 1889 the D-8 brake valve had appeared, designed to overcome the objectionable features of the preceding model. Instead of passing through the upper equalising piston chamber on its way to the brake pipe, main reservoir pressure was directly connected to the brake pipe (in FULL RELEASE position of the brake valve handle) via a port and cavities in the rotary valve and seat and a **direct-application-and-supply port**.

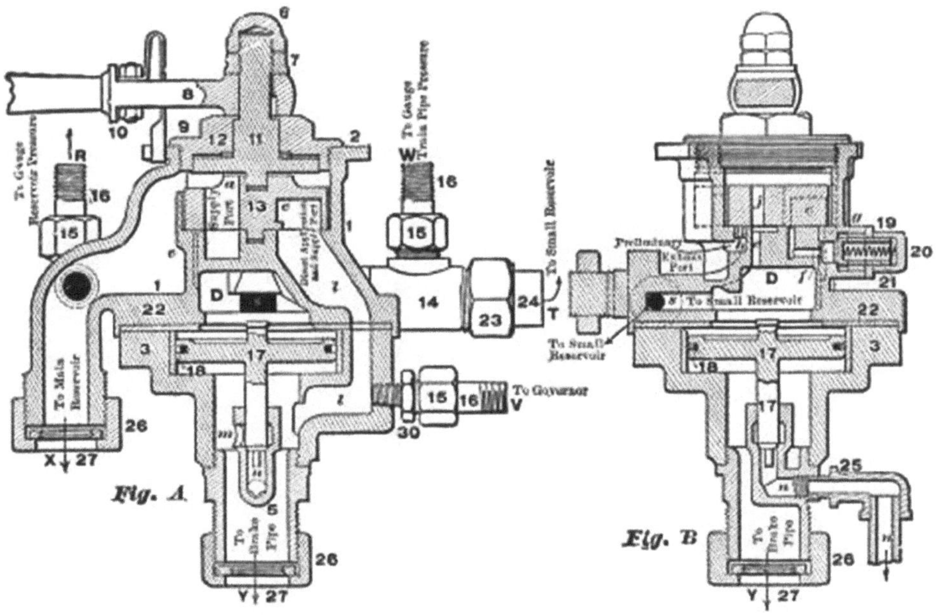

FIGURE 138: THE WESTINGHOUSE D-8 AUTOMATIC BRAKE VALVE – DIAGRAM

By 1892 the D-8 brake valve, beset by deficiencies that had become obvious as train lengths continued to increase, was superseded by the D-5 brake valve (also referred to—perhaps more logically—in consecutive Westinghouse catalogues as the E-6 and then the F-6). A **feed valve** or brake pipe governor assumed the function of the excess-pressure valve; port arrangements in the rotary valve and seat were altered; and the projection of the equalising discharge valve stem below the brake pipe exhaust valve seat—previously parallel—was now tapered.

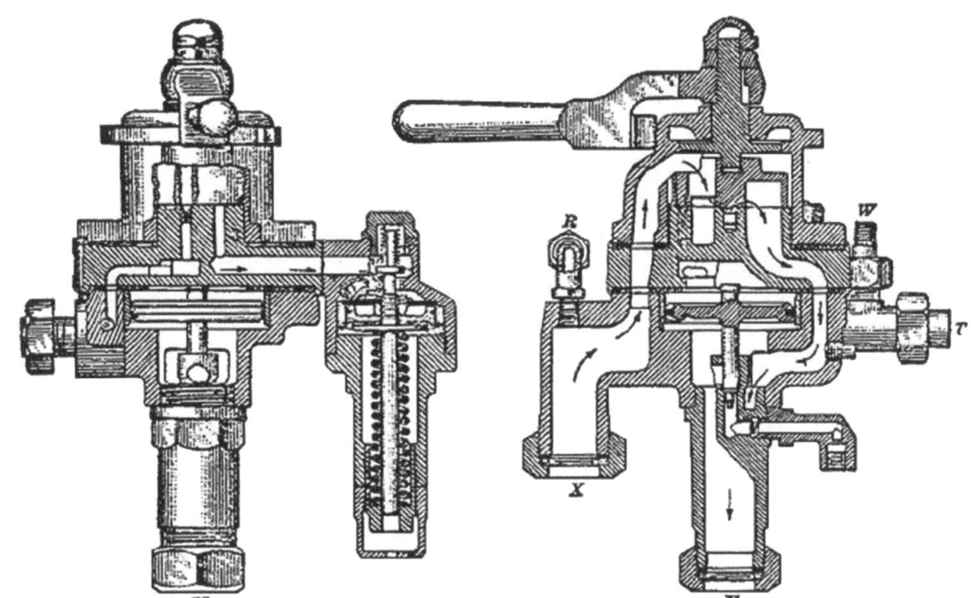

FIGURE 139: THE WESTINGHOUSE D-5 AUTOMATIC BRAKE VALVE – DIAGRAM

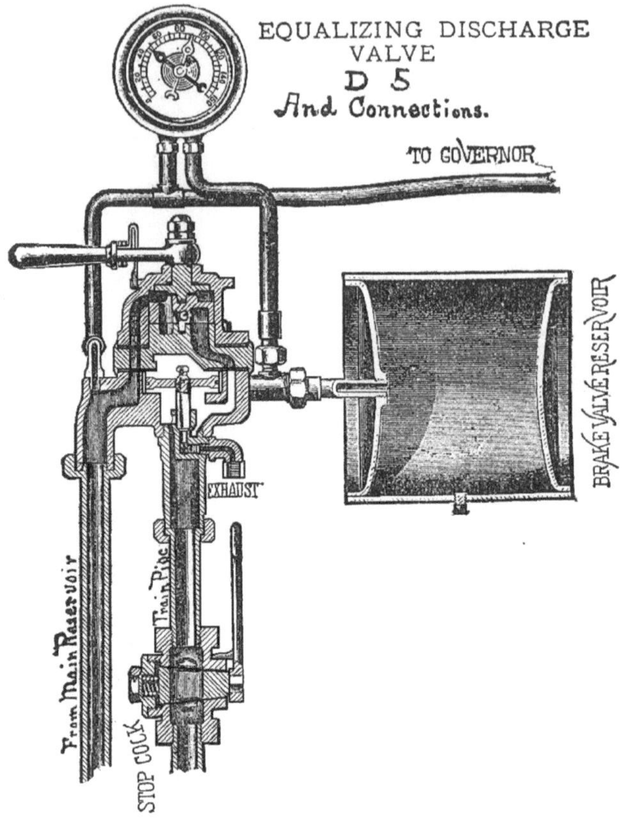

FIGURE 140: THE EQUALISING DISCHARGE VALVE. THIS EARLY DIAGRAM FROM AN AMERICAN AIR BRAKE HANDBOOK DEPICTS THE D-5 AUTOMATIC BRAKE VALVE AND PIPE CONNECTIONS. TO MAKE THE DIAGRAM CLEARER, THE EQUALIZING PISTON UPPER AND LOWER CHAMBERS HAVE BEEN WHITENED.

As mentioned, the D-5 brake valve was redesignated the E-6 and then the F-6, simply by listing it as such in successive Westinghouse catalogues. This process had effectively generated what looked like a logical sequence of successive models and the Westinghouse Air Brake Company utilised this to maintain the progression. The next model in this progression was the G-6 that differed from its predecessor only by the type of feed valve used - in the latter case, a slide valve. It was regularly furnished with the old-standard A-1, AD, and AG engine brake equipment.

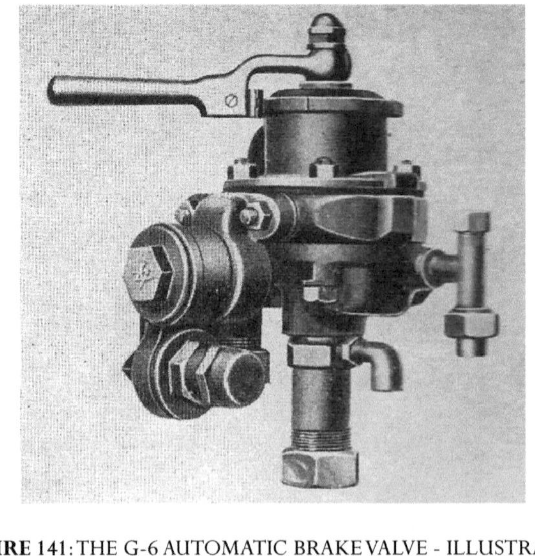

FIGURE 141: THE G-6 AUTOMATIC BRAKE VALVE - ILLUSTRATION

The H-5 followed, becoming the brake valve for the new 5-ET equipment and, although modelled on previous valves, differed significantly in detail in order to obtain the desirable operating features of the new **distributing valve** used in connection with it. Among other features, the H-5 introduced the HOLDING position of the rotary valve, which was so-named because with the brake valve handle in this position the locomotive brakes could be held applied while the train brakes were released and recharged to feed valve pressure. All ports but one—port ***l*** in the H-5 or ***p*** in the H-6—communicating via the Independent brake valve to the release pipe (Independent Release pipe **4**) and the application cylinder (Control Pipe **3** to the relay/application portion) register as in the RUNNING position. However, the latter port being closed, any application cylinder pressure that may have resulted from a previous service application was bottled, thus preventing the release of locomotive brake cylinder pressure. Therefore, the only difference between the two positions was that in RUNNING, the application chamber was open to atmosphere (and locomotive brakes therefore released), while in HOLDING position, it was not (locomotive brakes therefore held applied).

The № 4 Equipment

Although outstanding in their concept and design, these early equalising brake valves were relatively simple in both construction and operation. The first such device to become widely used in both Australia and New Zealand was the № 4 Equalising Brake Valve adapted from early US designs by the Westinghouse Brake and Signal Co. in England; the major difference being the positioning of the equalising portion *beside* the handle and rotary valve portion rather than beneath it in the same housing. Internally, the assembly is conventionally first-generation air brake valve, consisting in its upper part of a rotary valve with main reservoir pressure above it. In the usual fashion, the rotary valve is connected to the operating handle by a spindle such that movement of the handle turns the rotary valve on its seat, opening and closing access to various ports and passages drilled and cast in both parts. These ports and passages—when opened or closed by this movement of the rotary valve relative to its seat—in turn opened or closed connections between the main reservoir, brake pipe, and equalising reservoir; and brake pipe, equalising reservoir and atmosphere. Handle positions are RELEASE, RUNNING, LAP, SERVICE, and EMERGENCY.

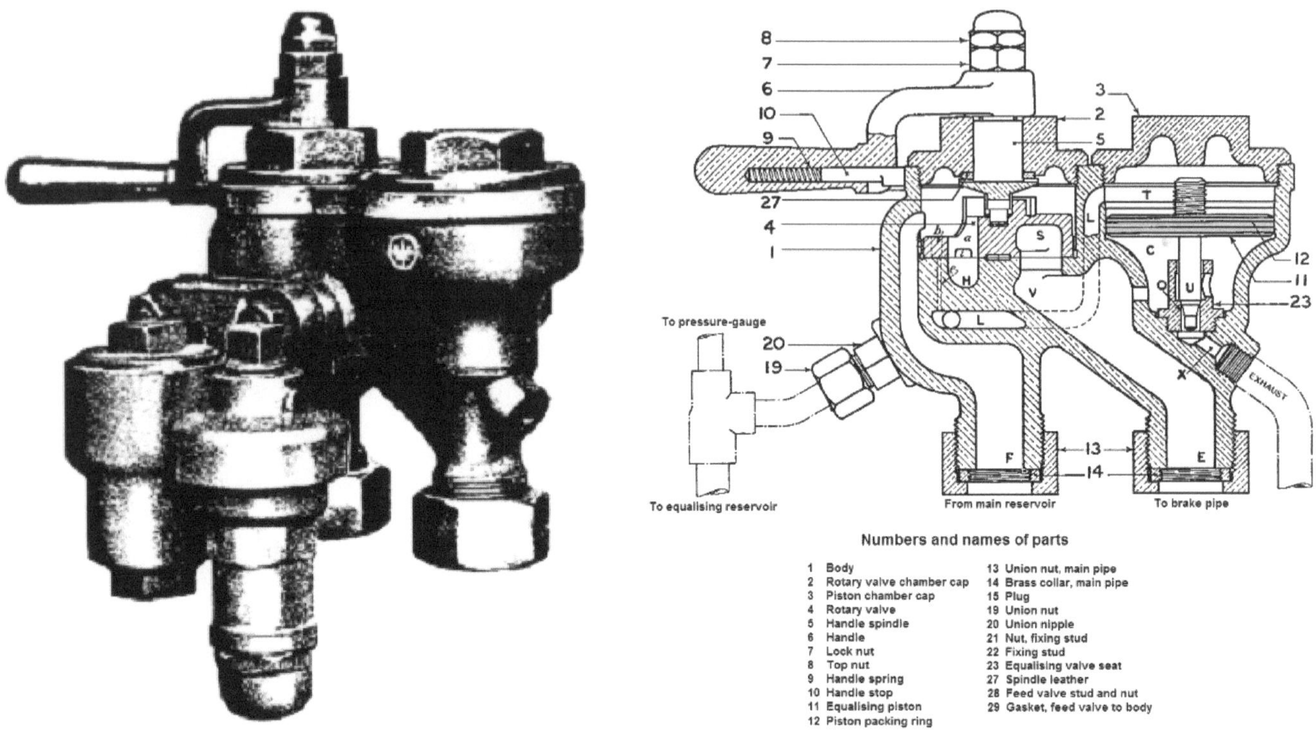

FIGURE 142: THE № 4 AUTOMATIC BRAKE VALVE – ILLUSTRATION AND DIAGRAM

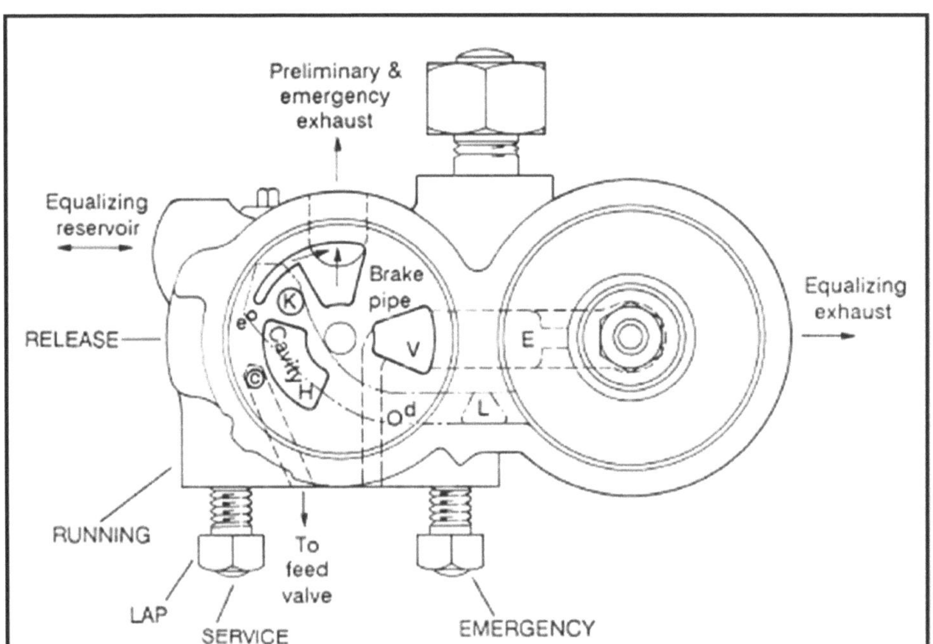

FIGURE 143: THE № 4 AUTOMATIC BRAKE VALVE, ROTARY VALVE PLAN VIEW

The equalising arrangement consists of the piston **11** with stem terminating in valve **U** that, in the position shown, closes exhaust port **O**. Movement of the brake valve handle to the SERVICE position to initiate a brake application provides for an exhaust of equalising reservoir pressure that is also in chamber **T** above the piston. Brake pipe pressure in chamber **C** being now greater will lift the piston and discharge valve so as to open the brake pipe exhaust port, permitting brake pipe pressure to flow to atmosphere until it equalises with that remaining in the equalising reservoir and chamber **T**. Brake pipe pressure, in fact, reduces to a value slightly less than equalising reservoir, and, as this occurs, the equalising piston and discharge valve are driven gently down to close the brake pipe exhaust. This arrangement for a driver's Automatic brake valve proved eminently suitable and would remain the design basis for such equipment until long after the demise of regular steam operations. The № 4 brake valve was still in service on many Australian and New Zealand steam locomotives as this era drew to a close, and it remains in use on operating steam locomotives to this day. (And not just steam locomotives! See Figure 146)

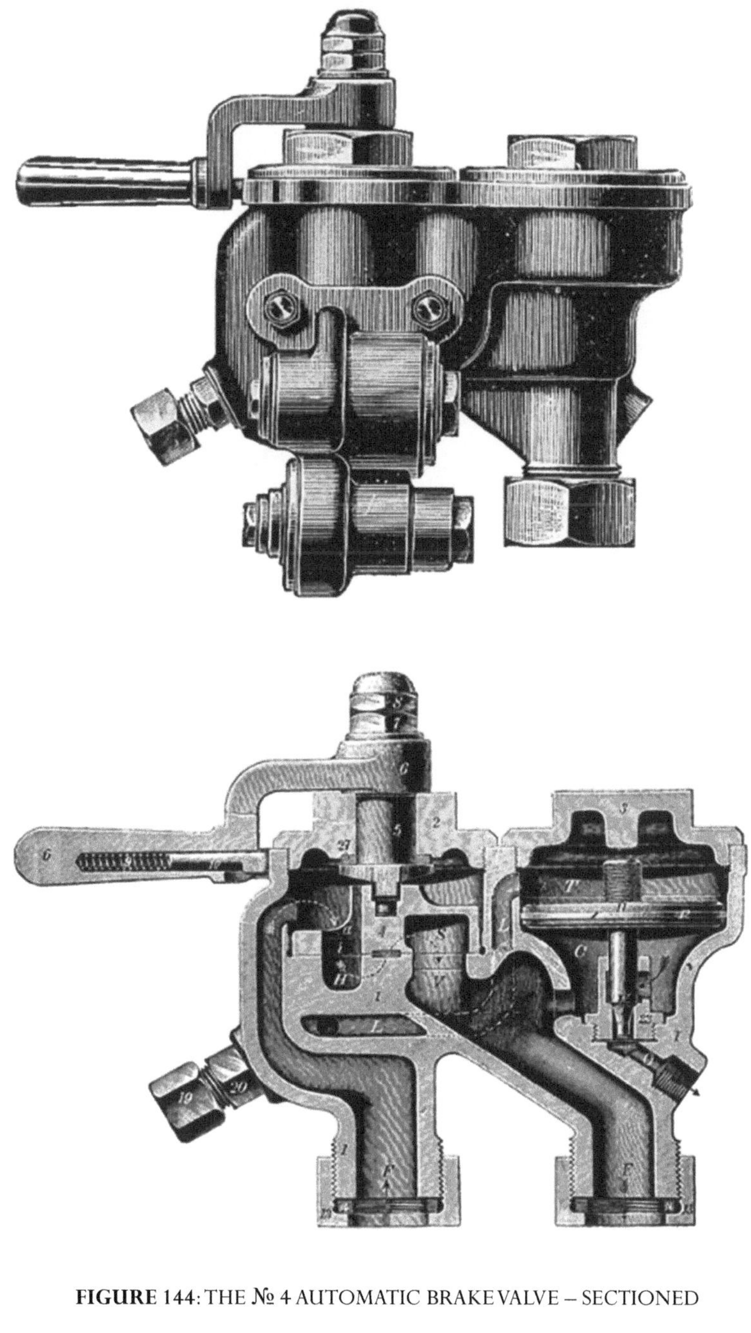

FIGURE 144: THE № 4 AUTOMATIC BRAKE VALVE – SECTIONED

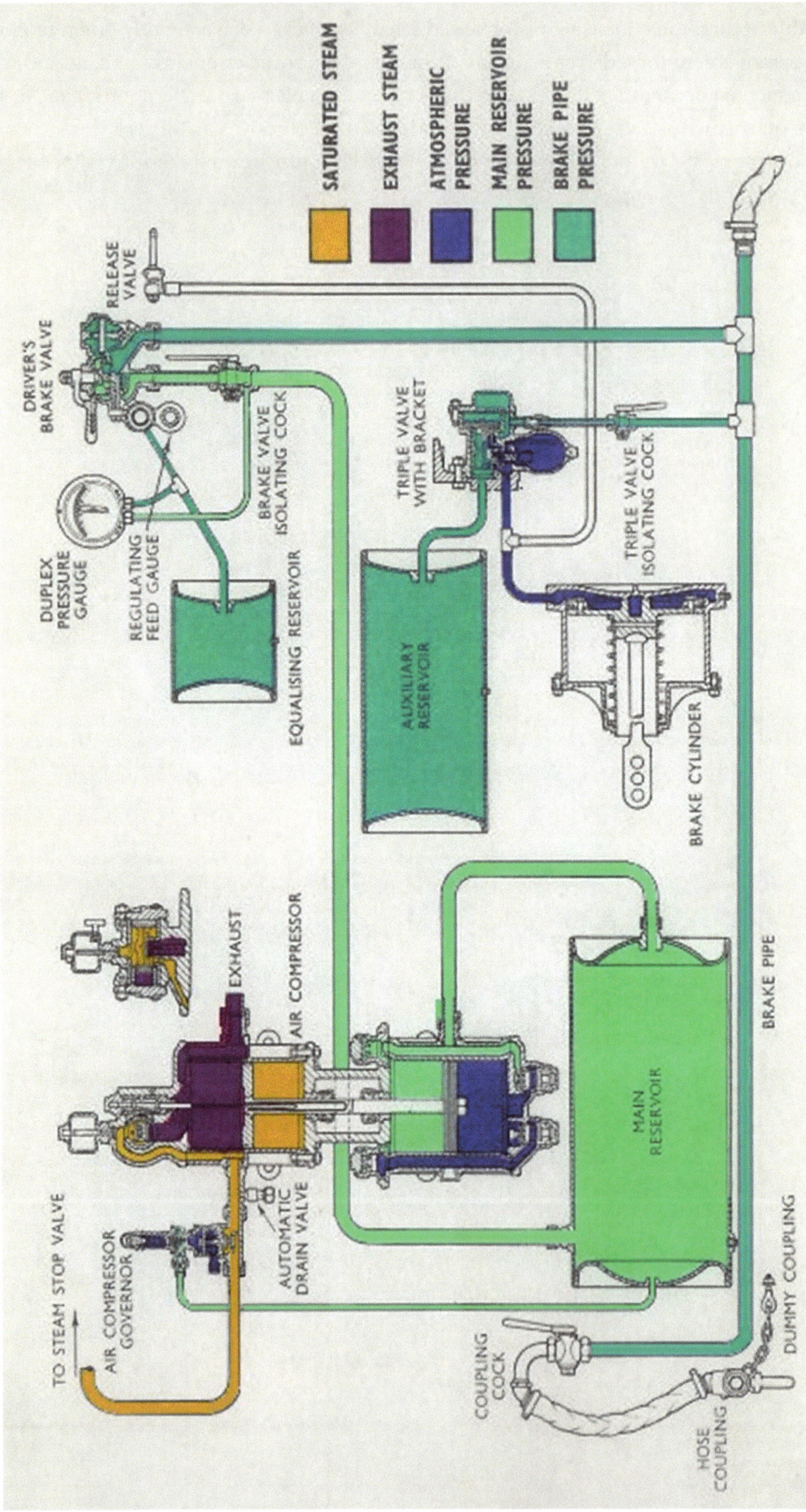

FIGURE 145: THE № 4 AUTOMATIC BRAKE VALVE – ILLUSTRATED LOCOMOTIVE SCHEMATIC DIAGRAM

FIGURE 146: THIS PHOTO OF THE DRIVER'S SIDE CAB INTERIOR OF A VENERABLE VICTORIAN Y-CLASS DIESEL LOCOMOTIVE SHOWS A RECYCLED № 4 AUTOMATIC BRAKE VALVE PAIRED WITH A W-TYPE STRAIGHT-AIR BRAKE VALVE. NOTE THE LEFT-HAND DRIVE CONTROL STAND AND THE MIRROR-IMAGE ORIENTATION OF THE EMD CONTROLLER PANEL, A CHARACTERISTIC UNIQUE TO AUSTRALIAN DIESEL LOCOMOTIVES, WHICH ARE OPERATED FROM THE LEFT-HAND SIDE (EXCEPT IN QUEENSLAND AND THE PILBARA REGION OF WESTERN AUSTRALIA), THUS REQUIRING BACK-TO-FRONT MOVEMENT OF ALL CONTROL HANDLES (AND WHICH—AMONG OTHER NEGATIVES—RENDERS THE THROTTLE NOTCH INDICATOR AT THE WRONG ANGLE FOR CONVENIENT OBSERVATION). *Bernie Baker*

The A6-ET Equipment

By the beginning of the twentieth century in the United States, the early equipment schedules, such as those depicted and described previously, were—in turn—being superseded. The Westinghouse Company had developed the ET equipment specifically for larger locomotives that required braking to both ***E***ngine and ***T***ender. The first incarnation of this schedule, the 5-ET dating from 1905, extended the numerical progression from the № 4 and introduced the new *triple valve/relay valve/auxiliary reservoir* combination known as the **locomotive distributing valve** and was somewhat prototypical. Significantly refined, it had evolved by 1907 into the 6-ET and would later be further developed into the 8-ET schedule providing for advanced brake valve and operational functions required for North American operations; in particular, the sophisticated streamlined passenger trains that were later introduced. In its 6-ET guise, the schedule utilised the H-6 driver's brake valve, much developed from the D-5/E-6/F-6 and H-5 models previously mentioned. The S-6 Independent brake valve and the № 6 distributing valve completed the schedule, and a competing version notwithstanding (the LT equipment from the New York Air Brake Company) became a standard across North America.

For Australia and New Zealand, the equipment was adapted by Westinghouse Australia for local requirements and—with local parts numbers—became the ***A***6-ET schedule with the ***A***H-6-P Automatic brake valve. This equipment was fitted to most modern, last-generation, air-braked steam locomotives in Australia and New Zealand and was exported to the United Kingdom for overseas contracts. The equipment provided for full independent operation of locomotive brakes (including brake cylinder pressure maintenance against leakage and piston travel variations); a new feature being the steam-operated Duplex air compressor governor which automatically controlled main reservoir pressure within a wide and useful range while minimising wear and tear on the compressor. It should be acknowledged here that the US 6-ET equipment (including H-6 brake valve) did find its way into Australia via some US-built steam locomotives; the Queensland Government Railways AC16 (one example of these wartime locomotives having now been rebuilt in Queensland and another undergoing restoration in New South Wales at the time of writing) and the New South Wales Government Railways 59 Class.

Distinctive features of the AH-6-P driver's Automatic brake valve ("P" indicating *Pedestal*-type as distinct from a direct pipe-bracket installation)—essentially a version of the American H-6 valve—included the following:

- With the handle in the RELEASE position, the engine and tender brakes released at a retarded rate (whereas with the US 6-ET equipment, they were held applied), thereby inhibiting the locomotive from gaining excessive speed before the brakes had released on the remainder of the train;
- With the handle placed directly to the LAP position, a minimum brake pipe reduction of around 50 kPa was achieved as a preliminary set, which snubbed the wagon or car blocks against the wheels to provide a gentle, initial brake application and was not a feature of the American H-6 valve;

FIGURE 147: № 4 BRAKE VALVE SHOWN ON A NZGR AB-CLASS LOCOMOTIVE. *E. J. McClare*

- With the handle in the SERVICE position, the engine and tender brakes applied at a retarded rate (thereby preventing train slack from running-in too rapidly), and the brake pipe was reduced at a rate that would not produce emergency Quick-Action;
- With the handle in the EMERGENCY position, the engine and tender brake application is not retarded and occurred promptly. In this case, the rate of brake pipe reduction delivered Quick-Action, propagating the reduction along the brake pipe as rapidly as possible;
- Deletion of the HOLDING position (locomotive brakes being held applied during train brake release and recharge). The Holding position had been deemed unnecessary in Australia and New Zealand due to differences in train size and operations (not least of which were lower overall speeds) and less complex rollingstock air brake equipment specifications.

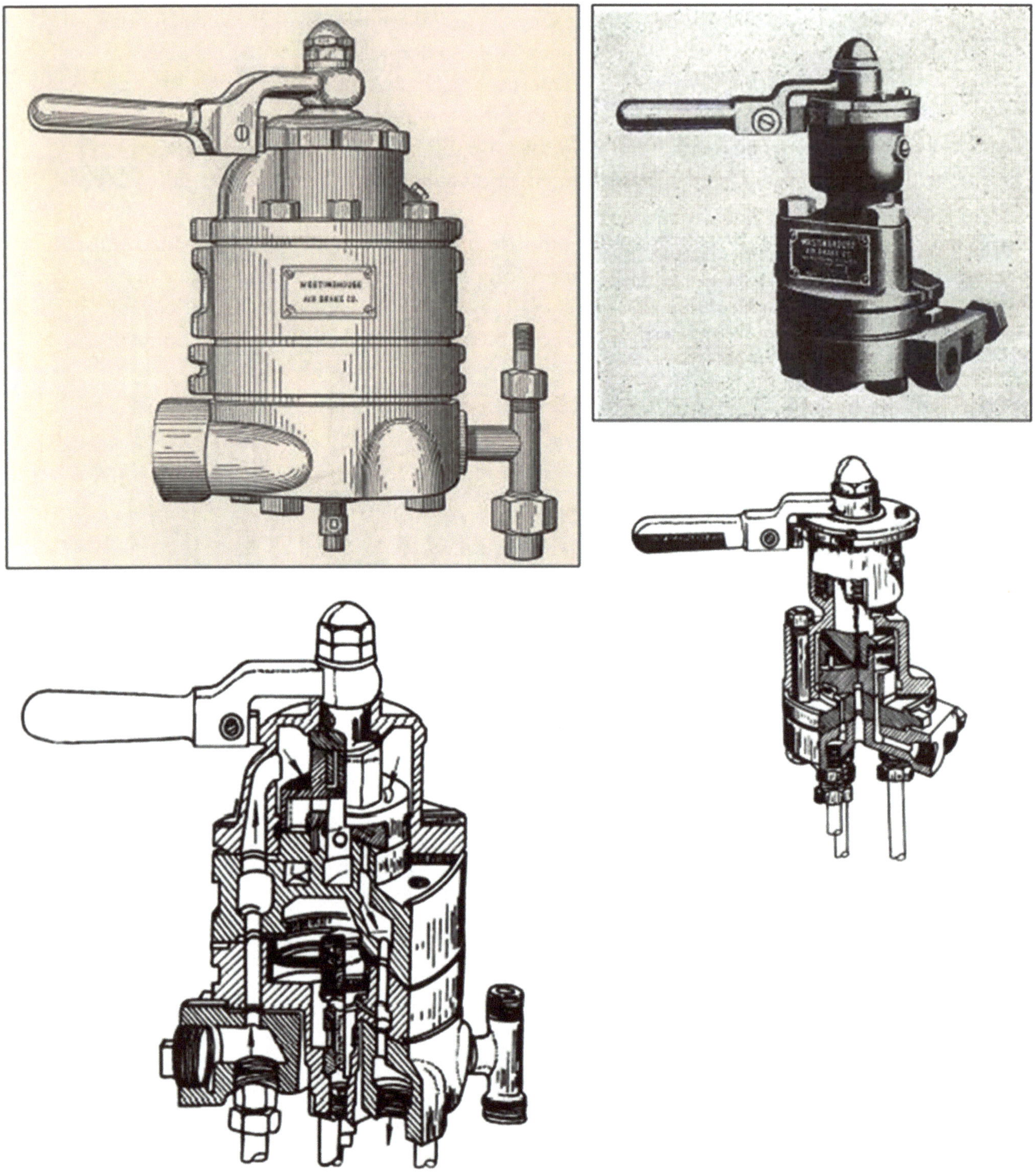

FIGURE 148: THE H-6 AUTOMATIC AND S-6 INDEPENDENT BRAKE VALVES – ILLUSTRATION

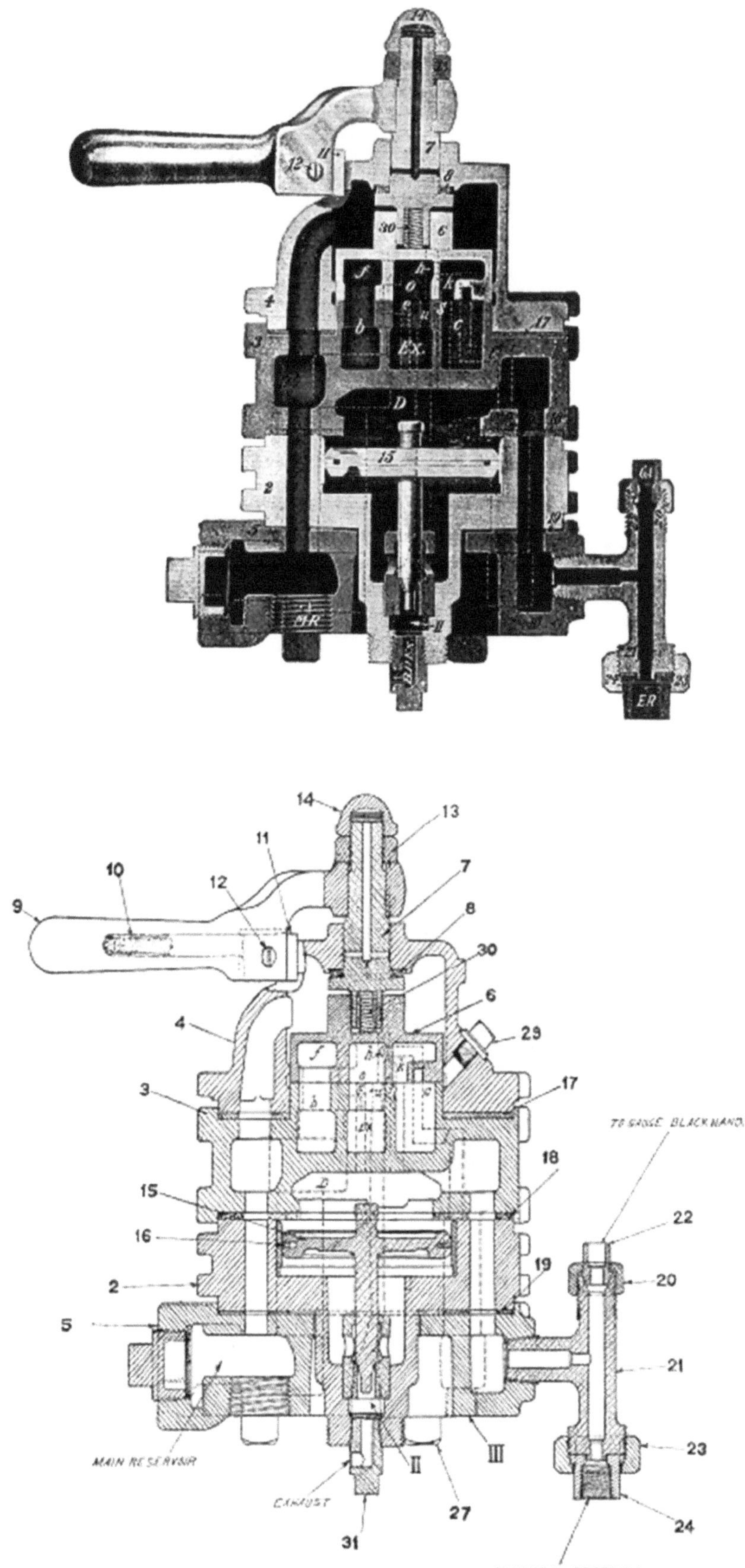

FIGURE 149: THE H-6 AUTOMATIC (WITH SOLID EQUALIZING PISTON) – SECTIONED VIEWS

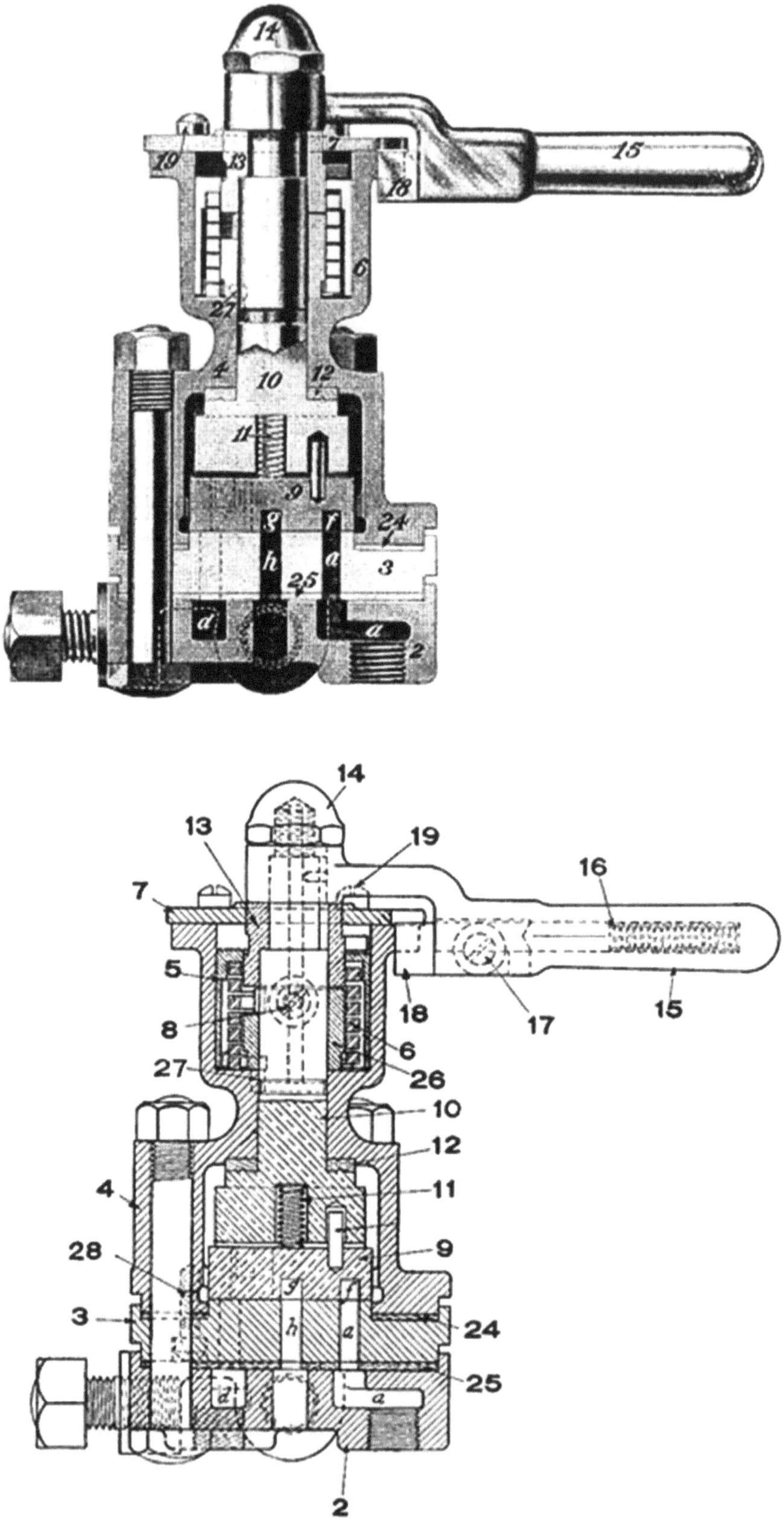

FIGURE 150: THE S-6 INDEPENDENT BRAKE VALVE – SECTIONED VIEWS

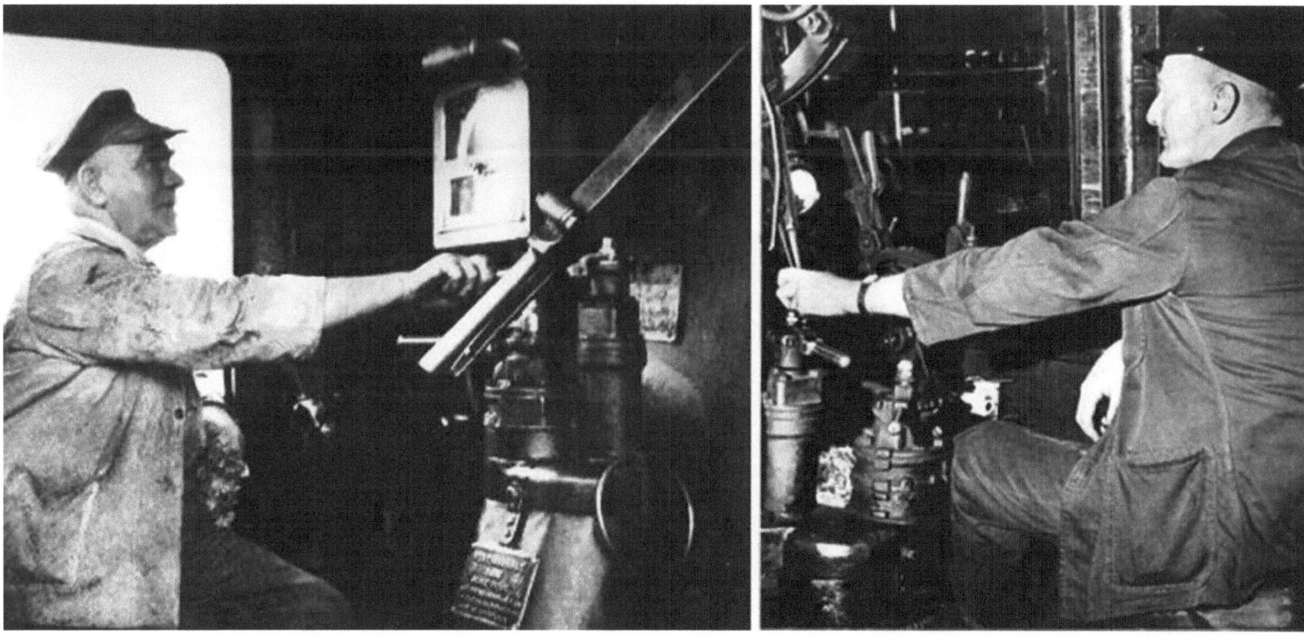

FIGURE 151: THE VICTORIAN RAILWAYS DRIVER IN THE LEFT-HAND IMAGE IS ABOARD THE SINGULAR H-CLASS LOCOMOTIVE. HIS HAND RESTS UPON AN ELONGATED S-6 INDEPENDENT BRAKE VALVE HANDLE. COMPARE THE HEIGHT OF THE BRAKE PEDESTAL WITH THAT ON THE NZR JA-CLASS LOCOMOTIVE SHOWN IN THE ADJACENT IMAGE. IN THE LATTER PHOTO, THE DRIVER HAS THE A6-ET HANDLES MORE READILY TO HAND - IN THIS CASE, THE INDEPENDENT HANDLE IN THE *LAP* POSITION; THE AUTOMATIC HANDLE AT THE *RUNNING* POSITION. L-H Image: *O'Neil Publishers P/L, Melbourne*; *I. R. Barkla, courtesy John Currey*. R-H Image: *Roy Sinclair, courtesy Whitcoulls Ltd., Christchurch, NZ*

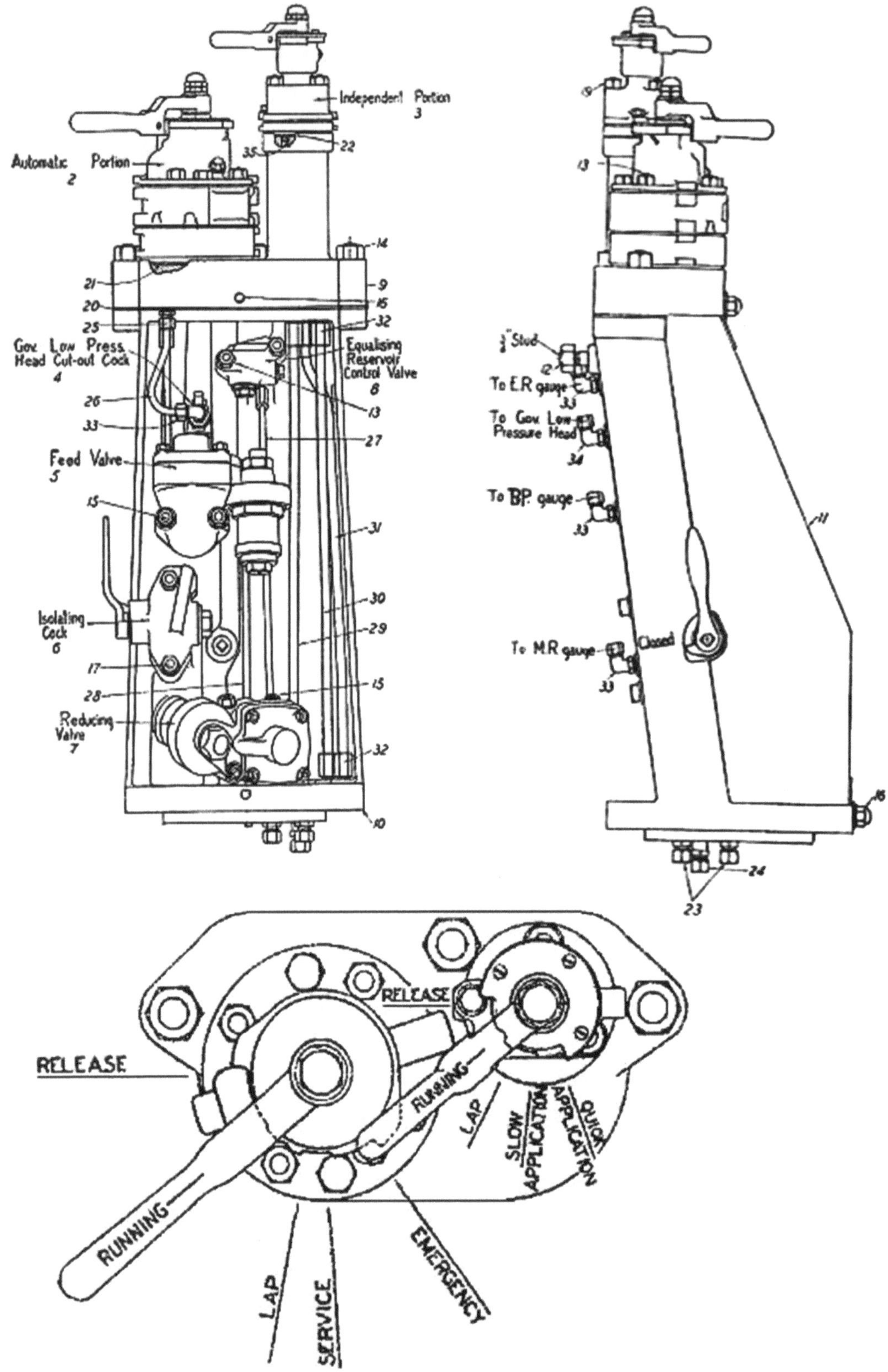

FIGURE 152: THE A6-ET/A7-EL BRAKE VALVE PEDESTAL ARRANGEMENT. BRAKE VALVE HANDLE POSITIONS ARE COMMON TO BOTH SCHEDULES

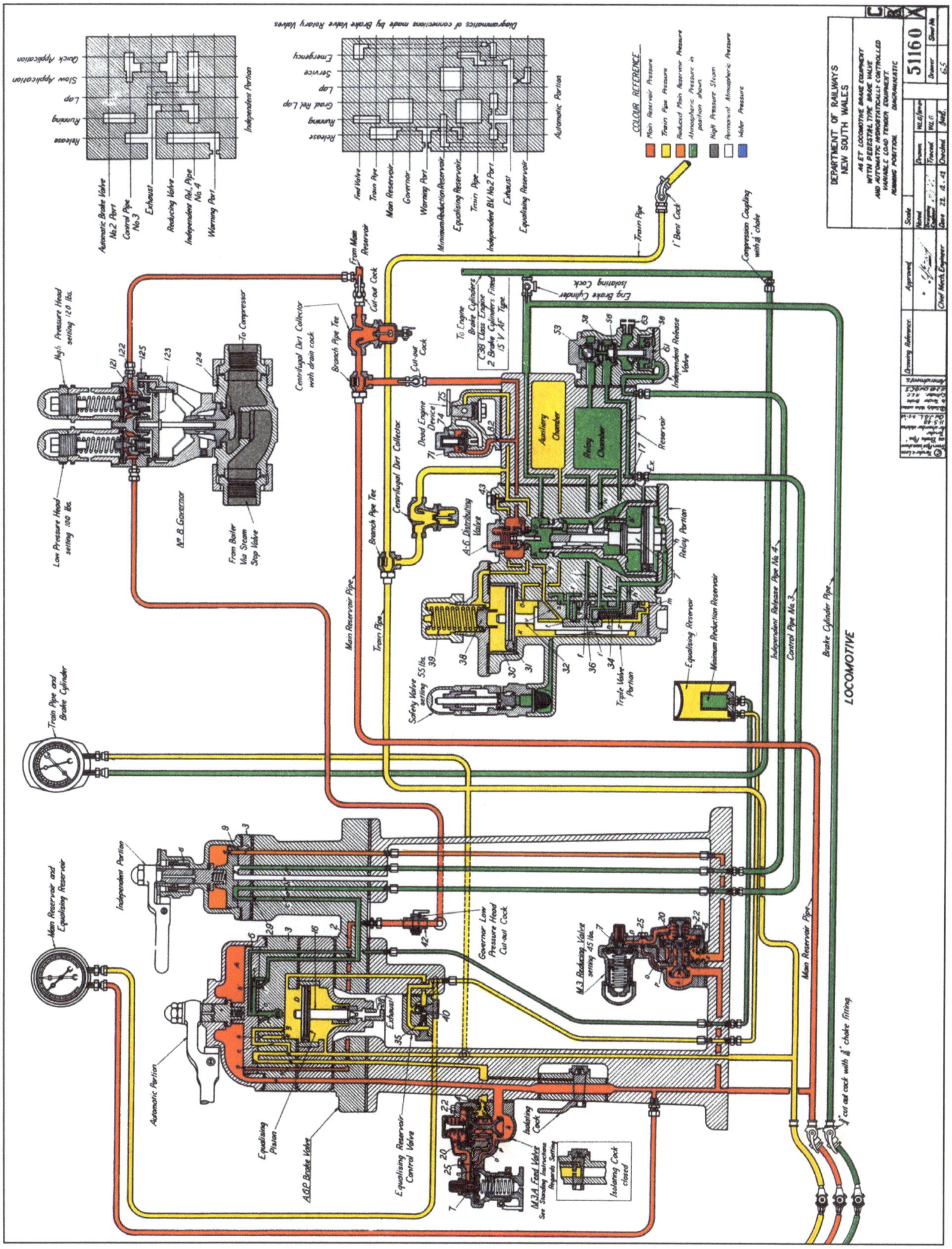

FIGURE 153: THIS IS THE A6-ET SCHEDULE AS APPLIED IN AUSTRALIA AND NEW ZEALAND. THIS DIAGRAM DEPICTS THE ARRANGEMENT FOR THE NSWGR C38-CLASS 4-6-2 STEAM LOCOMOTIVE. NSWGR CHIEF MECHANICAL ENGINEER AND DESIGNER OF THE C38 CLASS WAS HAROLD YOUNG. THIS SCHEMATIC DRAWN BY BILL GRIMSHAW. *Courtesy of Albert Taylor, Sydney*

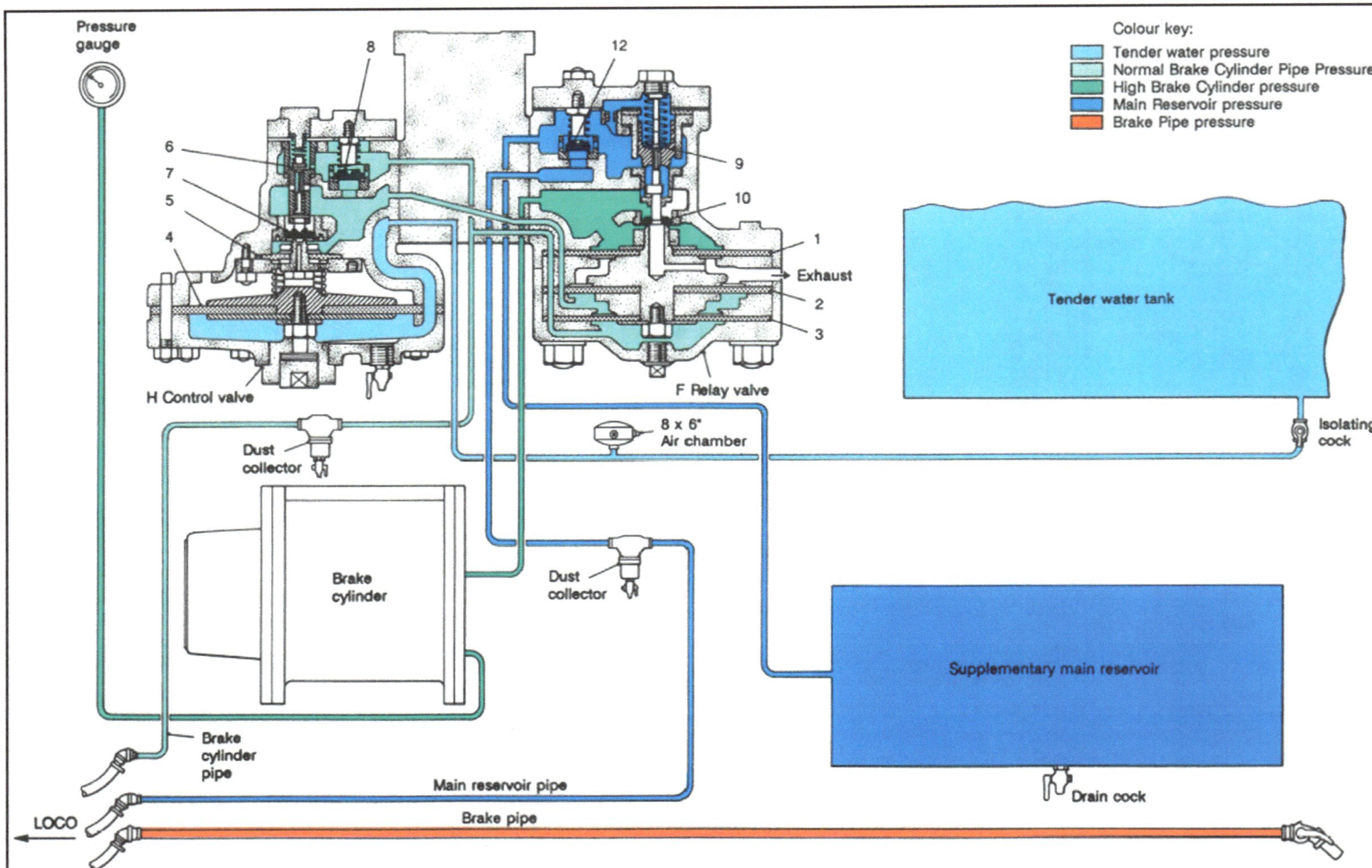

FIGURE 154: A UNIQUE ASPECT OF THE VR R-CLASS (4-6-4) AND NSWGR C38-CLASS (4-6-2) LOCOMOTIVE AIR BRAKE ARRANGEMENT WAS THE WESTINGHOUSE HF HYDROSTATIC VARIABLE LOAD EQUIPMENT FITTED TO THE TENDER. WITH A CONTROL VALVE OPERATED BY WATER PRESSURE (WHICH VARIED ACCORDING TO TENDER CONTENTS), A CONTROL AIR PRESSURE WAS SUPPLIED TO A RELAY VALVE, WHICH THEN FUNCTIONED TO SUPPLY BRAKE CYLINDER PRESSURE AT THE REQUIRED RATIO FROM A SUPPLEMENTARY RESERVOIR MOUNTED ON THE TENDER. THE EQUIPMENT DID NOT OPERATE RELIABLY OR ECONOMICALLY AND WAS EVENTUALLY REMOVED

As with the older № 4 brake valve, the A-6-P Automatic brake valve consisted internally of a rotary valve operated by the brake valve handle, having main reservoir pressure above it and a fixed valve seat against which the rotary valve turned. The porting within these two components was, however, considerably more complicated than for the № 4 valve. As well as providing a constant down-force to assist it to seal against the machined face of its seat, the main reservoir pressure above the rotary valve was available to flow directly into the brake pipe when the brake valve handle was placed to the RELEASE position, and to the air-compressor governor while the handle was in the RUNNING position. The normal main reservoir supply—via the **feed valve**—to the brake pipe was also facilitated when the brake valve was in the RUNNING position.

Similar in design to the Automatic brake valve, the Independent brake valve had **reducing valve** pressure above its rotary valve. With the handle in either of the APPLICATION positions, reducing valve pressure flowed via the rotary valve and its seat to the distributing valve which was thus activated to provide a controlled main reservoir supply to the brake cylinders of both engine and tender. The RELEASE position was the equivalent of our modern-day "bail-off" although somewhat more basic in operation. When held laterally (against spring tension) in this position, control air pressure that would otherwise be against the application piston in the distributing valve was able to exhaust to atmosphere, permitting a release of brake cylinder pressure. When hand tension on the Independent handle was released, the spring returned it to the RUNNING position. This spring also functioned in like manner to return the handle—when being held in the opposite direction—from the Quick- to the Slow-Application position.

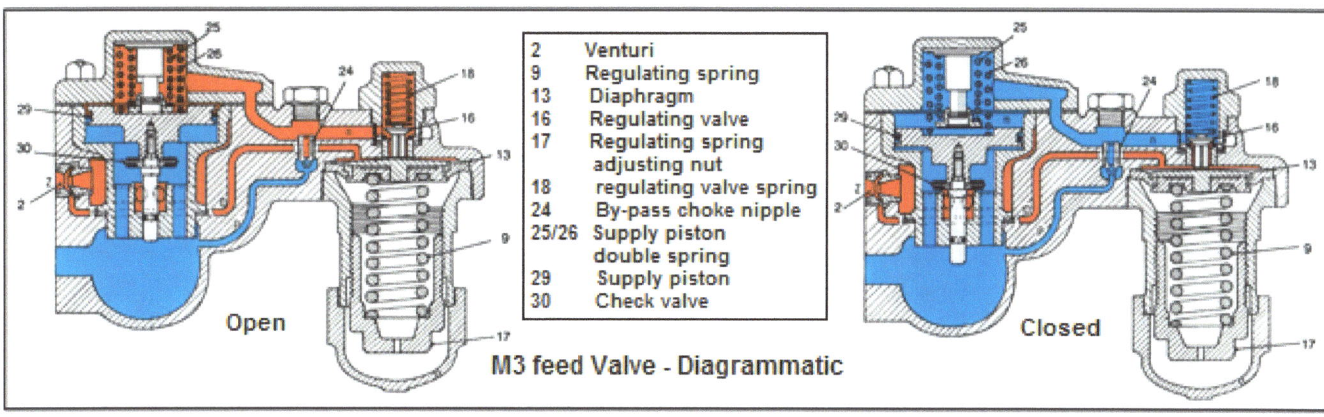

FIGURE 155: THE M3 FEED VALVE AND REDUCING VALVE FOR A6-ET/A7-EL AIR BRAKE SCHEDULE - DIAGRAM

The **feed valve** is a brake pipe pressure regulator that ensures the prescribed pressure is automatically maintained in the brake pipe, irrespective of the higher main reservoir pressure. It is attached to the brake valve main reservoir supply pipe and works in conjunction with the Automatic brake valve. The device is referred to as a **reducing valve** when it is used for the Independent brakes, and it consists of two portions: the *supply portion* and the *regulating portion*. The latter contains a spring providing a set force that will hold the regulating valve open to maintain a supply of main reservoir pressure into the brake pipe—via the supply valve—to the required value (in Australia at this time, typically 70 psi/500 kPa and in New Zealand 80 psi/560 kPa). The balancing force for the regulating spring is brake pipe pressure itself, and the feed valve is very sensitive to slight fluctuations in this.

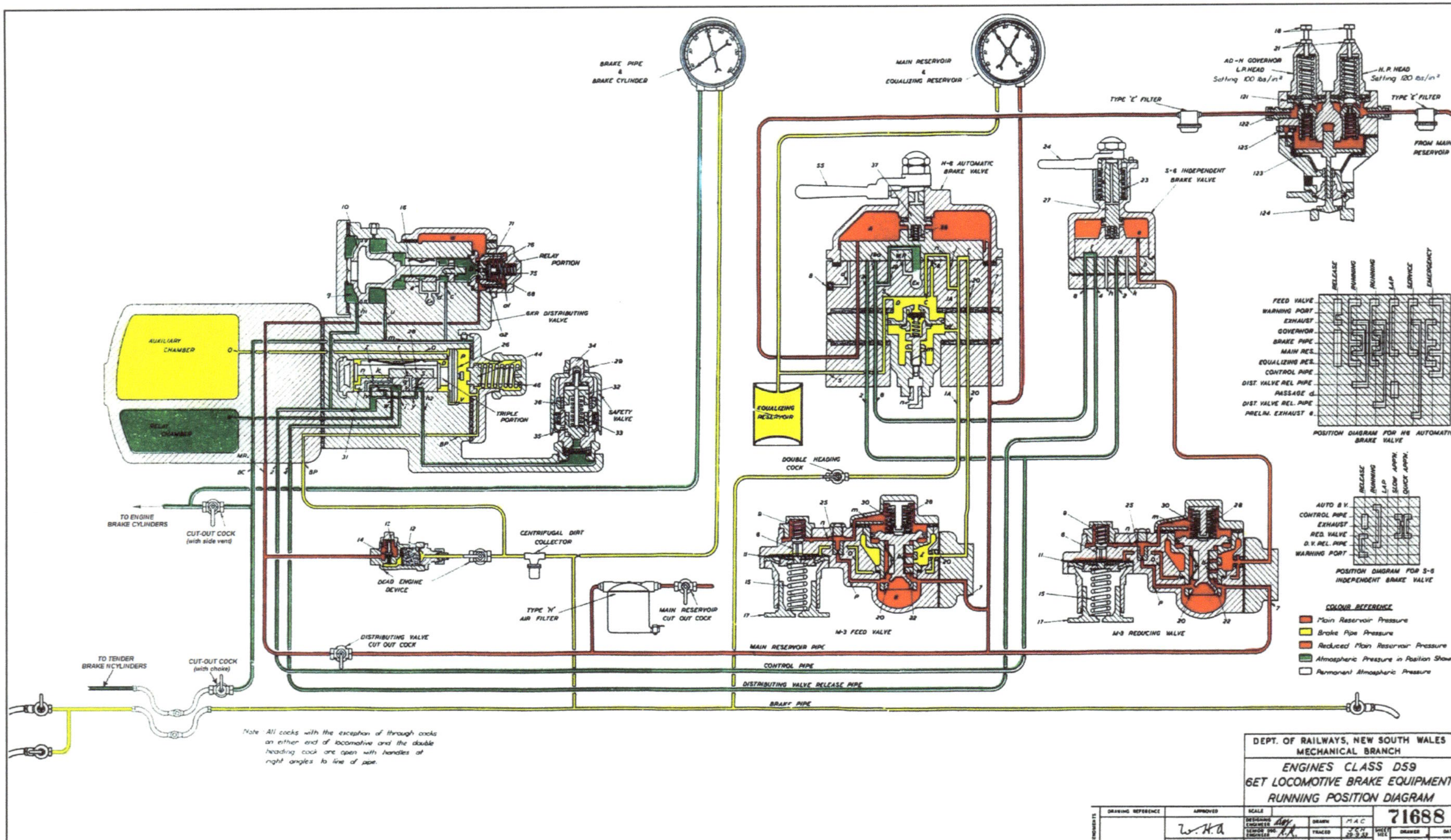

FIGURE 156: THIS AIR BRAKE SCHEMATIC DEPICTS ONE OF THE TWO KNOWN APPLICATIONS WITHIN AUSTRALASIA OF THE AMERICAN 6-ET AIR BRAKE SCHEDULE. THIS CAN BE COMPARED WITH THE A6-ET DIAGRAM IN FIGURE 153. NOTE THE COLLAPSIBLE EQUALISING DISCHARGE PISTON AND LACK OF A MINIMUM REDUCTION RESERVOIR AND EQUALISING RESERVOIR CONTROL VALVE. THIS SCEMATIC DRAWN BY MAURICE CHIN (NSWGR BRAKE ENGINEER'S OFFICE) AND APPROVED BY CHIEF MECHANICAL ENGINEER BILL ARMSTRONG. *Courtesy of Albert Taylor, Sydney NSW*

The **distributing valve** directly controls the application and release of the locomotive brakes according to the position of the Automatic and Independent brake valve handles. Consisting of both a **triple valve** and a **relay valve** in one casting, it is attached to a reservoir that contains an **auxiliary chamber** and a **relay chamber**. The auxiliary chamber equates to the auxiliary reservoir in a conventional air brake setup, and the relay chamber is, in effect, a dummy brake cylinder that—during brake operations—receives air from the auxiliary chamber and exhausts it to atmosphere according to the actions of the triple valve portion. The air pressure value in the relay chamber is duplicated in the locomotive brake cylinders by the action of the relay valve portion that receives its supply of air from the main reservoir. The relay valve action results in the same brake cylinder pressure being obtained and maintained for a given brake pipe reduction, irrespective of the length of piston travel or reasonable cylinder leakage.

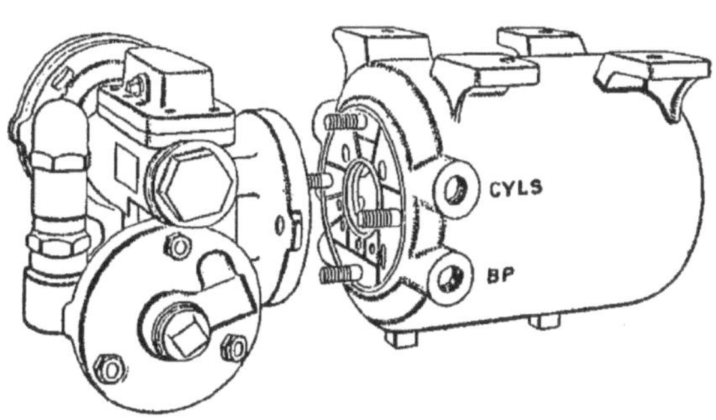

FIGURE 157: THE № 6 LOCOMOTIVE DISTRIBUTING VALVE - ILLUSTRATION

Note that the illustrations on this page (the two images showing the valve with different mounting options) depict the American № 6 Distributing Valve from which the Australian A-6 version depicted in Figure 158 is derived.

A separate Independent Release Valve—not part of the American version—was attached to the distributing valve, and an exhaust choke fitting (item 63 in Figure 160) was provided to assist with the exhaust of Relay Chamber pressure when the Independent Release Valve was actuated by a control pressure—provided through the № 4 Independent Release Pipe—from the driver's Independent Brake Valve. The NSWGR experienced problems with these Release Valves sticking, so the exhaust choke was removed and replaced with a square-head plug, meaning all of the Relay Chamber pressure had to flow back to exhaust at the driver's Independent Brake Valve via the No. 3 Control Pipe instead of some being able to exhaust directly at the Independent Release Valve. For drivers, this greatly increased the time required to obtain a locomotive brake release, and they began removing the plug for a quicker release, especially during shunting. The Department duly countered by replacing the square-head plug with an Allen-head fixture.

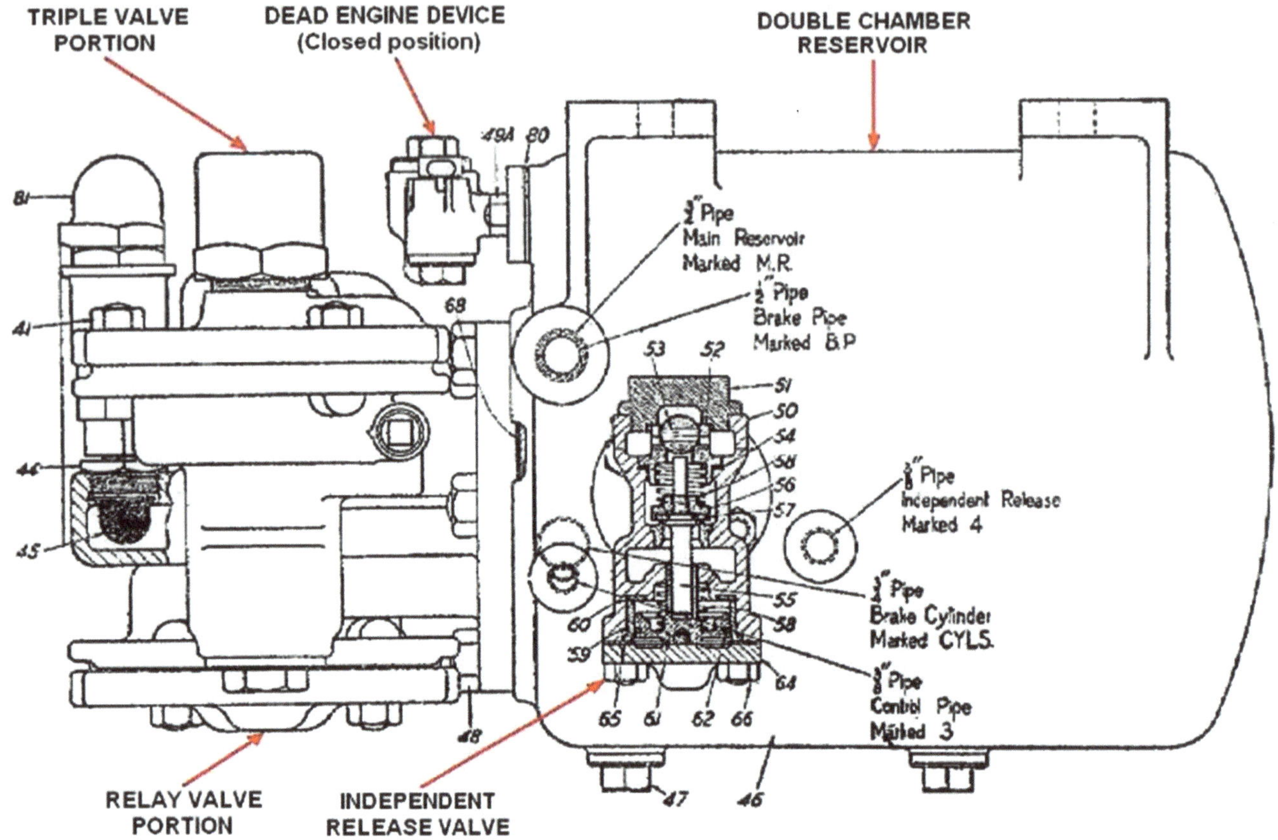

LEGEND

2. Body
3. Application-cylinder cover
4. Application-cylinder-cover
5. Application-piston-cage bush
6. Cage-bush seal
7. Application piston
11. Packing leather
15. Piston screw
17. Release-slide valve
19. Application valve
27. Equalising-cylinder cover
28. Equalising-cylinder cover gasket
31. Equalising-piston seal
32. Equalising piston
34. Equalising-slide valve
35. Equalising-slide-valve spring
38. Graduating-spring stem
39. Graduating spring
43. Charging choke
46. Distributing valve reservoir
50. Independent release valve body
52. Ball-valve seat
52. Ball valve
54. Ball-valve-seat seal
55. Valve spindle
56. Valve
57. Valve seal
58. Springs
59. Piston bush
60. Piston-bush seal
61. Piston
63. Exhaust-choke fitting
64. Cylinder cover
65. Cylinder cover gasket
66. Stud with nut
69. Dead-engine-device body
71. Valve
72. Valve seal
74. Valve spring
75. Cock plug
76. Cock plug handle
79. Choke plug (1/8in. drill)
81. Safety valve (set at 55 lb.)

FIGURE 158: THE A-6 LOCOMOTIVE DISTRIBUTING VALVE – GENERAL ARRANGEMENT

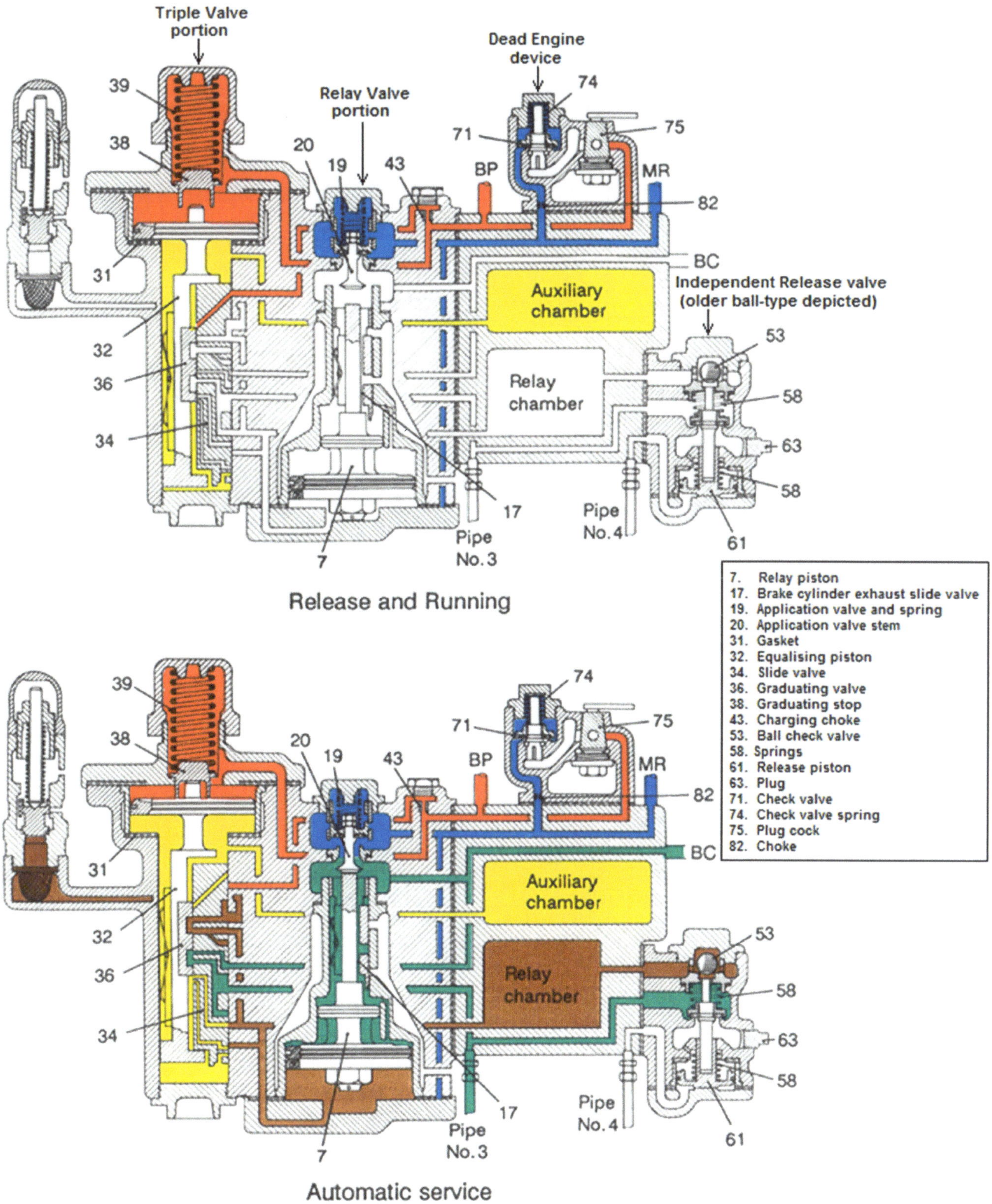

FIGURE 159: THE A-6 LOCOMOTIVE DISTRIBUTING VALVE – DIAGRAM

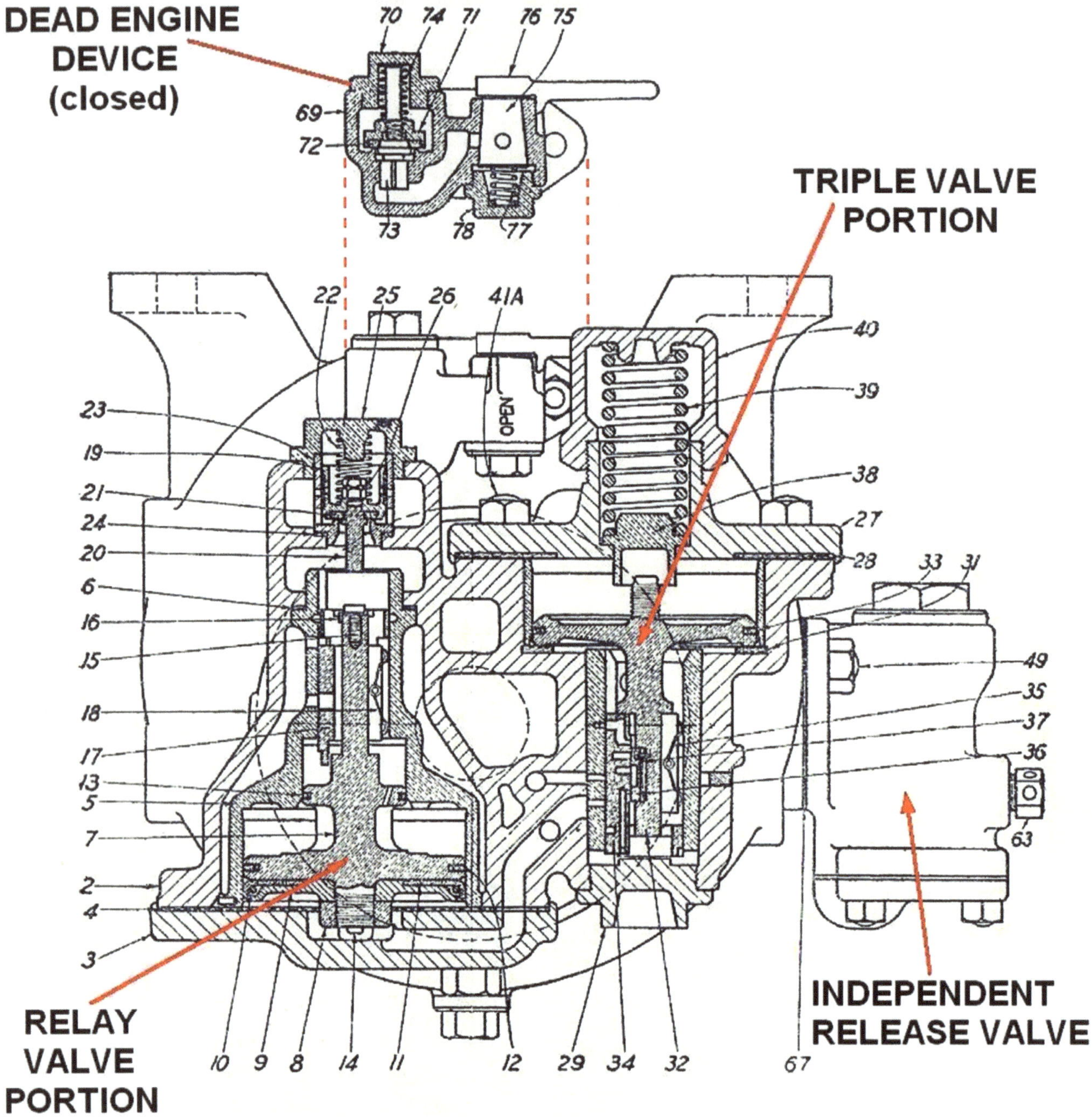

FIGURE 160: THE A-6 LOCOMOTIVE DISTRIBUTING VALVE – SECTIONAL FRONT VIEW

FIGURE 161: 6-SL BRAKE – INDUSTRY ADVERTISEMENT. *Railway Age*

The 6-SL Equipment

Designed originally for diesel ***S***witching ***L***ocomotives not requiring the multitude of extra locomotive and train operating features often finding their way onto air brake schedules, this equipment was made up in the United States from the 6-ET equipment, in its basic form being for single-end/single-unit operation. The equipment was to have only one application in Australia—the 40-Class units (Alco/MLW RSC-3 model) imported from Canada for the NSW Government Railways in 1951—and one in New Zealand, where it was supplied from 1955 to 1957 with the first forty units of the DA Class (EMD/GMD G12 export model). An H-6-B **Relayair valve** unit (Figure 164) was included on the Australian locomotives for multiple unit operation and the H-6-A version on the New Zealand locomotives.

The standard Automatic brake valve for this equipment was a pedestal-mounted H-6 derivative, the KH-6-P, which included a **bypass-type** (or 'collapsible') equalising discharge piston. This bypass feature was widely used in North American passenger operations and was designed to permit a balancing of pressure across the equalising piston at times when—with the brake pipe not fully recharged (i.e. at less than equalising reservoir pressure)—a further brake application was required. Since operating practices in New Zealand did not require this feature (although this type of equalising discharge piston was included with an early EMU electro-pneumatic brake), the NZR locomotives used the LH-6-P brake valve that had non-removable handles and a 'solid' equalising piston. New Zealand operating practice did not require the removal of brake valve handles on trailing diesel units.

The equipment included what was possibly the first self-lapping Independent brake valve to come from the United States. Its ease of operation and rapid response were particularly desirable features for operations, such as shunting, which required continual brake valve manipulation by the driver. Brake valve responsiveness was limited only by the capacity of the distributing valve to handle the installed brake cylinder volume. Sander and bell-ringer valves were included as options but not utilised on the New Zealand installation.

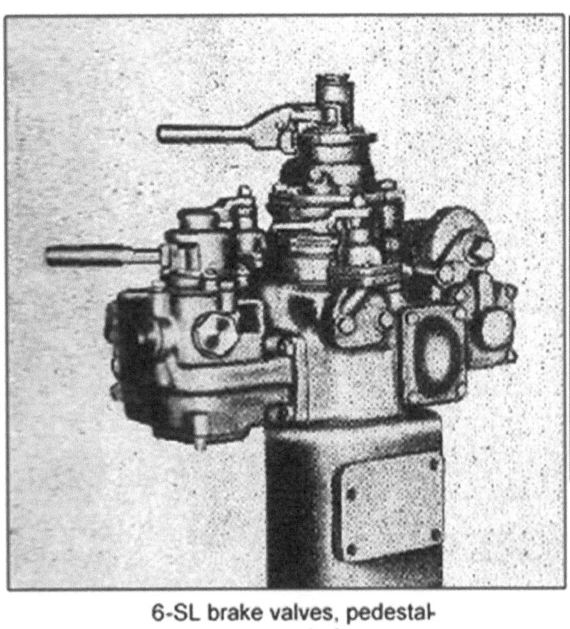

6-SL brake valves, pedestal-mounted

6-SL brake valves mounted directly to a pipe bracket on the control stand of an EMD G12 diesel. Apart from the absence of dynamic brake controls, this layout is identical to the NZR early-build DA class but is shown here on a South American left-hand drive model.

FIGURE 162: THE 6-SL LOCOMOTIVE BRAKE EQUIPMENT - ILLUSTRATION

The extremely bad effects of overcharging on modern brakes and the general use of large-capacity feed valves made the need for RELEASE position (providing a high-capacity main reservoir feed into the brake pipe and also referred to as FULL RELEASE) largely unnecessary and, for this reason, on the 6-SL Automatic brake valve a 'Full Release nullifier' was attached to the handle, which prevented its being placed to this position, a common modification in North America. In NSW, however, NSWGR brake engineer Geoff Lampard was concerned about this modification. He felt that denying the enginemen the use of the FULL RELEASE position also deprived them of the ability to optimise their brake pipe recharge time, potentially introducing an unsafe operating condition on the steep descents of the Great Divide and exposing the department to legal action in the event of a serious incident. As a result, the nullifier was removed from the Australian locomotives prior to their entering service.

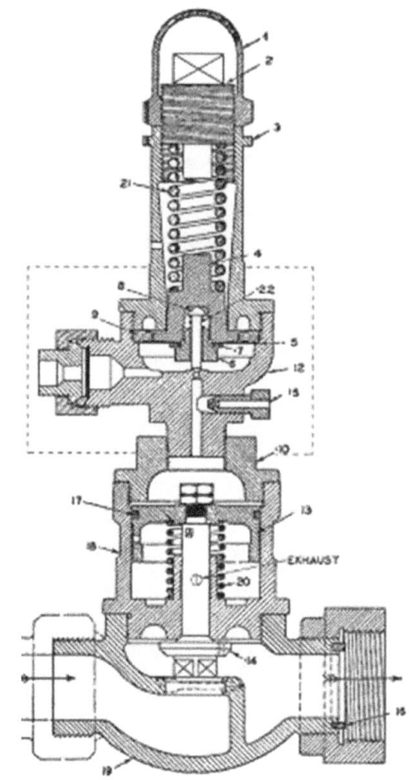

FIGURE 163: EARLY NSWGR BRAKE PIPE CHARGING FLOW INDICATOR (USING A № 7 COMPRESSOR GOVERNOR)

Lampard was also credited with inventing a very effective "brake pipe flow indicator" utilising the centre venturi portion (high-lighted in Figure 163) of a № 7 compressor governor. Initially, this device provided the engineman with an audible cue only; however, a gauge was later added. This was undoubtedly the first flow indicator to be used in Australasia and may even have predated the use of such a device in North America. As well, the HOLDING position—which was otherwise identical to the RUNNING position and held the locomotive brakes applied while the train brakes were released—was determined by the New Zealand Government Railways as not required and was disabled (by the removal of an internal plug, allowing distributing valve release pipe air to exhaust). Thus, the brake valve on the NZR locomotives equipped with 6-SL had two identical RUNNING positions with either being useable. The author is unaware if the above modification was ever applied to the Australian units.

There was no direct main reservoir air passage to the top of the rotary valve; instead, the feed valve supplied air to the rotary valve chamber. In RUNNING position, this pressure was at feed valve setting so that, in New Zealand, the rotary valve had 80 psi (560 kPa) acting on it and in New South Wales, 70 psi (500 kPa). The feed valve opening was controlled by pressure in a passage (1A in Figure 165) that was always connected to the brake pipe with the equipment set for Lead. So long as the Automatic brake valve handle was not in the RUNNING position, this pressure would be below the feed valve setting, and the feed valve **supply valve** would be open. In such a case, the feed valve then supplied air at main reservoir pressure to the rotary valve chamber so that in any position other than RUNNING, the rotary valve had main reservoir pressure acting on it.

Note:

- There was no Minimum Reduction feature in LAP position;
- The Independent brake valve was self-lapping;
- A three-position brake valve isolating cock was provided for operating in Multiple as well as Dead-in-Tow;
- A high-capacity D24 feed valve from the 24-RL schedule was utilised; and
- Automatic sanding was provided in EMERGENCY position.

The 6-IKR Distributing Valve

The normal distributing valve used with the 6-SL schedule was the type 6-KR. The NZGR units were fitted with the 6-IKR distributing valve, which included an FA-4 magnet valve mounted to the application cylinder cover, thus adding the dynamic brake ***I***nterlock feature. As with all diesel-electric locomotives, the DBI feature keeps locomotive brakes released during an Automatic (train) brake application when dynamic braking is in use, unless the Independent brake valve is used. The effect is to minimise the potential for locomotive wheelslide (causing flat spots that are expensive and time-consuming to rectify) as a result of excessive braking effort when rail condition is less-than-optimal.

This distributing valve had two main portions: the **equalising portion** (corresponding to the triple valve portion of the A-6 distributing valve) and the **application portion** (corresponding to the relay portion of the A-6). A double-chambered reservoir was incorporated, similar to the A-6 distributor. The chambers were the **pressure chamber** and the **application chamber**, serving the same purpose as the A-6 auxiliary and relay chambers respectively.

The H-6-A Relayair Valve Unit

The Relayair valve—used extensively in locomotive air brake schemes, particularly in North America—facilitates the control of certain air brake system functions through "piloting" when an air supply is connected to it at the desired time. A **Relayair valve unit** is several such valves connected to a common pipe bracket. In the 6-SL arrangement, this composite device incorporated three separate Relayair valves (Figure 164):

(1) *Cutoff Valve*: For break-in-two protection between locomotive units;

(2) *Transfer Valve*: On a Lead unit this device connected brake cylinder air to the equalising pipe, and on a Trail unit it connected the equalising pipe to the application cylinder pipe (and thus the Trail unit distributor), providing for the duplication on the Trail unit(s) of all Lead unit air brake functions, and;

(3) *Dynamic Valve*: With the three-position cut-out cock in LEAD position, this device connected the distributing valve application cylinder and the Independent brake valve to atmosphere via the distributing valve release pipe, enabling the distributing valve to exhaust brake cylinder pressure. With the three-position cut-out cock in TRAIL position, the Dynamic Valve connected the main reservoir to both the Dynamic and Transfer Relayair valve diaphragms. This permitted application cylinder air on Trailing unit(s) to exhaust (since brake cylinder on the Lead unit was exhausting) and caused the Trail unit distributing valve to also exhaust brake cylinder pressure.

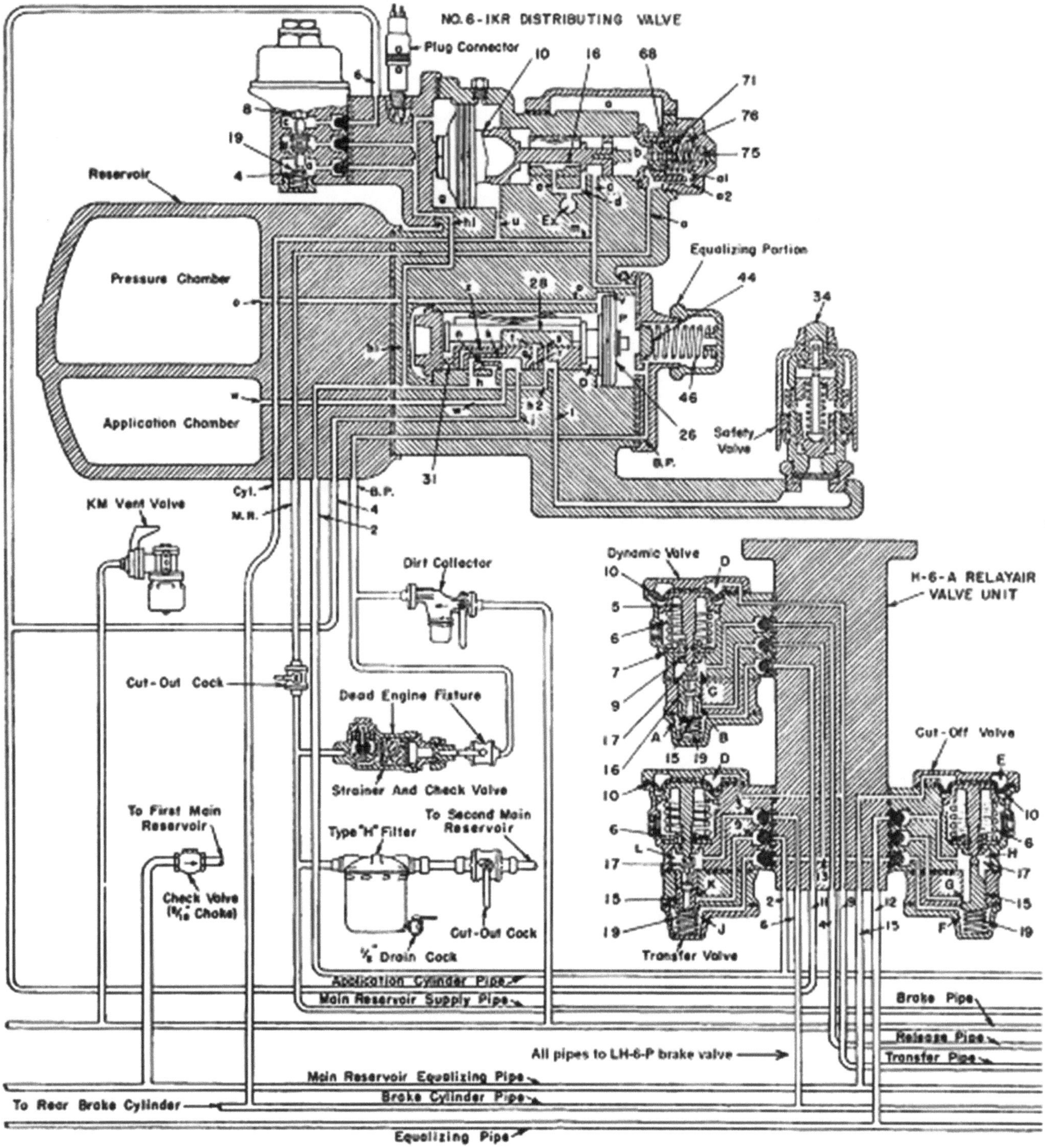

FIGURE 164: EARLY DISTRIBUTING VALVE AND RELAYAIR VALVE ARRANGEMENT WITH 6-SL SCHEDULE - SCHEMATIC

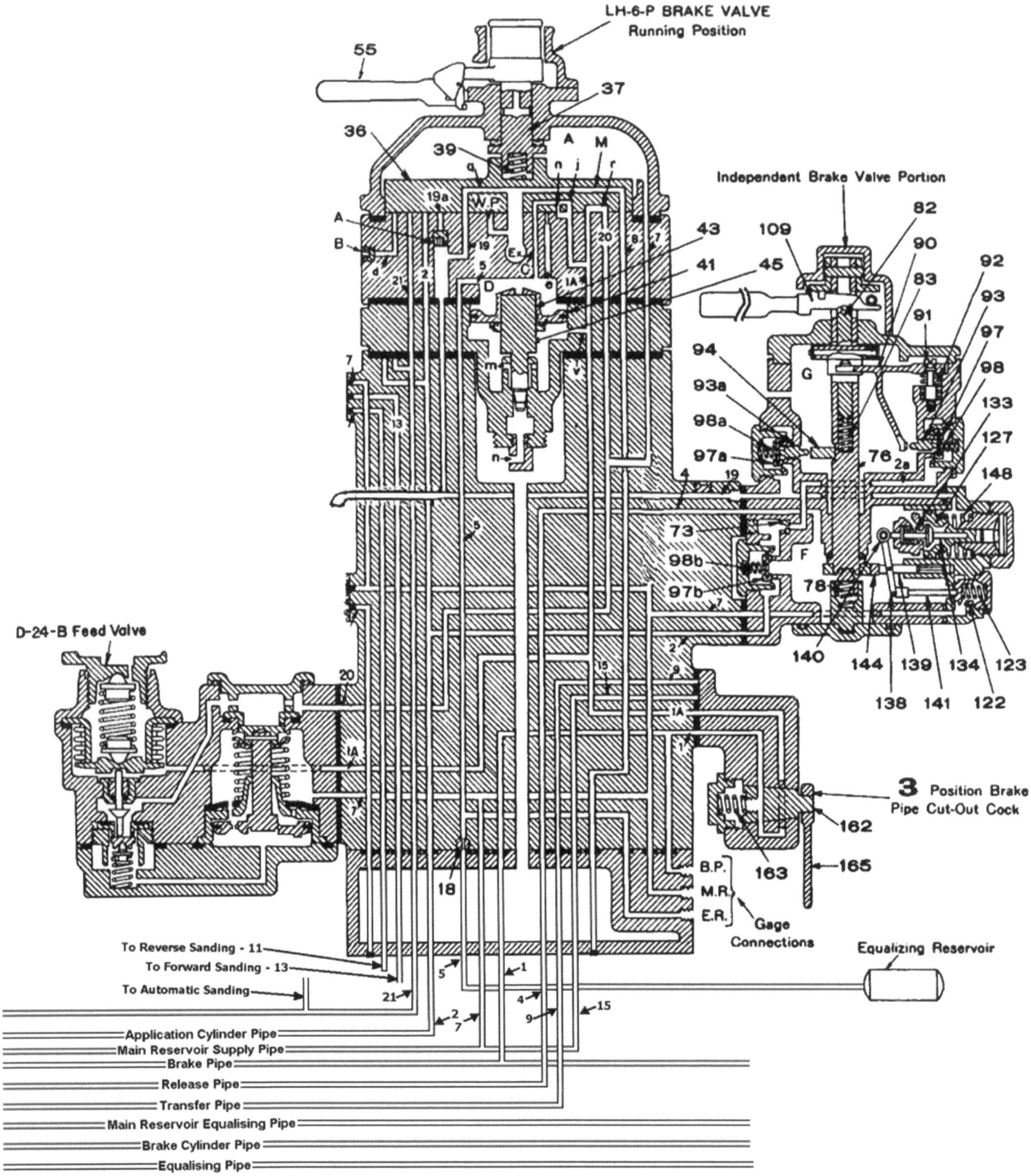

FIGURE 165: 6-SL SCHEDULE – BRAKE VALVE DIAGRAM

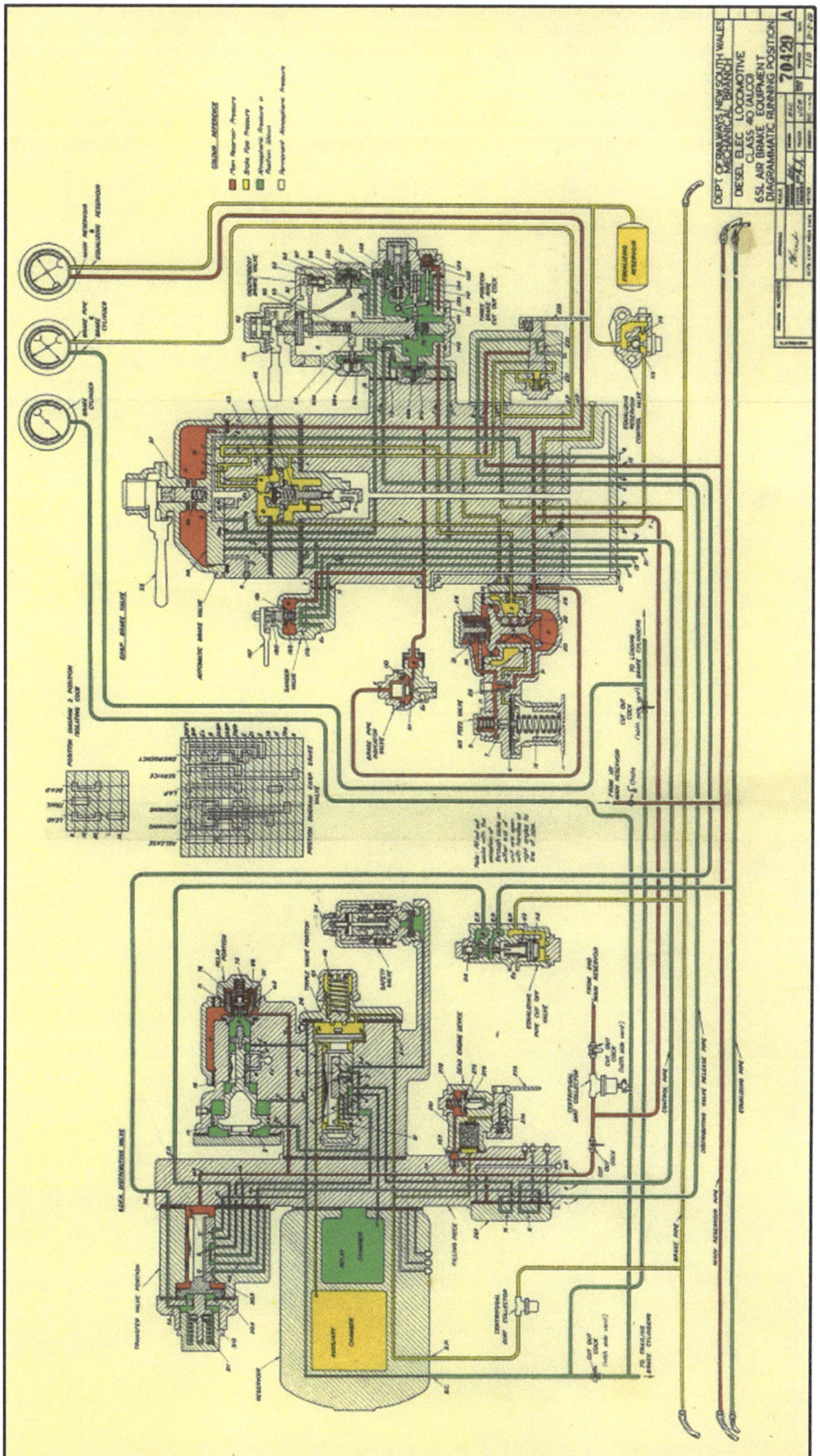

FIGURE 166: THIS SCHEMATIC DIAGRAM OF THE 6-SL EQUIPMENT ON THE FIRST NSWGR DIESEL-ELECTRIC LOCOMOTIVES, THE 40-CLASS CANADIAN-BUILT ALCOS, SHOWS THE BY-PASS EQUALISING DISCHARGE PISTON. I HAVE BEEN ADVISED BY ALBERT TAYLOR THAT THE DATE ON THE DRAWING IS INCORRECT.
Courtesy Albert Taylor, Sydney, NSW

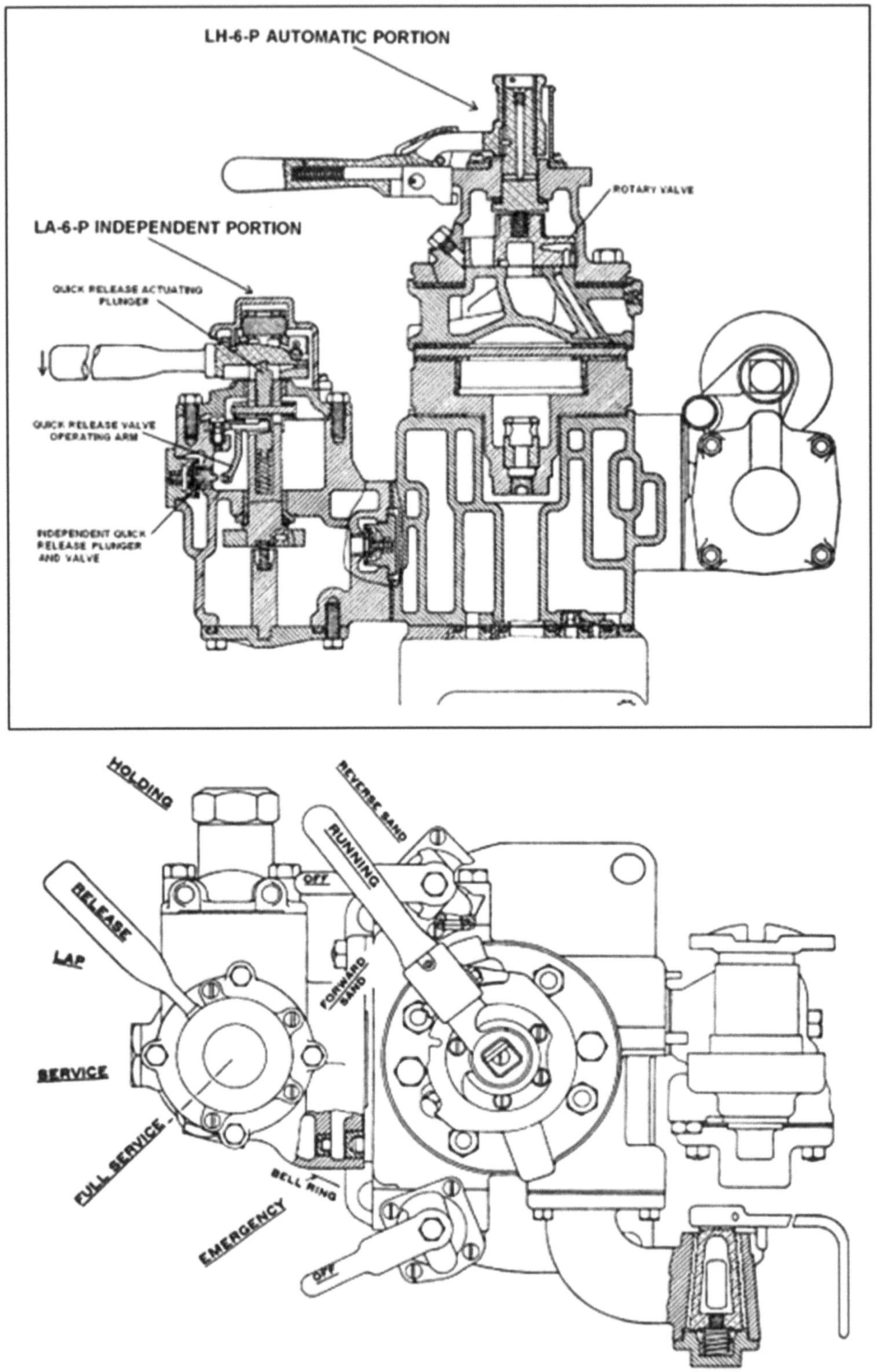

FIGURE 167: 6-SL AUTOMATIC AND INDEPENDENT BRAKE VALVE ARRANGEMENT – SECTIONED AND PLAN VIEW

THE A7-EL EQUIPMENT

This schedule is a development of the 6-ET and shares its principal features, including brake valve handle positions. Shortly after the successful introduction of the 6-ET equipment, the Westinghouse Air Brake Company was requested to adapt it for application to numerous classes of double-cab electric locomotives then being built for various railroads. These electric 'motors'—some of them extremely large and powerful machines—were the first American locomotives to require dual air brake stands. In turn, this made it necessary to compensate for the greater length of the application cylinder and distributing valve release pipes, which now had to connect the two brake stands with the single distributing valve. With two brake stands installed, removable brake valve handles were specified, and this evolutionary development is reflected in the subsequent '7' designation while 'EL' of course, denotes ***E***lectric ***L***ocomotive. The next major motive power development in the United States was dieselisation, and as this phenomenon progressed, the 7-EL equipment was soon being applied to diesel-electric locomotives. For this application, the 7-EL system was ideal because multiple-unit operation was now possible, requiring all air brake stands in the locomotive consist—except that on the Lead unit—to be 'cut out' (isolated from use) and the handles removable if desired by the railroad customer. Since no major design changes were required to accommodate the application of this equipment to diesels, no re-designation (e.g. to '7-DL') was required, and the 'EL' designation thus endured.

FIGURE 168: EARLY AMERICAN DOUBLE-CABBED ELECTRIC LOCOMOTIVE

The schedule was adapted by Westinghouse Australia for local requirements—principally the introduction of locally-built double-cabbed diesels, which required similar provisions to the early American electric locomotives with regard to the piping of two brake stands to one distributing valve—thus becoming the ***A***7-EL. Because of its commonality with the previous 6-ET/A6-ET equipment, and hence its familiarity to enginemen making the adjustment from steam locomotives, this new equipment was immediately accepted by railway administrations and was soon popular with engine service personnel. In Australia the A7-EL schedule began appearing on early diesels, such as the Commonwealth Railways GM-Class and the Victorian Railways B-Class from 1951, and in New Zealand it was specified for English-built electric and diesel-electric classes from 1954.

With 7-EL being essentially little more than 6-ET with a modified piping arrangement, it was inevitable that railway administrations would take the action, where they could, of providing brake valve equipment from retired steam locomotives to the manufacturers for recycling onto early diesel fleets. Victorian Railways—alert as ever against profligate expenditure—certainly did this with the purchase of its new B-Class Clyde-EMD fleet and the author has no doubt that other railways must have quietly done the same.

Early local versions of the equipment utilised an adaptation of the AH-6-P automatic brake valve; however, Wabco in the United States soon adopted the H-7, a development of the H-6 valve used with the 6-ET equipment, and—adapted by Westinghouse Australia—this became the ***A***H-7 and was thereafter used on Australian and New Zealand A7-EL systems. The AS-6 Independent valve was re-used as the AS-7.

The brake valve pedestal carries the Automatic and Independent brake valves, feed and reducing valves, pedestal isolating cock (with flow-meter venturi where fitted), and equalising reservoir control valve. The piping runs either to the base or to the back of the pedestal, and the various valves may be removed without the necessity to break any pipe joints. The system includes an electric compressor governor rather than one operated by steam pressure, two brake stands instead of one where the unit has two cabs or dual control stands, four instead of two duplex pressure gauges, and two instead of one combined equalising and minimum reduction reservoirs - again where the unit has two cabs or dual control stands.

Certain performance limitations of the AS-7 Independent brake valve soon became critical as Australian railways progressively began to utilise multi-coupled diesel locomotives. The time-lag, both of brake cylinder pressure development on the locomotive consist from an Automatic application and also for Independent quick-release (bail-off), became excessive on MU consists due to the increased total volumes of the control pipe, the Independent release pipe, and the distributing valve relay chambers, as well as greater pipe friction. For this reason, a faster-acting, self-lapping Independent brake valve was required and—under the urging of the NSWGR—was eventually sourced by Westinghouse Australia from the 6-SL schedule. With this improvement plus the replacement of the A7 distributing valve *Independent quick-release portion* with the newer B7 portion, the ***B*7**-EL equipment was created, significantly improving application and release times on multiple-unit locomotive consists.

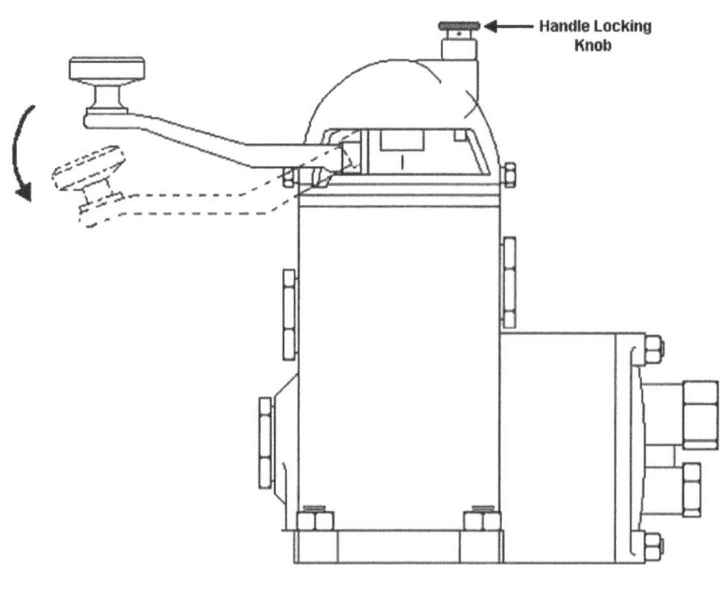

FIGURE 169: B7 INDEPENDENT BRAKE VALVE ARRANGEMENT – OUTLINE

This equipment continues in revenue service in Australia to this day on early-build Clyde-EMD and Comeng-Alco units. As the self-lapping Independent brake valve is supplied with air at main reservoir pressure (a principal reason for its faster operation), the reducing valve is eliminated and replaced with a ported blanking plate on the A7-EL brake valve pedestals. The B7 quick-release valve is diaphragm- rather than piston-operated and includes a local exhaust port. The larger diaphragm area enables the use of a heavier return spring, thus ensuring a more positive return to normal position after a brake release. A spring-loaded check valve replaces the ball check-valve of the A7 type. Apart from the manner in which the Independent brake valve is manipulated, the B7-EL equipment is identical in operation to the A7-EL.

The B7-EL schedule was not used in New Zealand (where A7-EL equipment only remains today on restored vintage English Electric diesel locomotives operated by railway heritage organisations); however, the B7 distributing valve quick-release portion was applied to many A-6 distributing valves on New Zealand A7-EL–equipped locomotives.

Some cultural differences between railway administrations regarding brake valve operability should be discussed here. Dating from the 5-ET schedule, American equipments were provided with the RELEASE position (sometimes referred to as FULL RELEASE) for the Automatic brake valve to permit the supply of recharge air at main reservoir pressure into the brake pipe to initiate the brake release. This was in an era predating the more advanced triple valves that would eventually incorporate features to enhance the propagation of the 'release' pressure signal down the brake pipe. This full-pressure main reservoir charge was intended to kick the brake release into action by boosting the brake pipe recharge for a brief period, albeit effective mostly at the head- rather than the rear-end of the train.

Use of the RELEASE position, though, could be a problem since the main reservoir pressure would overcharge the brake pipe if the handle was held there for too long. For this reason, RELEASE position—the valve porting being arranged to provide an audible airflow warning to the engineman—was intended to be used only briefly before the handle was returned to the RUNNING position. This passage from an early Westinghouse instructional pamphlet describes the complexity of the required brake valve manipulation:

Passenger Service – Direct Release:

In making the first release of a two-application stop [this being a technique with non-graduated release equipment intended to wash off high-speed energy before releasing and re-applying a more gentle application to make the eventual stop - the overall objective being smoothness and accuracy], *the brake valve handle should be placed to Release position and then quickly back to Running position, where it should be allowed to remain for an instant* [the duration of an "instant" not defined, and being at the judgment of the engineman], *(first, to permit the pressure in the equalising reservoir and brake pipe to equalise and second, to release part of the driver brake cylinder pressure)* ["driver" being a reference to locomotive driving wheels - this being an instruction dating from the steam era] *then moved to Lap position and from there again to Service position as required* [i.e. for the second application]. *In passenger service the time the handle is in Release position should be only momentary* [again, "momentary" being whatever the engineman thought it should be] *but the time in Running position should be governed by the conditions existing for each particular case, such as the length of the train, type of reduction made, time available and so on.*

In making the final release of a two-application stop with short trains, release [the brake] *shortly before coming to a standstill by moving the handle to Release position and immediately back to Running position, and leave it there. With long trains, the brakes should, as a rule, be held applied until the train stops.*

<u>Freight Service</u>:

If conditions permit a brake release while in motion, the brake valve handle should be moved to Release position and held there long enough [again, "enough" being a judgment call required of the engineman] *to move as many of the triple valves as possible to release without unduly overcharging the head-end of the train (the time in Release position should be governed by the length of train, amount of reduction made, etc.) then returned to Running position to release the locomotive brakes and complete the recharging of auxiliary reservoirs. A few seconds* [yet another judgment call] *after such a release, particularly on long trains, it is necessary to again move the handle to Release position and quickly back to Running position to 'kick off' any brakes at the head-end of the train that may have re-applied due to their auxiliary reservoirs having been slightly overcharged.*

The above piece of conversational instruction would have been appropriate in a classroom, where trainees could have asked questions, but it can nevertheless be seen that Automatic brake valve manipulation with the -ET (and thus also the -EL) equipments could be relatively complicated and required of the American engineman a significant degree of anticipation, prudence, and reasoning, including a situational awareness of the degree of brake pipe recharge and thus equipment state along the length of their train. Obviously, only an experienced, skilled, and knowledgeable locomotive driver would be adequately equipped to perform successfully as required.

Despite train handling requirements (including those pertaining to brake valve manipulation) within Australia and New Zealand being less complex by comparison to North American operations and equipment, and the AH-7 brake valve being slightly less complicated than its American progenitor (and thus, it could be argued, requiring a lesser degree of manipulatory skill to efficiently manage), Australasian engine-service personnel were still required to apply a critical degree of judgment in order to effect a smooth stop. The Victorians have long possessed a comprehensive air brake manual for operating and maintenance staff that is an excellent resource; in fact one of the best I've seen. On the subject of stopping passenger trains, it allows that:

Passenger trains equipped for direct release braking [which means all of the Victorian Railways locomotive-hauled passenger stock] *must be stopped at stations by two or more brake applications... The driver must aim at stopping the train where required with the brake cylinder pressures as low as practicable to avoid the tilting action of the bogies and the consequent reaction which causes severe jolts and discomfort to passengers.*

In the author's view, this is an important and prudent instruction. Too infrequently have I observed locomotive drivers make any specific attempt to stop passenger trains in a manner that will nullify the unsettling instant-of-stopping 'brake-snatch' phenomenon alluded-to in this instruction.

The VR instruction for the driver to stop "...*with the brake cylinder pressures as low as practicable*..." though, requires explanation. This was not a request that drivers merely make a light brake application to stop a passenger train; indeed, under the variable conditions that could apply to train handling, this might not be possible. It was a specific instruction for drivers NOT to come to a stand with a brake cylinder pressure that was high enough to cause brake-snatch and it might mean that after making whatever brake application the driver needed to get the stop under control, they would then release it at such a point prior to the stop that the train would come to a stand with the brakes releasing and brake cylinder pressure as "low as practicable". Getting this right continues to require precise judgement and takes great practice.

In some operations in New Zealand, using a light brake application to stop a passenger train simply was not possible, because it wasn't permitted. I recall informal advice from my learning of the requirements for running suburban passenger trains in Wellington, the specific instruction that—be it a locomotive-hauled carriage train (challenging) or an Electric Multiple Unit (more easily achieved)—the approach onto the platform should be at around 60 km/h and the train bought 'smartly' to a stand with the train brakes releasing (and thus a sufficiently low brake cylinder pressure to minimise any brake-snatch). This was a deliberate strategy to discourage passengers from either alighting or boarding before the train had stopped, and—with our direct-release triple valves of the day—required a brake release to be initiated at a well-judged prior point. Our colloquial term for this was a 'grand-stand' stop, and drivers were judged by both passengers and their peers as to their ability to smoothly achieve one. Often, control of speed during this phase required the assistance of a partial Independent brake application and we became quite adept at this as well. A perfect stop wasn't always achieved, but my view was that so long as one was always striving toward that goal then passenger train stops would always be better rather than bad. It stands to reason that grand-stand stops were NOT conducted at arrivals into Wellington's terminal station (and HQ) where all 9 platform tracks terminate.

The NZR did assist with valuable instructive advice regarding how to stop passenger trains[124]:

These trains are divided into several types. Suburban passenger trains are usually between 8 and 14 cars, and should be kept stretched at all times. Holding the locomotive brake off at the first reduction with a little power on will keep the slack out, then throttle back and allow the locomotive brake to apply at the second and subsequent reductions. This will keep slack stretched but not so tightly that the rear cars move up after the stop.

The automatic brake must be released prior to the stop so that brake cylinder pressures are low at the instant of stopping. The point-of-release requires some judgement but if a reduction of 100 kPa or greater has been made, the speed at release should rarely be less than 15 km/h and on level track could be 20 km/h. If the reduction has been less than 100 kPa then there is less brake cylinder pressure to exhaust, so speed-at-release can be less.

Releasing at the correct speed allows the drawgear tension to adjust; preferably to a lightly-stretched condition as the train comes to a stand.

A7-EL BRAKE VALVES ON A VR DIESEL LOCOMOTIVE
Photo: Bill Perrusich, from **The ML2 Story: The History of the Victorian Railways' Famous B Class Diesel-Electric Locomotive, *by Peter Bermingham***

B7-EL BRAKE VALVES ON A NSWGR ALCO LOCOMOTIVE
Photo: Maikha Ly

FIGURE 170: A7-EL AND B7-EL LOCOMOTIVE BRAKE VALVE CAB INSTALLATIONS

124. *Air Brake Handbook* – New Zealand Government Railways, 1970

Some Australian railways noticed an increased incidence of sticking brakes on freight trains, and the foremost reason for this appeared to have been the inappropriate use by drivers of the RELEASE position. Since air brake testing had revealed that adequate brake pipe charging and recharge performance via the feed valve (i.e. using RUNNING position) was achievable, some railways sought—in the impulsive management fashion customary of the times—to have the pesky RELEASE position on A7-EL and B7-EL brake valves either removed or nullified, rather than invest in re-educating drivers as required.

Commonwealth Railways in South Australia modified some brake valves by welding a 'stop' on the quadrant, thus blocking the RELEASE position from use. This resulted in complaints from drivers (and presumably some pressure from the union), so Westinghouse Australia was asked to design a special filling piece for the pedestal pipe bracket attachment that deleted the main reservoir porting in the RELEASE position, leaving the handle RELEASE position intact but rendering it superfluous. This modification remained prototypical due to production costs, and so the Commonwealth introduced a modification—with details of which I am unfamiliar—that was apparently not easily detectable and ostensibly designed to keep the troops happy by hoodwinking them into thinking they still had the use of a RELEASE position. I do not know if this proved to be an effective tactic. The BHP Company based at Whyalla sought—during the mid-1970s—a similar solution, and Victorian Railways looked at removing RELEASE position as late as the 1980s but did not proceed.

In New Zealand, RELEASE position on A7-EL equipment was left operationally intact; however, this did not carry through to the later introduction of 6-SL equipment on which, as has been discussed, a modification also common in the United States—the provision of a "Full Release nullifier" attached to the handle—was employed to prevent access to RELEASE position on the brake valve quadrant.

The NSWGR Metropolitan Vickers/Beyer-Peacock 46-Class electric locomotives that entered service from mid-1956 were fitted with a modified A7-EL schedule that included the AS-8 independent brake valve. This was called the A8-EL brake equipment, and it functioned similarly to the A7-EL; however, the AS-8 Independent brake valve was larger and had certain ports of a larger diameter than the AS-7, resulting in a more rapid application and release of the Independent brake. There was no difference in its operation.

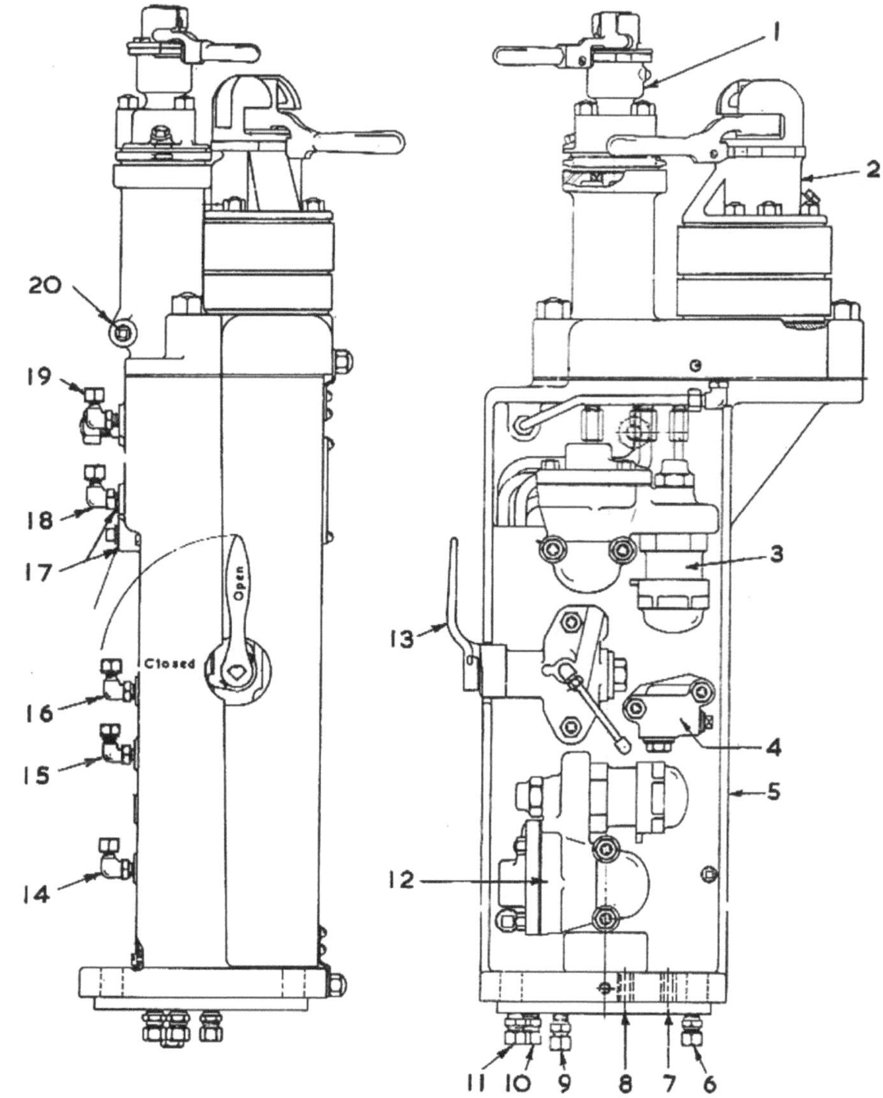

LEGEND

1. Independent brake valve
2. Automatic brake valve
3. Feed valve
4. Equalising reservoir control valve
5. Brake valve pedestal
6. To equalising reservoir
7. To brake pipe
8. To main reservoir
9. To minimum reduction reservoir
10. To independent release pipe
11. To control pipe
12. Reducing valve
13. Isolating cock
14. To eq. res. pressure gauge
15. To vigilance control
16. To main res. pressure gauge
17. To flowmeter
18. To brake pipe pressure gauge
19. To sanders (if required)
20. To magnet valve (if required)

FIGURE 171: THE A7-EL LOCOMOTIVE BRAKE PEDESTAL

B7–EL BRAKE VALVE PEDESTAL

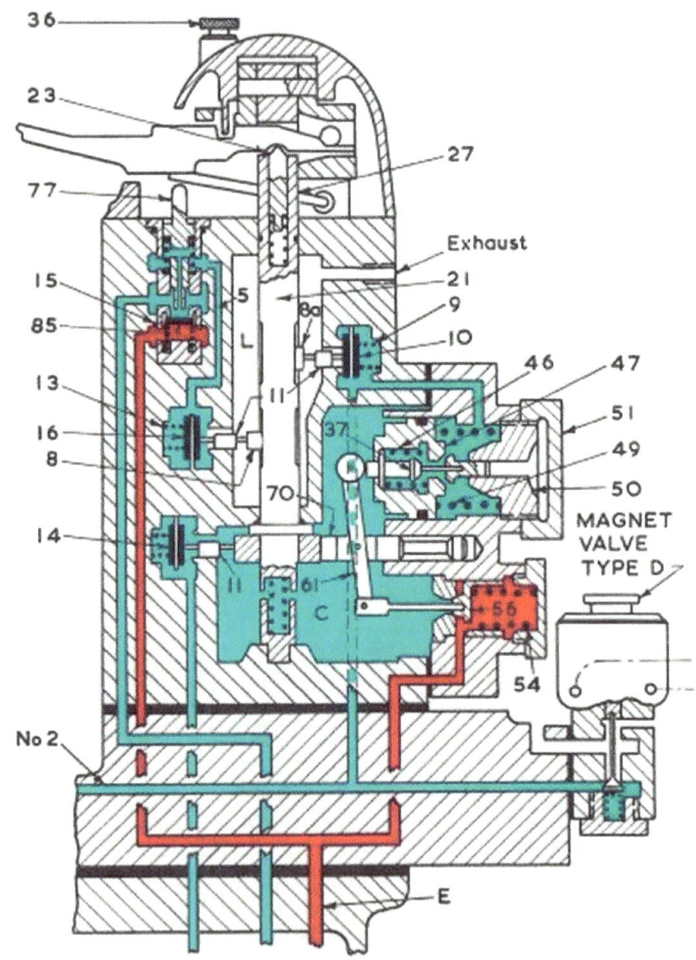

B7 INDEPENDENT BRAKE VALVE - Diagrammatic

FIGURE 172: THE B7-EL LOCOMOTIVE BRAKE PEDESTAL (ILLUSTRATION) AND INDEPENDENT BRAKE VALVE (DIAGRAM)

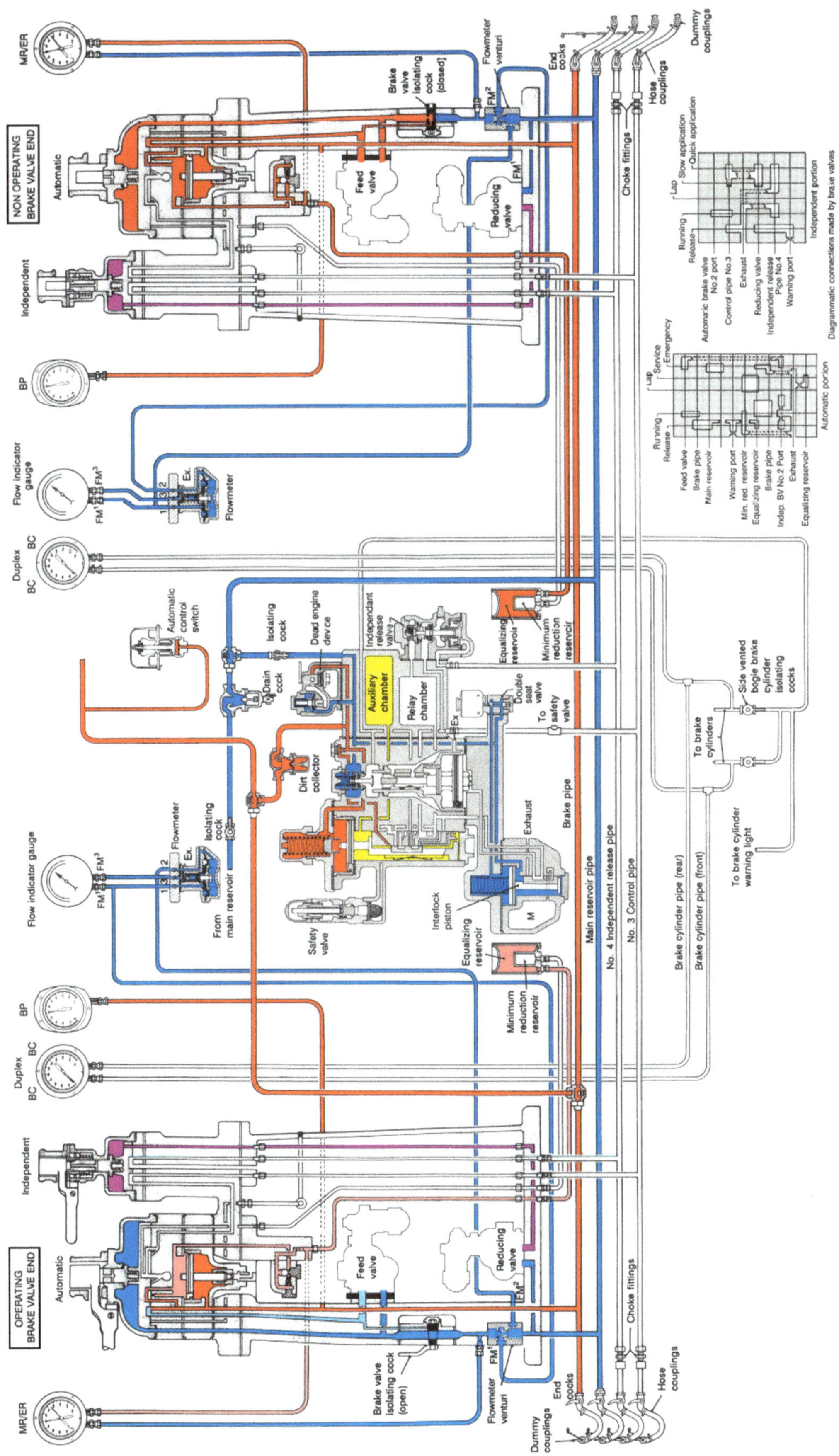

FIGURE 173: THE A7-EL AIR BRAKE SCHEDULE, 2-CAB ARRANGEMENT – DIAGRAM

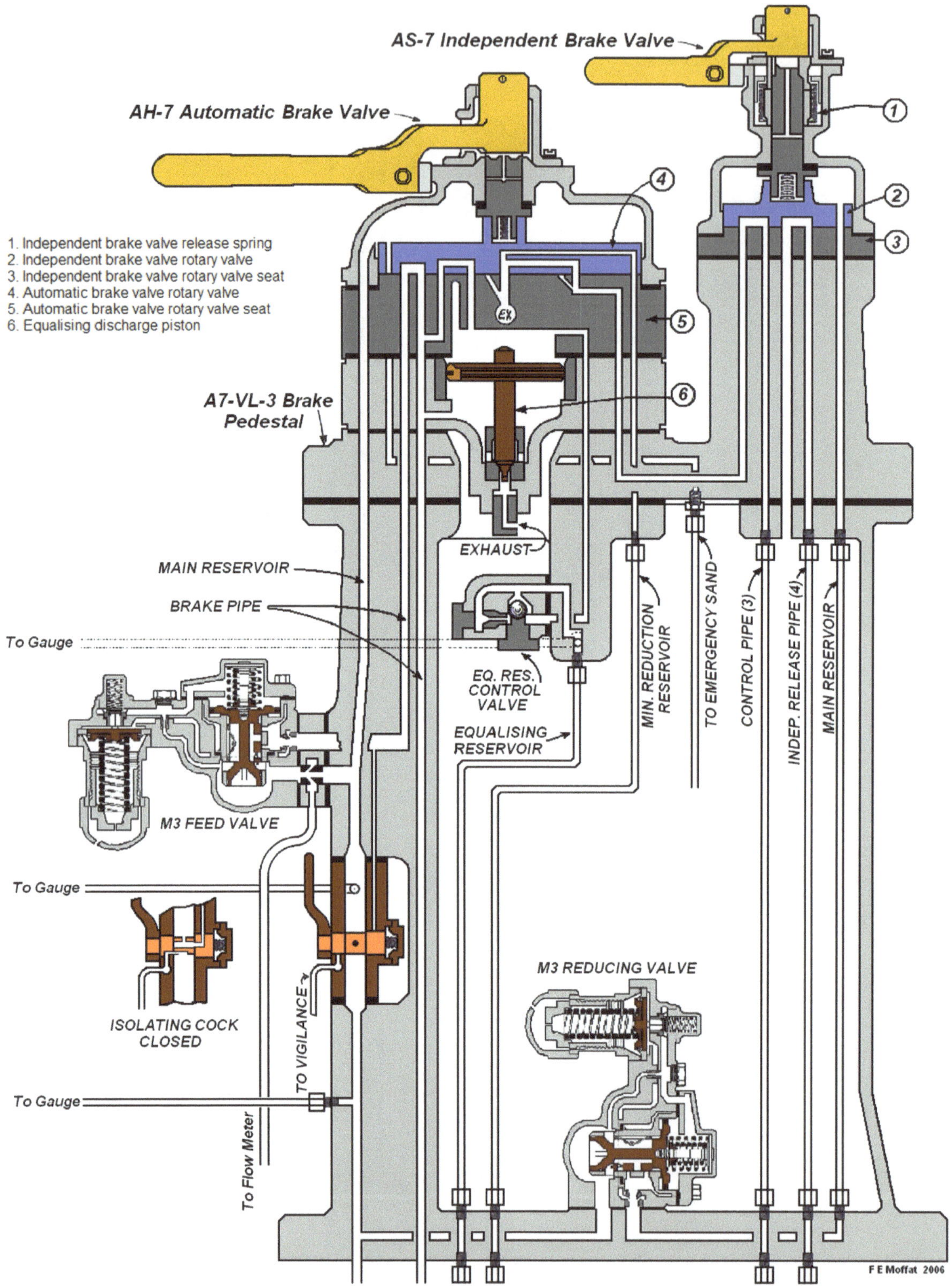

FIGURE 174: SIMPLIFIED BRAKE PEDESTAL DETAIL, A7-EL BRAKE – DIAGRAM

THE 14-EL EQUIPMENT

Dating from the 1920s, this US equipment followed the 7-EL and was offered to railroads for much the same reasons, being a further adaptation of and embodying the operative features and advantages of the 6-ET equipment (but including a *collapsible piston*[125], a *regenerative interlock* and *compressor governor synchronisation*) and applicable to any electric locomotive in any category of service without change or special adjustment. Its features included providing great flexibility in control of brake cylinder pressures and in operation, locomotive brake cylinder pressure maintenance against leakage or piston travel variations, emergency capability, and having a single common standard for all categories of locomotive service. These made the equipment an ideal fit-out for yard locomotives and, as with the 7-EL system before, allowed it to be easily applied to diesel-electric switchers.

Its only uses in Australia were on a lone 1946-vintage ex-Canadian National Railway GMD NW-2 model switcher imported for use by Comalco at Weipa in Queensland, and the Queensland Government Railways' first main-line diesel-electric units, the 1210 class (later reclassified 1300 then eventually 1150) built by General Electric in the United States in 1952 (Figure 175).

The standard brake valve for single-end 14-EL equipment was the K-14-B, and from photos supplied to me, this appears to be the valve fitted to the Comalco unit (Figure 176), as was the No. 6 distributing valve or variant (e.g. 6-DKR). The difference between both these devices and the usual K-14-A brake valve and № 14 distributing valve was simply that a *compensating port* within both was omitted due to the shorter *distributing valve release* and *application cylinder* pipes which resulted from the elimination of one of the two brake valves. The K-14-B differed from the H-6 in that both Automatic and Independent portions were on the same bracket to which all pipe connections were made. The Automatic brake valve handle was removable in LAP position, the Independent handle likewise in the RUNNING position. The difference in operation from the H-6 was that reducing valve air was connected to both the distributing valve release pipe and application cylinder pipe when the Independent brake valve was in the QUICK-APPLICATION position. With the equalising portion of the distributing valve in its CHARGING/RELEASE position, a more rapid build-up of application cylinder pressure occurred to provide a faster Independent application; a highly desirable feature for a shunting locomotive. A transfer valve assembly integrated with the distributing valve ensured the desired rate of Independent brake operation was obtainable and—along with a *brake cylinder equalizing* pipe—made the system multi-unit capable in air brake operation.

FIGURE 175: AUSTRALIAN LOCOMOTIVES THAT USED THE 14-EL AIR BRAKE SCHEDULE

125. See Collapsible or "By-Pass" Equalising Discharge Piston in the Glossary, chapter 8.

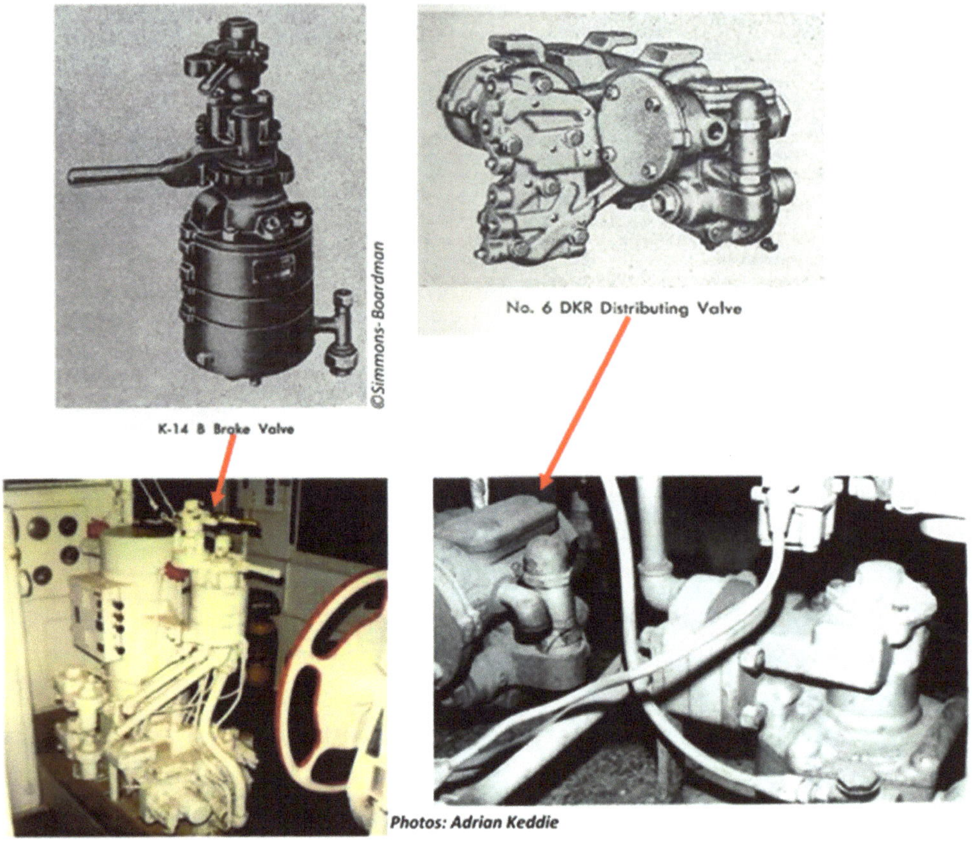

FIGURE 176: № 14-EL SCHEDULE EQUIPMENT AS INSTALLED ON COMALCO SHUNT LOCOMOTIVE

THE 24-RL EQUIPMENT

A brief description of this equipment will be included since it was used in a transitory way in Australia and was the immediate precursor to the celebrated 26-L schedule. But it was also responsible for a major advance toward brake pipe pressure maintenance during a brake application, and this characteristic is explored in some depth in chapter 6.

A unique air brake schedule with extensive application throughout North America, this equipment was developed in the United States to provide a generic system that could be applied to any steam, diesel, or electric locomotive in any type of service. It was not a new air brake scheme but rather a design that included, in one package, features widely used on other equipments then extant. These features were train control functions developed for the high-speed, heavyweight (and sometimes streamlined) passenger trains that had entered service from the late 1930s, and features on the 8-ET/8-EL and 14-EL equipments designed for steam, electric, and diesel freight and passenger locomotives (e.g. dead-man safety control, overspeed control, and train signal control). Because of the proliferation of these functional complexities, locomotive manufacturers and railroads had demanded basic brake equipment to which various features could be furnished by the addition of modular, interchangeable components. Wabco responded with a design for an affordable locomotive air brake schedule that would provide the flexibility of quick changeover to meet all ***R***oad ***L***ocomotive requirements.

The D24 Automatic brake valve was the basic component, and was a pedestal brake valve with external quadrant arranged for diesel locomotive installation, with an Independent brake valve mounted separately. Various portions could then be added in a clever 'sandwich stack' arrangement to provide the optional brake valve features. One such was the important new feature of pressure-maintenance against brake pipe leakage during service applications. It is often thought within Australasia that the feature of pressure-maintenance during Service applications originated with the 26-L, but this is not so (although the feature was much refined in 26-L).

The 24-RL equipment only appeared in Australia on two EMD F7 units imported second-hand from the Western Pacific Railroad in the United States during 1968 for construction of the Mount Newman Mining iron ore railroad in Western Australia. Both units then operated ore trains before being retired once the first Century-Series ALCOs arrived. One unit was restored to operating condition by a volunteer organisation and has been used infrequently for special trips on the Hamersley Iron track of Pilbara Rail (later Pilbara Iron), based in Dampier. The other unit resides (as far as the author knows)—minus prime mover, traction motors, and many other parts—as a static display in a Port Hedland park, having been internally refurbished as a model railway clubroom. I recall several challenging days in 1985 laboriously disassembling the air brake pedestal of this latter unit for shipment to the Pilbara Railway Historical Society as a spare parts source. The modular composition of the 24-RL brake pedestal was evident as each portion—one mounted atop the other—was removed for refurbishment.

The reader is referred to 'The development of Maintaining Braking' in chapter 6 for a further description of how the 24-RL air brake schedule contributed to the evolution of railway air braking.

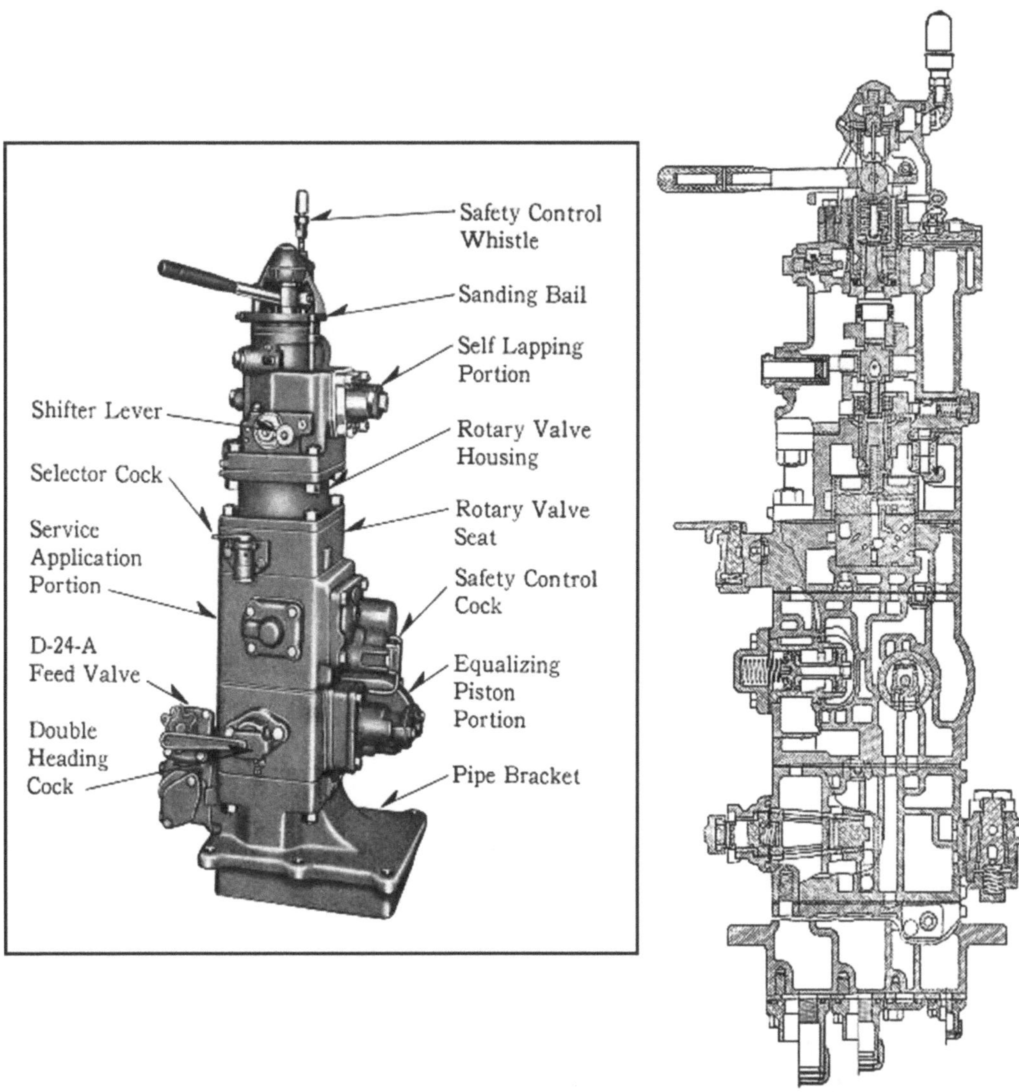

FIGURE 177: 24-RL AIR BRAKE SCHEDULE. THE D-24 BRAKE VALVE PEDESTAL (DSE-24-H DRIVER'S BRAKE VALVE) - ILLUSTRATION AND DIAGRAM

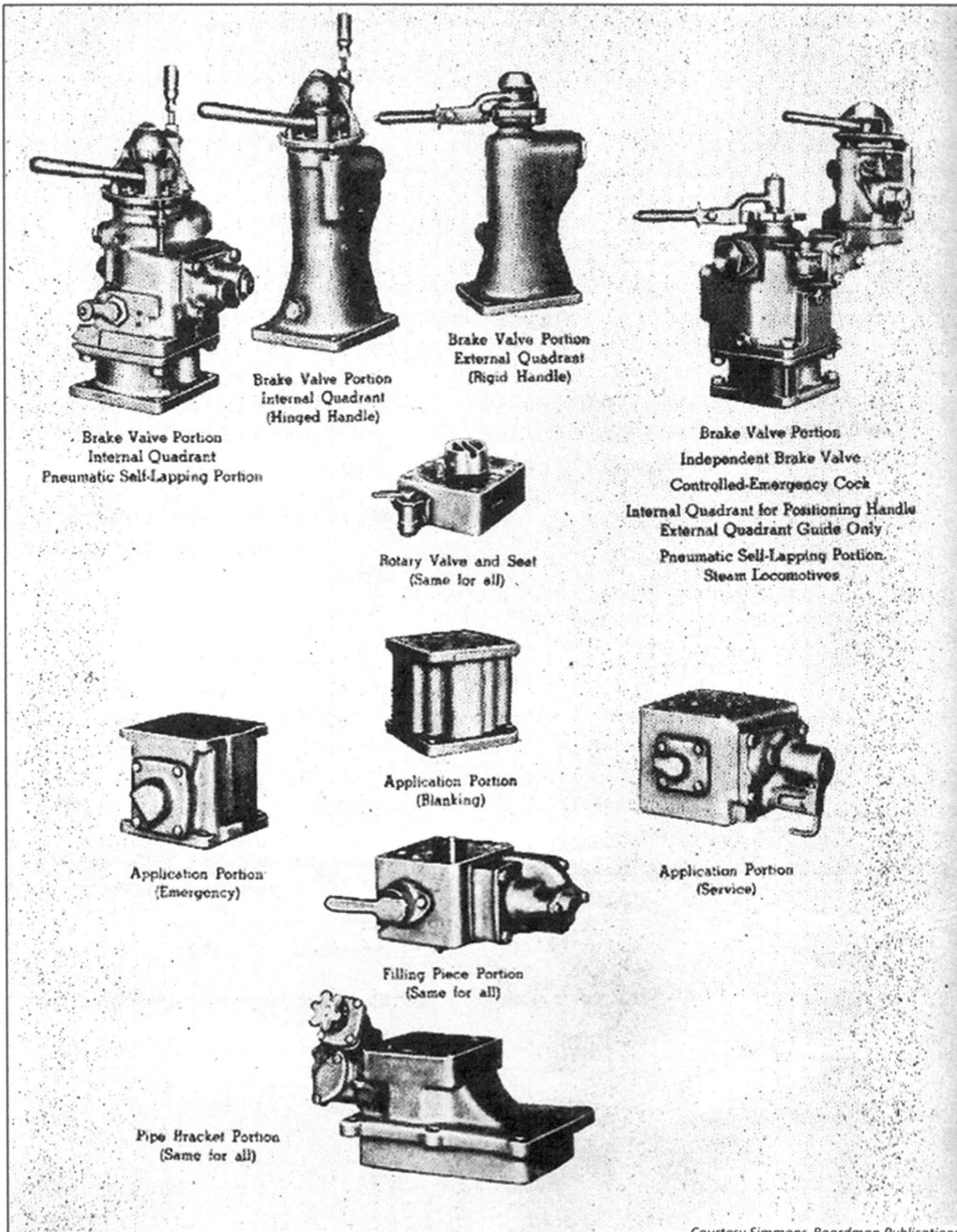

FIGURE 178: THE D-24 BRAKE VALVE PEDESTAL MODULE PORTIONS

THE 26-L EQUIPMENT

This schedule and its innovative arrangement became the definitive AAR-standard pneumatic system. Other fully pneumatic equipments to follow—for example, the 30-Series brake systems—would do little more than improve upon various engineering design or operator-interface aspects of this apparatus, and subsequent electronic systems—such as EPIC and CCB—would essentially replicate and adapt the pneumatic logic of the 26-L as well as its cab presentation and functionality. For this reason, the principles of operation of the Automatic and Independent portions—these remaining common to the ensuing pneumatic variants mentioned—will be explored in greater detail than with other equipments, hopefully keeping the basics easily understood.

The 26-L schedule was the inevitable result of the continuation of air brake engineering research and development driven by unrelenting pressure from railroad operators. This new system's distinctive attributes provided the ultimate in fully-pneumatic air brake operation and represented a quantum leap in locomotive-hauled train air braking technology. By the late 1950s, 26-L was in service with its first North American customers and its manifest benefits were being used and appreciated. New Zealand Railways had taken delivery of a fleet of 40 EMD G12 export model locomotives, commencing late in 1955 (equipped with the 6-SL brake equipment) and when this order was extended by a further 106 units with delivery commencing during 1961, NZR specified 26-L and became the first global customer outside of North America for this new equipment. The 26-L schedule was thereafter adopted for all diesel locomotive acquisitions in New Zealand until the Chinese-built DL-Class arrived in 2010 with 30-Series equipment.

The first appearance of the 26-L schedule in Australia occurred with three Alco Century-series C628 units and a centre-cab C415 imported between 1965 and 1968 from the United States (the C415 being an ex-Alco company demonstrator) for construction and line-haul operations on the Hamersley Iron railroad in the Pilbara region of Western Australia. West Australian Government Railways (Westrail) was the first Australian state rail system to acquire 26-L equipment when it was specified for the standard-gauge Clyde-EMD L-Class fleet (export model GT26CW, based on America's domestic SD-40) that entered service from late 1967. NSWGR introduced 26-L with the landmark Clyde-EMD 422-Class (J26C) in 1969; the Commonwealth Railways with their Clyde-EMD CL-Class (AT26C, a full carbody version of the Westrail L-Class) from early 1970; QR with the Clyde/Comeng-EMD 2100-Class (GL26C) of late 1970; South Australian Railways with the Goodwin-ALCO 700-Class (DL500G) in mid-1971; and Victorian Railways (VicRail)—doubtless wanting to first use up their stocks of salvaged A6-ET equipment—with the later series of their Clyde-EMD X-Class (G26C) of 1975/76.

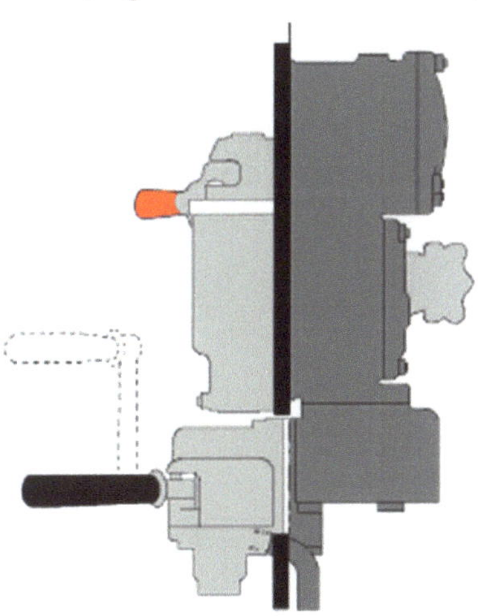

FIGURE 179: 26-L LOCOMOTIVE BRAKE VALVES SHOWING HOW MUCH OF THE ASSEMBLY SITS CONCEALED BEHIND THE CONTROL STAND FASCIA (THICK BLACK LINE)

Although the driver's brake valve and other components have been much modified and adapted since, the basic premise of the equipment remains and has been applied around the world in its various guises.

The design originated from the Westinghouse Air Brake Company, however arising from the commercial concerns of the railroad industry (around monopoly) and by US government regulations, the design was shared with, and the equipment was also manufactured, by the New York Air Brake Company. Note that although this action, presumably, allayed industry fears and certainly afforded NYAB a slice of the market, once they were part of the Knorr-Bremse Group and able to offer advanced brake systems of their own, they quickly—during 1993—ceased manufacturing the irksome Westinghouse product.

With 26-L the locomotive driver is still required to move the brake valve handle around an operating quadrant against which the conventional, spring-loaded handle locating latch is able to trace the handle operating positions with reference to detents machined into the quadrant profile. Gone, however, are pedestal-mounting and exposed piping, the rotary valve and seat, the equalising piston (and other ring-fitted pistons), and slide valves. In their place are a tidier, more compact panel-installation, a camshaft rotated by the brake valve handle (and by which various spool valves and diaphragm-operated pistons are moved or induced to move to provide the working functions), and a piloted, diaphragm-operated relay valve providing the equalising function.

The design provides for freight (direct-release) or passenger (graduated-release) service without altering the piping or component devices and is suitable for multi-unit operation with locomotives equipped with the same or older generations of brake equipment. The modification required to configure the Automatic brake valve for passenger operation is minor (see the Cut-off pilot valve detail in Figure 182).

The 26-L schedule introduced to American locomotive air braking the new design characteristics of cam-operated spool valves sealed by "O" rings and poppet valves in place of the rotary valve, slide valves, and piston rings, as well as two seminal performance features - reliable brake pipe pressure maintenance during brake applications in accordance with a self-lapping Automatic brake valve. Although the antecedent American 24-RL equipment had utilised a poppet valve—cam-operated from the rotary valve spindle—as an Emergency valve and even with a degree of brake pipe pressure maintenance being also provided to stabilise service reductions where leakage was excessive, it was 26-L that would launch these technical innovations into comprehensive use, and this critical advance cannot be overstated.

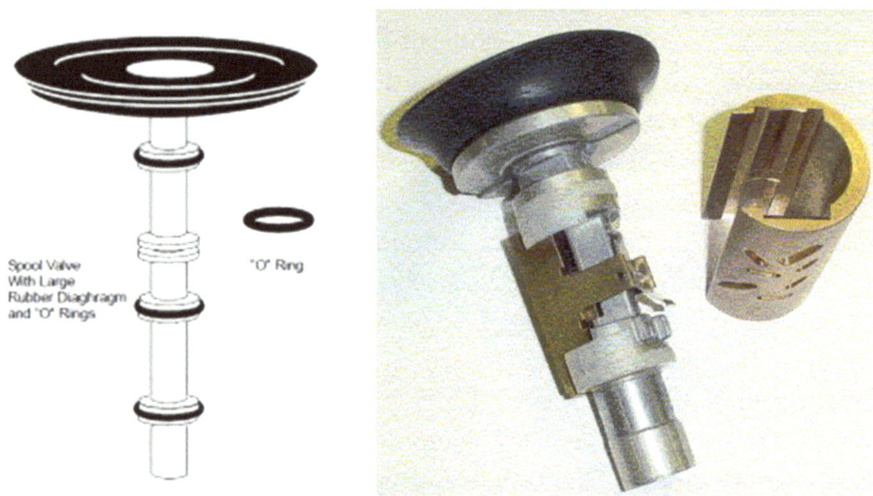

FIGURE 180: THIS ILLUSTRATION SHOWS (LEFT) THE DISTINCTIVE DIAPHRAGM, SPOOL, AND "O" RING SEALS THAT MADE THE 26-L AUTOMATIC AND INDEPENDENT BRAKE VALVES SUCH AN OUTSTANDING TECHNICAL ADVANCE ON PREVIOUS ROTARY VALVE EQUIPMENTS, AND (RIGHT) THE DIAPHRAGM SPOOL AND SLIDE VALVE ASSEMBLY OF MODERN WABTEC WAGON CONTROL VALVES. TOGETHER, THESE DEVICES REVOLUTIONISED TRAIN AIR BRAKING PRIOR TO THE ADVENT OF ECP BRAKING

A self-lapping Automatic brake valve, especially when coupled with an automatic (or 'programmed') minimum brake pipe reduction, provides precision in effecting brake applications and permits the locomotive driver to maintain their concentration outside the cab when necessary, focusing only on the gauges to the extent that making the required equalising reservoir reduction may require it. That said, in my professional opinion **self-lapping** has been largely responsible for the widespread—and lamentable—situation whereby many (perhaps even most) drivers that I have observed, these days make brake applications somewhat vaguely rather than by thinking about what they are doing and with scant reference to the gauges. Regrettably, one suspects this mediocre state of affairs will remain the norm.

The other vital function—of automatically maintaining brake pipe pressure (against reasonable leakage) during applications—is another matter altogether. This singular feature has enabled the smoother, more speed-consistent, and safer operation of larger trains in general operation (but most specifically on challenging downgrades) than was hitherto normally possible. It might be argued as well, that 26-L has also simplified the learning task of trainee locomotive drivers, significantly reducing the apprenticeship phase - excessively so, in my opinion.

It is of supreme irony, therefore, that having eventually accepted this remarkable equipment into service, some Australian railways took the bizarre action of finding a means to nullify its most exceptional attribute - automatic brake pipe pressure maintenance. Why this occurred, we shall visit later in this section.

The 26C Automatic and SA26 Independent brake valves comprise the two principal operating components of the 26-L equipment; both mounting to a convenient, common pipe bracket. The Automatic brake valve component itself consists of two integral valvular elements: the **Regulating** and **Relay** portions.

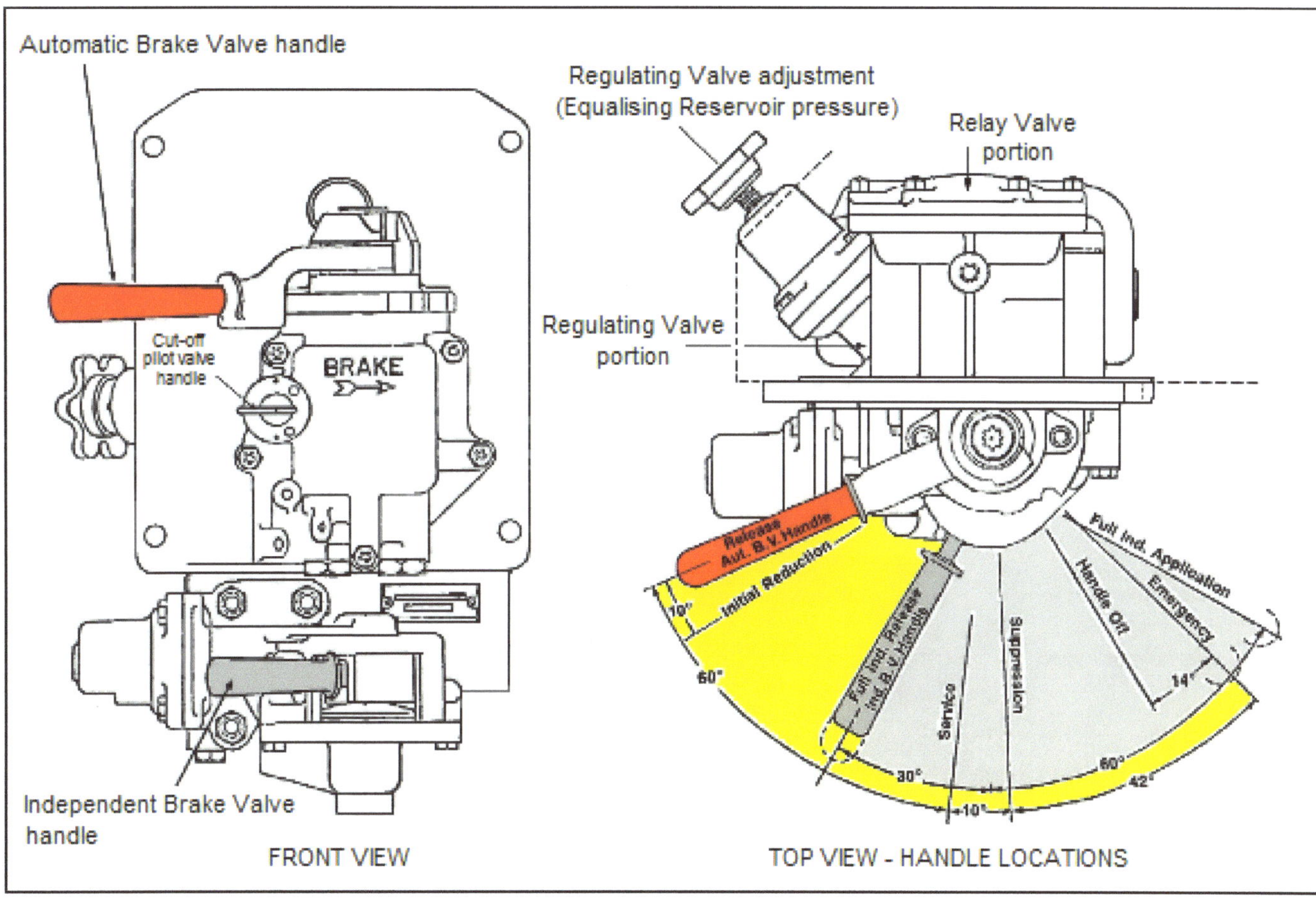

FIGURE 181: THE 26C AUTOMATIC AND SA-26 INDEPENDENT BRAKE VALVES DEPICTED IN THEIR CONTROL-STAND PANEL ARRANGEMENT

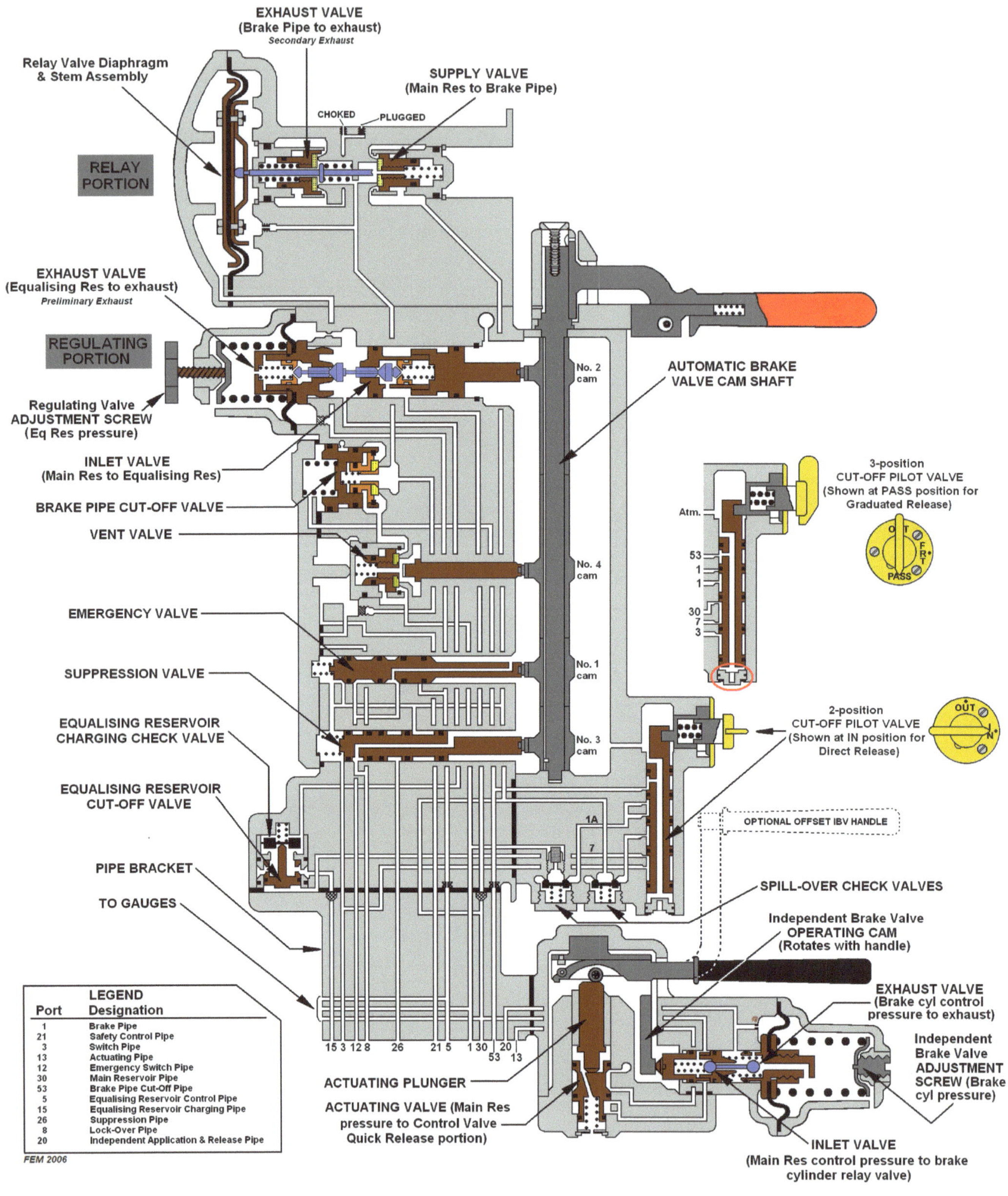

FIGURE 182: THE 26C AUTOMATIC AND SA26 INDEPENDENT BRAKE VALVES - DIAGRAM. THIS ILLUSTRATION REPRESENTS THE INTERRELATIONSHIP BETWEEN THE INTEGRAL PARTS AND PORTING OF THE BRAKE VALVE ASSEMBLY, RATHER THAN ITS PHYSICAL LAYOUT

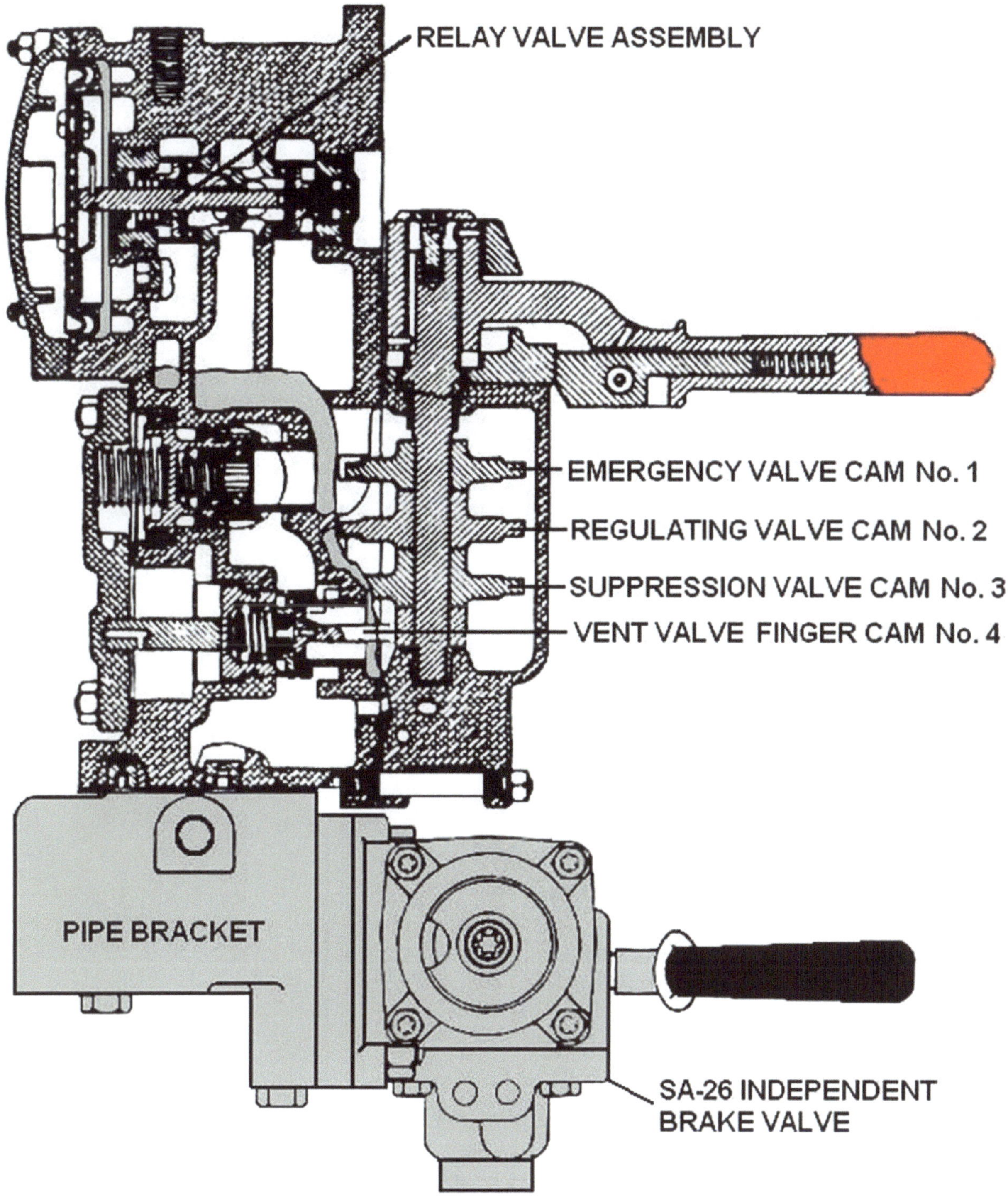

FIGURE 183: THIS ILLUSTRATION REPRESENTS THE ACTUAL CONSTRUCTION OF THE BRAKE VALVE AND CAN BE COMPARED WITH THE DIAGRAMMATIC VIEW SHOWN IN FIGURE 182

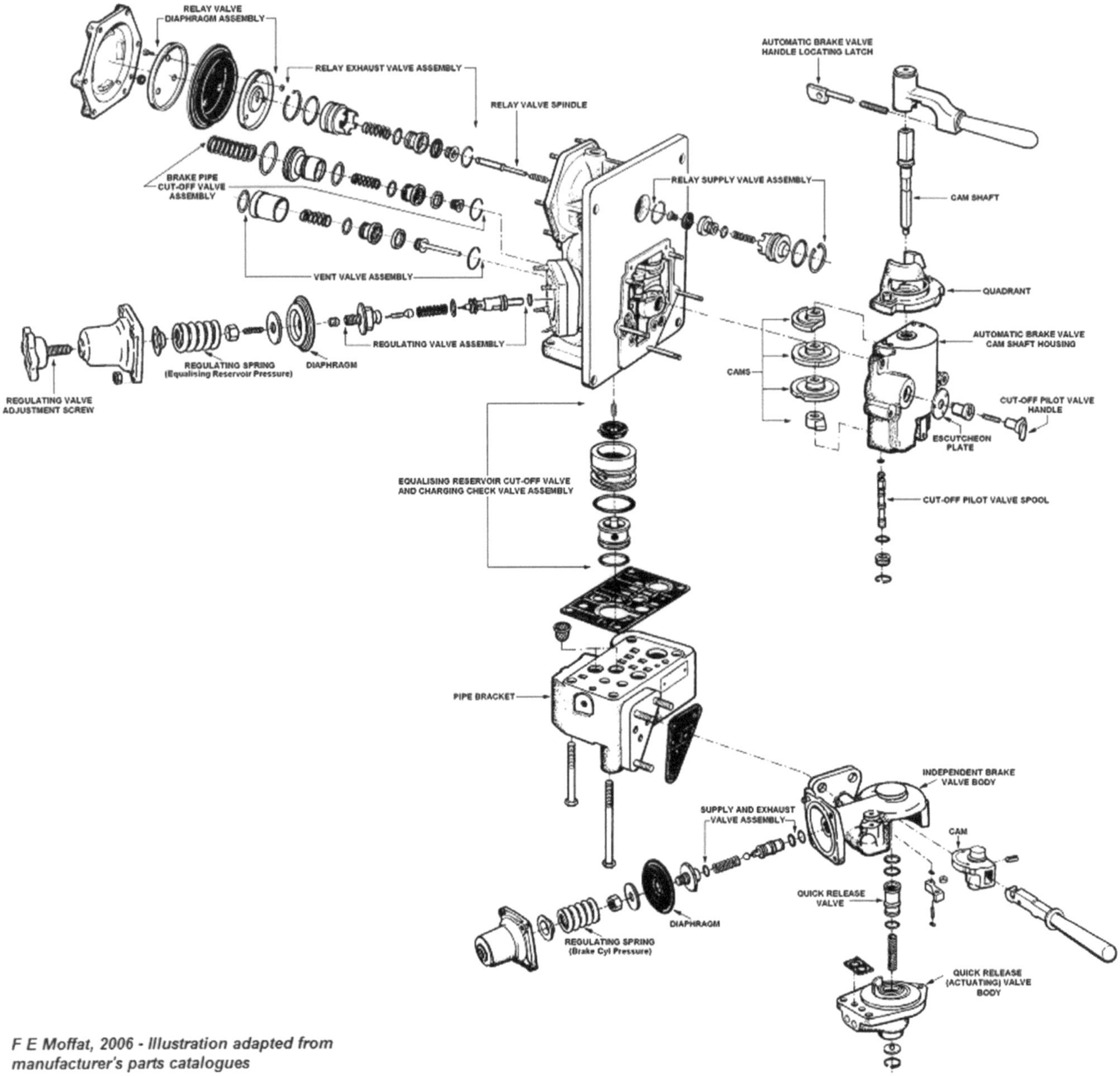

FIGURE 184: 26-L BRAKE VALVE EQUIPMENT – EXPLODED VIEW

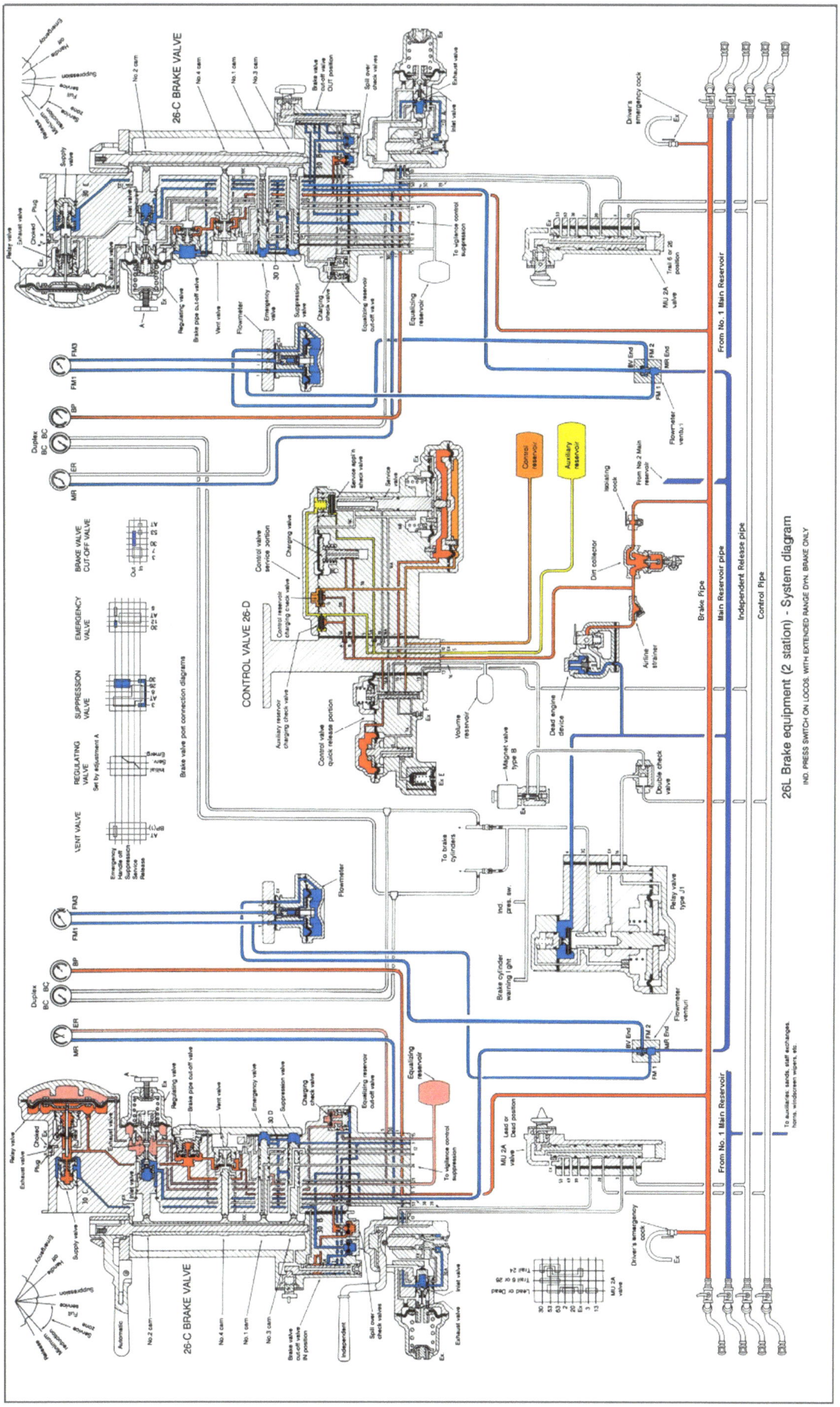

FIGURE 185: 26-L AIR BRAKE SCHEDULE (AUSTRALIA AND NEW ZEALAND) – SCHEMATIC DIAGRAM, 2-CAB ARRANGEMENT, F-1 SELECTOR VALVE NOT REQUIRED

26C Automatic Brake Valve

The following two series of drawings are provided to explain the internal functioning of the 26C Automatic brake valve. For those who hope for an easy way to grasp this learning, there is disconcerting news. A working comprehension of the driver's brake valve requires some study of the subject, and these diagrams are intended to facilitate that process. I created the first set of drawings many years ago, before the arrival of computer-based training and after seeing a Burlington Northern Railroad air brake training film that featured use of a large-scale, one-dimensional board-model of a 26-L air brake system. This training resource incorporated sliding valvular elements able to be moved by the instructor to demonstrate the internal working of the brake valve. I readily acknowledge the unsophisticated nature of these illustrations when compared to contemporary training resources that are available with advanced computer graphics, but I hope they will prove useful in conveying how the brake valve functions.

Regulating Portion

This is the equivalent of the *feed valve* in older equipments. It can be adjusted by a manual screw to provide the desired brake pipe pressure. Its two main components are an *inlet* and an *exhaust valve* - the former supplying the main reservoir feed to the equalising reservoir and the latter permitting the equalising reservoir to exhaust when the Automatic brake valve handle is moved away from Release position.

The Regulating valve has a self-lapping function and is the only one of the four valves operated by the camshaft to be held in tension when the brake valve handle is at RELEASE (Figure 186). The degree of tension that exists between the adjustment spring and the camshaft is established by the setting of the manual adjustment knob (or set screw). By holding the inlet valve open, it permits main reservoir pressure to charge the equalising reservoir and also flow to the outer side of the diaphragm in the Relay portion. Equalising reservoir pressure is also fed through a choke to the inner face of the Regulating valve diaphragm, and as this pressure develops, it begins to counter the regulating spring tension until the pressure called-for by the setting of the adjustment screw is achieved.

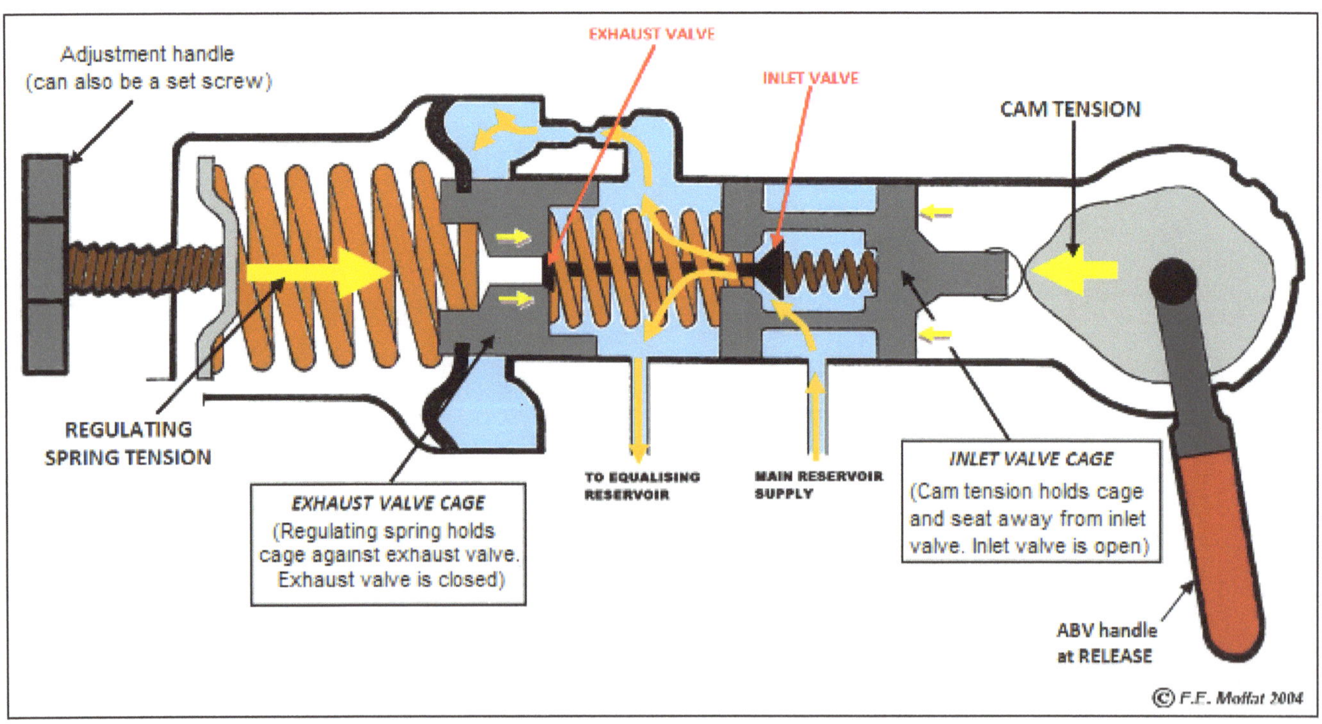

FIGURE 186: 26C AUTOMATIC BRAKE VALVE REGULATING PORTION – *RELEASE* POSITION

At this point, the assembly is balanced and the Regulating valve diaphragm assembly moves under the influence of minor spring tension to close the *inlet valve* - the Regulating valve being now 'lapped'. The equalising reservoir is thus charged and will pilot the Relay Portion to duplicate this pressure in the brake pipe.

When the aforementioned spring tension is relieved—by the regulating assembly coming 'off-cam' when the brake valve handle is moved from RELEASE—the exhaust valve is able to open and permit equalising reservoir pressure to escape to atmosphere by an amount proportional to the degree of handle movement (Figure 187). This is referred to as the brake application ***preliminary exhaust,*** and it occurs via the Regulating spring chamber. The preliminary exhaust is of extremely short duration since the equalising reservoir system is of small capacity (consisting of a small length of piping, some ports and passages within the brake valve, and a reservoir about the size of a basketball) and is barely discernible, as the more-obvious brake pipe exhaust follows very quickly. When equalising reservoir pressure reduces to a value consistent with the now-reduced regulating spring tension (caused by brake valve handle movement), the diaphragm assembly moves to close the equalising reservoir *exhaust valve* and the Regulating valve assembly again assumes a 'lap' condition.

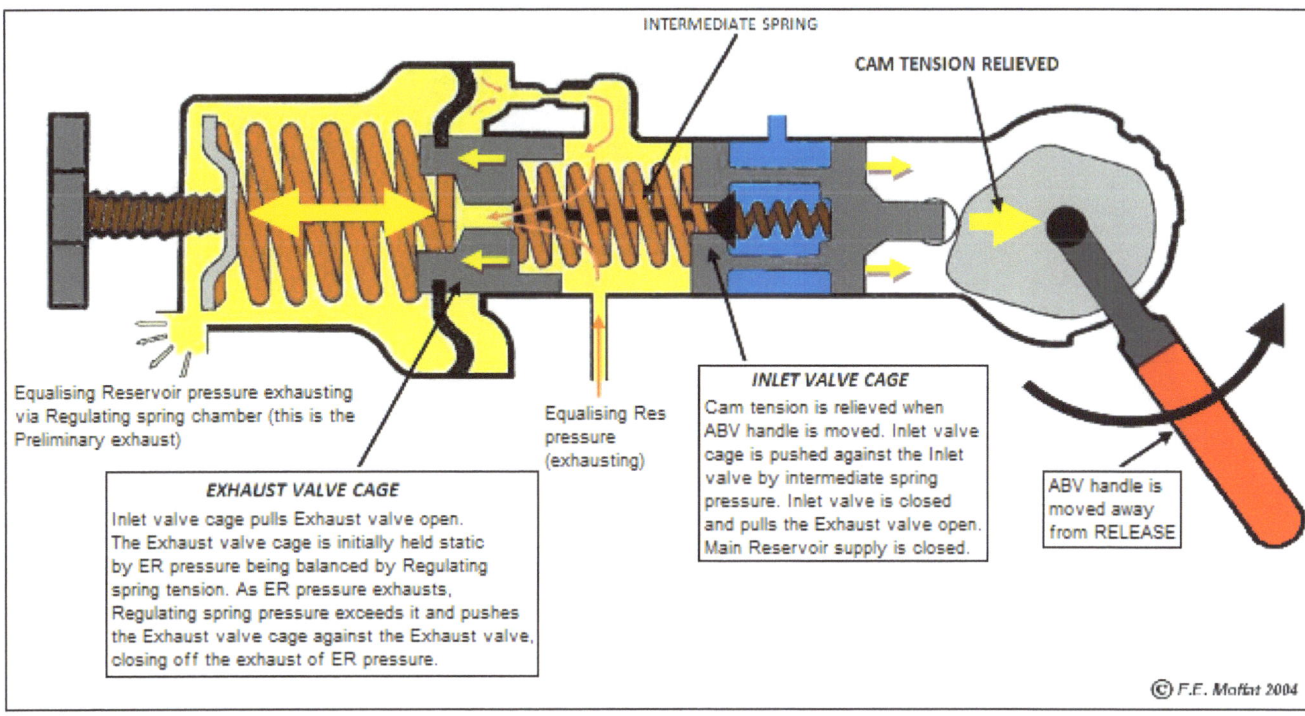

FIGURE 187: 26C AUTOMATIC BRAKE VALVE REGULATING PORTION – SERVICE APPLICATION POSITION

Relay Portion

The functional equivalent of the equalising piston in older equipments, the Relay Portion consists of a large diaphragm and spindle, a *supply* (or inlet) *valve* designed to supply main reservoir pressure to the brake pipe, and an *exhaust valve* designed to release brake pipe pressure to atmosphere. The spindle attached to the diaphragm assembly passes through and extends beyond the *exhaust valve* to touch against the *supply valve* (equivalent to the Feed Valve in older equipments). No part of the Relay valve assembly is attached to or is directly operated by the brake valve handle or camshaft.

As mentioned in the description of the Regulating valve operation, one side of the Relay valve diaphragm—its outer face—has equalising reservoir pressure against it. During charging (Figure 188), this pressure moves the diaphragm assembly away so that the spindle pushes the *supply valve* open and permits main reservoir pressure to flow into the brake pipe. At the same time, this pressure is able to develop against the opposite inner face of the Relay valve diaphragm such that when it attains equalising reservoir pressure, the assembly is balanced and the diaphragm stem moves away from the *supply valve*, which is then closed by its operating spring, thus 'lapping' the Relay Valve. By this action, the brake pipe is charged to the same pressure as the equalising reservoir.

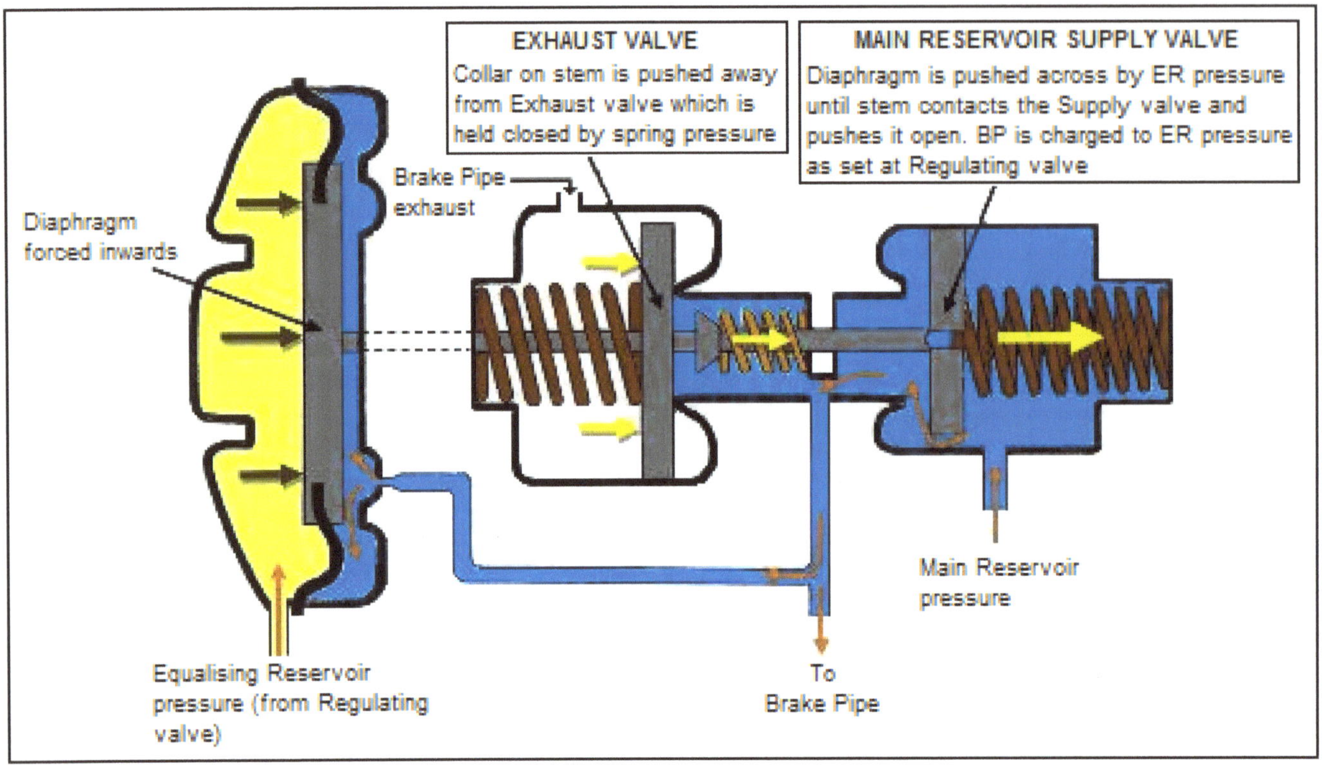

FIGURE 188: 26C AUTOMATIC BRAKE VALVE RELAY PORTION – *RELEASE* POSITION

When a brake application is made, equalising reservoir pressure is reduced (Figure 189). Since this reduces pressure against the outer face of the Relay valve diaphragm, the greater brake pipe pressure on the other side moves the assembly such that a collar on the spindle pulls the *exhaust valve* open. Brake pipe pressure now exhausts to atmosphere in what is referred-to as the brake application ***secondary exhaust***. This exhaust is the airflow most familiar to locomotive drivers when making a brake application. It results in the audible brake pipe exhaust noise that drivers recognise in terms of its 'length and strength' when we are discussing making an *effective* brake pipe reduction. This exhaust is usually ducted through the floor to reduce cab noise levels.

When brake pipe reduces to match equalising reservoir pressure, the Relay valve diaphragm is balanced and the *exhaust valve* is then closed by its operating spring. In this way, the Relay valve assumes its 'lap' condition.

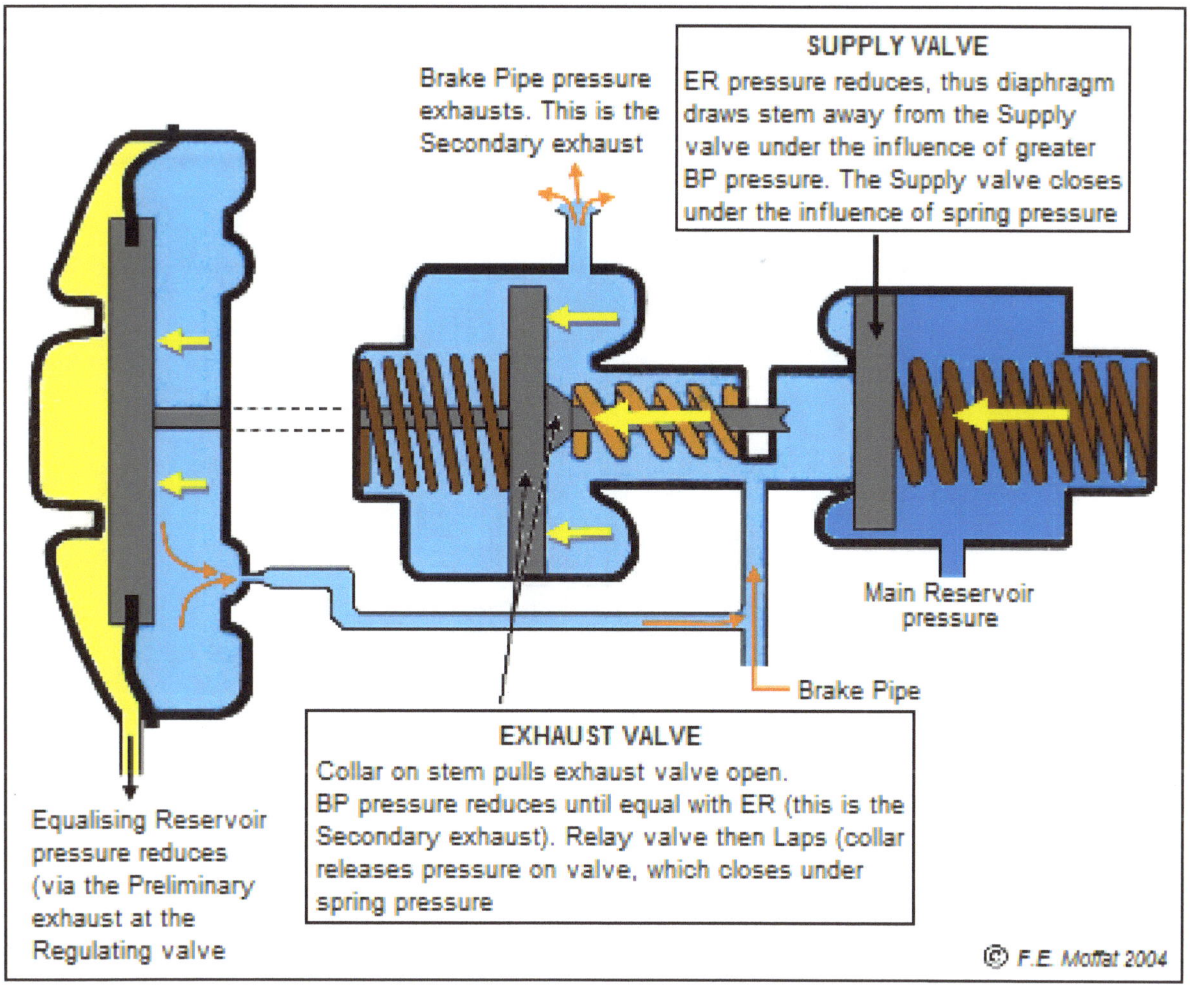

FIGURE 189: 26C AUTOMATIC BRAKE VALVE RELAY PORTION – *APPLICATION* POSITION

Any reasonable leakage in the equalising reservoir will be automatically compensated-for by the Regulating valve (under the influence of the spring setting), while reasonable leakage in the brake pipe will be replenished by the Relay valve (under the influence of equalising reservoir pressure). This is the 'pressure-maintaining' feature of the 26C Automatic brake valve.

Summary

The following series of diagrams depicts the 26C Automatic brake valve in extremely simplified diagrammatic form. The illustrations are intended to provide the reader with a broader comprehension of brake valve functionality.

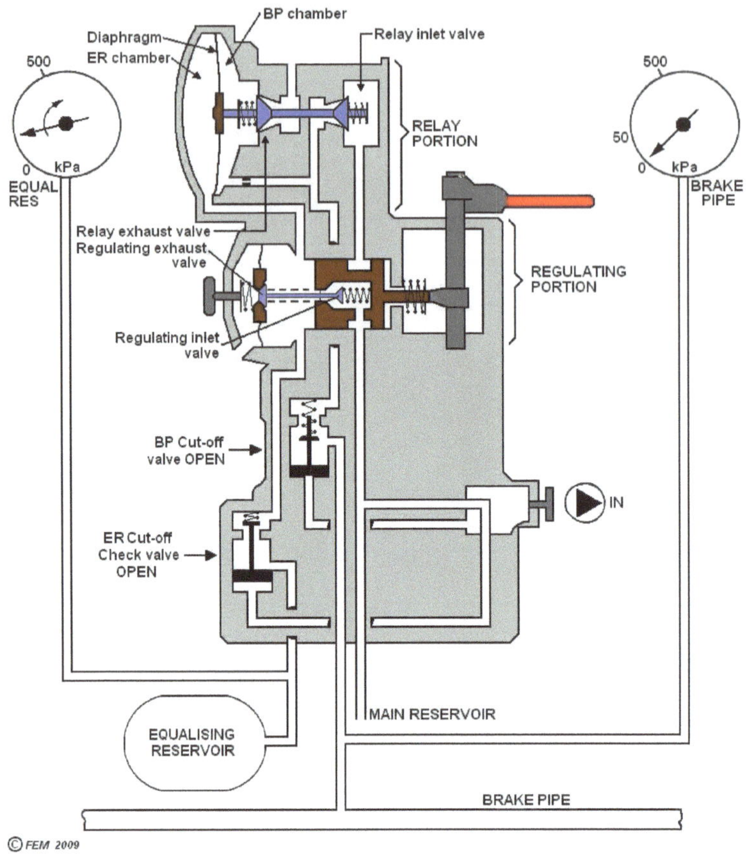

FIGURE 190: 26C AUTOMATIC BRAKE VALVE SIMPLIFIED – INITIAL EQUALISING RESERVOIR CHARGING

Initial Equalising Reservoir charge. This is brake pipe 'dry charge'. The train is in the initial stages of brake pipe charging. Main Reservoir pressure is flowing via the Regulating valve portion to both the ER side of the Relay valve diaphragm and to the ER itself. This causes the Relay valve portion to admit Main Reservoir pressure into the brake pipe as depicted in Figure 191.

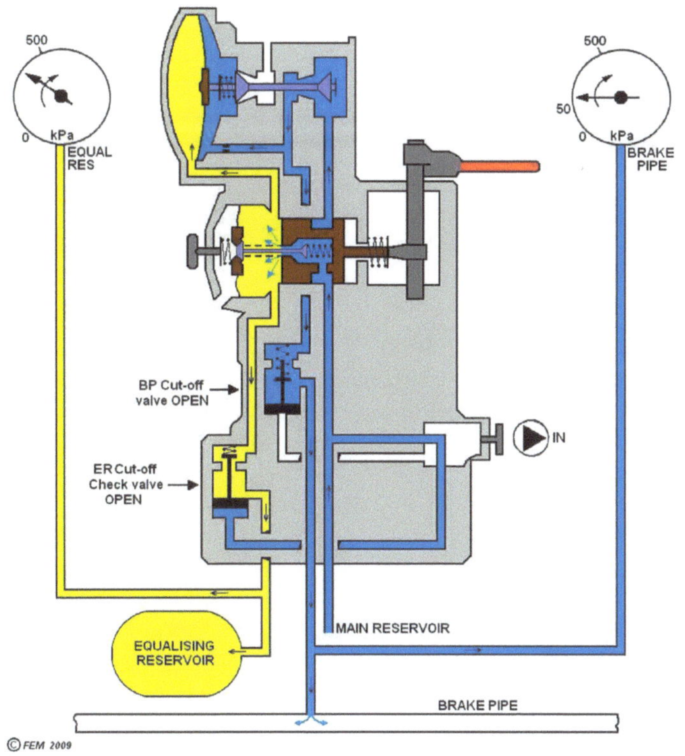

FIGURE 191: 26C AUTOMATIC BRAKE VALVE SIMPLIFIED–EQUALISING RESERVOIR IS CHARGING, BRAKE PIPE IS CHARGING

Equalising Reservoir charging. The brake pipe begins to charge.

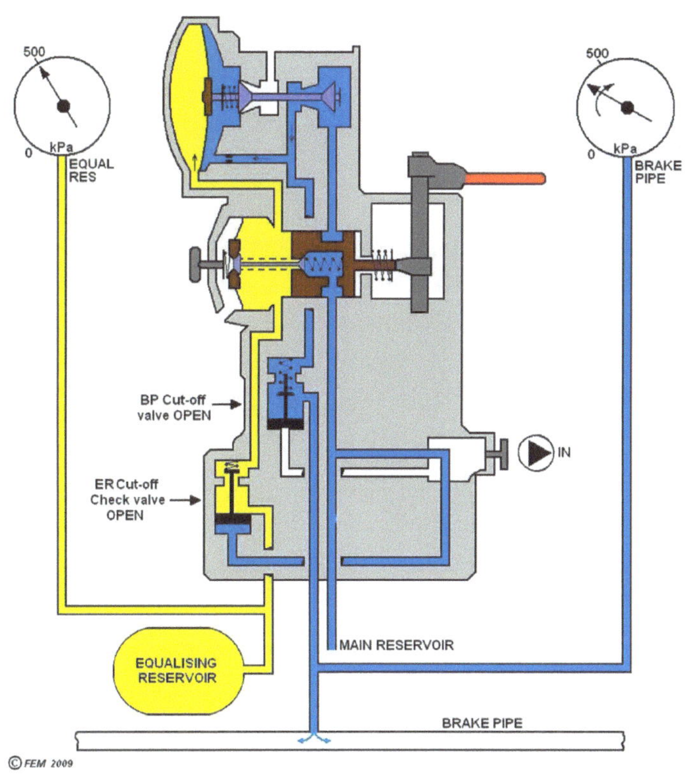

FIGURE 192: 26C AUTOMATIC BRAKE VALVE SIMPLIFIED – EQUALISING RESERVOIR IS CHARGED, BRAKE PIPE IS CHARGING

Equalising Reservoir charged. The ER is fully-charged and the Regulating valve has lapped. The brake pipe continues to charge.

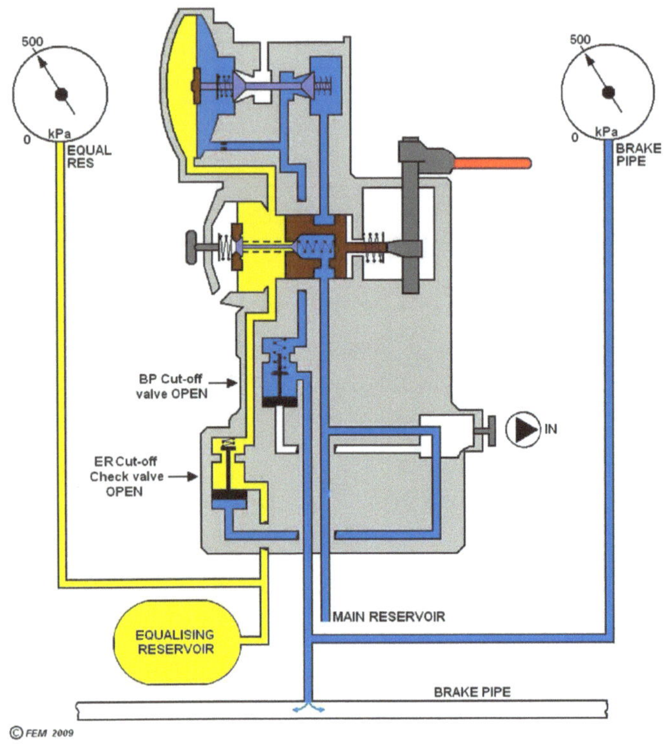

FIGURE 193: 26C AUTOMATIC BRAKE VALVE SIMPLIFIED – SYSTEM CHARGED

Brake Pipe charged. BP charging is complete. The Relay valve has lapped. The system is fully charged, assuming no leakage.

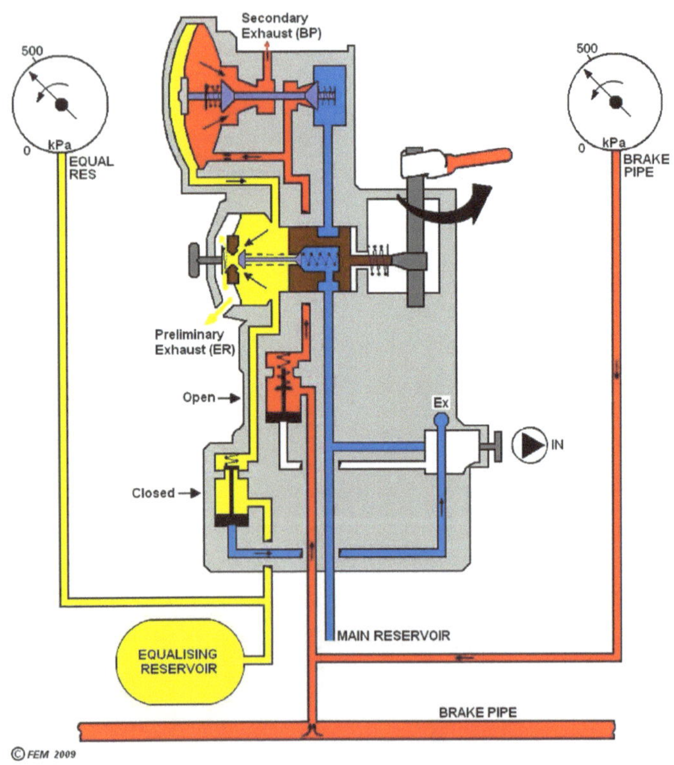

FIGURE 194: 26C AUTOMATIC BRAKE VALVE SIMPLIFIED – SERVICE APPLICATION

Service application. An ER pressure reduction is made to a value consistent with the degree of brake valve handle movement. This is the 'preliminary exhaust' and it is very brief. The Regulating valve will lap when this reduction is attained. The BP reduction commences along with the ER reduction. This is the 'secondary exhaust' that will directly cause a brake application. The brake pipe is reducing.

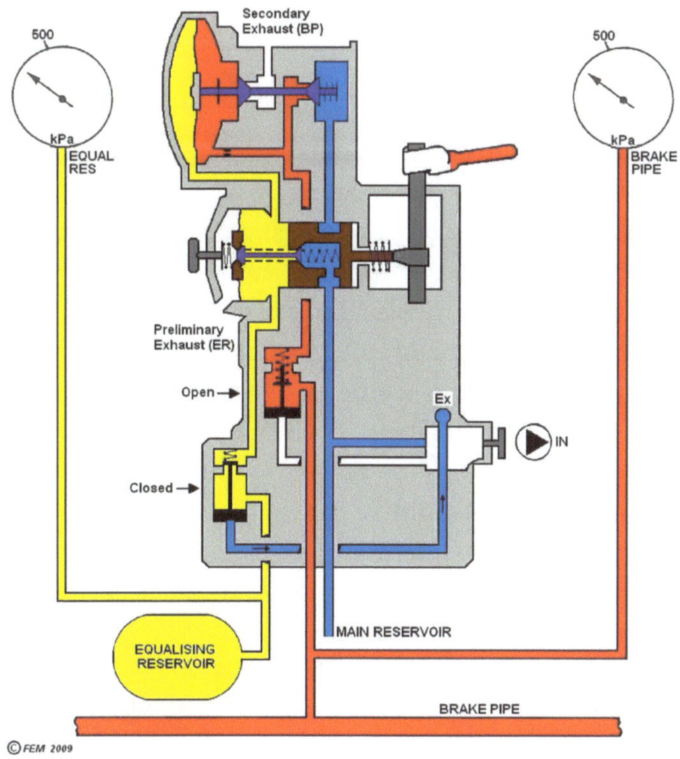

FIGURE 195: 26C AUTOMATIC BRAKE VALVE SIMPLIFIED – SERVICE LAP

The ER reduction is complete, and the Regulating valve has lapped. The resultant BP reduction has also been attained and is completed by the lapping of the Relay valve. The system has stabilised at a BP reduction that is consistent with the degree of Automatic brake valve handle movement.

SA26 Independent Brake Valve

All Independent brake functions are carried to trailing multiple units by the relevant Multiple-Unit air connections. The SA26 Independent brake valve assembly mounts as a modular device to the front of the brake valve pipe bracket (Figures 183 and 184). It is also self-lapping and pressure-maintaining, and is infinitely variable in operation through its APPLICATION zone. Construction and operation of the Independent portion is similar to the Regulating valve portion described previously, and the basic design premise of the SA26 borrows heavily from the original Westinghouse Controlair reducing valve.

An adjustable regulating spring acts against the outer face of a diaphragm assembly to hold the *exhaust valve* open in the RELEASE position. Movement of the Independent brake valve handle into the APPLICATION zone adjusts this spring tension by increasing it (as distinct from **reducing** it, as occurs with the 26C Regulating Valve). Moving the handle rotates a circular cam face that seats the *exhaust valve* and pushes the *inlet valve* open to feed main reservoir air as a control pressure to the locomotive brake cylinder relay valve via Independent Application & Release pipe 20 (Control pipe in 3-pipe systems).

The locomotive brake cylinder relay valve, in turn, supplies main reservoir pressure to the brake cylinders to a value consistent with the position of the Independent handle within the APPLICATION zone, maintaining this pressure against leakage and also compensating for a reasonable increase in brake cylinder piston travel due to brake shoe wear. As pressure in the Independent Application & Release pipe develops, it also develops against the inner face of the Independent brake valve diaphragm until it is sufficient to move the diaphragm assembly back against the spring pressure and close the *inlet valve*.

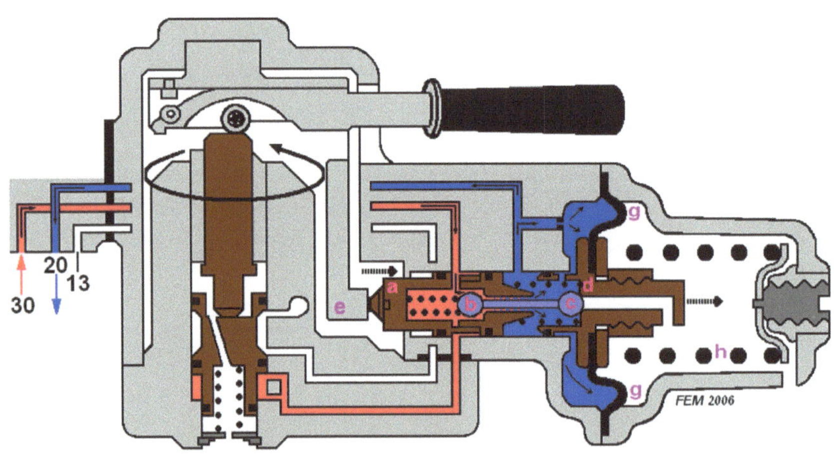

APPLY

- Independent Brake Valve handle into Application Zone
- Operating cam 'e' pushes inlet valve cage 'a' (to right in this diagram).
- Inlet valve spring bears against inlet valve 'b' forcing exhaust valve 'c' to close.
- Inlet valve spring is further compressed so that inlet valve 'b' is moved off its seat as inlet valve cage 'a' is pushed further out.
- Main reservoir pressure (pipe 30) flows to Indep. Appl. & Rel. pipe 20 as a brake cylinder control pressure.
- This pipe 20 brake cylinder control pressure also develops against the pressure-regulating diaphragm 'g' until it overcomes the force of regulating spring 'h' sufficiently to be able to move exhaust valve cage 'd' (to the right in this diagram) such that inlet valve 'b' seats in the inlet valve cage eventually closing off the main reservoir pressure supply.
- At this point the independent brake valve has 'lapped'.

FIGURE 196: SA26 INDEPENDENT BRAKE VALVE. *APPLICATION* – DIAGRAM

Movement of the Independent handle back to or toward RELEASE position reduces cam pressure against the seated *exhaust valve* until—with the assembly held in balance by regulating spring and brake cylinder control pressures across the diaphragm—the *exhaust valve* opens and discharges the control pressure until it equalises with the now-reduced regulating spring tension, at which point the *exhaust valve* closes and the assembly assumes a lap condition. If the Independent handle is moved directly to RELEASE position, all cam pressure is removed from the regulating valve and its diaphragm assembly so that the *exhaust valve* is again held open by the regulating spring, permitting a complete discharge of control air pressure. This depletion of control pipe pressure permits the locomotive brake cylinder relay valve to exhaust locomotive brake cylinder pressure.

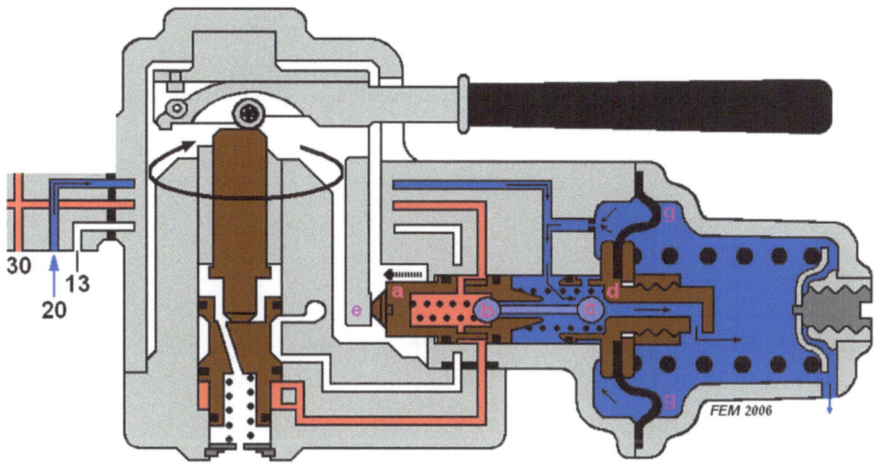

RELEASE

- Independent Brake Valve handle to (or toward) Release position.
- The receding operating cam profile 'e' permits inlet valve cage 'a' to retract (move to left in this diagram) and drag with it the valve spindle, closing inlet valve 'b' and opening exhaust valve 'c'.
- Brake cylinder control pressure (Indep. Appl. & Rel. pipe 20) escapes via exhaust valve cage 'd' and diaphragm spring chamber.
- This same pressure also recedes from the inner face of the pressure-regulating diaphragm 'g' until the diaphragm and regulating spring assembly are able to move the exhaust valve cage back (to the left in this diagram) until exhaust valve 'c' seats, closing off the exhaust of brake cylinder control pressure and the independent brake valve 'laps'.

FIGURE 197: SA26 INDEPENDENT BRAKE VALVE. *RELEASE* – DIAGRAM

Depression of the Independent brake valve handle (in the RELEASE position) following an Automatic application will release that application on the locomotive. Under certain operating circumstances, though, the driver may desire to allow the locomotive brake to partially apply along with the Automatic application to assist with the slowdown or stop. Depression of the handle **at any position within the APPLICATION zone** (other than the RELEASE position)—and then leaving the handle in that selected position—will create an Independent application that is limited to a reduced value determined by the handle position (per the driver's judgement). This latter point is quite important since—due to the relative mass of a diesel locomotive by comparison to that of some passenger train rollingstock—the braking ability of the locomotive might be required to assist with the stop. This operating practice can also be useful if urgent circumstances necessitate a rapid, unplanned slowdown or stop requiring a heavy Service brake application.

By permitting the locomotive Independent brake to partially apply, coupler slack (relatively less in most passenger trains but of critical importance to passenger comfort) can be gently bunched and the heavy locomotive braked to augment the train brakes and assist the slowdown in a manner that minimises the danger of sliding the locomotive wheels. Normal operation of the Independent brake is available at all times, and locomotive brake cylinder pressure can be adjusted as required regardless of the partial bail-off originally initiated.

Depressing the Independent brake valve handle pushes down on the internal quick-release plunger (also called the 'actuating plunger') and thus also the *quick-release valve* (also referred to as the 'actuating valve') and opens a main reservoir feed to the Quick-Release portion of the locomotive brake operating valve (control or triple). In this way, the Quick-Release portion is triggered to interrupt any flow of air to the brake cylinders and exhaust any pressure already therein. This is the process that locomotive drivers know as 'bailing-off'; however, it is technically referred-to as *actuation* or *quick-release*.

The main reservoir flow via Actuating pipe 13 to the Quick-Release portion of the control valve is indicated on the driver's flow meter as a high initial flow that quickly tapers off to zero as the Actuating pipe rapidly fills to main reservoir pressure. On locomotives with four-pipe systems, this feed is piped to trailing units, and the driver's bail-off action must be of sufficient duration to fully charge the pipe throughout a locomotive consist so as to properly *actuate* all trailing unit Quick-Release valves. It is generally accepted that the Independent brake valve handle should be held depressed for at least four seconds per trailing multi-coupled unit.

Note that the obsessive action—sometimes observed—of drivers 'pumping' the Independent brake valve handle up and down during an Automatic brake application does not serve to pressurise Pipe 13 correctly and is quite ineffectual since Pipe 13 is not being charged to any useful degree by the momentary downward action of the handle is then exhausted each time the handle is released.

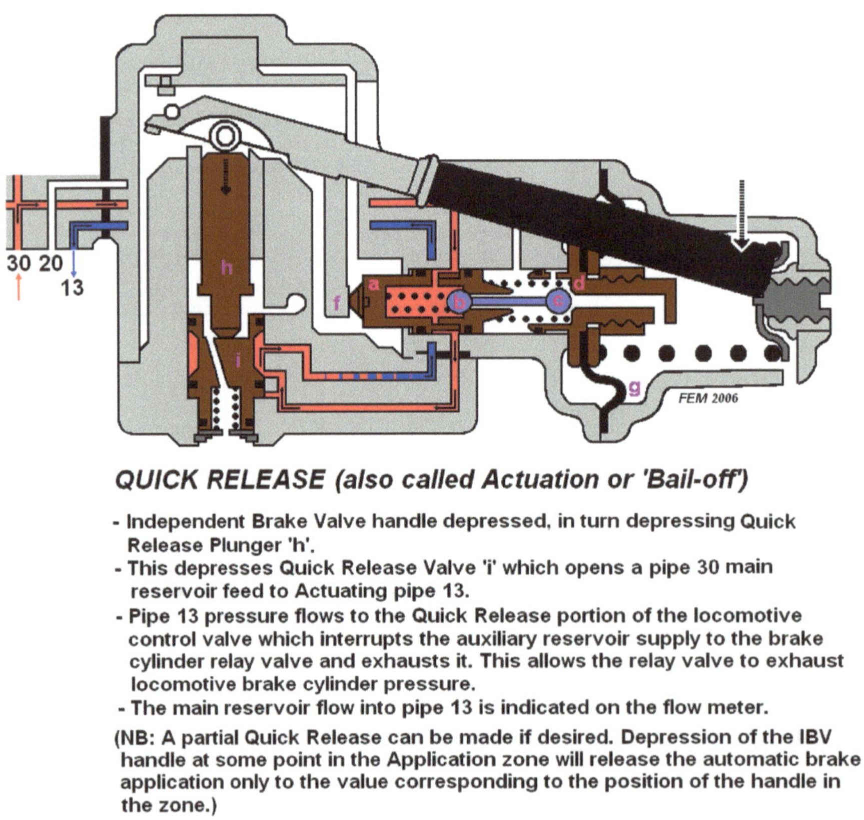

FIGURE 198: SA26 INDEPENDENT BRAKE VALVE. *ACTUATION* – DIAGRAM

Although air brake equipment manufacturers recommend the use of non-removable Independent brake valve handles, some Australian operators who utilise dual control stands or double cabs—and have developed familiarity with the practice of removing all brake valve handles on older equipments—insist on retaining a removable handle. In this case, the Independent handle is removable in the RELEASE position by pulling it directly out against the action of a spring-loaded retention bearing. Replacement is the reverse process.

It was originally commonplace for 26-L brake valve equipment to be mounted quite high on the cab air brake stand, with the Automatic brake valve handle located adjacent to the driver's left shoulder and with the Independent handle conveniently to hand at midriff height. Later, as the equipment was progressively applied to Australian diesel locomotives with left-side driving stations (and therefore with 'back-to-front' AAR control stands), the air brake stand migrated to a location either ahead of the control stand (rather than aft of it as applied to the conventional American-style, right-side driving station) or shoe-horned into the front-left corner of the driver's cab space. In these latter Australian-style layouts, the brake and control stands could not be permitted to obstruct the view through the cab windscreen or centre front windows of the many classes of Australian forward-control diesel locomotives that lacked a short hood structure. The brake valve and control stands were thus positioned lower, relative to the seated driver.

FIGURE 199: SA26 INDEPENDENT BRAKE VALVE. THESE ILLUSTRATIONS DEPICT TWO VERSIONS OF THE 26-L BRAKE EQUIPMENT. ALTHOUGH LITTLE HAS CHANGED IN THE PHYSICAL PRESENTATION OF THE VALVES, THE PHOTO ON THE RIGHT SHOWS AN ARG 3100-CLASS (EX-WAGR L-CLASS) INSTALLATION THAT INCLUDES THE OPTIONAL "OFF-SET" INDEPENDENT BRAKE VALVE HANDLE. (THE BOX-LIKE METAL SHROUD FASTENED AGAINST THE AUTOMATIC BRAKE VALVE CAMSHAFT HOUSING PROTECTS A MICRO-SWITCH FOR VIGILANCE SYSTEM CANCELLING BY HANDLE MOVEMENT)

Additionally, in later circumstances—for example, arising from operational experience or new regulations and other stipulations mandating improved visibility—other operators sought to have the height of air brake and control stands reduced, lowering their 26-L installations. Thus was introduced a problem: the Independent brake valve handle was now in close proximity either to the driver's left thigh (for a R-H drive control stand) or right knee (Australian L-H drive control stand), would intrude into the seating space (again, a consequence of the back-to-front L-H drive control stand), and would interfere with the driver's freedom of movement, especially in and out of the seat. The solution was the 'cranked' or vertically offset handle that raised the handgrip to a more convenient position and that is nowadays commonly used.

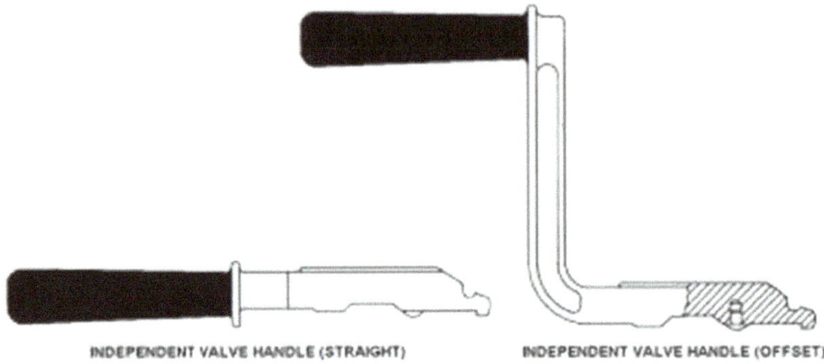

FIGURE 200: SA26 INDEPENDENT BRAKE VALVE HANDLE STYLES – CONVENTIONAL AND OFF-SET (OR 'CRANKED')

A third Independent brake valve handle style deserves mention. In 1986, the Mt Newman Mining Co P/L (now BHP Billiton Iron Ore) had some of their ALCO Century-Series locomotives rebuilt into an enhanced General Electric Dash-7 model[126], taking delivery from January 1987. These units retained their 26-L brake systems but the new driver's control stand (incorporating a conventional-height brake stand) was located further away from the seat to provide easier access/egress for the driver. This created a considerable problem for the driver who was barely able to reach the Independent brake handle if required to manoeuvre the locomotive while leaning out of the cab-side window. Union members on the MNM Cab Design Committee demurred, and builder A. Goninan & Co—keen to establish themselves in the Pilbara heavy-haul market and thus ever-accommodating—provided a clever solution by designing a handle with a hollow hand-grip that contained an extension of the handle shank. This hand-grip could be unscrewed to permit it being slid out along the shank and thus its reach extended. I expect this is the only occasion such a modification was ever made to the SA26 brake valve handle.

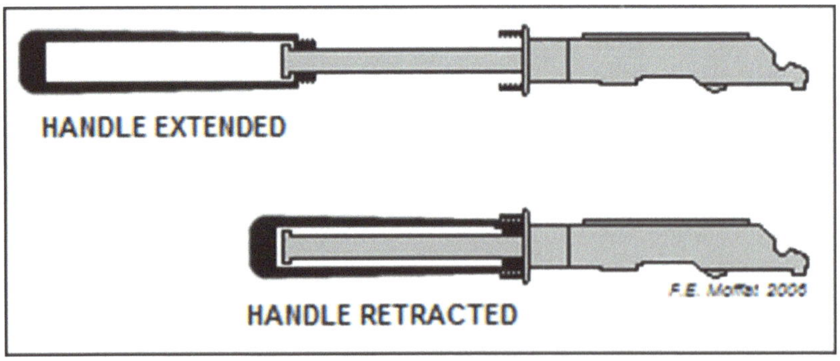

FIGURE 201: SA26 INDEPENDENT BRAKE VALVE HANDLE – EXTENDABLE (MT.NEWMAN MINING GE DASH 7 LOCOMOTIVE)

126. The project to rebuild these Alco locomotives into modern GE units was a bold sales move by the General Electric Transportation Systems' Australian licensee, A. Goninan & Co., who were aware that both Mt. Newman Mining and Hamersley Iron were facing the imminent requirement of locomotive fleet renewal. Convinced they had the right product at the right time, GE offered MNM a prototype product at an acceptable price. MNM wanted the new Dash 8 model, but GE convinced them to settle for an 'Enhanced Dash 7'. It was a clever strategy to prove their technology in a cost-effective manner. MNM went on to purchase a fleet of GE Dash 8 locomotives, although the first four of these—patterned on a sub-model then being built for the Union Pacific Railroad—were also ALCO rebuilds. From this strategy, GE gained a powerful foothold in the Pilbara during the last decade of the closing century and the first of the new one. It would not be until BHPIO hired senior railroad mechanical decision-makers from the United States who had first-hand experience of the latest General Motors products that EMD would return with a vengeance to the Pilbara.

Brake Pipe Pressure Non-Maintaining Valve

Early in this section, I commented on a rather perverse action by some Australian railways to amputate the most beneficial feature of the 26-L schedule - its pressure-maintaining ability. It is probably relevant at this point to discuss this unique peculiarity that has so incredulised those outsiders who have become aware of it, in particular American air brake industry observers.

Mention has been made in a previous section on Australian triple valves that Victorian Railways was renowned for having persisted with the use of outmoded air brake apparatus long after more advanced equipments became available. Certainly, those other state railway systems with which VicRail interchanged their rollingstock were well aware of the outdated triple valves and less-than-perfect condition in which VR wagon air brake equipment was maintained. Typical examples were the continued use of recycled Improved Triple Valves long into the 1990s and the ill-advised use of grease for triple valve lubrication where the manufacturer specified oil. This latter immoderation was for no other reason than to stretch the interval between examinations and then to mask the leakage that inevitably resulted from this deferral of maintenance, the result being the porous air brake systems, especially with regard to triple valves, for which the Victorians were infamous.

Perhaps the Queensland Government Railways, meanwhile, had also been somewhat less than diligent in the maintenance of certain air brake equipment (although it must be acknowledged that QR has been more disposed over the years to take up the new triple valves as they became available) as here, too, auxiliary reservoir leakage had been prevalent. All of this was quite manageable in the decades prior to the 1970s as train sizes were modest and the predominant A7-EL equipment coped adequately.

This languid state of affairs changed abruptly, however, with the introduction of 26-L, which—courtesy of its ability to maintain brake pipe pressure with the brakes applied—would expose the auxiliary reservoir leakage (mostly from ill-maintained triple valves and pipework joints) that had become commonplace on these railways. Simply put, the threat was that if any particular brake pipe pressure obtained during a brake application were maintained while auxiliary reservoir continued to reduce through leakage, then brake pipe pressure would eventually exceed that in the auxiliary reservoirs, and the triple valves would detect this as a 'release' signal, albeit an uncommanded one. This, of course, was (and is) quite unacceptable from a safety perspective. But rather than fixing their leaking equipment, these railways sought to eliminate the new equipments' most advanced feature—the very reason for its development—so that the leakage could be accommodated! When news of this reached the United States, one flummoxed Wabco executive questioned why a railway would "*...drag modern equipment down to the level of the obsolescent stuff.*"

The Queenslanders' original solution was crude but effective. To artificially diminish 26-L to the inferior performance level of A7-EL, they added a choked isolating cock to the 26C brake valve equalising reservoir circuit and referred to it by the grandiose euphemism "Maintain/Non-Maintain Lever Control Valve". This allowed the locomotive equalising reservoir—when the train brakes were applied—to be able to artificially 'leak' at a rate of up to 5 psi (35 kPa) per minute (the maximum allowable leakage rate under ROA[127] standards), which—of course—allowed the brake pipe to also reduce at this rate. Thus, any leakage occurring along the train was not maintained, permitting the brake application to increase in intensity as traditionally occurred with older equipments, and preventing an unintentional release (and of course, meaning that the brake pipes of <u>all</u> trains hauled by 26-L locomotives would 'leak' regardless of whether or not there were actually leaky wagons on the train!). The following passage from a QR Chief Mechanical Engineer's instruction to enginemen refers:

127. Railways of Australia (now the Australasian Railway Association). Refer to Glossary, chapter 8.

This valve is provided on locomotives fitted with type 26 brake so that the driver may convert the 26C automatic brake valve from a pressure maintaining valve to a non-maintaining valve, depending on the type of brake equipment fitted to the hauled vehicles.

If the maintaining position of the lever control valve is used on trains having slide valve triple valves [i.e. ITV and AF], *because the brake pipe pressure is maintained, the moment the auxiliary reservoir pressure leaks down to a pressure approximately 15 kPa below the brake pipe pressure, the brakes on the vehicle concerned will release.*

Note that in its terminology the instruction is presented as an entirely normal requirement; use of the device simply "...*depending on the type of brake equipment fitted to the hauled vehicles.*"

The Victorians, on the other hand, pursued a more elegant answer. No doubt, equipment orders from Victorian Railways were a novelty, but as with the American Wabco executive, Westinghouse Australia personnel must have also been shaking their heads in wonderment as they set about providing the requested solution. This was duly offered as the "Brake Pipe Non-Maintaining Valve" which was to be fitted to 26-L–equipped locomotives to enable suppression of the brake pipe maintaining feature "as required" (meaning when hauling VR freight trains; the Victorians being as adept as the Queenslanders at making it all sound 'normal'). This more advanced device was later also adopted by QR.

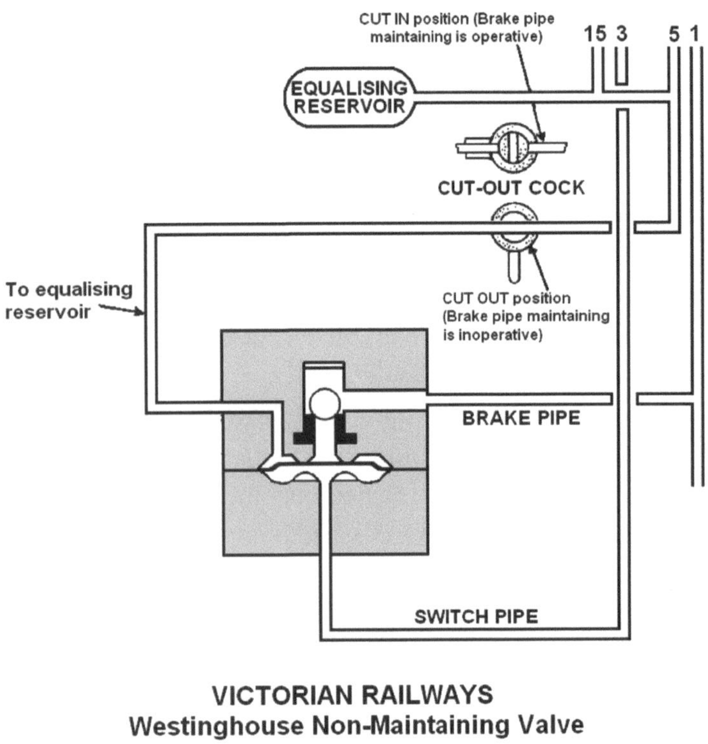

FIGURE 202: BRAKE PIPE NON-MAINTAINING VALVE – DIAGRAM

With the Automatic brake valve at RELEASE (Figure 202) and the Non-Maintaining Valve Cut-Out Cock at its CUT OUT position, main reservoir pressure was admitted via the brake valve Suppression valve to port 3, raising the diaphragm to seal the check valve port and thus disconnect communication between brake pipe and equalising reservoir. Confusingly, the Cut-Out Cock is *not provided to cut the Non-Maintaining Valve IN or OUT* but rather to cut the 26C brake valve maintaining feature IN or OUT (which is not the same thing), thus when this cut-out cock is IN, the Non-Maintaining Valve is 'Out' (isolated from operation) and vice versa. When the Automatic brake valve handle is moved away from RELEASE position, port 3 and the switch pipe are exhausted at the suppression valve per normal 26C brake valve functioning, and within the Non-Maintaining Valve, this permitted its diaphragm to withdraw under the greater influence of equalising reservoir pressure. The resulting admission of equalising reservoir pressure into the port below the ball-valve unseated it and a charging flow was thereby supplied to the brake pipe.

Should leakage now cause a reduction in brake pipe pressure, equalising reservoir will reduce accordingly. Since the 26C relay portion maintains brake pipe at the equalising reservoir value (which is now reducing along with the brake pipe), it will not detect any brake pipe leakage and will not, therefore, operate to maintain it. During initiation of the brake application—while equalising reservoir pressure is briefly lower than the brake pipe—the check valve prevents the backflow of brake pipe pressure to equalising reservoir. Once the brake pipe reduction is effective, equalising reservoir pressure will flow into the brake pipe at a rate determined by the amount of brake pipe leakage. The 26C relay portion—via its exhaust valve—will still function to ensure the brake pipe pressure cannot exceed equalising reservoir.

It would appear that once Wabco heard of their Australian division's production of the non-maintaining valve (and presumably having overcome their initial indignation), they decided to design one themselves, probably to secure patent rights. The US version made use of a 26C Automatic brake valve equalising reservoir cut-off valve, cost more than the Australian device, and—since the latter was perfectly adequate for the task—was not used in any Australian application.

This auxiliary reservoir leakage phenomenon and the requirement to nullify the brake valve's brake pipe maintaining ability to 'allow' for it is the reason for the Maintain/Non-Maintain switch mounted on the eyebrow panel in NR-Class locomotive cabs (and perhaps other family classes built since and with which I am technically unfamiliar), the NR being designed for universal operation across mainland Australia. Via customised software, this switch modifies EPIC brake functionality to achieve the same result as detailed on the previous page and for the same reasons. It is probably little-used nowadays (if at all) for its intended reason—that is, conditioning the locomotive's brake system to be *non-maintaining* when hauling leaky Victorian and Queensland rollingstock—except that some drivers were under the misapprehension that it can be utilised when carrying out brake pipe leakage tests.

In fact, a reliable brake pipe leakage assessment cannot be achieved by means of this switch, since differences between brake pipe and equalising reservoir transducer calibrations, mechanical friction within the relay valve, and any internal leakage present within the EPIC brake pipe portion would all contribute to a possible invalid flow reading if the brake pipe portion were not cut out via IFC[128] as required. For this reason, company operating procedures—at least as I recall them from my time operating these units between 1998 and 2001—expressly prohibit its use in this context (although without the rationale provided here).

Any reader sharing my jaundiced view of the practice of suppressing the maintaining feature of the 26-L system as being somewhat bizarre would be interested to know of South Australian Railways' action—taken in the late 1950s—to conceal excessive system leakage (which the locomotive driver interprets during brake testing as the rate of brake pipe charging flow) by the cunning expedient of modifying their flow meters with a heavier diaphragm spring. Requiring twice the flow rate to produce a given reading, the customised flow meters thus provided the driver's flow indicator gauge with a reading at half the brake pipe charging flow rate that was actually occurring!

Was there something peculiarly 'Australasian' about air brake equipment that made the actions referred-to above strictly necessary? It would seem not, since both NSWGR and NZGR happily operated trains equipped with both 26-L on the locomotive <u>and</u> ITV-equipped rollingstock without having to eliminate the brake valves' pressure-maintaining functionality.

In Victoria, this aberrant modification to 26-L equipment was removed in about 2003–04 and the brake valves restored to their design functionality.

Other Features of the 26-L Schedule

As with previous air brake equipments, the 26-L schedule provides, through the use of numerous optional devices, for more operating features than are used in Australia and New Zealand (with the exception

128. Refer to Glossary, chapter 8.

of the Pilbara iron ore railroads that use the North American version of the schedule). In addition to its self-lapping and pressure-maintaining abilities, this system provides the following features:

- *Safety control.* Initiation of air brake applications through a "dead-man" feature utilising a foot-pedal valve or reset alerter system in US applications and similar vigilance systems in Australasia;
- *Overspeed control.* Initiation of a Full Service penalty air brake application if locomotive speed exceeds a predetermined limit. Temporary or permanent suppression of the penalty is available as an option;
- *Split or limited reduction penalty air brake applications*; As above, except that a preliminary reduction (leading to Full Service) or a single lighter brake application can be automatically made.
- *Power and dynamic brake knockout during penalty applications.* (PCS operation via the P-2-A Brake Application Valve or other device);
- *Brake pipe charging cut-off and/or sanding, plus power and dynamic brake knockout* from an emergency brake application caused by brake pipe fracture or separation;
- *Multiple unit operation between 3- and 4-pipe air brake systems.* (NB: This occurred extensively in New Zealand where older 4-pipe English-built classes and a more recent Japanese-built class were later required to multiple with 3-pipe EMD units. This state of affairs arose from the South Island network having originally been equipped with English Electric classes having A7-EL with 4 pipes (later being joined by the fleet of Japanese-built locomotives having 26-L and 4 pipes) whereas the North Island became predominantly American, all having 6-SL or 26-L with 3 pipes. This distinction was later blurred when some North Island locomotives were transferred to the South Island; some being equipped with an extra pipe on one side of the headstock. The few remaining operable English Electric locomotives in New Zealand are products of restoration by rail heritage groups and have been modified with 26-L equipment, which provides the added benefit of simpler MU functionality when these units are mixed on Tourist & Heritage trips.)

Characteristic of air brake practice in Australasia, the application of locomotive multi-unit piping nomenclature can be inconsistent and perhaps even unconventional. In some cases UK practice was established and endures. In others, American labeling prevails (particularly on the Pilbara railroads). And in yet others, the headstock labeling system appears to be the hybrid product of individual customer preference and/or entrenched local state-rail custom and practice.

This sometimes leads to confusion by locomotive staff (especially those who have moved between states and systems – something that was never traditional but has occurred increasingly since the mid-1990's) when coupling units, especially when Dead-In-Tow operation is required. However, since much current air brake instructional documentation is American in origin, the comparative terminology for multiple unit air-pipe connections as used in North America, Australia, and New Zealand is herewith presented by the author (as Figure 203). Note that in North America the Brake Pipe and Main Reservoir multi-unit connections are considered to be a 'given' and are not counted; therefore only the other pipes are referred-to - hence the American reference to *2-pipe* (our 4-pipe) and *single-pipe* (our 3-pipe) systems.

4-PIPE SYSTEM *(2-pipe in Nth America)*

	PIPE NAME	HEADSTOCK LABEL
1	Brake Pipe	BP
This is the #1 pipe. It conveys the regulated pressure supplied through the driver's brake valve along the train for operation of car brakes.		
2	Main Reservoir Equalising	MR or MREQ
This is the #30 pipe. It connects the main reservoirs of all operating locomotives in consist to ensure a common pressure and for maximum brake pipe supply capacity and it supplies MR pressure to dead-attached locomotives for auxiliary purposes (e.g. to hold the park brake released).		
3	Indep Application & Release	IND (when a transfer valve is used) IAR (when transfer valve not used)
or	Brake Cylinder Equalising	BC (Nth America/Pilbara)
or	Control Pipe	CP (Australia and New Zealand)
or	No. 3	3 (Australian state systems)
or	Application and Release	A & R (recently-built locomotives)
This is the #20 pipe. It conveys independent brake (control) pressure from Lead to Trailing units for operation of brake cylinder relay valves.		
4	Independent Release	IR or IRP (Australia and NZ)
or	Actuating	ACT (Nth America/Pilbara)
or	No. 4	4 (Australian state systems)
This is the #13 pipe. It conveys MR pressure to Trailing units when the Independent BV handle is 'bailed' to actuate the locomotive control valve for Quick Release.		

3-PIPE SYSTEM *(single-pipe in Nth America)*

	PIPE NAME	HEADSTOCK LABEL
1	Brake Pipe	BP
2	Main Res Equalising	MR
3	Equalising Pipe	EP (New Zealand)
or	Control Pipe	CP (New Zealand)
or	Straight Air Brake	SAB (West Australia)
This is the #20 pipe. It conveys independent brake (control) pressure from Lead to Trailing units for operation of brake cylinder relay valves.		
NOTE: ***3-pipe locomotives***, when multi-coupled as Trailing units, do not develop their own brake cylinder pressure from a brake pipe reduction (i.e. via control valve action). In this case their brake cylinder control pressure is automatically supplied from the Lead unit's control valve (and exhausted thereat during 'bail-off') via the 3rd pipe. Thus, such locomotives do not require the 4th pipe to accept and convey a Quick Release (bail-off) pressure signal from the Lead unit.		

FIGURE 203: DESCRIPTION OF LOCOMOTIVE MULTI-UNIT PNEUMATIC CONNECTIONS

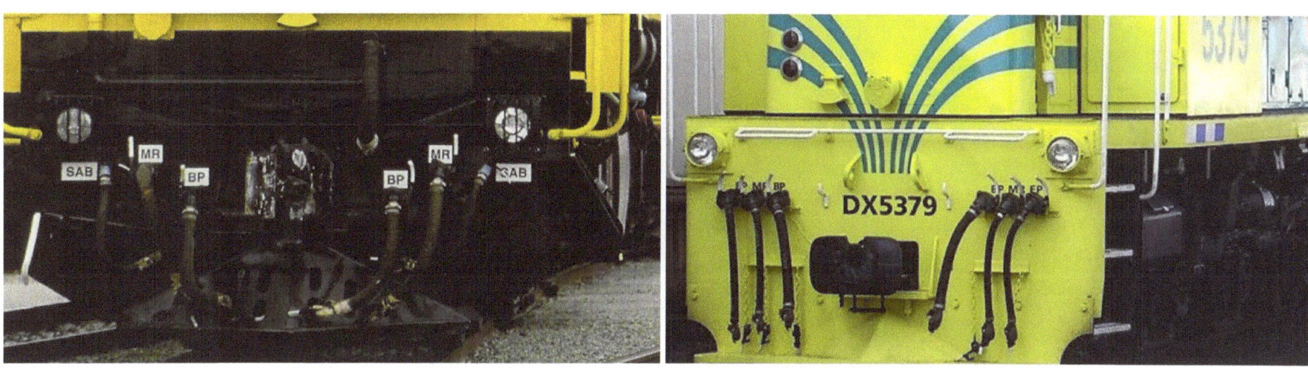

FIGURE 204: DIVERSE LABELING TREATMENTS FOR 3-PIPE LOCOMOTIVES. THE WESTERN AUSTRALIAN NARROW-GAUGE DUAL-BRAKED UNIT IS STILL BRANDED WITH THE ANTIQUATED STRAIGHT-AIR BRAKE LABEL WHILE THE NEW ZEALAND UNIT FOLLOWS CLASSIC 3-PIPE PRACTICE.

FIGURE 205: DIVERSE LABELING TREATMENTS FOR 4-PIPE LOCOMOTIVES. THE 'PNQ UNIT DISPLAYS THE PONDEROUS 'MREQ' LABEL, WHILE THE BIG PILBARA RAIL GE EXHIBITS THE ULTIMATE IN DUMBING-DOWN EXERCISES. THIS LATTER CONTINUES THE SIMPLISTIC AND UNLAMENTED AUSTRALIAN STATE RAILWAY PRACTICE OF LABELING THE 'BRAKE CYLINDER EQUALISING' AND 'ACTUATING' PIPES WITH THE ANONYMOUS NUMBERS '3' AND '4' RESPECTIVELY, THUS CONVEYING NOTHING IN THE WAY OF UNDERSTANDING TO LOCOMOTIVE PERSONNEL USING THEM

Multi-Unit Operation of Air Brakes

An examination of the differences of multiple unit pneumatic couplings between locomotives gives rise to reasonable questions about *how* and *why*, thus a brief look at this subject is in order.

The No. 6 brake equipment dates back to steam locomotives and their single-unit operation (acknowledging multi-heading, where individual drivers are running each locomotive and manipulating the engine brakes). Dieselisation introduced the necessity to control the brakes on each unmanned trailing unit in a multiple-unit consist by transferring brake cylinder pressure from Lead-to-Trailing unit(s) to operate the control valve, and—to satisfy this requirement—equipments derived from the 6-ET and modified to be MU-capable had been carried across to some diesels.

This resulted in the appearance of (1) the **transfer valve**, (2) a 3-position cut-out cock, and (3) either one or two additional pneumatic connections between units. The 3-position cut-out cock (which is double-ported and does not always have all three positions utilised) controls the proper functioning of the brakes on each unit via the transfer valve and, therefore, must be positioned for Lead or Trail according to the location of the locomotive within the consist. Since the transfer valve on the Lead unit connects its brake cylinder pipe to the brake cylinder equalising pipe between units (this latter being one of the new multiple-unit pneumatic connections and also referred to as the **control pipe**), any brake application—Automatic or Independent—will produce an air flow from the brake cylinder pipe of the Lead unit through the transfer valve and on to the Trailing unit(s) via the brake cylinder equalising pipe and its MU connecting hoses.

At the Trailing unit(s), this pressure flows to the transfer valve and the 3-position cut-out cock in TRAIL position, then to the distributing or control valve that functions to produce a brake cylinder pressure equal to the pressure value received from the Lead unit. Releasing the locomotive brakes reverses this sequence. In later years, H-5 Relayair valves (see the section on the 6-SL equipment) have been used as transfer valves, while the F-1 Selector valve—an auxiliary device that is part of the 26-L schedule—is another.

In North America, the 24-RL equipment appeared in the 1940s, utilising **Independent Application & Release** and **Actuating** pipes for transferring smooth and rapid independent brake pressure signals direct from the Independent brake valve to Trailing units. These pipes were carried over to the new 26-L scheme, to which was added (for those circumstances where locomotives with different types of brake systems are required to multi-couple) the F-1 Selector as a transfer valve, plus the MU-2 or MU-2A valve to pilot it and for the control of Independent brake functions.

Locomotive drivers will be familiar with the US position-labelling on the MU-2A valve escutcheon, which—on units equipped with an F-1 Selector valve—reads LEAD OR DEAD, TRAIL 6 OR 26, and TRAIL 24. NZR enginemen read the latter two positions as: TRAIL 6 OR 26 (GM); and TRAIL 24 (EE).

The type of brake equipment on the lead unit determines the position of the MU-2A valve on the trailing unit(s). In New Zealand, TRAIL 6 OR 26 position is used when a 3-pipe unit trails another 3-pipe unit (or a 4-pipe unit trails another 4-pipe unit) equipped with either the 6-SL or 26-L schedule. TRAIL 24 position was used when a 4-pipe GM unit trailed an English Electric unit with A7-EL. Either TRAIL position could be used when two English Electrics having A7-EL were multipled. This demonstrated how the 26-L schedule was designed to facilitate the mixing of locomotives in consist by being capable of connecting dissimilarly-piped air brake systems for effective operation. The key to this mixing of coupled air brake systems is the F-1 Selector valve, whose job is two-fold:

1. To transfer the correct Independent brake pressures from the lead unit to the control valves on trailing unit(s). Where trailing units are 26-equipped and utilise the F-1 Selector, these signals are received by the Selector valve, which transfers them as the relevant control pressure to the brake cylinder relay valve. Note here that the F-1 Selector valve is not required where all locomotives to be multi-coupled have 4-pipe air brake systems, which is the predominant case throughout Australia.

2. To protect the air brake integrity of trailing 3-pipe locomotives in the event of a separation between units, by automatically placing the trailing unit(s) into a Lead condition so they can develop brake cylinder pressure as a result of the loss of brake pipe pressure. The F-1 Selector valve is not physically manipulated but is piloted or "conditioned" by air pressure supplied from the MU-2A valve (which IS physically manipulated) in its various positions.

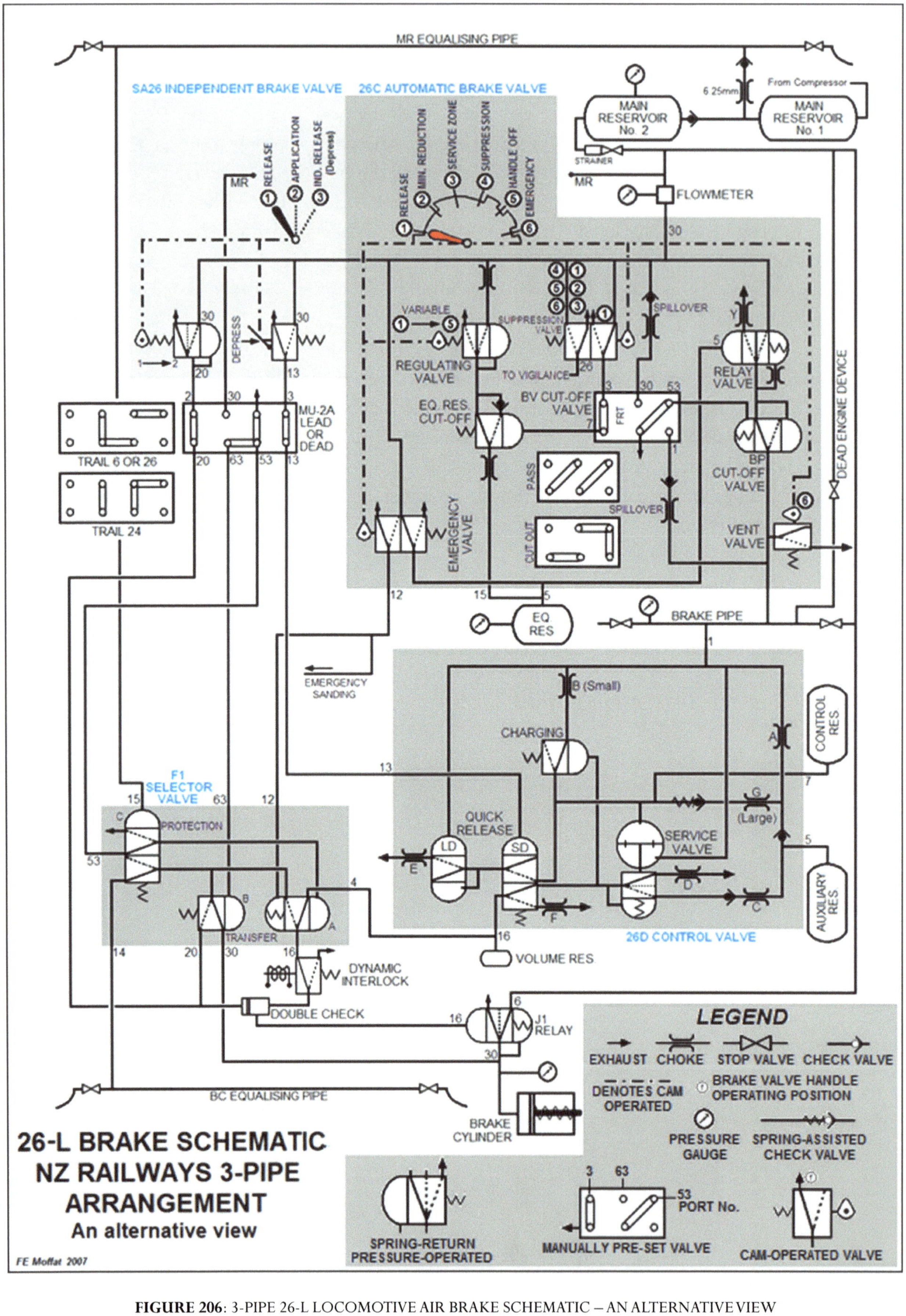

FIGURE 206: 3-PIPE 26-L LOCOMOTIVE AIR BRAKE SCHEMATIC – AN ALTERNATIVE VIEW

THE 30-SERIES EQUIPMENT (WABCO)

The WL Schedule

If the 24-series equipment (24-RL with D24 brake valve and versions) was improved-upon logically by the advancements offered by the 26-series equipment (26-L and derivatives), then it was inevitable that further evolution would extend this logical numeric progression. Thus, the next advancement from the 26-series became the 28-series equipment, wherein the two integral elements of the Automatic brake valve—the Regulating and Relay portions—were physically split from each other with only the Regulating portion and associated operating handle being in the cab, while the Relay portion and its related (and noisy) brake pipe exhaust function were banished to a convenient out-of-cab position. Whereas the 26-L design had dispensed with much in-cab piping and related complexity, the 28-L configuration reduced this yet again as the automatic brake Relay portion—now called the C2W Relay (or "module portion")—and attendant pipe work was also moved out of the driver's cab. The remaining panel-mounted cab device was, in reality, nothing more than the Regulating portion of the Automatic brake valve, controlling the low-capacity equalising reservoir. The benefits were a more compact control stand, and a quieter cab.

In my view, though, the downside with this transformation is that the audible brake pipe exhaust was engineered out of the driver's realm of awareness, and this has ramifications with regard to train handling practices.[129] Reducing cab noise levels is generally desirable for crew health reasons and is one thing, but completely isolating the driver of a lengthy freight train from that ancient perceptible cue long considered vital to effective brake manipulation—the *length and strength* of the brake pipe exhaust—is something else altogether.

I have reluctantly come to accept that the days of locomotive drivers being required to "handle the air" with judgemental finesse and expertise are at an end. The contemporary, commercially-focused rail industry can no longer indulge itself at taxpayers' expense in permitting locomotive drivers to develop their aptitude over time as they once did, honing the vital air brake and other train-handling skills required of their calling. The next generation of engine service personnel, even now among us, will qualify in a much shorter period than has historically been the case, and technical knowledge will be much less of a priority for the railroads; and indeed, already is!

Locomotive drivers today seldom scrutinise their air brake gauges for the equalising reservoir indication when making an air brake application; in fact, it is not easy to assimilate this information on some of the ill-designed and unrefined screen-based displays currently installed in some modern locomotives (the EMD/Rockwell ICE displays on Clyde-EMD units, such as the Q and S Classes in WA, the FQ Class in SA, and the solitary V Class from Victoria, being classic cases in point). Drivers today are likewise seldom encouraged by their instructors to look for **brake pipe equalisation** (in other words, an *effective* application) prior to initiating brake release, and many have learned to operate by 'feel'; simply moving the handle of the electronic brake controller until it clicks into the next détente, and *presto!*, a pre-programmed Service reduction without reference to any gauges or displays.

I did, though, have some experience with a 28-LAV (dual ***A***ir/***V***acuum) schedule in the course of a rail project in India; the brake valves being the A-9 Automatic and SA-9 Independent. The A-9 was not a particularly high-capacity valve, and the action around the quadrant—while positive—was extremely light. (The process by which

129. A Wabco instruction pamphlet (5071-6) has this warning: *"An awareness that air is exhausting at the brake valve when an Automatic brake application is called-for by the operator is one of the first indications that the air brake equipment is functioning. A direct audible exhaust in the cab is the most desirable arrangement. Failure* [by railway organisations] *to comply* [with specifications for piping the exhaust out of the cab to reduce noise levels] *can result in improper air brake operation and/or the inability to adequately control the retardation of the vehicle or train..."*

a brake pipe reduction on the locomotive was relayed—via a vacuum brake application—to create locomotive brake cylinder pressure, while logical, is interesting and somewhat complex![130]) With the Regulating and Relay portions physically separated, the "28" equipment was an undoubted evolutionary improvement, but the driver's brake valve was still panel mounted in a vertical control stand. This was to become an issue when railroads in North America began requesting brake valves suitable for their new "desk" or console control layouts.

In the late 1970s, the US passenger train operator, Amtrak, was adapting a Swedish AC electric locomotive for high-speed operation on their North East Corridor route. As part of the improved driver's cab layout, an air brake valve able to be incorporated into a control desk in front of the driver was specified. A similar request had also been made of US air brake equipment manufacturers by the Canadian National railway, which was redesigning the locomotive driver's cab workstation to conform to the new AAR Clean Cab[131] principles. The result, from Wabco, was the 30-Series equipment. The new 30C brake valve arrangement, incorporating both Automatic and Independent brake valve elements within one compact housing, provided a modern appearance, was space efficient, and simplified cab piping arrangements in a similar fashion to the 28 equipment. Presented as the 30-CDW (***D***esk ***W***idecab), the new brake valve utilises a common cut-off pilot valve that combines the functions of isolating both Automatic <u>and</u> Independent brake valves from operation.

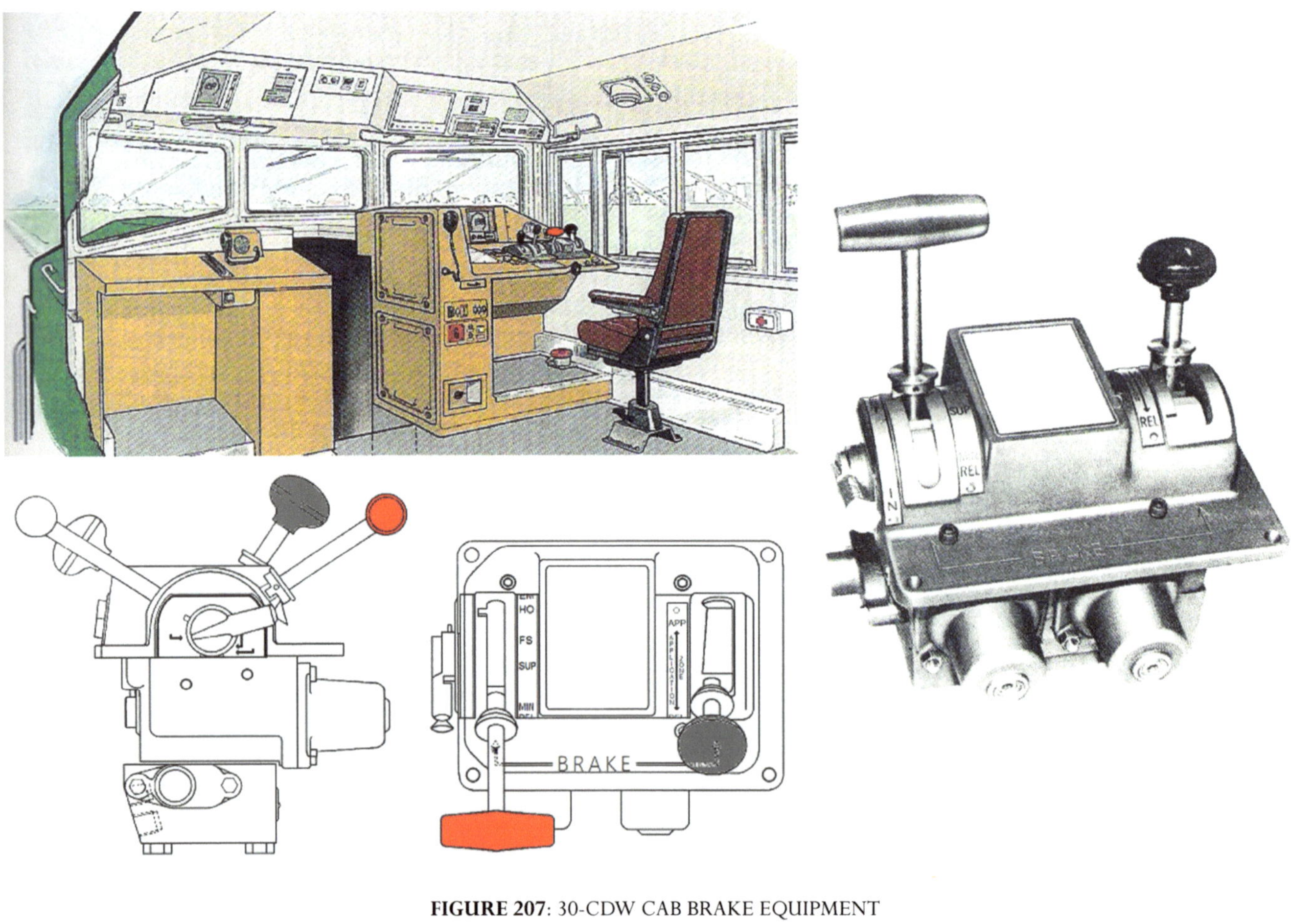

FIGURE 207: 30-CDW CAB BRAKE EQUIPMENT

130. The 28-LAV locomotive brake application process: (i) Auto brake handle moved away from Release → equalising reservoir reduced; (ii) C-2 <u>brake pipe</u> relay valve (module portion) responds → brake pipe reduced; (iii) VA-1B vacuum control valve responds → vacuum train pipe increased (vacuum proportionally destroyed); (iv) 28-VB control valve responds → control pressure supplied to C-2 <u>brake cylinder</u> relay valve; and (v) C-2 brake cylinder relay responds → main reservoir feed to locomotive brake cylinders.

131. Refer to Glossary, chapter 8.

As with the 28-L equipment, only the Regulating portion of the Automatic brake valve is installed within the cab environment. The Relay portion and associated larger high-capacity components, such as the integral brake pipe cut-off, vent, and emergency valves, are all mounted remotely and called the **30-CW module portion** incorporating the same **C2W relay valve** previously mentioned. In operation, the 30-CDW brake valve and associated 30CW module portion function identically to the 26C and SA26 Automatic and Independent portions of the 26-L schedule (within which the 30-CDW equipment is designed to operate).

The 30A-CDW brake valve

Because the 30-CDW equipment was designed specifically for US passenger service, the cut-off pilot valve, when positioned to IN, provided for *graduated release* of the Automatic brake. Westinghouse Australia, on behalf of Australian clients, asked its US parent to provide this brake valve with a *direct release* function only, to suit most operating practices in this country. The resultant ***A***ustralianised version is the 30A-CDW (author's note: a more logical designation following an established naming convention might have been *A30-CDW*) with two usable cut-off pilot valve positions—OUT and FRT (***FR***eigh***T***)—and one blocked position, PASS (***PASS***enger). This version is now also in use within North America.

When supplied with the Australian WL (***L***ocomotive) or WLG (***L***ocomotive ***G***raduable) triple valve and WI ***I***ndependent release valve, it constitutes Westinghouse Australia's incarnation of the 30-Series - the WL equipment. The WL locomotive triple valve is actually the ubiquitous 'W' triple valve without either the

FIGURE 208: 30A-CDW BRAKE VALVE INSTALLATION ON KIWIRAIL DL-CLASS LOCOMOTIVE (1)

inshot or **retarded recharge** portions, these being required on a freight wagon but not a locomotive. Different choke-plates are used that permit recharging of brake pipe and auxiliary reservoir more rapidly than on a hauled vehicle. This more rapid charge rate ensures that the auxiliary reservoir will always be ready for the next Automatic brake application. The brake cylinder build-up rate is deliberately slower than on a freight wagon so as to reduce the chance of severe slack run-in. The separate WI **Independent release valve** functions during Independent brake valve **actuation** (bail-off) to provide Quick-Release.

This equipment has been fitted in Australia to several classes of Clyde-EMD and Goninan-GE locomotives (now operated by Aurizon and Pacific National), two fleets of Clyde-EMD units built for FreightCorp (NSW) and now also operated by Pacific National, and a group of ex-QR English Electric units rebuilt in South Australia by the Morrison-Knudsen Company. (Note: Some of these latter units, classed MK, were operated during the mid-1990s in "Ready Power" lease service in Malaysia, then back in Australia with Pacific National Queensland, and are now operating in Tasmania). 30A-CDW has recently been introduced to New Zealand for the first time on KiwiRail's 26-L-equipped DL-Class fleet built at Dalian, China.

FIGURE 209: 30A-CDW BRAKE VALVE INSTALLATION ON KIWIRAIL DL-CLASS LOCOMOTIVE (2)

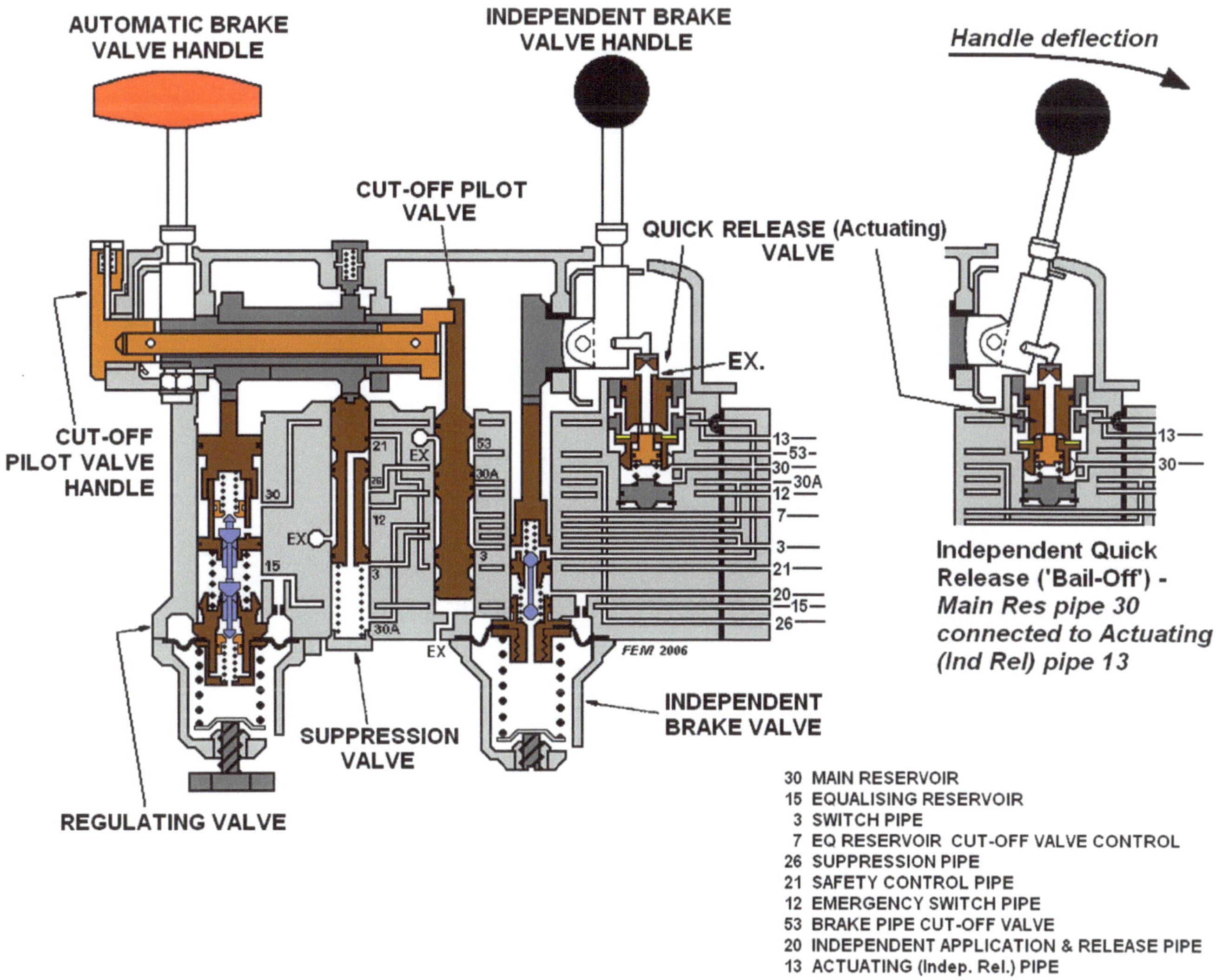

FIGURE 210: 30A-CDW BRAKE VALVE DIAGRAM – 'RELEASE POSITION' (AUTOMATIC AND INDEPENDENT)

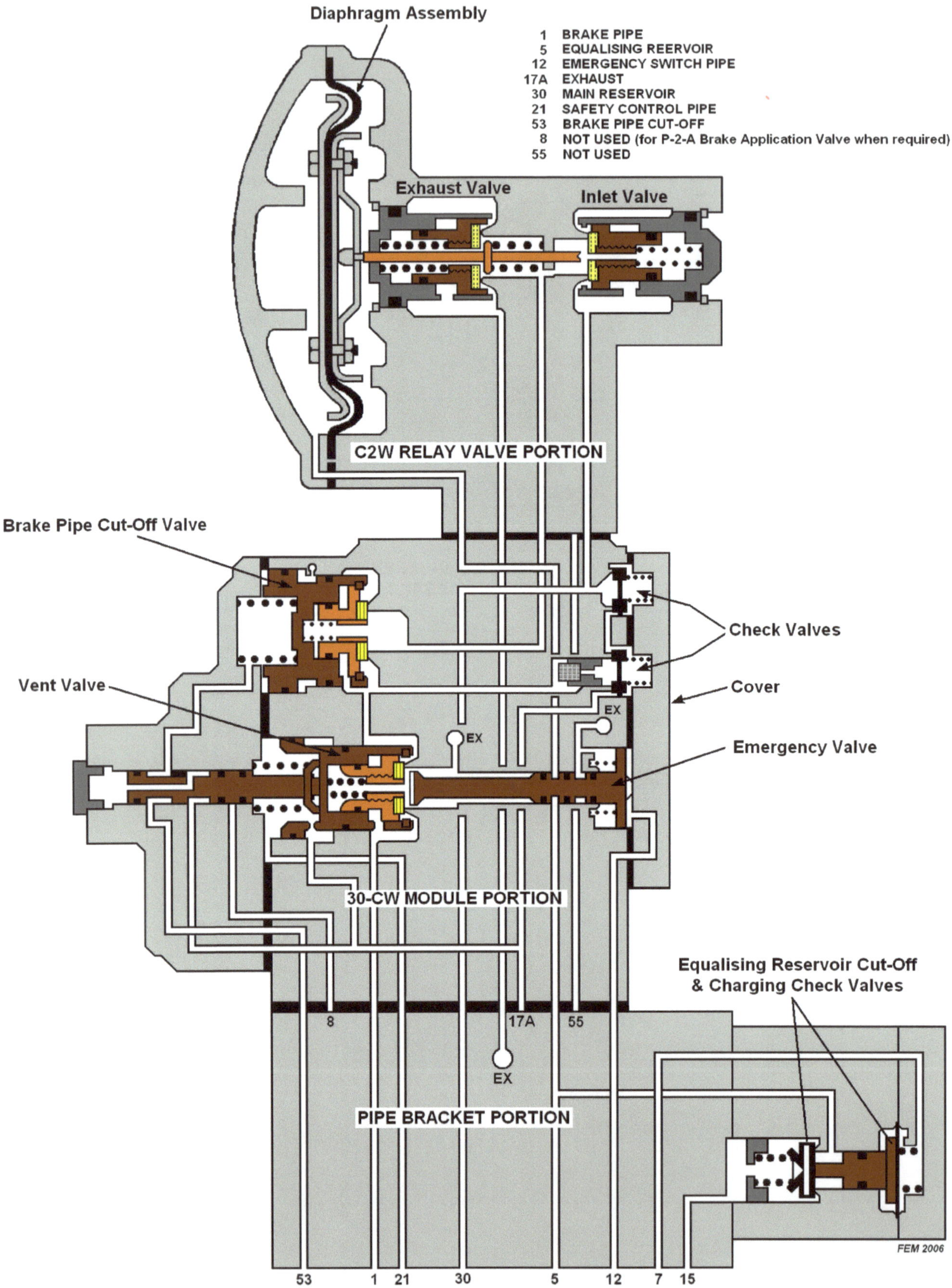

FIGURE 211: 30-CW MODULE AND C2W RELAY PORTIONS – DIAGRAM, 'RELEASE/CHARGED' CONDITION

The W30 Brake Valves

Early in 1986, the Mount Newman Mining Company in the Pilbara region of Western Australia—at the behest of their enginemen's union—set about designing a customised driver's cab layout for their forthcoming GE C36-7M locomotives to be rebuilt from their Century-Series ALCO units by A. Goninan & Co. As part of the driver's cab design phase, the use of the 30A-CDW brake valve—mounted vertically in an AAR-style control stand and oriented similar to the 26 equipment—was investigated (Figure 215). Although deemed practicable by Westinghouse Australia, the proposal was eventually dropped by MNM due to budgetary constraints, and refurbished 26-L equipment was used.

Two years later, the company was completing specifications for their new Goninan-GE CM39-8 units. Again, a driver's cab design project was undertaken to incorporate the US AAR Clean Cab standards and contemporary ergonomic principles as well as local requirements, and again the 30A-CDW brake valve was under consideration. By this time, however, Westinghouse Australia was offering an innovative local redesign of this valve by which the Automatic and Independent portions were physically separated and able to be mounted apart from each other if desired for greater cab layout flexibility. Convertible for either **push-to-apply** or **pull-to-apply** operation, this new version was offered to MNM and enabled their cab layout requirements to be met. In this MNM application, the W30 equipment was fitted as part of the advanced-specification 26-LUM unitised rack-mounted arrangement rather than the more modest WL schedule. This was the first use of the new W30 equipment.

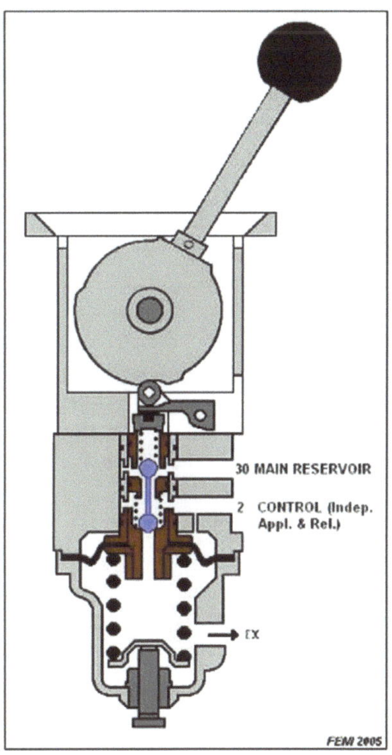

FIGURE 212: W30-IND INDEPENDENT BRAKE VALVE - DIAGRAM

Designed principally by the late Mr Noel Mather[132]—retired New Zealand Railways Chief Brake Inspector, who was working at the time for Westinghouse Australia—the new equipment was conceived initially for NZR who had originally intended to install it on new Brush electric locomotives. Given the "W" designation apparently to reflect its Westinghouse Australia design and manufacture, this new 30-Series locomotive equipment consists of the W30D-G Automatic (providing ***D***irect or ***G***raduable release) and W30IND ***IND***ependent elements mounted in the driver's cab, plus the 30CW relay module portion mounted remotely. The MU-2A valve is deleted since both the 13 and 20 pipe Independent brake functions—**actuation** (or "bail-off") and **control** respectively—are carried through the brake valve cut-off pilot valve. The cut-off pilot valve itself has three positions: OUT, D (***D***irect release), and G (***G***raduated release). A *G-stop* fitted to the exterior face of the brake valve or, alternatively (and more permanently), a plug specially inserted into the cut-off valve cavity can be provided to block the G position from use. Westinghouse Australia decided to use the labels "D" and "G" on the cut-off pilot valve instead of the usual "FRT" and "PASS," respectively, since PASS refers to US passenger train *graduated release* operation. Because of this, the manufacturer felt that a cut-off pilot valve position labelled PASS might be

132. During my railway service in New Zealand I had never had occasion to meet Noel Mather, but in 1995, while I was working in a technical role with a Southeast Asian luxury tourist train company, I had the privilege of meeting this elderly but vital gentleman and hosting him on board for a day. As our train rolled through the verdant scenery of the southern Malaysian peninsular, I enjoyed a long, wide-ranging discussion with one of Australasia's leading specialists on railway air braking, and I consider the encounter a career highlight.

problematic if an Australian driver were to unintentionally use it when conditioning the brake valve on a locomotive in passenger service, since almost all passenger brake gear used in Australia is *direct release* (or freight-style) equipment.

Note that the danger in setting the cut-off pilot valve for 'graduated release' operation when the train is fitted with 'direct release' triple valves is that, whereas a locomotive driver in Australia would expect the Automatic brake valve handle to require to be placed fully to the RELEASE position in order to effect a brake release, with the brake valve configured for graduated release, only a slight—perhaps unintentional—movement of the handle back **toward** RELEASE will result in a complete and possibly unplanned brake release on the train. In such circumstances, the result could be at best a train handling miscalculation and at worst a runaway.

I was puzzled that in Western Australia, ARG (later QRN, now Aurizon) continued to permit its W30-equipped locomotives (the Goninan-GE P-Class, now re-classified 2500, that I was operating at the time) to be operated—purely in freight service and against well-established international convention—with the brake valve cut-off pilot valve set in the G (GRADUATED RELEASE) position. The potential danger described above clearly exists, and this policy appeared inconsistent not only with this railways' avowed intention (at the time) to be the safest in the nation, but also with any notions of excellence or normal industry custom and practice, let alone "World's Best". The following passage is from a Wabco equipment pamphlet (5071-6, page 3- the emphasis is theirs);

IMPORTANT:When operating in Freight service, a brake release is accomplished by manually moving the controlling brake valve handle directly into the Release position WITHOUT HESITATION. A GRADUATED BRAKE RELEASE IS NOT PERMITTED IN FREIGHT TRAIN OPERATION.

Meanwhile, other niche train operators in Australia are proposing to import second-hand locomotives from the United Kingdom and Europe (indeed, some have arrived as this is written; refer to the section on the Knorr-Bremse Type D-2-b driver's brake valve), predominantly for freight haulage, that are equipped with European graduated release equipment. It remains to be seen how they intend to operate them in this context and what 'risk mitigation' leeway the rail regulator might see fit to extend.

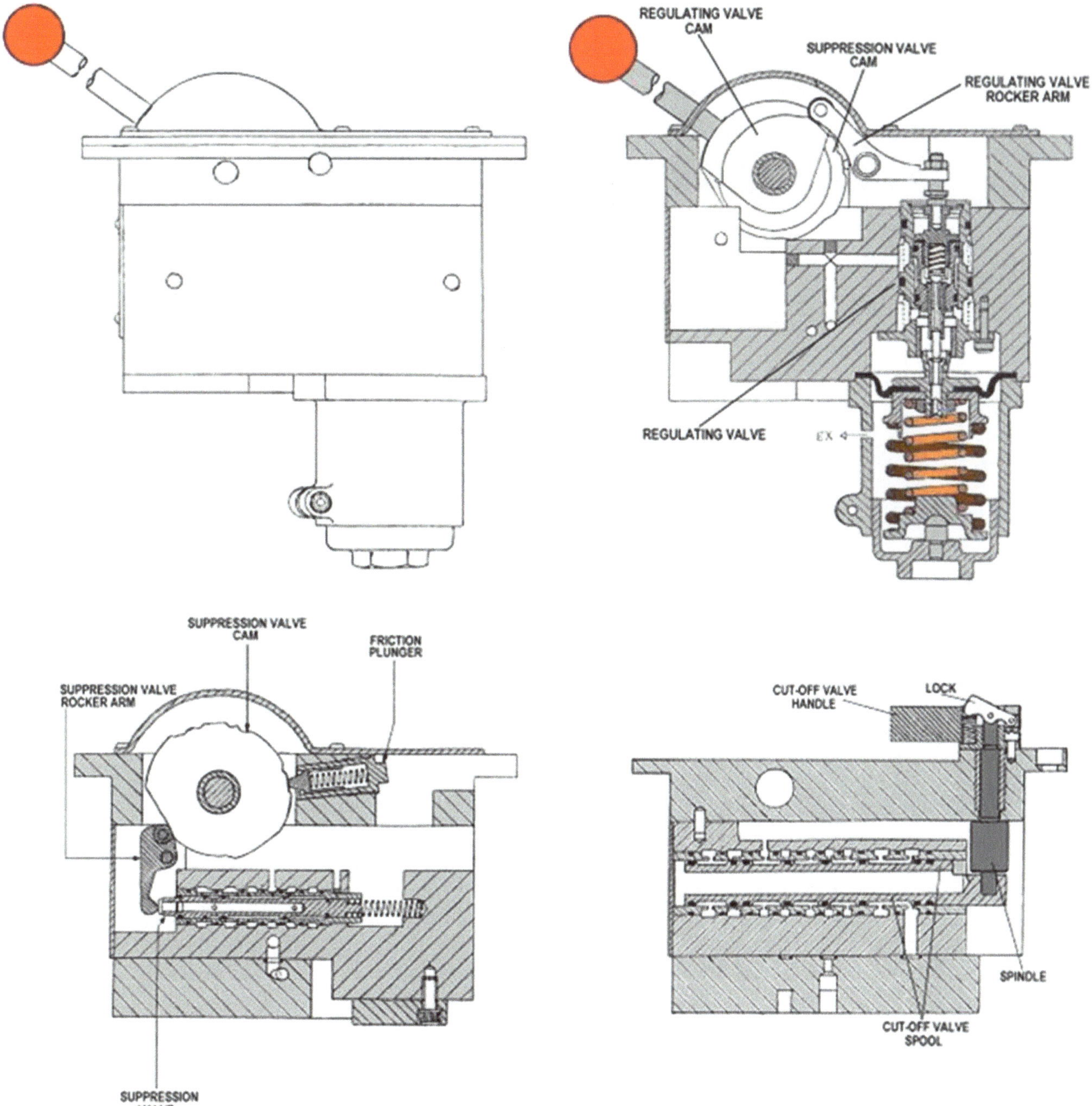

FIGURE 213: W30-DG AUTOMATIC BRAKE VALVE – DIAGRAMMATIC SIDE VIEWS THROUGH COMPONENT SECTIONS

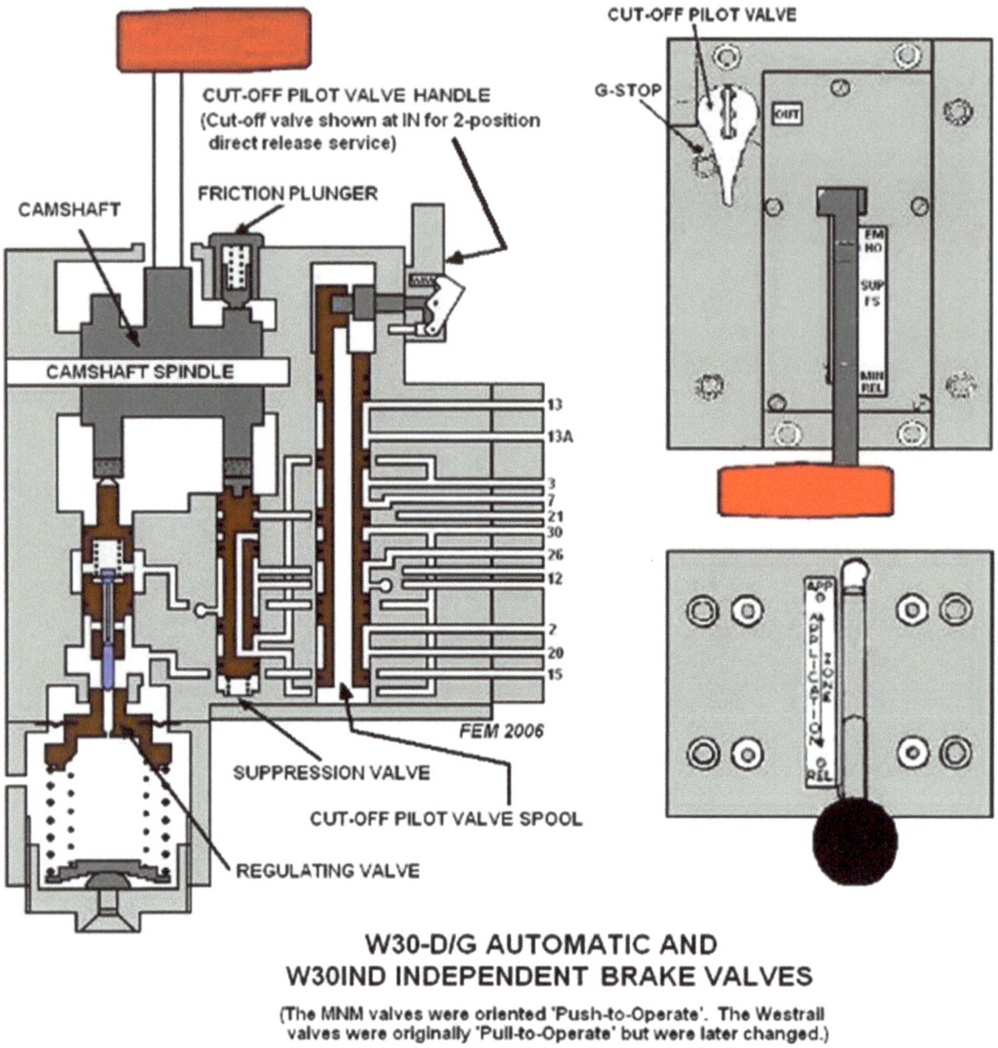

FIGURE 214: W30-D/G AUTOMATIC AND W30-IND INDEPENDENT BRAKE VALVES – DIAGRAM AND GENERAL ARRANGEMENT PLAN VIEWS

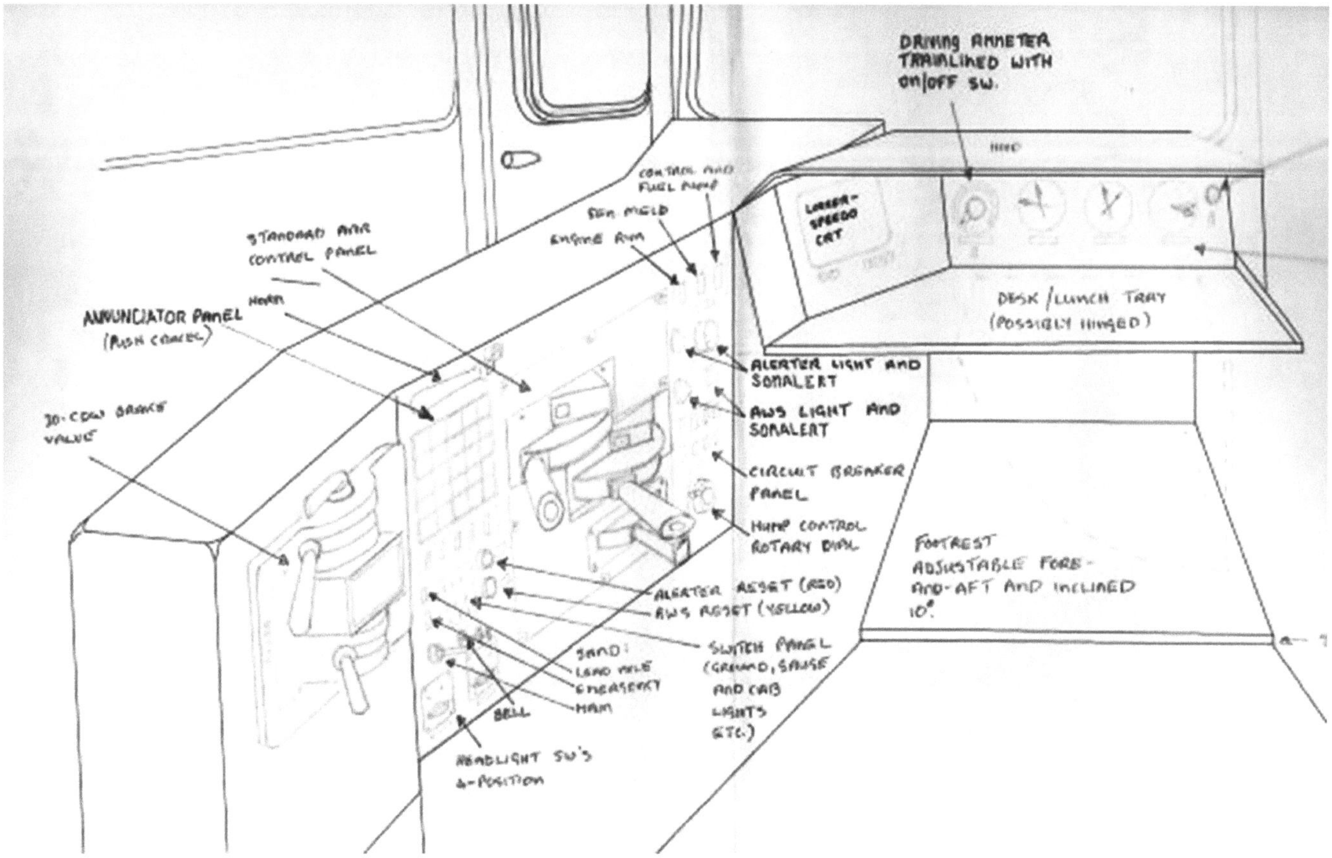

FIGURE 215: EARLY DRIVER'S CONTROL DESK DRAFT CONCEPT FOR GONINAN-GE C36-7M LOCOMOTIVE REBUILD

The 'back-of-an-envelope' concept sketch (Figure 215) was submitted by the author on behalf of the Mt Newman Mining (later BHP Iron Ore) locomotive Cab Design Committee during planning in 1986 for the rebuilding of Goodwin/MLW-ALCO C636 locomotives into modernised GE Dash 7 units. This sketch depicts the 30A-CDW brake valve mounted vertically on the control stand, utilising 26-style handles. The concept was not adopted - refurbished 26-L equipment eventually being re-used (Figure 216).

An interesting feature of this eventual cab layout was that the control stand was deliberately located to provide generous access between it and the driver's seat. As a result of this distance, the SA26 Independent brake valve handle was not conveniently to hand for a driver leaning out of the cab-side window during shunting operations. Consequently, Independent brake valve handles for these rebuilt locomotives were custom-manufactured that could be screwed to 'unlock' and then pulled out to significantly extend their length (Figure 201).

FIGURE 216: EVENTUAL CONTROL LAYOUT FOR MNM GONINAN-GE C36-7M LOCOMOTIVE REBUILD

During the cab design phase preceding manufacture of the GE CM39-8 locomotives that followed the Dash 7s in 1988, the driver's work-group representatives provided a request to builders A. Goninan & Co. for the GE controller panel to be adapted for a horizontal desk-style application. The resulting solution proposed by Goninan was subsequently created as part of a cab mock-up (Figure 217). The desk layout proposed by the builder was altered in order to better utilise the flexibility of the new W30 brake valve arrangement. The full desk-style concept was eventually rejected in favour of a version of the more conventional control stand as seen to the left in Figure 218. A more conservative version of the desk layout depicted in this sketch was later accepted by Westrail for their new P-Class fleet.

As originally produced for BHP Iron Ore, the W30D-G Automatic brake valve utilised a Regulating valve portion borrowed from the United Kingdom. This appeared to be satisfactory for the requirements of this first customer; however, when the brake valve was later applied to the Westrail P-Class locomotives, an interesting problem emerged. The British component utilised soft-seated valves that permitted a sympathetic grooving of the valve seat over time and the eventual development of a hysteretic reaction by which the equalising reservoir setting and the required brake pipe value could become contradictory. This was problematic since some of the P-Class fleet had been fitted with Locotrol distributed power equipment, which is rather intolerant of such inconsistency. The solution for Westinghouse Australia was to replace the UK Regulating valves with the standard 26C component, which has a harder valve seat. This modification produced the W30D-G-H (***H***ard seat) version that—since the BHP Billiton Iron Ore CM39-8 units have been modified for EPIC electronic operation—is now used universally. Apart from the Western Australian P-Class units, W30 brake valves are used in the QR 2800-Class (also Goninan-GE Dash 8 units), and the rebuilt Chicago Freight Car Leasing Australia GL-Class (Goninan-GE), as well as some track maintenance machines that are capable of hauling freight wagons.

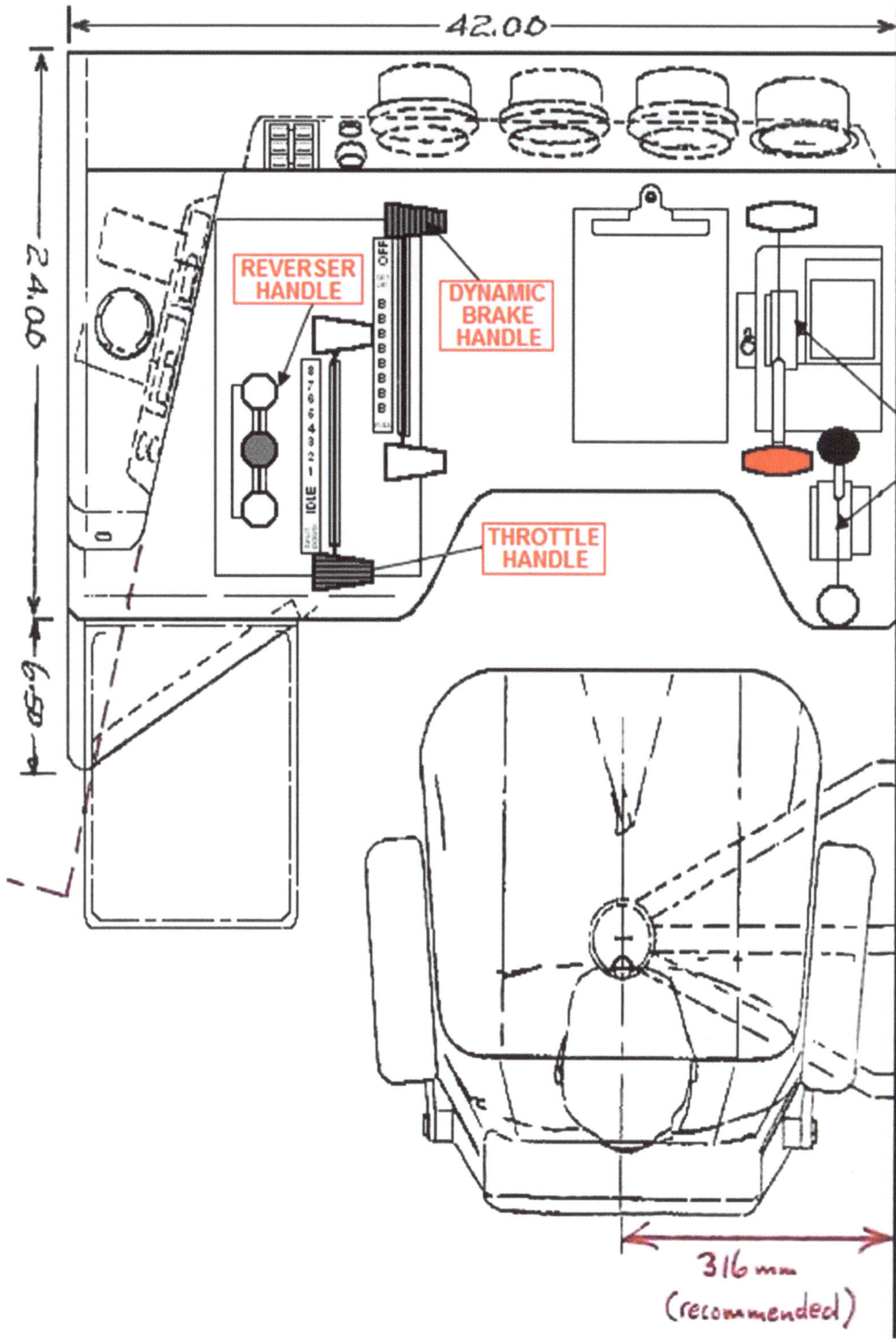

FIGURE 217: DRIVER'S CONTROL DESK CONCEPT FOR MOUNT NEWMAN MINING CO. GONINAN-GE CM39-8 LOCOMOTIVE REBUILD. THIS (A. GONINAN & CO) DRAWING DEPICTS AN EARLY CAB MOCK-UP PROPOSAL RESULTING FROM MNM CAB COMMITTEE AND GONINAN COLLABORATION

The author's photo (Figure 218) shows the original brake valve layout in the driver's cab of the CM39-8M locomotives. These units have since been upgraded with EPIC 3102 equipment. They were the first GE Dash 8s to be placed in service outside North America and were the forerunners to other GE Dash 8, Dash 9, and AC 6000 units that have powered the Pilbara iron ore railroads since 1989. They were also the first

FIGURE 218: EVENTUAL CONTROL LAYOUT FOR MNM GONINAN-GE CM39-8 LOCOMOTIVE REBUILD. THE DESK-STYLE CONTROL STATION CONCEPT HAS BEEN REJECTED. THE W30DG AND W30IND BRAKE VALVES HAVE BEEN INSTALLED IN A TANDEM CONFIGURATION WITH THE INDEPENDENT BRAKE VALVE HANDLE CONVENIENTLY TO HAND FOR ALL OPERATIONS. THE LOCOTROL PUSH/PULL-BUTTON AIR BRAKE CONSOLE IS LOCATED PROXIMATE TO THE W30 BRAKE VALVE HANDLES, KEEPING ALL AIR BRAKE CONTROLS CONVENIENTLY GROUPED. THESE LOCOMOTIVES HAVE SINCE BEEN REBUILT WITH THE EPIC BRAKE SYSTEM, A REVISED VERTICAL CONTROL STAND, AND SCREEN-BASED DISPLAYS

locomotives in Australia to have the W30 brake valves, which can be seen arranged in tandem alongside the cab wall in close proximity to the LOCOTROL II push-button air brake console. The independent quick release (bail-off) push-button—incorporating a blue light to indicate the presence of brake cylinder pressure—is located adjacent to the lower right-hand corner of the LOCOTROL air brake console. The tandem arrangement of the Automatic and Independent brake valves illustrates the extreme layout flexibility offered by the W30 design.

Notably, with the Westrail application (Figure 219), the Automatic and Independent elements—although physically separate—were located side-by-side on the driver's control desk. With this arrangement, the design flexibility—the essence of the W30 concept and the very reason for its existence—has not been fully exploited, although it *has* enabled the two portions of the valve to be co-located to suit the operator's requirements - in this case, with the Independent to the right of the Automatic.

Located to the right of the Automatic handle and being in an extreme-forward position on the desk, the Independent portion is not exactly within easy reach of the driver. The button on the desk in front of the Automatic brake valve handle is Independent Release "actuate" (bail-off). The panel to the right of the brake valve handles, containing the locomotive control handles, is the eventual manifestation of that depicted as a proposal to MNM drivers in Figure 217. It is essentially the standard General Electric AAR controller (Figure 216) moved from a vertical control stand position to a horizontal desk and modified with T-handles.

FIGURE 219: THIS CONCEPT WAS CHOSEN BY WESTRAIL FOR THEIR P-CLASS LOCOMOTIVES. W30IND INDEPENDENT BRAKE VALVE IS POSITIONED BESIDE THE AUTOMATIC BRAKE VALVE AND IN A FORWARD POSITION ON THE CONTROL DESK. THIS REQUIRES AN INCONVENIENT DEGREE OF REACH FOR OPERATION AND IS NOT AN OPTIMUM LOCATION. *The author*

FIGURE 220: W30 BRAKE VALVES AND LAYOUT ON THE CONTROL DESK OF A QR 2800-CLASS LOCOMOTIVE. NOTE OPPOSITE ORIENTATION OF AUTOMATIC AND INDEPENDENT PORTIONS AND DIFFERENT PRESENTATION OF THE VALVE ENCLOSURES TO THAT DEPICTED IN FIGURE 219. *Author's collection*

Figure 220 depicts the W30 installation to the QR 2800 Class locomotives and clearly demonstrates the flexibility offered by this design. The driver's station is on the right-hand side of the cab and the Independent portion has been located to the left of the Automatic portion. The physical shape and design (Western Australian versus Queensland version) of these two brake valve housings and handles as well as the cut-off pilot valve handles is also worthy of comparison.

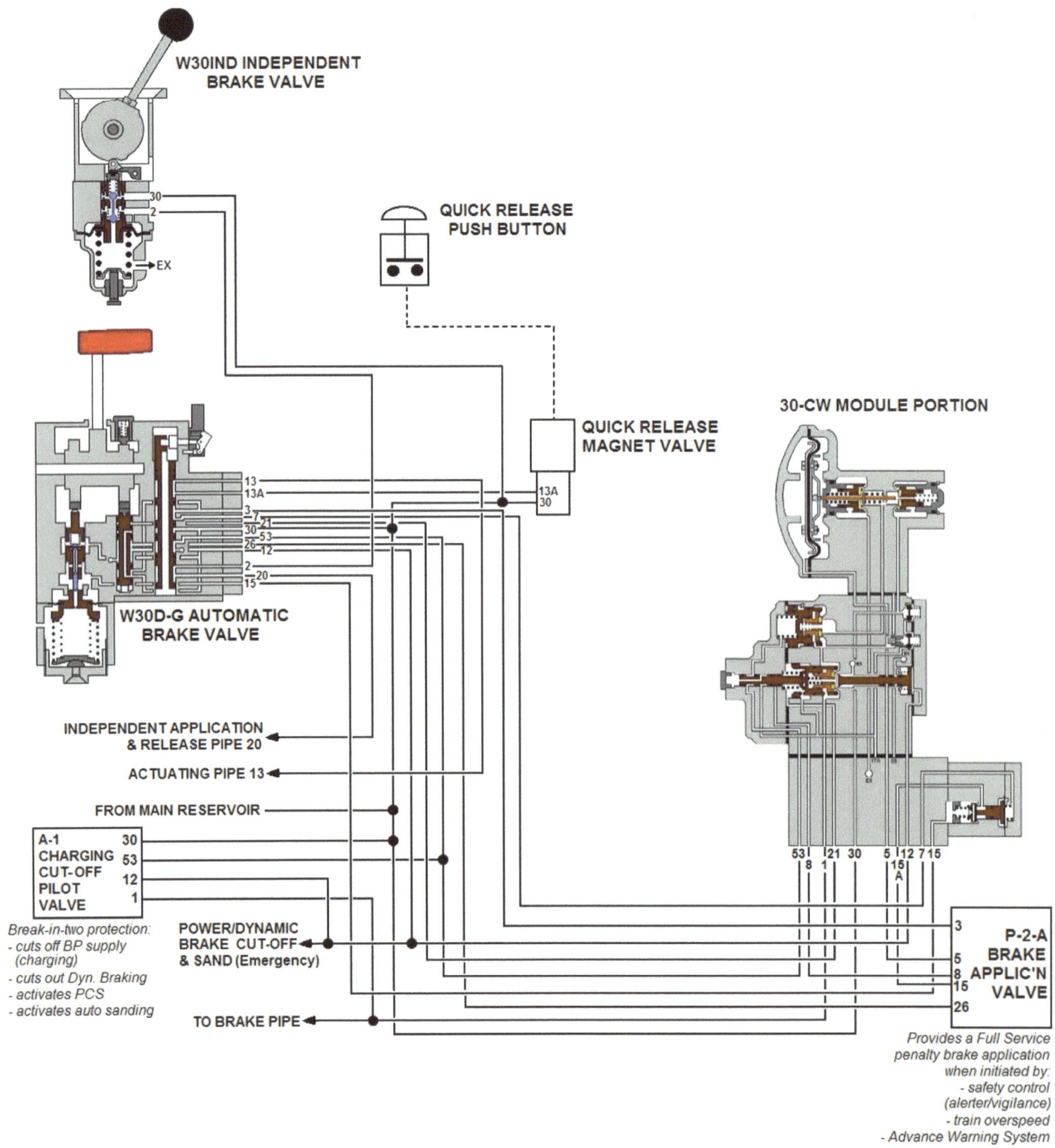

FIGURE 221: W30 BRAKE VALVES AND 30-CW MODULE PORTION – SCHEMATIC DIAGRAM. THIS IS A SIMPLIFIED PIPING DIAGRAM FOR A BHP IRON ORE GE CM39-8 LOCOMOTIVE AS ORIGINALLY EQUIPPED. THE INSTALLATION IS TYPICAL FOR AN AMERICAN 4-PIPE LOCOMOTIVE

THE 31-SERIES EQUIPMENT (WABTEC)

EPIC™

By the mid-1980s, industry observers realised that the 30-Series equipment would probably be the ultimate expression of the fully pneumatic locomotive air brake format, and that the inevitable next step in the advancement of locomotive brake systems would be the integration of electronics and electrical control to replace some or all of the conventional pneumatic logic. Since the late 1970s, the Locomotive & Rubber Products Division of Westinghouse Air Brake had been active in the design, manufacture, and supply of computer-controlled brake equipment related particularly to the Rail Transit industry. In 1986, development began in earnest on the concept of a new electronic/pneumatic braking arrangement for locomotives that would become the company's versatile and innovative ***E**lectro-**P**neumatic **I**ntegrated **C**ontrol air brake scheme. US railroads were extensively consulted during the design phase, as were the Federal Railroad Administration and the Association of American Railroads' Research Centre[133]. The new equipment's first incarnation—EPIC 3101—was applied in 1989 to a 1 050 hp Norfolk Southern Railroad switching and yard transfer locomotive. To ensure user convenience and maximum unit availability, this prototypical system was installed parallel with the existing 26-L equipment; cut-out cocks being strategically located to isolate the latter equipment when running under EPIC control.

The microprocessor-based system proved to be stable in operation and rapid in its response to control inputs. The arrangement reduced the relatively complex pneumatic logic functions (previously accomplished by numerous valvular hardware devices) to the more uncomplicated preserve of a software program. In turn, this permitted individual pieces of electrical and pneumatic equipment to perform multiple functions and for the functionality of some of the 26-L pneumatic devices to be emulated in the software, while the total equipment package—being lighter, simpler, and easier to maintain—was expected to be inherently more reliable. EPIC 3101 saw the introduction by Wabtec of the *Pneumacard* modular equipment-mounting design feature permitting rapid change-out of the brake system as well as the international-standard *Eurocard* format with DIN edge-connectors[134], allowing quick change-out of electronic equipment.

The system's programmable flexibility permits operational changes via software modifications rather than hardware changes, and the equipment is designed to be physically interchangeable and MU-compatible with the 26-L schedule and congruent variations in Lead, Dead, and Trail service. Mechanically, this compatibility was achieved using proven components such as industrial S-type relay valves and Super Spool valves as well as the proven C2W relay module, J-type brake cylinder relay valves, and other pneumatic devices from the N-1 and MC-30 families. Electronic control is achieved with solenoid valves and transducers. As alluded to in the previous paragraph, and in a manner reminiscent of the use of printed card slots in a personal computer, the pneumatics were mounted on three mechanical plates that slid into channels on a cage unit, with electrical and pneumatic connections being made when a plate was fully positioned and secured.

Further operational benefits include the following:

- Cab air piping is eliminated;
- Air devices are incorporated into three primary *removable modules*, two secondary *all-pneumatic backup air brake elements*, and a number of miscellaneous items, including pressure transducers and switches, all of which were mounted—as described above—on sliding plates (for 3102 this became a single laminated manifold), providing simultaneous disconnection of air and electrical interfaces from the modules);

133. See Glossary, chapter 8.

134. Ibid.

- Equalising reservoir pressure is maintained at all times against even moderate leakage. This is an important factor for freight train brake control in the modern-day environment, particularly for long-train and wireless distributed power operations (which—in the absence of competitor systems—means the GE LOCOTROL equipment, being notoriously intolerant of equalising reservoir leakage on the controlling locomotive.);
- An all-pneumatic *backup* feature provides critical brake pipe pressure protection in the event of a complete electrical power loss (and computer shutdown) as well as when the locomotive is being hauled Dead-in-Train;
- Familiarity for engine service personnel—in appearance and operational 'feel'—to 26- and 30-Series equipments as well as increased feedback information and improved accuracy in braking control.

The 3101 equipment subsequently evolved into the 3102 version via a mechanical upgrade that saw stainless steel protective covers provided for some of the components and major control portions being machined rather than cast, as well as the three Pneumacard plates being combined into a single laminate. Further evolutionary improvements, herein described, saw the system become EPIC II.

The EPIC equipment currently operated in Australia is either the 3102 version or the later EPIC II development. This latter version is functionally identical to 3102; however, both electronics and computer construction have been upgraded (including a better printed circuit board rack), diagnostics significantly improved (it became possible to test either individual portions or the complete brake system *in situ*; record data for the various penalty, emergency, and diagnostic events; and review real-time brake system functions), and software is now able to be flash-loaded. By comparison, the 3102 system uses EPROMs, for which software modification is comparatively laborious. The individual 3102 and Epic II operating portions, though, remain interchangeable.

Australian railways using this equipment currently include Rio Tinto (Hamersley Iron and Robe River) and BHP Billiton Iron Ore, Pacific National (including the EPIC II-equipped ex-Freight Australia V-Class unit and some refurbished ex-FreightCorp 48-Class units), Pacific National Queensland, FreightLink, and QR National/QRN West (Aurizon).

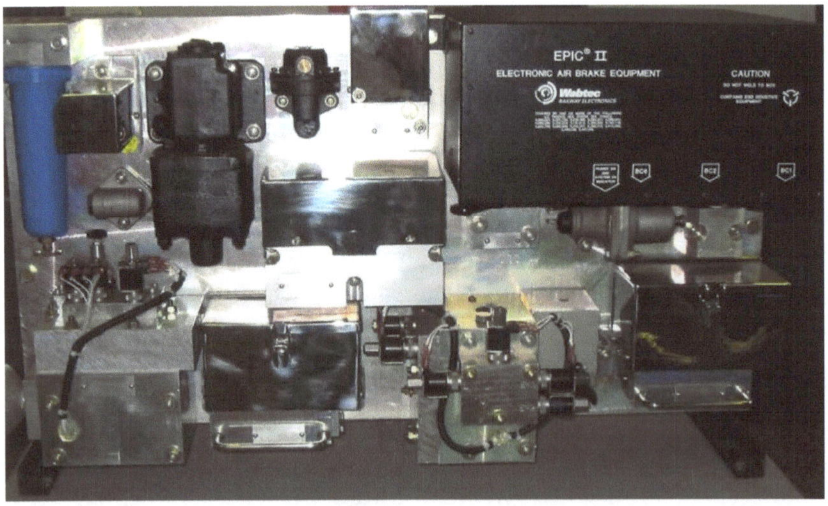

FIGURE 222: EPIC II PNEUMATIC OPERATING UNIT. *Wabtec, courtesy Ian Livingstone*

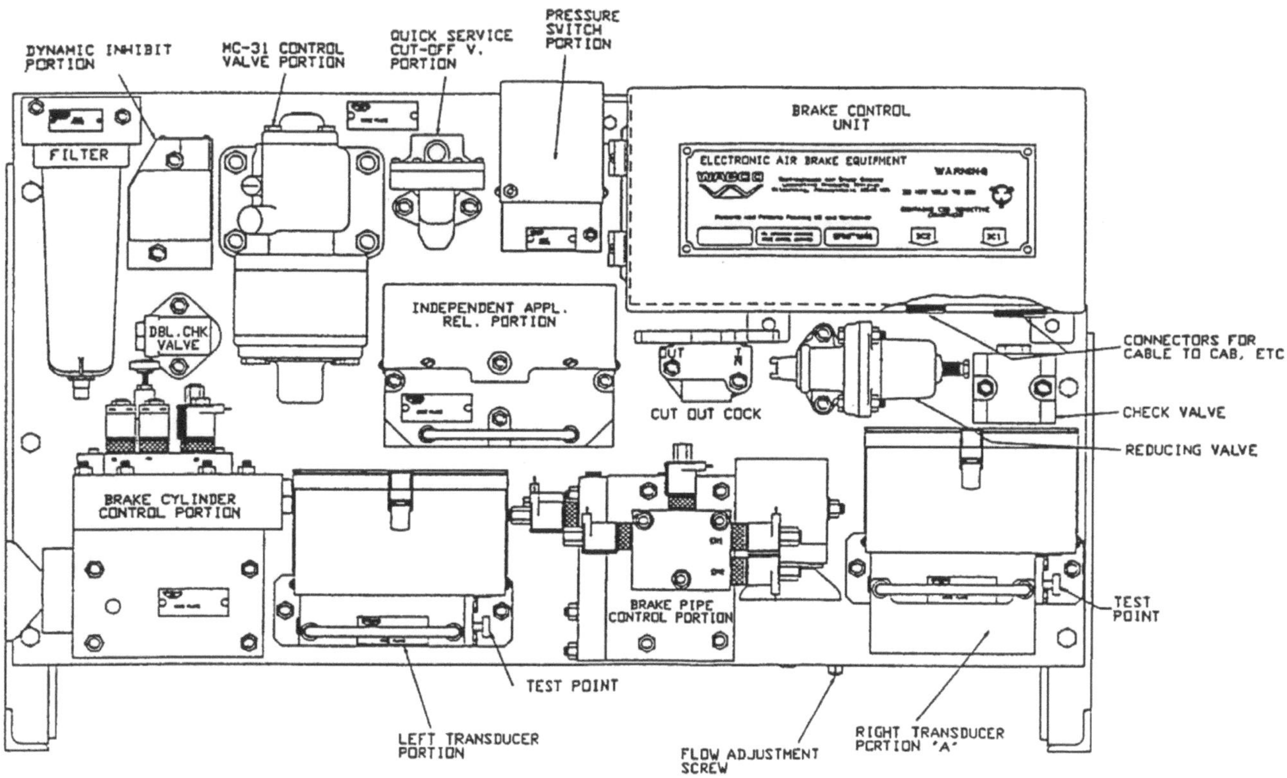

FIGURE 223: EPIC II PNEUMATIC OPERATING UNIT – GENERAL ARRANGEMENT

Major Components and their Functions

The *Cab Control Unit* (CCU), also referred-to as the Cab Control Portion, is the locomotive driver interface and is an electromechanical switch providing the same functional commands as the conventional locomotive brake valve. The unit has the two familiar handles—Automatic and Independent—with the same handle positions as the 26C and SA26 or similar Automatic and Independent valves but has no internal pneumatic function. Despite being simply a mechanical braking switch, it continues to be universally referred to by industry as a "brake valve."

The *Cab Control Computer* (CCC) processes the active cab status, Automatic and Independent brake handle positions, and setup parameters and sends the information to the Brake Control Computer. It also processes feedback information from the Brake Control Computer to the cab display.

The *Pneumatic Operating Unit* (POU) is the equivalent (and more) of the 26-L unitised rack. On its laminated panel are mounted all of the air-handling devices as well as the Brake Control Unit in which resides the system software. The rear face of the manifold mounts several removable canister volumes and a fabricated reservoir volume for various operational purposes. There are eight required air connections and two electrical connections:

- The *Brake Control Unit* (or "computer") contains the electronics that process the signal information for control of all intended electronically-controlled air brake functions as required by the customer;
- The *Brake Pipe Control Portion* controls equalising reservoir pressure (and thus brake pipe), permits cut-in/cut-out of the Automatic brake, and contains the emergency vent valve for emergency applications;

- The *Independent Application & Release Portion* controls both 20-pipe pressure for the application and release of the Independent brake, 13-pipe pressure for quick-release (bail-off), and allows the Independent brake to be cut In (LEAD) and Out (TRAIL);
- The *Brake Cylinder Control Portion* controls brake cylinder pressure for application and release of both Automatic and Independent brakes;
- The *MC-31 Control Valve* and the *Quick-Service Valve* provide the pneumatic backup signal to apply and release the Automatic brake in response to brake pipe pressure changes;
- The *Double Check-Valve Portion* selects which of the backup signals—either Independent brake 20-pipe pressure or control valve 16-pipe pressure (developed from an Automatic brake application)—gets passed to the Brake Cylinder Control Portion;
- The *Dynamic Inhibit (or Interlock) Portion* interrupts the development of the control valve-developed pneumatic backup signal when dynamic braking is in effect;
- The *Dead-In-Train feature* permits charging of the No. 2 main reservoir from the brake pipe. This is only to be activated when the locomotive is hauled shut down and remote from the locomotive consist (Dead-in-Train) as distinct from shut down and coupled to the locomotive consist (Dead-in-Consist). Maximum brake cylinder pressure is thus limited to that set for the No. 2 main reservoir;
- The *Transducer Portions* provide feedback of 'system pressures and flow' to the Brake Control Unit, the cab display, and for diagnostics;
- The *Air Filter Portion* provides a second step of moisture removal and filtration to reduce potential manifold component contamination and to enhance reliability of the magnet valve assemblies.

Pneumatic Backup

The pneumatic backup arrangement consists of the MC-31 control valve and Quick-Service valve. Spring- and diaphragm-operated, the MC-31 control valve is initially charged with brake pipe pressure and provides backup protection and braking where total loss of electrical power (and computer shutdown) occurs, or should the locomotive require to be hauled Dead-in-Train. The Quick-Service valve is used in conjunction with the pneumatic backup valve and ensures an adequate inshot pressure of 55–69 kPa to the brake cylinders during Automatic brake applications.

During charging of the brake pipe, this pressure flows through the charging and dissipation valve (Figure 224), the inner and outer control valve diaphragm chambers, the filtered choke, and into the 220-cubic-inch control reservoir. Brake pipe pressure is also supplied against the closed service valve. During a brake pipe reduction, pressure in both the inner and outer diaphragm chambers is also reduced, but due to the restriction of the filtered choke and the charging and dissipation valve, it reduces faster from the inner chamber than from the outer. The resultant pressure differential across the diaphragm causes the valve assembly to close the charging and dissipation valve.

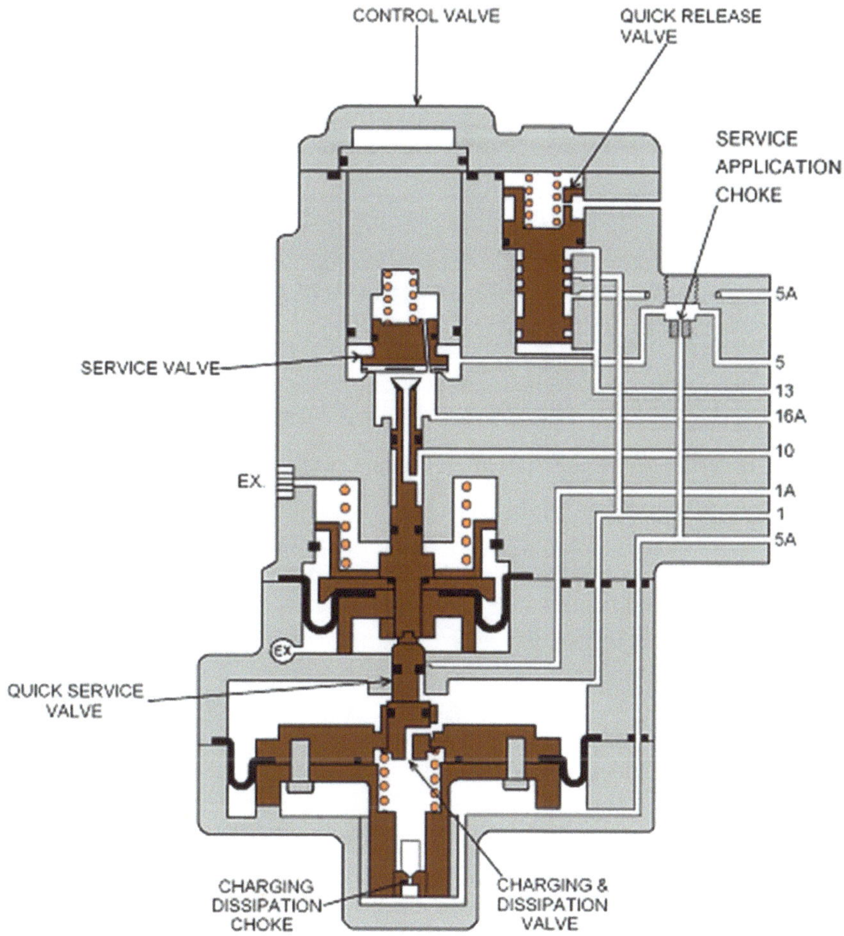

FIGURE 224: EPIC II MC-31 CONTROL VALVE – DIAGRAM

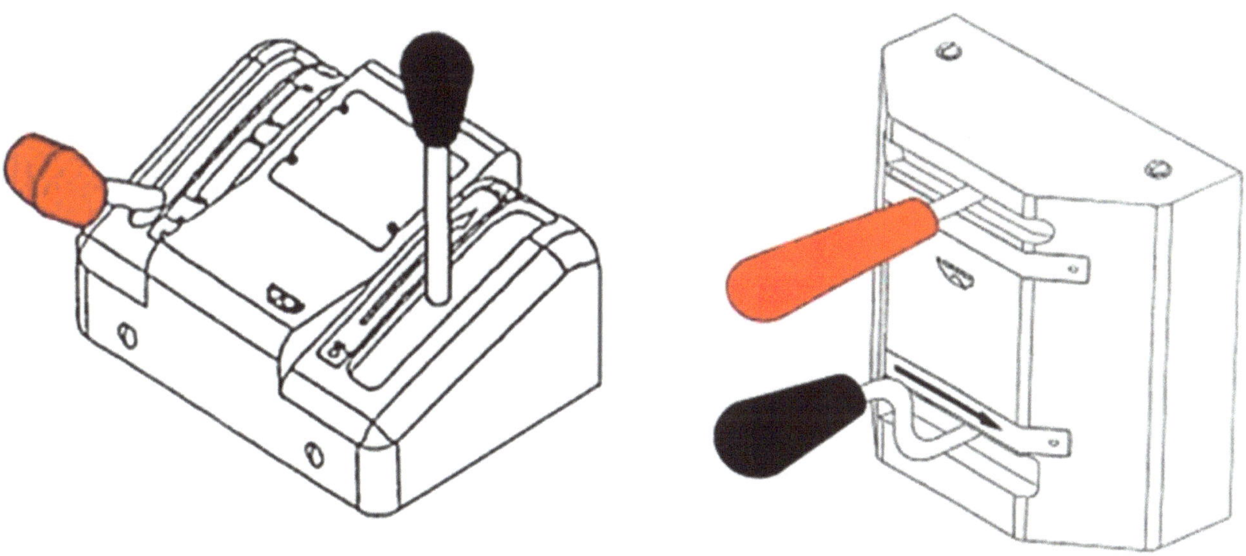

FIGURE 225: EPIC II 30- AND 26-STYLE CAB CONTROL UNIT OPTIONS

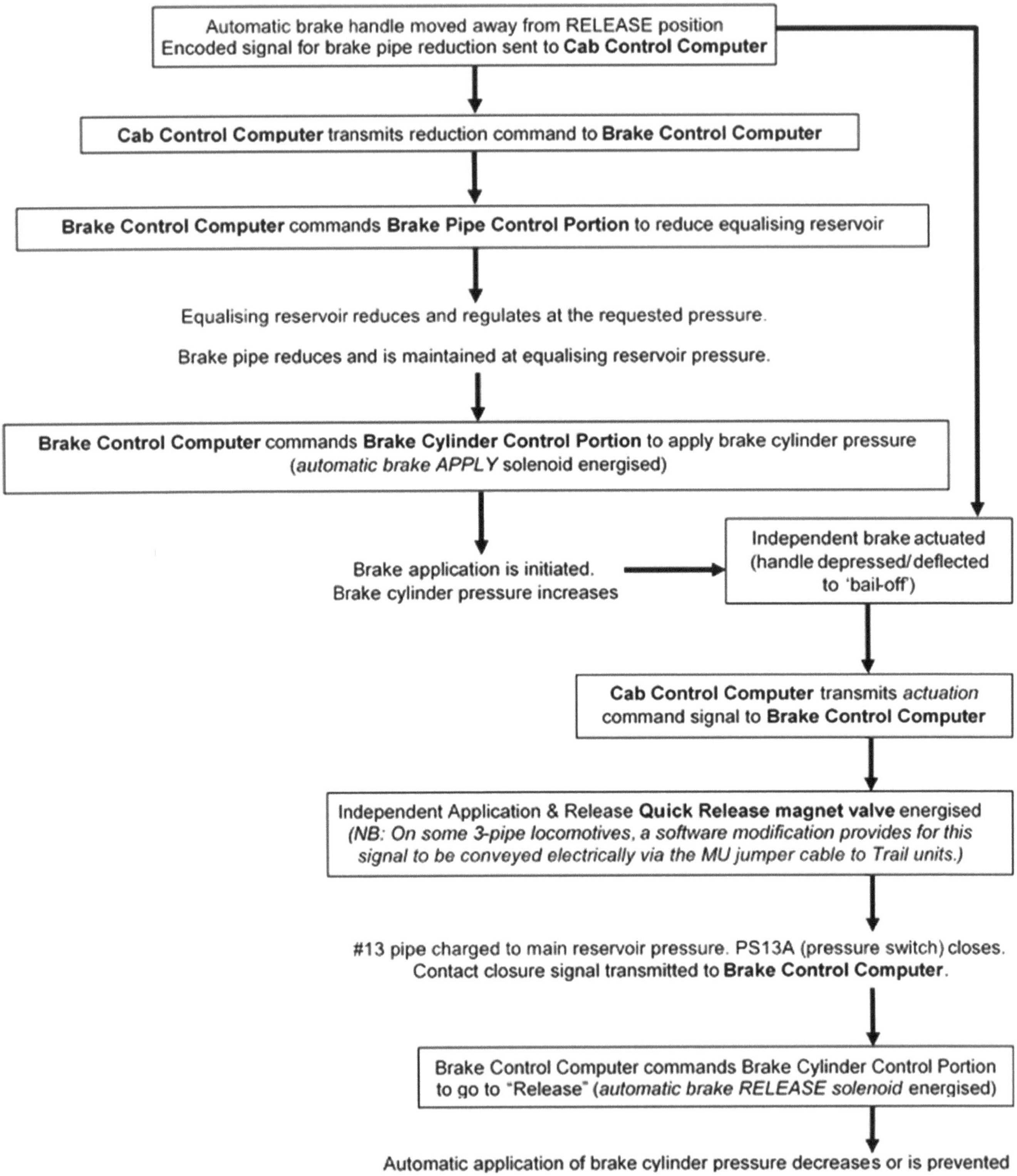

FIGURE 226: EPIC II AUTOMATIC BRAKE APPLICATION PROCESS - FLOWCHART

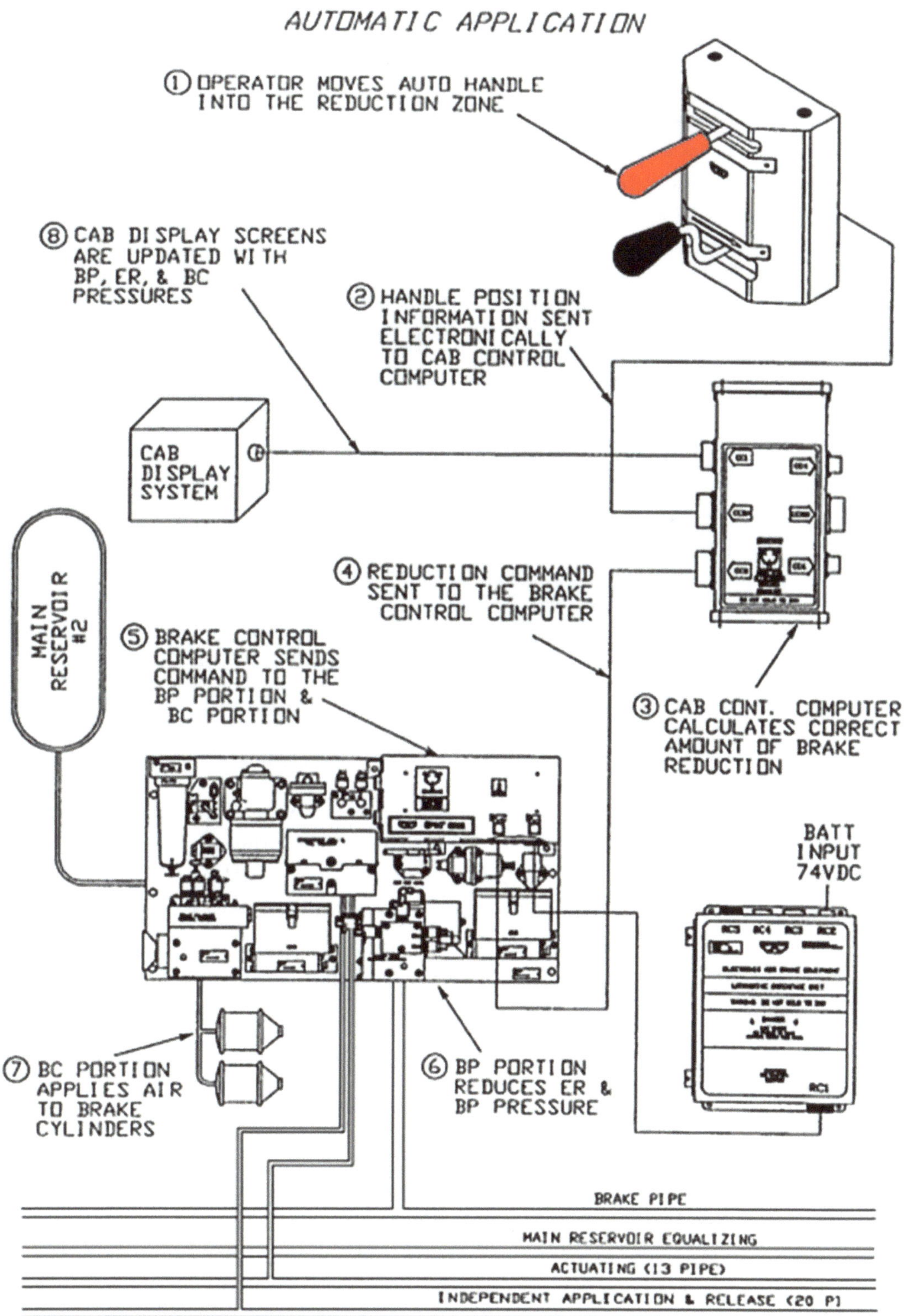

FIGURE 227: EPIC II AUTOMATIC BRAKE APPLICATION PROCESS - SCHEMATIC

FASTBRAKE®

This more recent Wabtec development is an evolutionary enhancement of EPIC 3102 and is quickly becoming the standard brake equipment for EMD locomotives, including the SD70ACe locomotives recently supplied ex-Canada to BHP Billiton Iron Ore. The 11 EDI Rail-EMD GT46C-AC locomotives supplied to SCT Logistics also have this equipment.

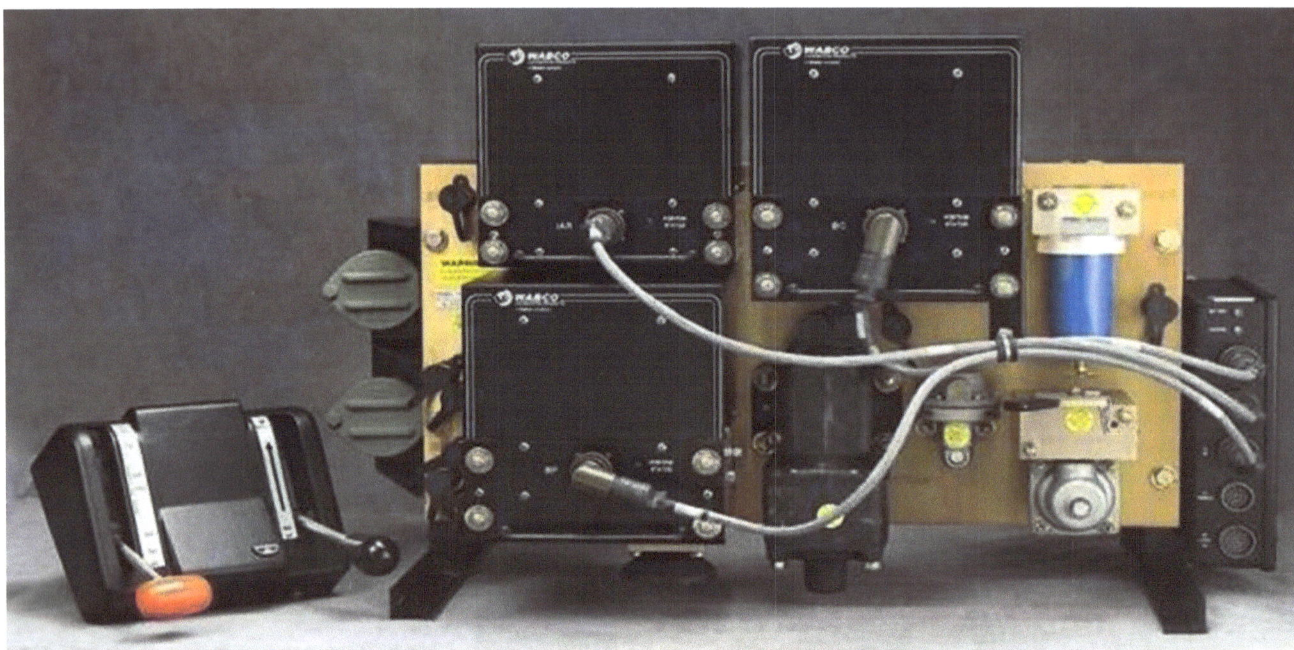

FIGURE 228: FastBrake ELEMENTS - CAB HANDLE UNIT AND PNEUMATIC OPERATING UNIT. *Wabtec, courtesy Ian Livingstone*

With fewer components in a lighter, more compact arrangement, FastBrake differs from EPIC II.

The system consists of two basic components: a Cab Handle Unit (CHU) and the Pneumatic Operating Unit (POU). The CHU is the driver's brake handle and is basically identical to the EPIC Cab Control Unit (CCU) [one wonders why—just when we were getting used to 'CCU'—the name had to be changed], while the POU is the brake rack. The Epic II Brake Control Computer (BCC) and Cab Control Computers (CCC) have been removed, with the logic function now being internal to the four electronic portions on the POU.

The FastBrake system uses a Controller Area Network (CAN) bus for communications between the electronic components. The CAN bus is divided into two levels: the *Railcan* bus for communications between the locomotive and its FastBrake system (via the CHU) and the *Wabcan* bus for communications between the FastBrake components. Mounted on the brake rack are the four main **operating portions**:

- A Power Supply unit and three **control portions**—the Brake Cylinder Control Portion (BCCP), the Brake Pipe Control Portion (BPCP), and the Independent Application and Release Portion (IARP)—and the air-handling devices associated with the pneumatic backup brake components;
- Air filters; and
- The dead-engine device.

FIGURE 229: FastBrake INSTALLATION. THIS IMAGE OF A BHP BILLITON IRON ORE EMD SD70ACE LOCOMOTIVE CONTROL STAND FEATURING WABTEC 'FastBrake' EQUIPMENT IS INTERESTING. WITH THE MODERN BRAKE CONTROLLER—ORIGINALLY DESIGNED FOR DESK-MOUNTING—INSTALLED VERTICALLY IN THE CONTROL STAND (NOTE DIFFERENT STYLE OF CAB HANDLE UNIT FROM THAT SHOWN IN FIGURE 225), IT REPRESENTS THE EVENTUAL EMBODIMENT OF THE 30A-CDW ARRANGEMENT ORIGINALLY SUGGESTED BY THE AUTHOR AS A CAB COMMITTEE PROPOSAL BUT NOT INSTALLED ON THE MOUNT NEWMAN MINING ALCO-TO-GE DASH 7 REBUILT LOCOMOTIVES IN 1987 (FIGURE 215). *Jim Ward*

Each of the three control portions consists of the magnet valves, transducers, and micro-controllers, along with the pneumatic relay valve used to control the brake system. Brake control software is now loaded via the FIRE computer[135] instead of the Brake Control Computer as for EPIC systems.

The diagnostics for FastBrake have been improved, allowing maintenance personnel to run on-board brake tests down to the 'portion level'. These tests can be accessed via the driver's screen after entering the required passwords.

THE KNORR-NYAB CCB II EQUIPMENT

As the use of electronics and computer control expanded rapidly within the mass transit industry—particularly in the area of power and braking controls—during the late 1970s and into the '80s, New York Air Brake identified its applicability to locomotives. The company recognised the opportunity provided by the nascent microcomputer technology to develop equipment that would conform to the new Clean Cab standard recently developed within North America under the auspices of the Association of American Railroads. This standard had been proposed to modernise and, where possible, standardise the locomotive cab layout and render it safer and more efficient as a work environment. It was realised that the application of electronics would allow the cumbersome air brake valve and other controls to be reduced in size, removing bulky piping and wiring and permitting a desktop arrangement with ample clearance for the operators'

135. See Glossary, chapter 8.

legs as well as providing an ergonomic control layout designed for balanced, two-handed use. The New York Air Brake Corporation's initial development in 1987 was the proof-of-concept Computer Controlled Locomotive (CCL) equipment providing electro-pneumatic control of braking as well as integrated control of throttle and dynamic braking functions, plus horn and bell. Targeted features included the following:

- *Clean Cab Concept*: three-handled controller with an interactive display;
- *Ease-of-maintenance and improved reliability*: consolidation of air brake valve devices onto a unitised package and a reduction of components through the use of computer logic;
- *Onboard diagnostics*: providing rapid fault identification for improved safety and simplified maintenance.

The first installation was undertaken for the CSX Railroad on a GE U36B locomotive, and this was soon followed by a second installation on a Burlington Northern Railroad EMD SD60M. CCL equipment and throttle controls, complete with display screens, were also installed on eleven remanufactured slug units[136] that the BN operated for a decade in hump yard service.

The North American locomotive manufacturers—General Electric and the Electro-Motive Division of General Motors (now Electro-Motive Diesel)—were developing their locomotive cab display systems; placing most of the operator's selective controls and indicators on screen-based displays, which led to their interest in having an integrated electronic air brake system.

NYAB became a part of the Knorr-Bremse AG group in 1991, and since Knorr had introduced an AAR-compliant electronic air brake system to South African Railways, the combined talents of the companies were put to work to redesign the electronic air brake portion of CCL to best meet the needs of the locomotive builders and the demanding requirements of the railroads. With priority being given to safety in system design and fault identification with corrective actions through microcomputer logic, the Computer Controlled Brake (CCB) emerged. The initial installations for CCB were on GE Dash 9 locomotives for Amtrak and EMD SD70MAC units for the Burlington Northern Santa Fe Railroad in 1993. More than 2 500 locomotives, including export units, have been equipped with first-generation CCB.

A key development in 1994 that strongly advanced the practicality of and need for electronic braking was the integration of Harris Locotrol distributed power equipment[137] within the architecture of the cab display system. By application of the new technology, conventional pneumo-mechanical air pressure gauge displays were no longer required, and air braking and throttle control connections and linkages were likewise rendered redundant, these utilities now able to be accomplished through serial data communications. NYAB has been a leading participant in the advanced development of this "systems integration" with nearly half of the freight locomotives equipped with first-generation CCB operating in 'distributed power' service.

In 1996—in an agreement with (then) GE-Harris Corporation—NYAB embarked on a redesign and repackaging project. The second-generation CCB (now known as CCB II) was developed with two major objectives: (1) to increase reliability, reduce failures and maintenance intervals, and improve or simplify maintenance processes; and (2) to combine Locotrol distributed power (DP) logic and electronic braking logic into a single microcomputer. General Electric Transportation is the prime lead in the development and distribution of the Locotrol Electronic Brake (LEB) product. In the context of "reliability", NYAB's design intention was to maximise *mission reliability*, including extra components to provide redundancies and backup modes to enable the system to continue operating and complete a run when a component or primary device might fail.

136. See Glossary, chapter 8.

137. The 'Locotrol' brand is now owned by GE Transportation.

To accomplish this objective, NYAB repackaged the pneumatics into functional modules relative to the braking controls of a locomotive. All the components for brake pipe charging, cut-off, and Emergency brake initiation were integrated into a single Line Replaceable Unit (LRU). All the components for the development and control of Independent braking were likewise contained into a single LRU for each primary function. A microcomputer was placed on each LRU for 'distributed intelligence', each LRU controlling its own particular function. A central computer provides the high-level intelligence and coordination, and communicates the command and status information from the locomotive's central processor with the intelligent Distributed Power LRUs.

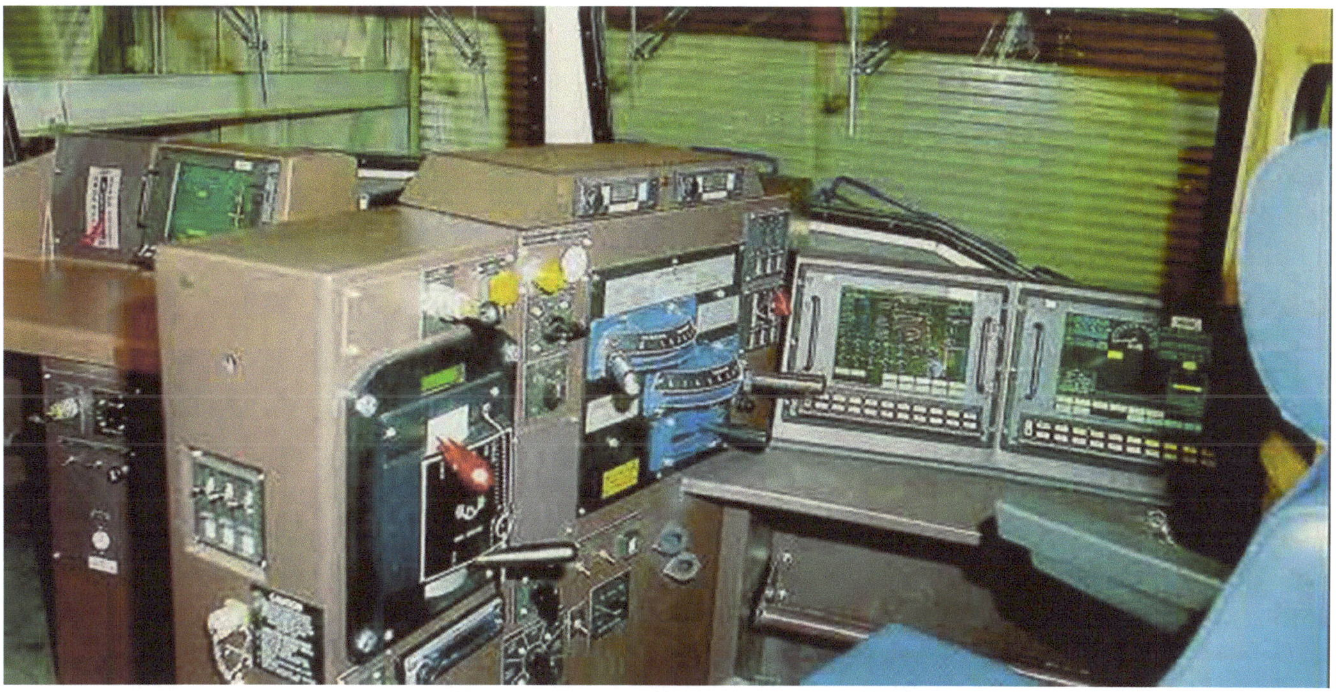

FIGURE 230: NYAB CCB II BRAKE CONTROLLER AS FITTED TO BHPBIO GE AC6000 LOCOMOTIVES. THE VERTICAL INSTALLATION SHOWN HERE PREDOMINATES WITHIN OTHER AUSTRALIAN RAILWAYS, EITHER FITTED TO A STANDARD AAR CONTROL STAND OR TO A CAB WALL (LEFT- OR RIGHT-HAND SIDE OF CAB). NO DESK-STYLE DESIGN UTILISING THIS BRAKE VALVE HAS YET EMERGED THAT SATISFIES AUSTRALIAN LOCOMOTIVE DRIVER'S VIEWS ON ERGONOMIC DESIRABILITY. *Jim Ward*

Installations of CCB II braking equipment started in 1998. At the time of writing there are more than 4 000 locomotives in service with approximately 2 600 being the above-described LEB units. There are about 500 units internationally in South America, South Africa, and Australia, plus 230 units under manufacture for mainland China.

The flexibility of CCB—when incorporated with a modern ECP control valve (in NYAB's case, the EP-60)—allows a locomotive driver to control an ECP brake system using the Automatic air brake handle (rather than dedicated buttons and switches used to signal devices on an air brake equipment rack) and to apply distributed locomotive consist braking in response to ECP trainline brake commands, all the while maintaining the brake pipe charge throughout the train.

CCB II installations in Australia include the QR 2160 and 4000 Classes, Pacific National Queensland 5000 Class, and the BHP Billiton AC6000 units (now scrapped), as well as a prototype 80-tonne branch-line unit with dual driving stands and displays built in 2006 by the Junee Railway Workshops.

Major CCBII Components and their Functions

Electronic Brake Valve (EBV)

The ergonomically designed EBV is the interface between the brake system and the locomotive driver and features either a desktop or AAR control stand installation. An easy-to-read digital display provides instantaneous information on the equalising reservoir target pressure. This feature avoids feedback delays the manufacturer claims are inherent in other systems and permits the locomotive driver to exercise precise brake control. As a fail-safe design, the EBV provides direct-acting emergency venting of the brake pipe.

Integrated Processor Module (IPM)

The IPM manages the electronic interfaces between the brake system elements and the driver's display. It communicates with the locomotive on-board computer and interfaces with electrical trainlines. The IPM also communicates with a portable test unit (PTU) for troubleshooting and system diagnostic tests. With the appropriate hardware and software conversions, the IPM serves as the computer in a Locotrol distributed power application.

Electro-Pneumatic Control Unit (EPCU)

The EPCU manages the pneumatic interfaces between the brake system and the locomotive consist. It controls the locomotive's brake cylinders, brake pipe, Independent Application and Release pipe, and the Actuating pipe. The heart of an EPCU's pneumatic circuit is the Knorr-Bremse patented *Analogue Converter*. This utilises Pulse-Width Modulation[138] to provide highly accurate and stable control of the equalising reservoir, brake cylinder pilot, and Independent Application & Release pipe pressures. A pneumatic backup system is integrated into the EPCU, providing redundancy for partial loss of computer control in the lead locomotive or functions in the Trail mode when the system is unpowered.

Networked Modules

Fundamental to the CCB II design is the application of Echelon LonWorks network technology to interconnect electro-pneumatic modules (Line Replaceable Units) that communicate constantly with one another, actively controlling a number of functions. Collectively, these networked modules provide faster response time, simplified troubleshooting, and easier train setup. The LRU systems approach is intended to minimise spare parts inventory, reduce operating and maintenance costs, and increase train availability.

---

138. See Glossary, chapter 8.

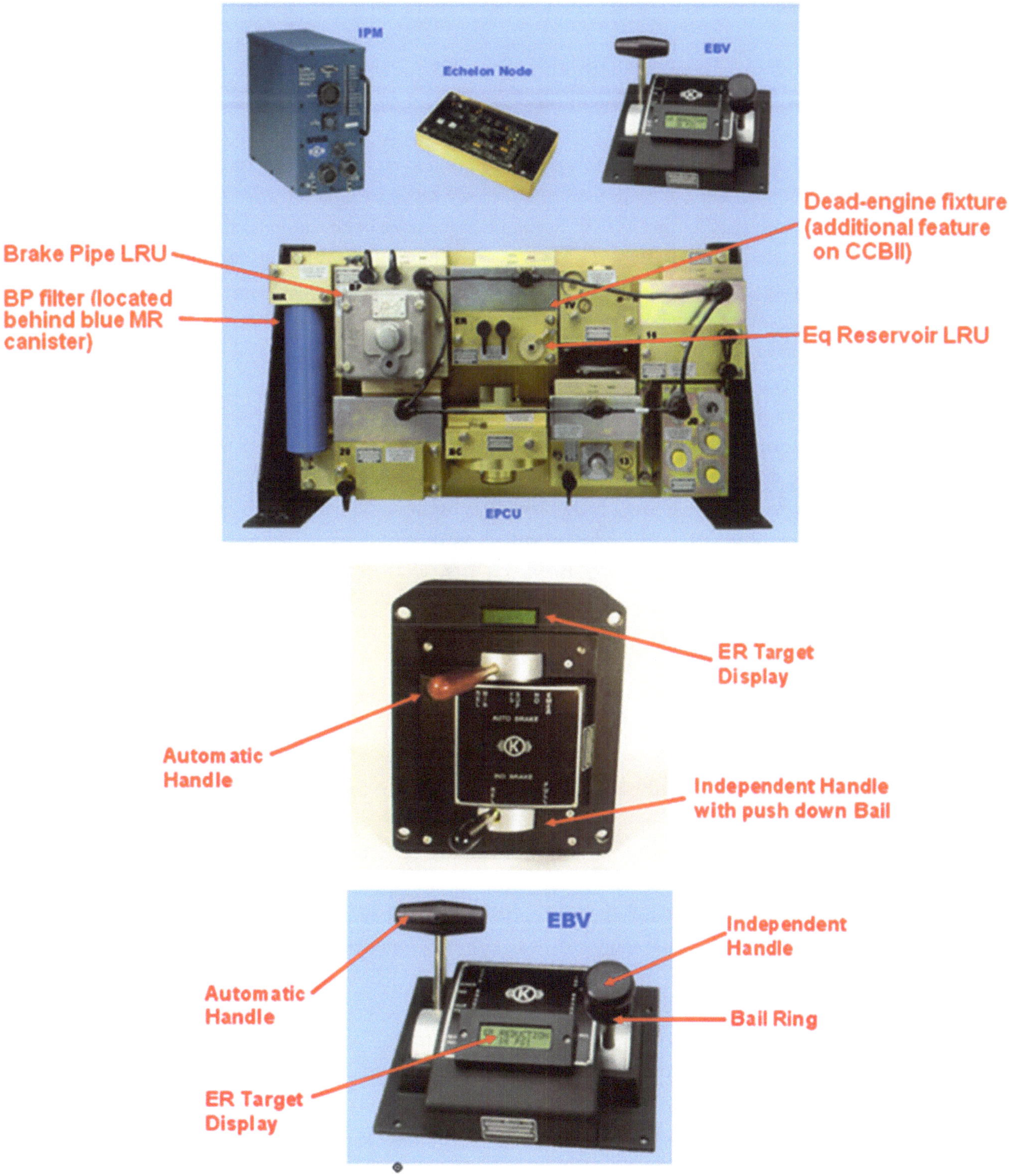

FIGURE 231: CCB II BRAKE EQUIPMENT. ELECTRO-PNEUMATIC CONTROL UNIT AND ELECTRONIC BRAKE VALVE OPTIONS

The CCB-26 variant[139]

By 2001, following the success of its CCB products, NYAB identified a global business opportunity to furnish a new-generation electronic brake technology at an affordable price for retro-fit to existing 6- and 26-equipped as well as other 'low-tech' road and shunting locomotives. This initiative has provided a version of ***CCB*** that emulates *26*-L and is an adaptation of the CCB-II system designed to be used on locomotives without integrated cab electronics. CCB-26 uses many of the same subassemblies of pneumatic valves, electronic controls, and software (LRUs) as CCB-II. Some changes have been made to simplify the system while maintaining or increasing the level of safety. For example, the penalty brake interface has been changed to mimic the 26-L system interface, allowing for a fully pneumatic penalty brake application. Also, the brake cylinder pilot pressure development has been simplified from an electronic control to a fully-pneumatic version. Much of the software and diagnostic logic that detects critical failures and takes appropriate action to effect a safe stop has been carried over from CCB-II.

In Australia, the privatisation and ongoing reorganisation of the rail industry has seen some operators seeking not only to continue established services with their existing and often ageing motive power fleets, but also to create new business opportunities. In some cases, significant locomotive refurbishing or rebuilding programs have been undertaken, and it is within this environment (not to mention the importation by some operators of foreign locomotives with decidedly 'foreign' brake equipment) that the Australian arm of the organisation, Knorr-Bremse (Australia) P/L, hopes to find a market; indeed, a new "lite" version is being sold into Australia. In Melbourne, Avteq [140] installed CCB-26 in the 130-ton EMD-powered VL Class units built for rollingstock lessor CFCLA. We stand to see more of this system in Australasia in the future.

CCB-26 has been fully developed and field-tested in both the United States, for 4-pipe AAR air brake systems, and Brazil for 3-pipe MU systems as found in that country and others such as China, South Africa, New Zealand, and Western Australia. NYAB describe CCB-26 as:

- modern CCB II technology for the non-integrated locomotive or for locomotives without cab electronics or displays;
- a stand-alone derivative of CCB II that does not require screens or LSI communication and is designed to functionally imitate and be cost-competitive against 26-L;
- derived from the well-proven CCB II pneumatic and closed-loop pressure control technology, and includes internal diagnostics to support an extended COT&S[141] interval.

The corporation's design objectives for CCB-26 are the creation of an electronic air brake system that:

- Is cost competitive with 26-L and less expensive to operate;
- Is reliable (using proven electronic and pneumatic components);
- Requires no communication or integration with other on-board equipment;
- Possesses similar pneumatic and man-machine interfaces as 26-L;

139. For further information, refer to *CCB-26 Computer Controlled Brake 26-L Emulation, Service and Troubleshooting Guide IP-209* (New York Air Brake, February 2008).

140. Avteq P/L was an Australian engineering company based in Sunshine, Victoria. Between 2007 and 2009 the company manufactured the 12 VL Class diesel locomotives for Chicago Freight Car Leasing Australia, then ceased trading and was in liquidation by June 2012.

141. Clean, Oil, Test & Stencil is a standard AAR-mandated (i.e. North American) scheduled maintenance process for rollingstock. Within Australia this standard is currently utilised only on the Pilbara iron ore railroads.

- Is easily upgraded to Locotrol-EB[142], Wire-based Distributed Power, or EP integrated braking;
- Introduces an alternate penalty methodology allowing direct exhaust of Equalising Reservoir as per the new US FRA 238.105 requirements[143];
- Introduces a J-1.6-16 relay valve (this is provided by the addition of a second relay);
- Retains diagnostic capability.

Although a product of the field-proven CCB II system, CCB-26 is much simplified, using fewer electronics and more pneumatic valves. Compared to CCB II, CCB-26 uses three fewer electronic modules, a pneumatic triple valve as the primary brake cylinder pressure development device, a J-1 type relay valve, a purely pneumatic penalty brake interface, and a 26-L-style man/machine interface (using original gauges and selector knobs).

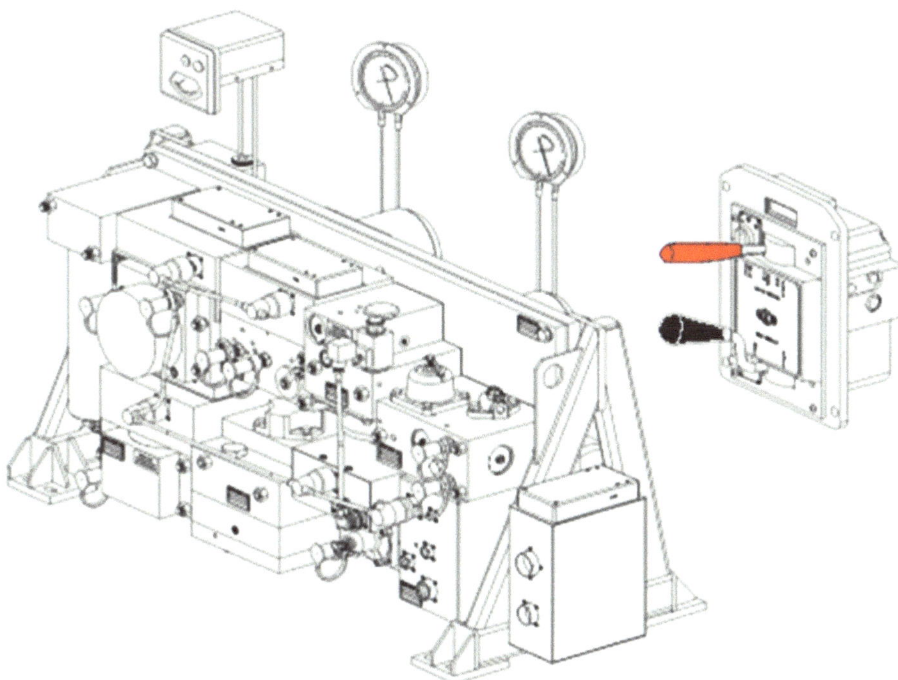

FIGURE 232: CCB-26 ELECTRO-PNEUMATIC CONTROL UNIT AND ELECTRONIC BRAKEVALVE

In terms of the design goal of "reliability," CCB-26 differs in a subtle way from its CCB II parent. Since minimising the parts count and keeping the locomotive available for use were primary objectives, *mission* reliability is considered by NYAB to be less crucial than *component* reliability, hence their focus on the latter. Having less in the way of redundancies and backup modes, the strength of CCB-26 is in its use of fewer, more highly reliable parts.

One last feature of CCB-26 is worthy of note: its ability to be easily upgraded at a later date, should the operator desire it. Promotion from CCB-26 to full CCB II requires the following:

- The addition of an Integrated Processor Module and cabling. The IPM—a primary component of CCB II—provides advanced brake control logic. It communicates with the

142. See following section on Locotrol.

143. http://edocket.access.gpo.gov/cfr_2004/octqcr/pdf/49cfr238.105.pdf. (retrieved 2012)

driver's display screen and the Trainline Integration Module (TIM) and Radio Interface Module when required;
- The addition of an integrated display screen (ICE, IFC, FIRE, etc.);
- Change-out of the 16-pipe Control Portion (contains Service and Emergency limiting valves as well as Emergency brake cylinder assurance valves to charge brake cylinders from the main reservoir during Emergency);
- Change-out of the 13-pipe Control Portion (controls the Actuating Pipe "bail-off" function);
- Replacement of the Lon cable and a modification of the system software.

A further upgrade to LOCOTROL-EB requires the relatively minimal addition of a TIM (see above; used by the LOCOTROL system to drive trainlines in a non-integrated system and driven in turn by the IPM) as required, plus radios, antennae, and cabling, and a change to both system and screen software as required. No piping changes or major removal are required. CCB-26 may also be provided as part of an integrated package combining ECP train brake functionality with electronic air brake.

The Electro-Pneumatic Control Unit

The Electro-Pneumatic Control Unit is mounted in the locomotive brake bay and consists of LRUs that control the development of all pneumatic control pressures. Four of the LRUs are "intelligent" and communicate via the network. They are as follows:

1. The *Brake Pipe Control Portion* (BPCP): Functions to supply, exhaust, maintain, and cut off the trainline brake pipe. The BPCP includes the brake pipe relay valve, Emergency magnet valve and vent valve, and brake pipe cut-out function as well as break-in-two detection and brake pipe pressure sensing.

2. The *Equalising Reservoir Control Portion* (ERCP): Controls the brake pipe relay and equalising reservoir pressure. The pneumatic and electrical control portions of the ERCP include emulation of the #3 (Switch Pipe), #10 (Safety Control), and #26 (Suppression) pipes. Optional Penalty magnet valves can be piped to the manifold #3 port for suppressible penalties and to the #10 port for non-suppressible penalties. In this manner, the penalty interface is identical to a 26-L pneumatic braking system.

3. The *20 Control Portion* (20CP): Provides Independent Application & Release Pipe pressure.

4. The *Relay Control Portion* (RCP): Contains the system relays and provides discrete signal interface to locomotive controls and sanding equipment.[144]

The EPCU also contains the following:
- A *Brake Cylinder Control Portion* (BCCP): Provides brake cylinder pressure based upon the level of 16-pipe and 20-pipe pressures. Various BCCP portions are available depending upon the required brake ratio of the applied locomotive;

144. On some applications, the RCP may be mounted remotely from the EPCU.

- A *DB Triple Valve Portion* (DBTV): Develops brake cylinder pilot pressure during Service brake applications sensed by reduction of brake pipe pressure;
- A *Power Supply Junction Box* (PSJB): Contains the EPCU power supply;
- A *13 Control Portion* (13CP): Provides Bail-Off (Actuating) pipe pressure;
- A *Penalty Control Portion* (PCP): Used on Single Pipe Systems ('3-pipe' in Australasia) instead of a 13CP to provide 'split penalty' functions;
- A *16 Control Portion* (16CP): Provides brake cylinder limiting and brake cylinder assurance in Emergency.

EBV

The locomotive driver commands the computer through the ***E***lectronic ***B***rake ***V***alve controller. The EBV is also on the network and signals the handle positions for Automatic and Independent braking. An exception is the initiation of an Emergency brake application, which is propagated mechanically through a vent valve by placing the Automatic handle into the EMERGENCY position.

Equalising Reservoir pressure is set via a normally-centered switch with 'momentary' positions on each side of center. Lead/Trail Mode and Cut-In and Cut-Out functions are provided via a three-position selector switch. These switches are located on the EBV.

The CCB-26 system performs several diagnostic functions. Faults are enunciated to the locomotive driver via a small LCD screen mounted on the EBV that is also used for routine advisory instructions such as Penalty and Emergency reset.

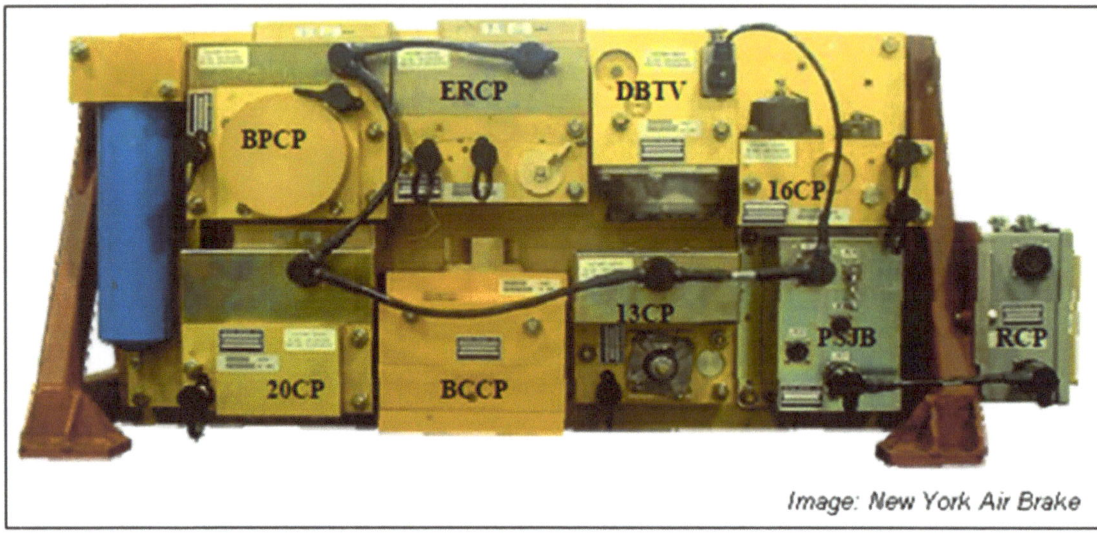

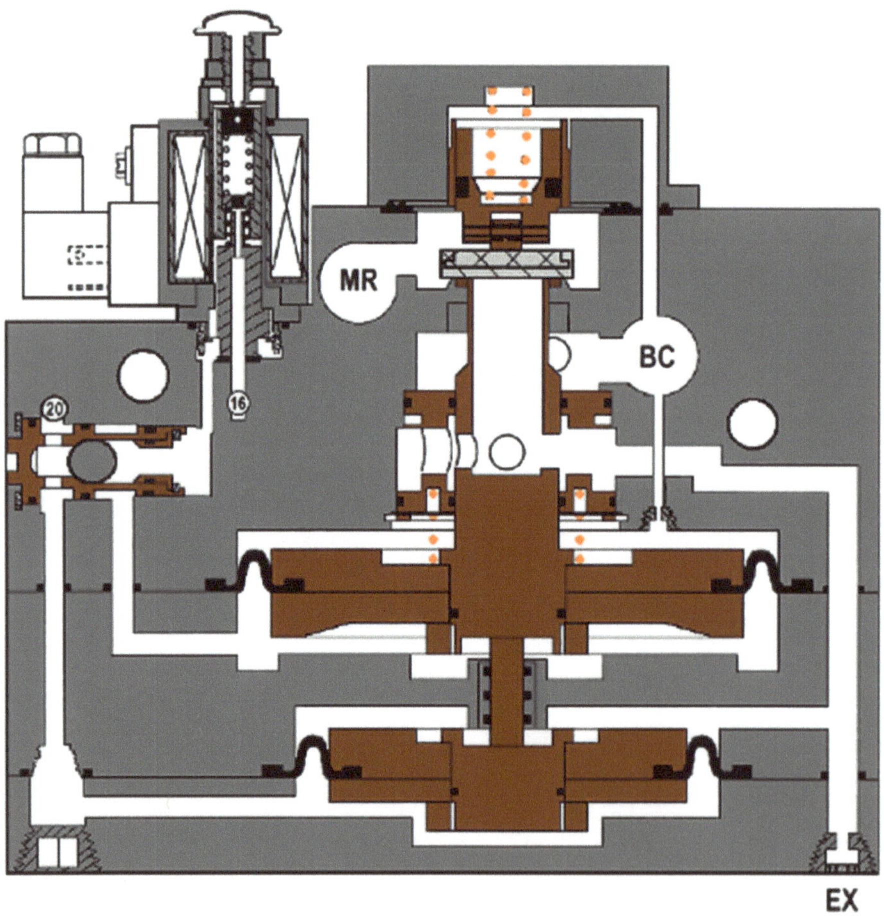

FIGURE 233: CCB-26 EPCU (PHOTO) AND CCB-26 RELAY VALVE (DIAGRAM)

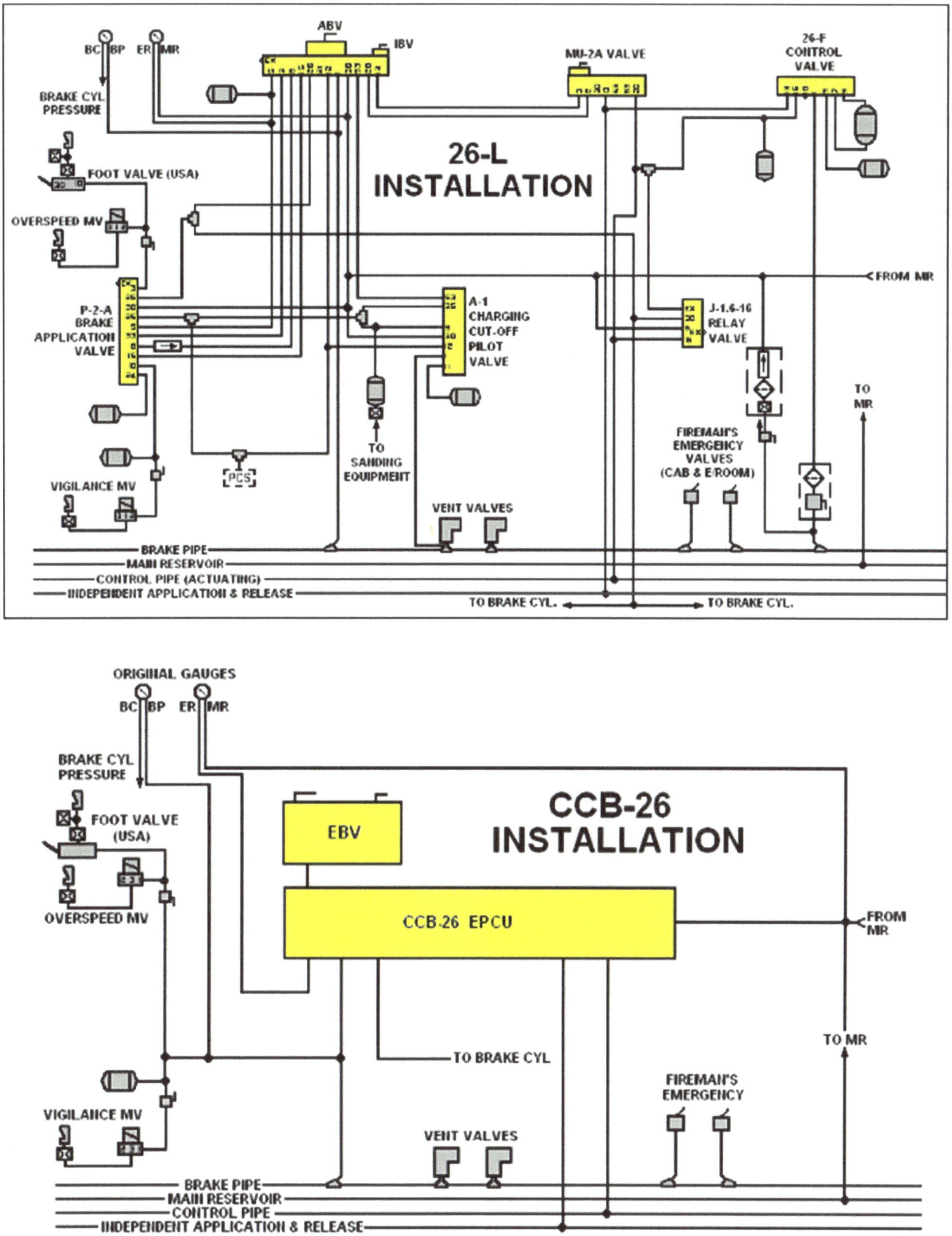

FIGURE 234: CCB-26 AND 26-L INSTALLATION SCHEMATIC COMPARISON. THE ABOVE TWO DIAGRAMS COMPARE THE LOCOMOTIVE AIR BRAKE PIPING TYPICALLY REQUIRED FOR A CONVENTIONAL 26-L INSTALLATION (US & PILBARA IRON ORE LOCOMOTIVES DEPICTED) VERSUS THAT REQUIRED FOR CCB-26. *Adapted from diagrams supplied by Knorr-Bremse (Australia) Pty Ltd*

THE WESTINGHOUSE (UK) WX AND WXC INDEPENDENT BRAKE VALVES

The type WX driver's brake valve from Westinghouse UK was designed to provide sensitive control of locomotive Straight-Air brake systems. The brake valve operation is graduable, self-lapping, and pressure-maintaining. A high brake cylinder pressure is possible in an Emergency application, in which the valve does not lap but brake cylinder pressure continues to build to the feed valve setting (or where no feed valve is used, to main reservoir pressure). The WX equipment is also designed for locomotives equipped with vacuum-controlled Independent brakes, in which case the locomotive is fitted with both a vacuum driver's brake valve and the type WX. The vacuum valve controls the Automatic vacuum train brakes plus the locomotive's vacuum-controlled Independent brakes while the WX brake valve is used—in conjunction with a double-check valve—to operate the Independent brake.

The WXC version differs only by the inclusion of a set of electrical contacts and an operating trigger that are used to control the brake cylinder cut-out magnet valve. In dual brake systems, such as the vacuum-controlled Independent brake referred-to above, the brake cylinder cut-out magnet valve enables the locomotive brake to be released independently of the train brakes. Depression of the brake cylinder cut-out magnet valve trigger moves a push rod down against a pivot that, in turn, moves a set of contacts to complete the electrical circuit required to energise the magnet valve. To remove the handle from either type, an Emergency application must first be made, and the brake cannot be released until the handle is replaced.

The WXC valve is fitted to the Aurizon 1900 Class (ex-Westrail D Class) Clyde-EMD narrow-gauge locomotives. In New Zealand, at least two other versions of this family of brake valves were used. The type WV was fitted to the twin-engine DSC Class diesel-electric shunt locomotives and the Vulcan single-unit diesel-mechanical railcars (the latter air brake scheme featuring the interesting inclusion of two identical air compressors; one being engine-driven, the other axle-driven). The Vulcan railcars also utilised a type WA valve for throttle control, whereby the output of the WA valve was supplied to a fuel-rack operating piston. A throttle control version was also used in some Australian Rail Motors (railcars).

Release:

Main reservoir air passes via an external reducing valve to chamber **A** (Figure 236) but is prevented from entering chamber **E** by closed inlet valve **B**. Brake cylinder (from port **J**) is open to chamber **E** which, in turn, is exhausted via open exhaust valve **O** and port **R**.

Application:

With handle movement from RELEASE toward the FULL SERVICE position, cam **K** engages cam follower **H**, which—along with rollers **L**—is held by the balance levers **G** and anchored by yoke **M** at fulcrum pin **N**. The two rollers **L** engage with inlet valve **B** and exhaust valve **O**. Since exhaust valve spring **P** is lighter than inlet valve spring **F**, the first movement of cam **K** closes exhaust valve **O**. Piston **Q** is prevented from moving by the heavy spring **S**, and the increasing tension produced by the cam—as it is rotated by handle movement—unseats inlet valve **B**, allowing a main reservoir feed to flow to the brake cylinder.

This pressure also develops in chambers **E** and **C** (which are connected by port **D**) until there is sufficient to move piston **Q** against spring **S** by an amount dependent upon the pressure admitted by inlet valve **B**. The pressure continues to move piston **Q** until it is balanced by the increasing resistance of spring **S** and, at this point, the brake valve 'laps'.

Exhaust valve **O** moves with piston **Q** and remains seated due to the effort of inlet valve spring **F**—acting through levers **G**—exceeding that of exhaust valve spring **P**. Simultaneously, the movement of levers **G** permits the inlet valve **B** to close when piston **Q** stops moving. Maximum brake cylinder pressure depends upon the pre-set load on piston spring **S**.

<u>Graduated Application</u>:

Further movement of the handle toward FULL SERVICE increases a partial application. In this case, inlet valve **B** will again be opened until brake cylinder pressure in chamber **C** has increased sufficiently to overcome spring **S** and move piston **Q** far enough to permit inlet valve **B** to again close. The brake cylinder pressure developed is always proportional to the extent of handle movement.

FIGURE 235: WESTINGHOUSE W-TYPE STRAIGHT-AIR SELF-LAPPING INDEPENDENT BRAKE VALVE

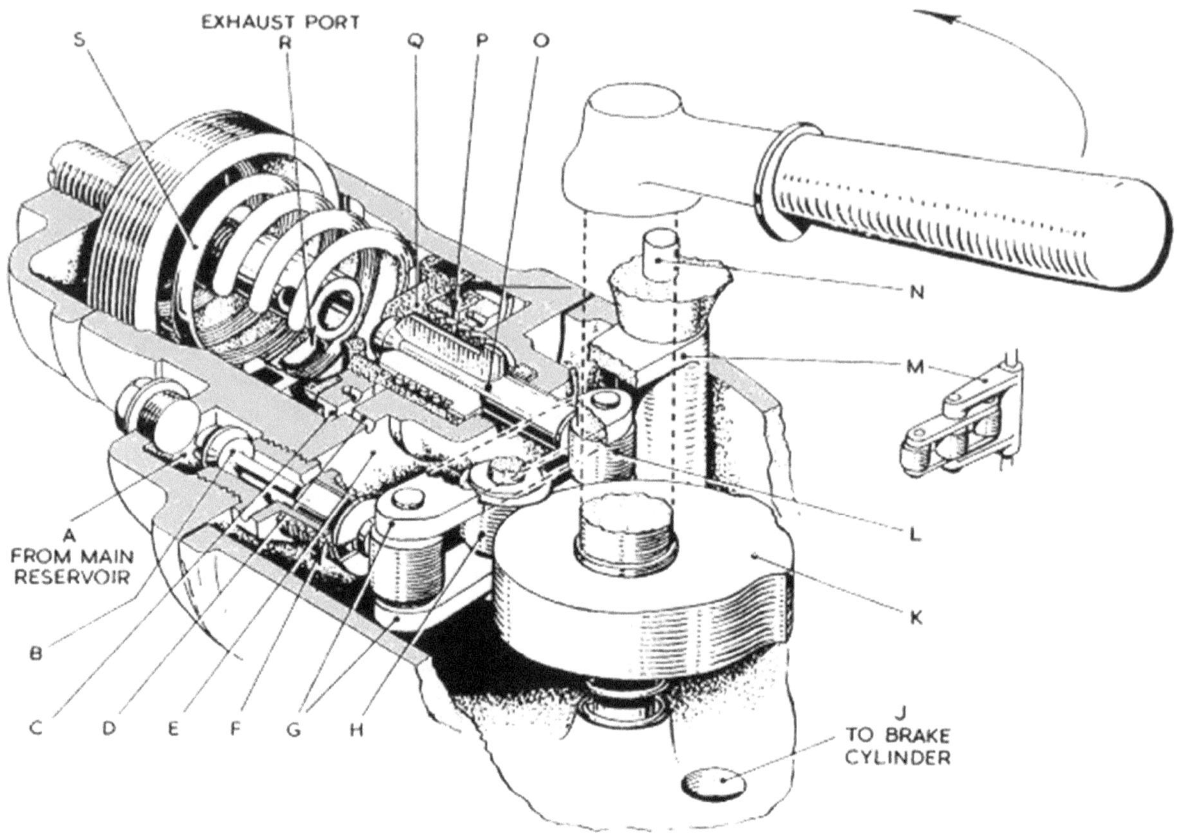

FIGURE 236: W-TYPE STRAIGHT-AIR SELF-LAPPING INDEPENDENT BRAKE VALVE – SECTIONED VIEW

<u>Graduated Release</u>:

Handle movement back toward RELEASE position reduces cam pressure on cam follower **H**, permitting spring **P** to open exhaust valve **O** and reducing brake cylinder pressure in chamber **C** to a value that permits spring **S** to move piston **Q** to close the exhaust valve. This sequence is repeated with each backward movement of the handle until the brake is fully released, at which point the piston is at the end of its stroke, the exhaust valve remaining open and the inlet valve being closed.

<u>Emergency</u>:

At this handle position, the extreme cam contour holds the inlet valve open, admitting the maximum available pressure to the brake cylinder, and the valve does not 'lap'.

<u>Pressure Maintenance</u>:

Loss of brake cylinder pressure (e.g. leakage) will allow spring **S** to move piston **Q** and the exhaust valve, thus opening the inlet valve until pressure increases sufficiently for the brake valve to again 'lap'.

Other variants of this equipment include the WSC used on early NSWGR suburban EMU stock and the W2E used on first-generation NZGR internal combustion rail cars.

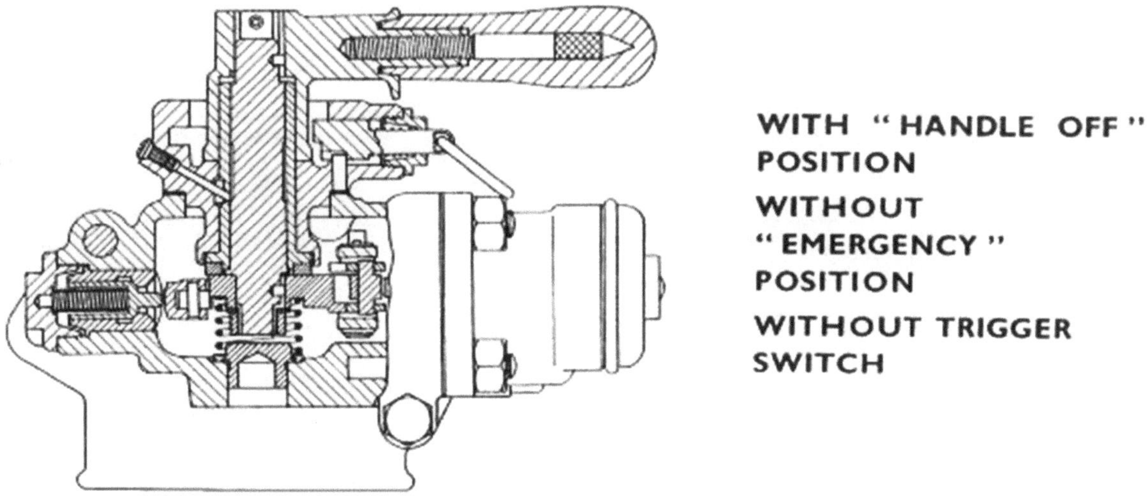

FIGURE 237: WV STRAIGHT-AIR SELF-LAPPING INDEPENDENT BRAKE VALVE

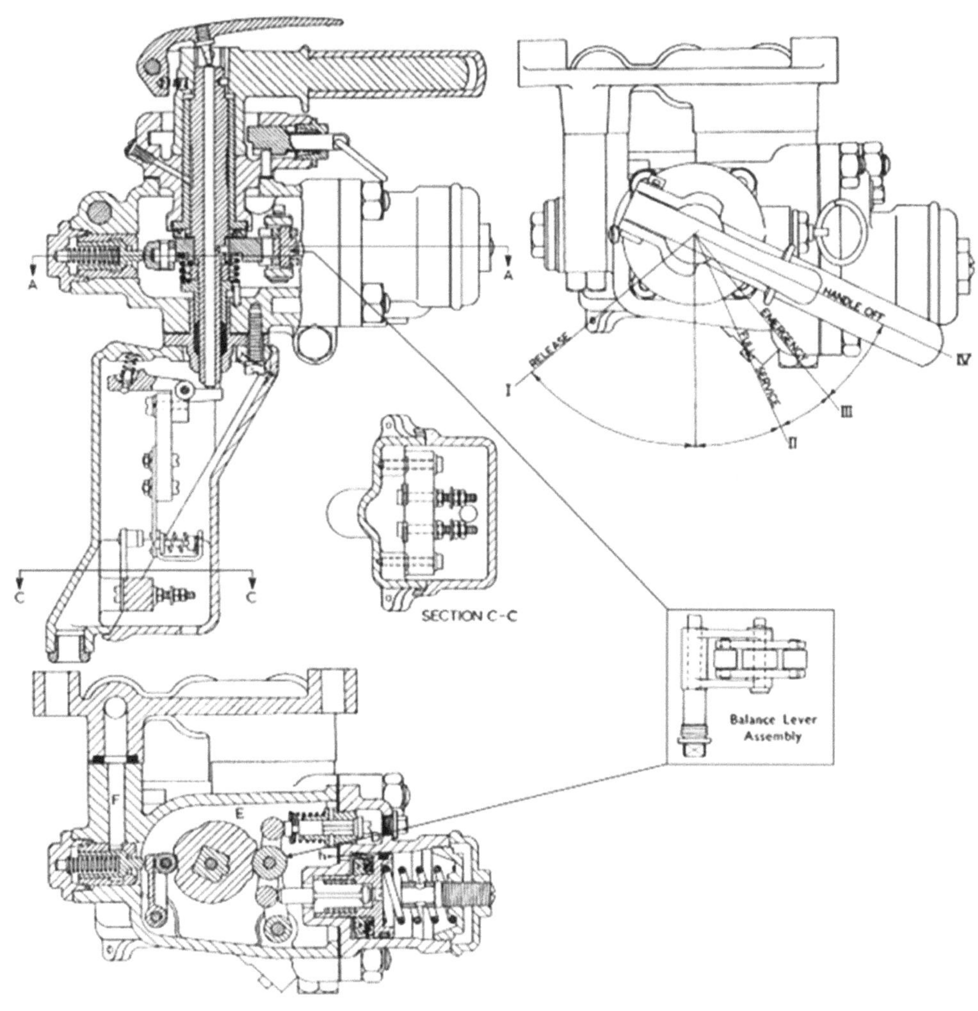

FIGURE 238: WXC STRAIGHT-AIR SELF-LAPPING INDEPENDENT BRAKE VALVE WITH TRIGGER SWITCH

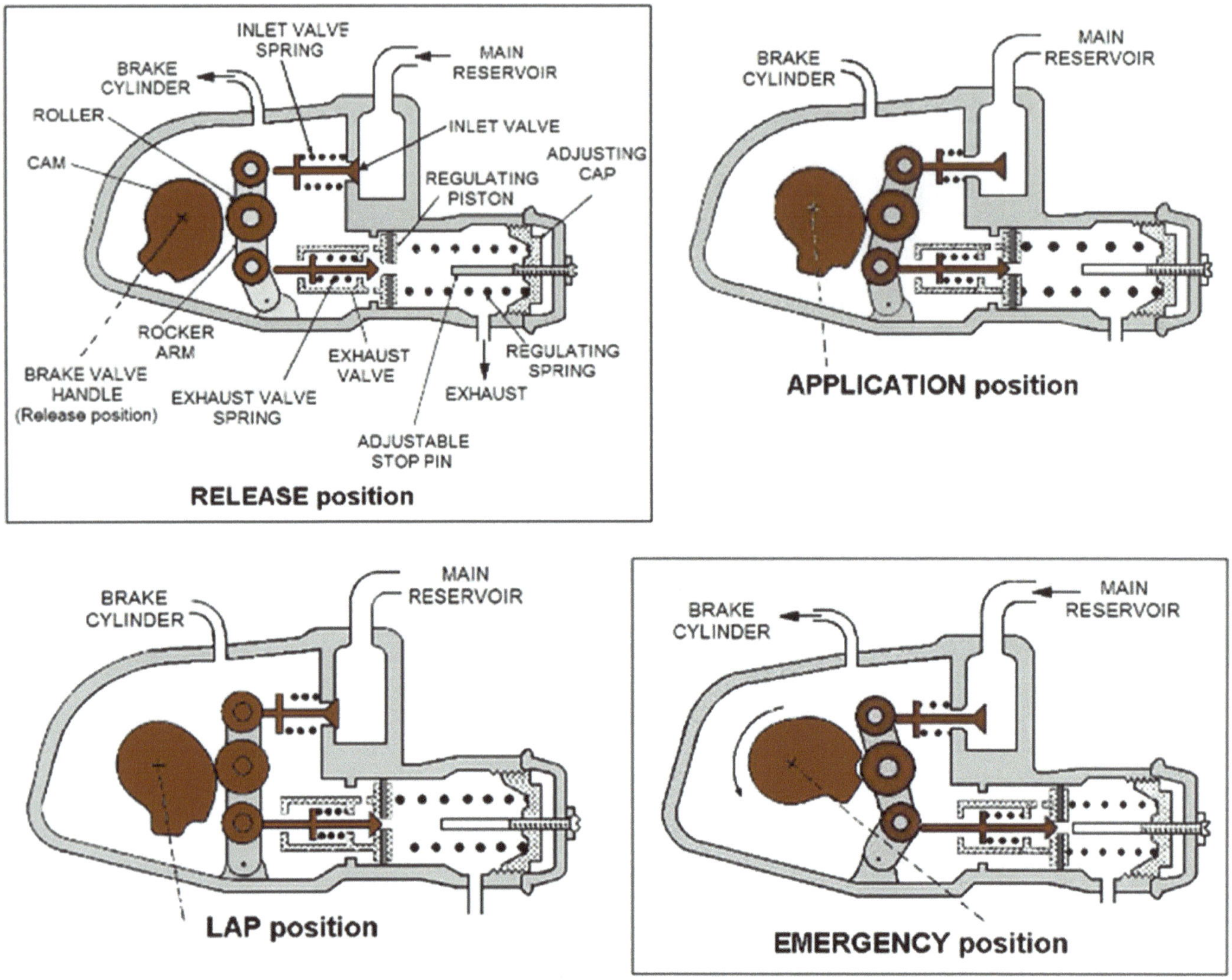

FIGURE 239: W-TYPE STRAIGHT-AIR SELF-LAPPING INDEPENDENT BRAKE VALVE OPERATION – SIMPLIFIED

THE AUSTRALIAN WESTINGHOUSE/NZR PUSH-BUTTON BRAKE

This equipment—developed in New Zealand—was based on a system designed in France and used there since 1965, originally utilising five push-buttons. In 1971, New Zealand Railways Chief Brake Inspector Noel Mather, who was overseas in connection with the design of new railcars, inspected this equipment. Riding a French locomotive fitted with their PBL brake[145], he was suitably impressed, felt that the equipment compared more than favourably with the standard arrangement in New Zealand, and realised that—if modified to suit direct-release trains—might have application within Australasia. A major part of this reasoning was due to comparative costs; a 'normal' pneumatic driver's brake valve being a relatively intricate, precision-built, and thus costly device. Furthermore, the piping and connections require a considerable amount of space and result inevitably in excessive noise in the driver's cab. All of these factors could be alleviated, and cab equipment layout potentially improved, by the use of a small push-button control system for the driver.

Aware at the time that NZR was intending to order some new provincial railcars, Westinghouse Australia also knew that they (NZR) did not desire to use the W2E brake valve employed on earlier models and could not justify the cost of a Westcode installation. It seemed to Westinghouse engineers that a simple push-button arrangement along the lines of that inspected by Noel Mather in France and previously discussed with them might be suitable, and this was suggested. Accordingly, and under the direction of Mather, NZR drafted a scheme and forwarded it to Westinghouse Australia for comment. Since Managing Director Mr L. Swan was about to depart on a visit to Westinghouse UK, he took the NZR proposal to seek their opinion. The British were apparently unenthused, although they could find no fault with the scheme. Shortly thereafter, a change of government in New Zealand resulted in the proposed acquisition of new railcars being abandoned, thus rendering redundant intentions for the new brake scheme.

By this time, Mather had retired from NZR and joined Westinghouse in Sydney. In 1977 the opportunity arose for more detailed study of the scheme, and an arrangement to replace the old No. 4 brake valve still used by some railways on smaller locomotives was drafted and a prototype installation set up at McKenzie & Holland (NZ) in the Hutt Valley suburb of Naenae, near Wellington. The NZR Deputy Chief Mechanical Engineer, Ian McGregor, was enthused by a demonstration of the equipment and requested further development to make it suitable for installation on main-line locomotives with multiple-unit operation. At the time, a rebuild program for a group of the English Electric DG Class locomotives fitted with the A-7-EL brake schedule was commencing in NZR workshops. The rebuild was to include a redesigned driver's cab and nose-hood structure plus the 26-L schedule; however, the new push-button brake arrangement was eventually requested to simplify the cab layout. Enthusiastically accepted by drivers, the new brake scheme provided excellent control during service tests on a 70-wagon freight train and was adopted by their union—the NZ Locomotive Engineers Association—as its preference for all future locomotives.

For ease of maintenance, the equipment was unitised; all except cab valves being mounted on a frame that could be fork-lifted into position beneath the cab. Pipe unions were via accessible couplings rather than a manifold system, as it was considered that for the number of pipes concerned, the added complication of a manifold was unwarranted.

---

145. A widely-used UIC (European) braking system designed to allow vehicles with no electro-pneumatic brake controls to operate in a train with E-P control available on the locomotive or power car. The system consists of a Brake Control Panel (currently type PBL-90) mounted inside the equipment compartment, usually on the brake frame. This panel performs the function of the driver's brake valve on a locomotive equipped with a conventional air brake system and is standard equipment on the French and Belgian railways. The principle-of-operation of the Brake Control Panel is that it converts the driver's braking commands from electrical signals into pneumatic signals to control brake pipe pressure. If the rollingstock is so configured, it can also transmit the signals electrically along the train to provide electro-pneumatically assisted brake control on each vehicle. The Brake Control Panel is designed to respond automatically to unplanned or emergency losses of brake pipe pressure to ensure that such incidents cause the train to stop as quickly as possible.

In 1978 the scheme was offered to BHP for their Port Kembla (New South Wales) locomotive rebuild program but not taken up, apparently on the grounds of cost.

The equipment located in the driver's cab included:

- a Brake Operating Unit (on which were mounted Automatic Brake APPLY and RELEASE as well as Independent RELEASE push-buttons);
- an electro-pneumatic (E-P) Independent Brake Valve Operating Unit, incorporating a miniature joystick control lever;
- a "get home" valve and associated cut-out cock; and
- a push-button Emergency valve.

Automatic Brake Operation:

Operation of the equipment was somewhat similar to a Locotrol air brake console. An Automatic brake *Minimum Reduction* was effected by a momentary push of the APPLY button causing the *stick-in* relay solenoid in the E-P Control Unit (mounted in the short-hood nose) to connect the equalising reservoir with the minimum reduction reservoir. By this action, the relay's two normally-closed contacts were momentarily opened, while the normally-open contact was momentarily closed. The relay coil thus lost its retaining feed, dropped out, and remained de-energised even after its contact on the Application switch closed. Opening of the other n/c contact de-energised the Release and Minimum Reduction magnet valves and extinguished the RELEASE indicating lamp.

The minimum reduction reservoir being an empty volume, the resultant in-flow from the equalising reservoir caused an equalising reservoir pressure reduction of about 50 kPa. A J-1 Relay valve was used as a *relay portion,* and this then reacted to provide the brake pipe reduction of like proportion. The simple equalising reservoir piping scheme meant that leakage was not an issue, thus the J-1 Relay was able to maintain the brake pipe at the reference pressure. For a further brake pipe reduction beyond Minimum Reduction, the APPLICATION button was held depressed, operating the Service solenoid to exhaust equalising reservoir pressure at the Service rate (the Service magnet valve being choked to provide a 125 kPa Full-Service reduction in 8–10 seconds). The brake pipe exhaust ceased, and the brake system 'Lapped' when the APPLICATION button was released.

Depressing the RELEASE button momentarily caused a direct release of the brake application by enabling the equalising reservoir to be recharged to its regulated *limiting valve* setting (550 kPa in New Zealand). This, in turn, provided an increased reference pressure signal to the J-1 Relay, which then operated to recharge the brake pipe to the equalising reservoir setting.

An Emergency application resulted when the EMERGENCY button was depressed, opening two vent valves. The large-capacity vent valve directly opened the brake pipe to atmosphere, while the other vented the equalising reservoir. A lever-reset lock-in facility retained the valve assembly 'in' until it was manually reset. Operation of the button resulted in E-P Control Unit solenoids being de-energised, the sanders being activated, and the Emergency solenoid being energised to provide a main reservoir feed to suppress the vigilance device and to provide a reference signal to the brake cylinder relay valve. This latter function resulted in a very rapid build-up of brake cylinder pressure to the safety valve setting, thus providing extra Emergency brake cylinder pressure on the locomotive.

A type-WL triple valve provided the standard operating valve functions, although not—as is mentioned above—those for an Emergency locomotive brake application. The WL triple was choked such as to provide a rapid auxiliary reservoir recharge but a more measured build-up of locomotive brake cylinder pressure.

Attached to the locomotive brake system volume reservoir was a B-7 Independent release valve that operated under the control of the Independent RELEASE button. Depressing this button provided Independent Quick-Release (Actuation/Bail-off), by which main reservoir pressure was supplied to Independent Release pipe 13 via the Multiple Unit isolating valve. This caused the B-7 Independent Release valve to exhaust volume reservoir pressure and thus release the locomotive brakes while leaving the train brake applied. Since pipe 13 is a multiple unit connection, all locomotives in consist would have their brakes released regardless of what brake system they had, and this feature was an NZR requirement.

For Trail operation, the brake valve isolating cock was turned to OUT position, supplying main reservoir pressure to pipe 53 for the following results:

- Suppression of the vigilance device;
- Isolation of the connection between brake pipe and relay valve;
- Isolation of the connection between pipe 13 and the Independent Release solenoid;
- Provision of a feed to the E-P Straight-Air Brake Control Unit to isolate it from its supply air pressure. (This function was provided for the Dead-Engine condition and was not strictly necessary for Trail conditioning.)

It is significant that with only this one control requiring to be positioned when setting a locomotive up for MU operation, the consequences of overlooking the MU-2A valve (as sometimes occurred when setting up 26-L-equipped locomotives) were avoided. The OUT position was also used during the brake pipe leakage test.

For towing Dead, an isolating cock above a Super Spool valve was closed, the Dead-Engine cock opened, and the brake valve cut-out cock closed. The Dead-Engine cock supplied brake pipe pressure into the main reservoir. The straight-air brake isolating valve incorporated a spring-loaded spool that would not permit the valve to close until main reservoir pressure had developed to about 300 kPa. During this time, air would flow to the brake cylinders but would be released upon closure of the isolating cock. This was a desirable feature, as a locomotive should not be moved—even when Dead—until main reservoir pressure has developed sufficiently to ensure efficient brake operation.

Independent Brake Operation (E-P Straight-Air Brake):

This portion of the proposed NZR Push-Button system consisted of an Operating Unit (incorporating a small joystick control handle) and a Control Unit. The Operating Unit contained two switches, Release (normally-open) and Application (normally-closed), while the Control Unit contained two solenoids: Application (n/o, so that in the event of an electrical failure, the Independent brake would apply) and Release (n/c). The brake was not self-lapping; brake cylinder pressure being varied by moving the Operating Unit handle between the desired position and LAP.

With the Independent Brake handle at RELEASE, the Release switch was closed and the Release solenoid energised, thus connecting Control pipe 20 to atmosphere. With the handle at the APPLY position the Application switch was opened, de-energising the Application solenoid and permitting it to supply air rapidly to pipe 20. In LAP position, both the Application and Release solenoids were closed, thus holding any pressure in pipe 20. The Independent Brake handle was spring-loaded to return to the central or LAP position when released; however, a gate was provided at both RELEASE and APPLY to hold the handle in those positions if required (e.g. when parking the locomotive). However, as eventually applied to the rebuilt DG

Class units, this E-P Independent brake was not used, the W-type straight-air brake valves recycled from decommissioned railcars being fitted instead.

FIGURE 240: RE-CABBED NEW ZEALAND GOVERNMENT RAILWAYS DG CLASS LOCOMOTIVE CAB. *G.J. Alecock (top) and G. McClare, courtesy Triple M Publications, NZ*

Control pipe 20 fed the brake cylinder relay valve via a double-check valve, providing a reference pressure up to 350 kPa for the relay valve during Independent brake operation. So long as this pipe was maintained airtight, this reference pressure enabled the relay valve to maintain brake cylinder pressure against cylinder leakage. A Dynamic Brake Interlock was provided to inhibit Independent air braking while dynamic braking was in effect.

Figure 240 shows the Push-Button control box mounted in the control stand of the rebuilt NZR DG Class locomotive. The Automatic brake APPLY and RELEASE buttons plus the Independent brake indicating lamp are on the front of the box, while the Independent RELEASE button is located on the top. This juxtaposition of Automatic brake APPLY and Independent brake RELEASE buttons facilitated good finger-and-thumb coordination when making Automatic brake applications and simultaneously 'bailing-off.' The W-type straight-air brake valve used instead of the originally proposed joystick Independent brake control is located next to the control box, and the protected Get-Home valve handle is behind. The mushroom-shaped Emergency valve button is located above and to the right on the switch panel.

A simplified version of the system was proposed for diesel-electric shunting locomotives (and, in fact, the Independent brake portion was trialed on one DE Class unit—which the author operated on several occasions—engaged on yard shunt and Roll-On/Roll-Off rail ferry duties in Wellington). With this version, the valves for MU operation were deleted, and since the scheme was intended for yard shunting and similar duties—with only occasional main-line working—the Get-Home Valve was also deleted; the Cut-out and Dead-Engine functions being provided-for by manual valves (but able to be 'piloted', if desired). Two smaller ELS brake cylinder relay valves—one per bogie—were used in place of a single J-1, providing the rapid brake cylinder pressure development desired for shunting work.

The brake pipe relay valve used was also an ELS type—being considered capable of providing adequate capacity to supply the brake pipe of trains hauled by main-line locomotives—despite the perceived drawback of a possible comparative lack of sensitivity due to its having a smaller-diameter diaphragm than the J-type. (Note that whereas good brake pipe relay valve sensitivity is necessary on a road locomotive operating down long gradients using *maintaining braking*, this duty is not required of a shunt/transfer locomotive not fitted with dynamic brake equipment.)

The Get-Home Valve

The push-button scheme provided 'fail-safe' functionality whereby an electrical failure or open circuit would cause a Service brake application and—to enable the locomotive to continue under these conditions—a Get-Home facility was provided. This originally included an open-ended pipe terminating at the equalising reservoir tee-piece. In Get-Home mode, the pipe from the Service solenoids to the tee was disconnected (and pushed aside) and the "loose" pipe coupled instead. The locking pin was then removed from the Get-Home valve handle on the control stand and the handle placed to RELEASE position to charge the equalising reservoir and release the brakes. The scheme was later improved by the provision of an isolating cock (**H** in Figure 241) interlocked with the Get-Home Valve such that in normal operation, the cock would be open and the valve "Lapped." Removing the interlock allowed the isolating cock to be closed, isolating the equalising reservoir from the Service solenoids and permitting the Get-Home Valve—that had RELEASE, LAP, and APPLY positions—to be operated manually to control equalising reservoir and thus brake pipe pressure. Consistent with this principle, the brake pipe emergency application valve was held closed by air pressure in the line to the Emergency push-button; any failure of the integrity of this line or severe leakage from it causing an Emergency application.

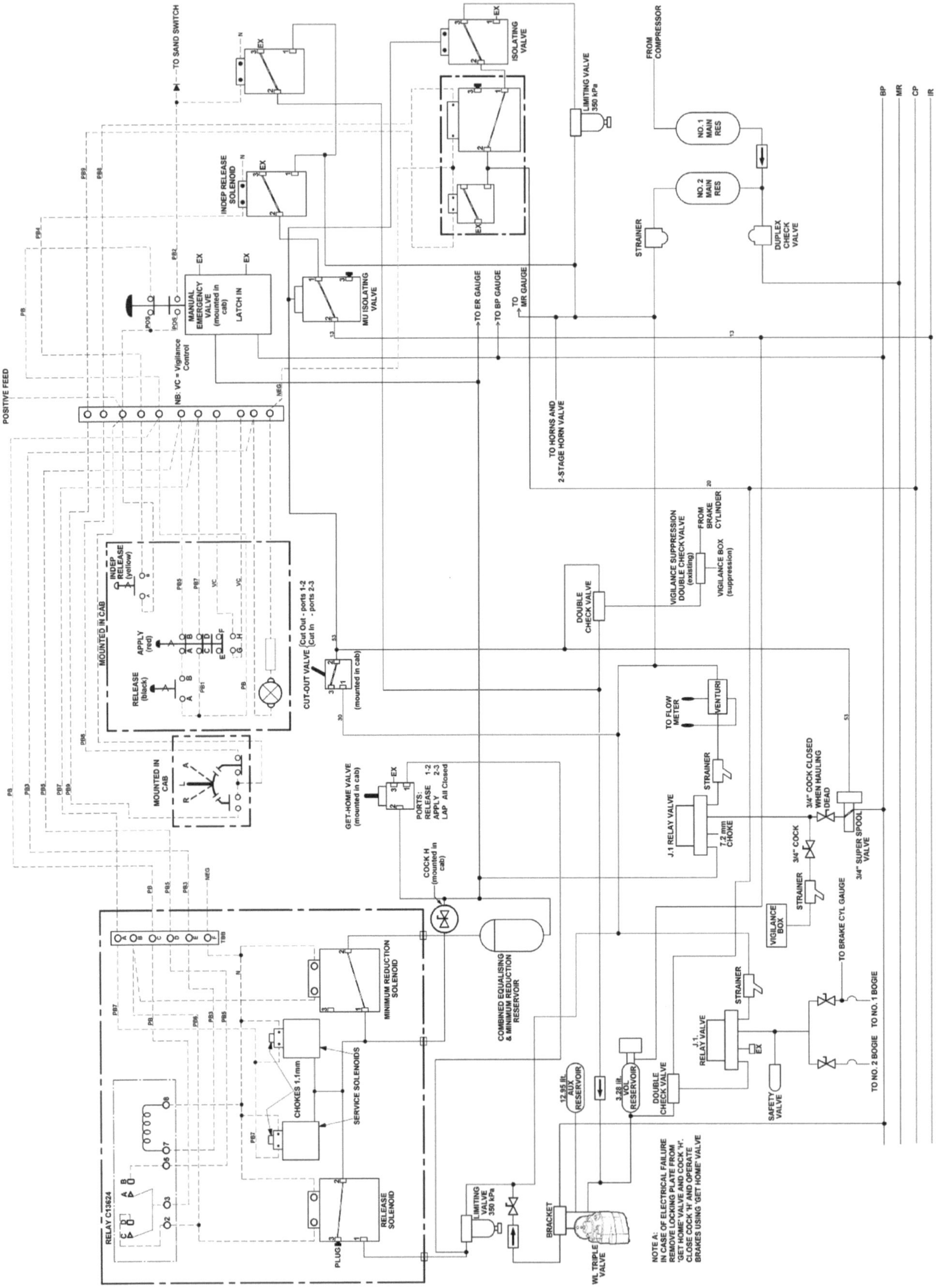

FIGURE 241: SCHEMATIC DIAGRAM OF THE WESTINGHOUSE/NEW ZEALAND RAILWAYS PUSH-BUTTON BRAKE SYSTEM WITH THE EP INDEPENDENT OPERATION AS ORIGINALLY PROPOSED. ALSO DEPICTED IS THE ISOLATING COCK 'H' AS PART OF THE *GET-HOME* FEATURE.

THE DAVIES & METCALF (METCALF-OERLIKON) FV4A AUTO BRAKE VALVE

The English manufacturer's claim for the presentation of this automatic brake valve assembly is that it is *"a modern design, with a simple and attractive outer shape, making it eminently suitable for mounting in modern locomotive driving cabs."* Whether or not these claims are valid must be left to the opinions of the locomotive drivers in Western Australia who operate the locomotives to which they are fitted in a forward-facing, desk-mount arrangement. In fact, the valves are quite "deep" in proportion and extend vertically down below the control desk surface, thus limiting knee-room beneath the desk and I have personal experience of these knee-room constraints.

The FV4a valve (Figures 243 and 244) is a *removable-handle* development of the FV3 (1950) and larger-capacity FV4 (1954) European brake valves and is used on Western Australian narrow-gauge locomotives fitted with dual-brake (air-over-vacuum) equipment. It is self-lapping and pressure-maintaining, and uses self-sealing synthetic rubber diaphragms and oil-resistant rubber-seated valves. Specifically designed for effective operation of graduated release triple valves on 'long' trains (by European standards), it is capable of handling large volumes of air—both in Application and Release—and provides an automatically-controlled high-pressure *releasing charge*. In the latter case, main reservoir pressure can be passed directly into the brake pipe by holding the Automatic brake valve handle in the RELEASE (or FILL) position. When the handle returns (by spring pressure) to the RUNNING position, brake pipe pressure is automatically reduced to a "low-pressure overcharge" that is then reduced at a controlled rate to normal Running pressure without re-applying the brakes (an important aspect of this valve's operation). Cam-operated switches are provided for control of electrical circuits that may be associated with braking.

Although being a much more capable brake valve than was required by West Australian Government Railways' early air-braked freight train operations, it was presumably selected for its graduated release capacity applicable to vacuum braking.

Handle positions are as follows:

- RELEASE/FILL: Described above. Puts main reservoir pressure directly into the brake pipe;
- RUNNING: Equivalent to Release position on other equipments. Recharges the brake pipe to regulating valve setting;
- INITIAL: Provides a 50 kPa (7 psi) minimum brake pipe reduction;
- FULL SERVICE: Provides a maximum 150 kPa (22 psi) service brake pipe reduction. The Service Application Zone is located between these last two and provides for the usual selected brake pipe reduction according to handle position;
- EMERGENCY: Brake pipe is vented directly via the emergency valve and its exhaust valve, and vacuum train pipe is opened to atmosphere via the two-inch valve;
- NEUTRAL: Brake valve is isolated from the brake pipe, and the brake pipe from the main reservoir. The handle can be removed if required. *(Note that this is not a brake valve "Cutout" position providing for isolating the valve following a brake application. The FV4a driver's Automatic brake valve is unique in modern-day train operations within Australasia in not having a setting by which a normal brake pipe leakage test can be accomplished. Narrow-gauge freight trains in Western Australia conveying hazardous materials and hauled by locomotives with this brake valve are regularly operated by ARG (Aurizon) down extensive 1:45 grades without having undergone the standard safety test to confirm brake pipe integrity regarding leakage; indeed, when the author operated trains in this region, network regulations—that otherwise mandate the brake pipe leakage test—specifically exempted its application in these cases simply because it could not be accomplished! This was quite an extraordinary state of affairs, and given this operator's emphasis on safety one could only assume that it resulted from a general corporate ignorance of air brake operation 'best practice' and an associated complacency arising from a historical vacuum brake culture.)*

Brake applications with the FV4a brake valve occur from a *control air* (not equalising reservoir) pressure reduction at the regulating valve. As with American air brake equipments, this action causes a replicate brake pipe reduction via the diaphragm-operated *application valve*, similar to the 26-L Relay Portion. As mentioned, the FV4a brake valve was required to pilot vacuum brake operation on a train so-equipped, this being a normal feature of vacuum brake operations. A unique aspect of the graduated release function in this brake valve—and one of the reasons for the previous comment that it offered greater capability than could be used by the West Australian Government Railways—is that at the commencement of each step of graduation, a high-pressure main reservoir releasing charge, supplied for a time period proportional to the degree of brake pipe recharge being called-for, is supplied to the brake pipe. In this manner, the advantages of the high-pressure release are obtained automatically at each stage of graduable release, propagating a high-pressure wave down the brake pipe that ensures the rapid attainment of the brake pipe pressure increase required, even at the rear of longer trains. Additionally, this 'release impulse' assists in ensuring the reliable operation of older wagon operating valves that may not be as sensitive and responsive as more modern types.

THE DAVIES & METCALF (METCALF-OERLIKON) FD1 INDEPENDENT BRAKE VALVE

Depending on which document or diagram one views, this device (Figures 245, 246, and 247) is variously referred-to by the manufacturer and its then-client railway—the West Australian Government Railways—as a "Straight-Air Brake Valve," a "Direct-Air Brake Valve," or (per the labelling on the valve escutcheon) an "Independent Brake Valve." This is a classic example of the confusion the air brake industry and railway organisations, between themselves, are so often able to create when naming and referring to equipment.

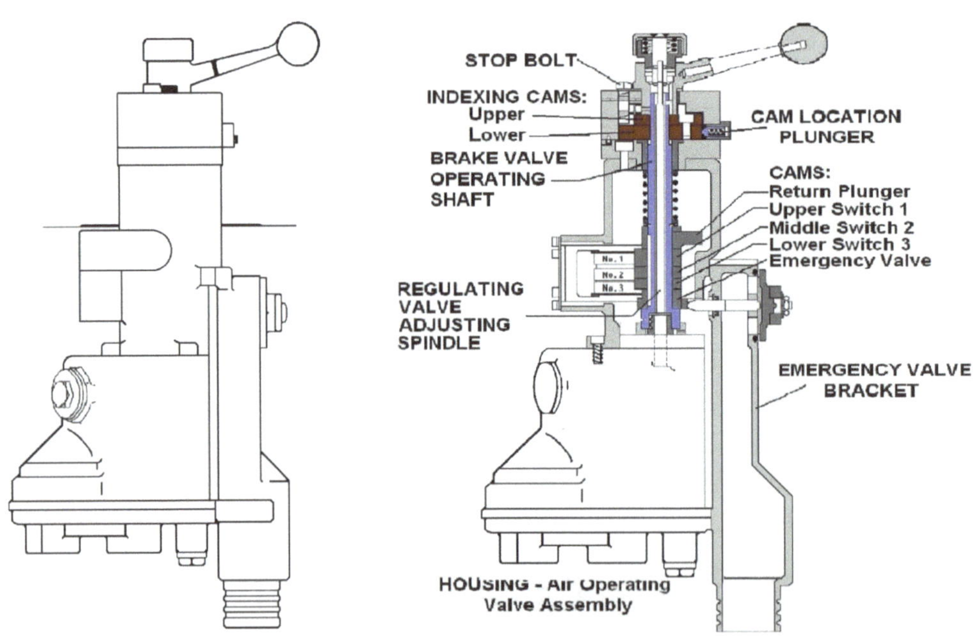

FIGURE 242: FV4A AUTOMATIC BRAKE VALVE HANDLE AND OPERATING SHAFT ASSEMBLY – CUTAWAY DIAGRAM

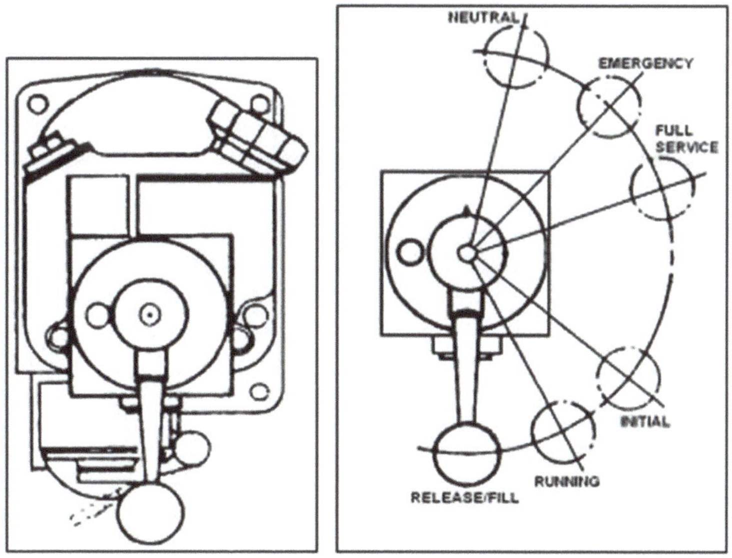

FIGURE 243: FV4A AUTOMATIC BRAKE VALVE HANDLE OPERATING POSITIONS - PLAN VIEW

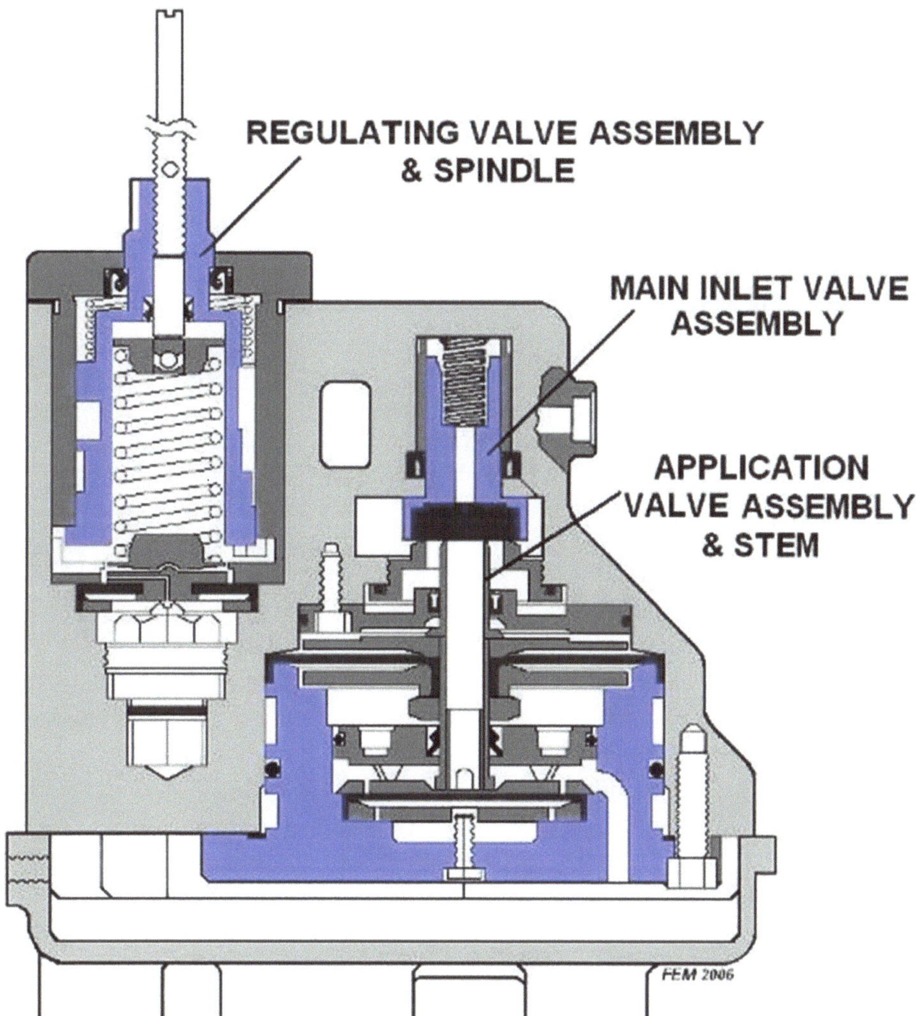

FIGURE 244: FV4A AUTOMATIC BRAKE VALVE AIR OPERATING VALVE ASSEMBLY – CUTAWAY DIAGRAM

D&M describes this valve as "simple and efficient"- a fair assessment. Handle movement rotates a 'cam disc' so that two vertically oriented helical cam faces around the periphery of the disc are presented to both *Regulating valve* and *Isolating valve* assemblies in such a way as to either increase or decrease operating pressure on these valves (Figure 245). As the handle is rotated into the Application Zone, increasing tension is

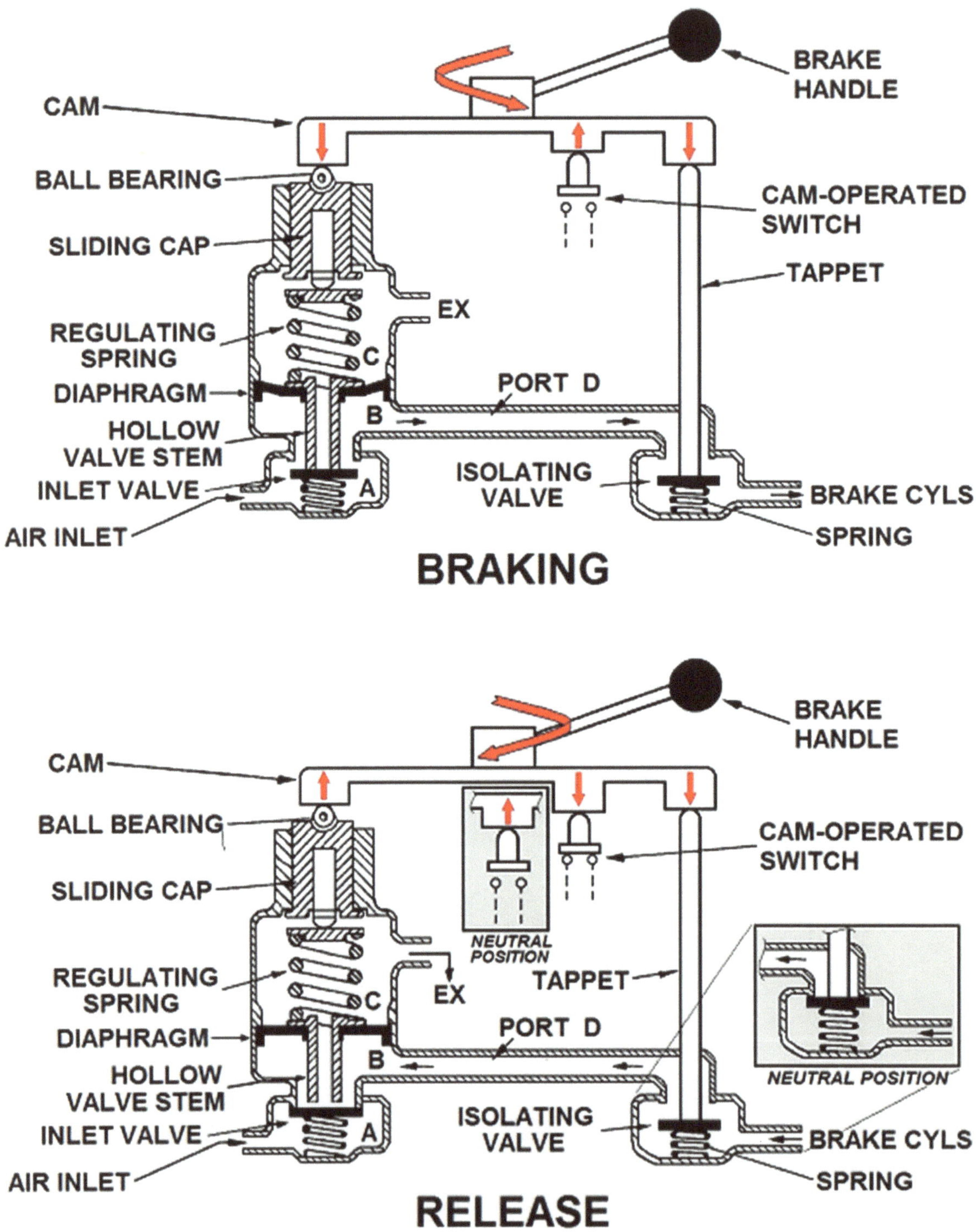

FIGURE 245: DAVIES & METCALF FD1 INDEPENDENT BRAKE VALVE - SIMPLIFIED OPERATING DIAGRAM. *Adapted from a WAGR drawing*

imparted by one of the helical cam faces to the vertical Regulating valve assembly, permitting main reservoir pressure to flow to the locomotive relay valve. As this pressure increases to a value set by the position of the handle, Regulating valve spring and diaphragm pressures are eventually balanced in a conventional manner to achieve the self-lapping function. Another cam profile imparts appropriate movement (via a tappet) to an Isolating valve that is able to close when the handle is placed to the NEUTRAL position.

A third cam face integral to the cam disc operates an electrical switch when the handle is moved to the RELEASE position such that a *brake cylinder cut-off magnet valve* is energised to exhaust locomotive brake cylinder pressure that might otherwise be developed from an Automatic brake application (i.e. the 'bail-off' function).

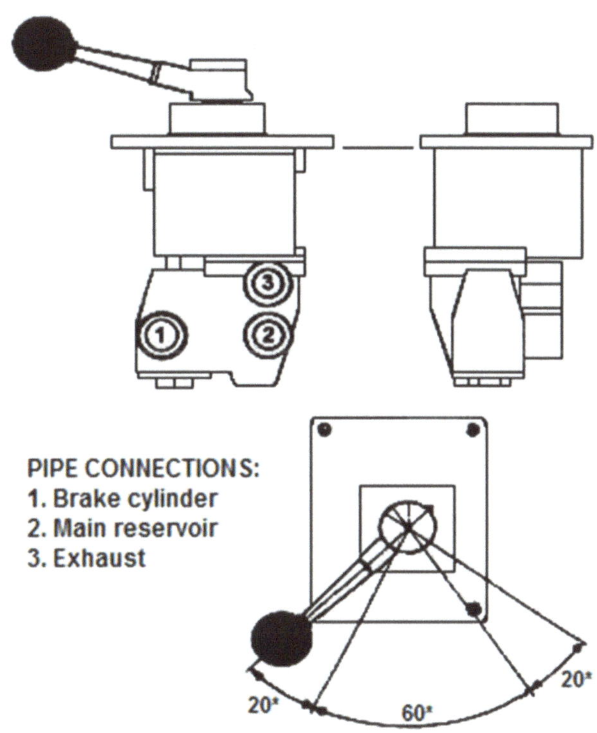

FIGURE 246: DAVIES & METCALF FD1 INDEPENDENT BRAKE VALVE – HANDLE OPERATING POSITIONS

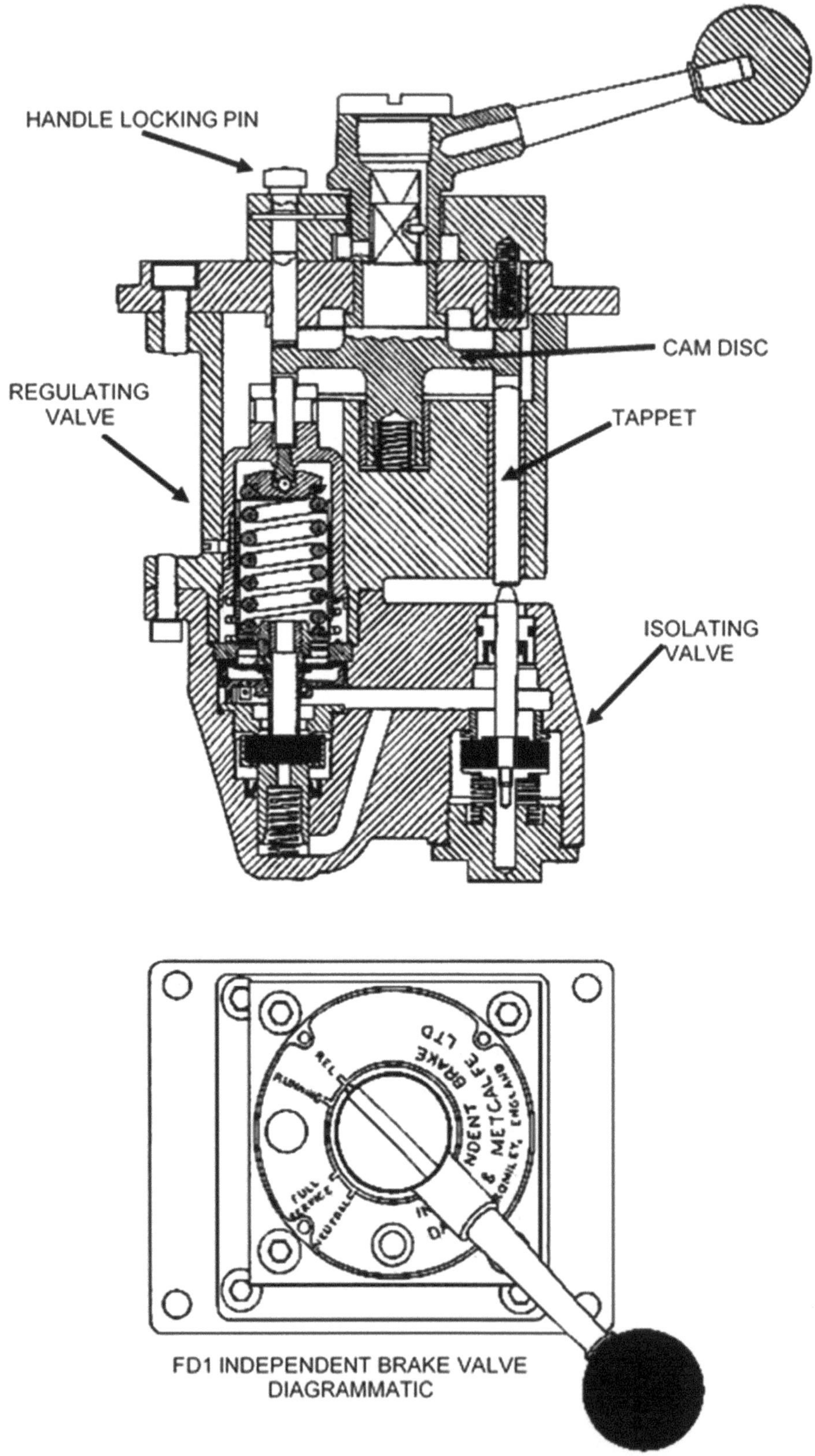

FIGURE 247: DAVIES & METCALF FD1 INDEPENDENT BRAKE VALVE – DIAGRAM AND PLAN VIEW

THE METCALF P85 EQUIPMENT

This analogue-controlled direct or graduable-release system—from Davies & Metcalf of the United Kingdom—was specified by the New Zealand Railways Corporation for their 30 Class (later to become EF Class) 25 kV triple-bogie electric locomotives delivered in 1986 from the Brush Traction Division, and by Queensland Rail for several groups of 30-Series 25 kV triple-bogie electric locomotives built by Walkers and ABB. The NZR units were configured for 3-pipe air brake operation, while those for QR employed the 4-pipe system.

The P85 was designed as a British alternative to the 26-L and similar equipment with which it is operationally interchangeable. This design premise arose from a peculiar reality confronting Davies & Metcalf following the development of their first electronic brake control system, the E70, to become—when fitted to the British Railways High Speed Train in 1974—the first electronically-controlled brake pipe control system in railway service. This reality was that some railway authorities, including within Australia and New Zealand, were nervous about the existence and use of electrical connections between the driver's brake controller and a central brake control unit and tended to specify conventional AAR 26-L systems for new motive power.

Despite this perceptual conservatism, D&M was aware of significant colonial interest in the central brake unit concept of their E70 (***E***lectronic, 19***70***) system and its E75 and E80 derivatives for graduated application/direct release and vacuum schemes. For these reasons, the P85 (***P***neumatic, 19***85***) locomotive brake system was developed to be the equivalent of and an alternative to the 26-L, and to be based on the demonstrated success of D&M technique and experience, including the use of poppet valves in preference to spool or slide valves. Thus, D&M adopted the E70 concept of a central brake control unit housing the large-bore valves (except the Emergency valve as described below) with the driver's Automatic and Independent controllers being the signaling devices, except that the communication between them would be pneumatic rather than electrical. The P85 concept incorporates the following features:

- All valves are of modular construction and manifold-mounted;
- Driver's Automatic and Independent brake valves are separate portions to facilitate cab layout flexibility in a desk-mounting arrangement;
- AAR practice is followed with respect to the Automatic brake handle positions whereby HANDLE-OFF comes between the Service braking range and Emergency, thus differing from standard European practice by which the equivalent NEUTRAL position is at the end of the handle movement (usually beyond Emergency, as seen with the FV4A brake valve described in a previous section);
- Locomotive brake Quick-Release (bail-off) is separated from the Independent brake handle to prevent inadvertent operation (considered possible due to the fore-and-aft plane of handle movement);
- The Independent brake is provided with a relay valve to give constant brake application times on a single unit or an MU consist and to provide a high-capacity auxiliary Independent brake function for specially-equipped wagons coupled behind the locomotive/s, this latter feature being used by QR;
- Computer-controlled drafting and machining technology has permitted D&M to develop the "Tri-plate Manifold" concept (Figure 248) - essentially a pneumatic version of an electronic printed circuit board. This involves sandwiching a thin sheet of aluminium—the separation plate, having air holes drilled strategically through it—sealed between two thick slabs that have air passages routed from their inner faces and individual valves mounted to their

outer faces. Serpentine 'O' rings that seal onto the thin centre sheet are located in grooves machined around the routed passages. (A similar laminated concept having been utilised by Comeng for the 26-LUM unitised air brake rack fitted to their Alco/MLW locomotives built during the 1960s and 1970s for Pilbara iron ore railroads. Unlike the D&M arrangement, though, these laminated assemblies were permanently glued together.);
- Deletion of equalising reservoir pipe-work. Due to the total length of equalising reservoir piping required for a dual-cab locomotive having a single brake control unit (with the necessity for the driver to be able to view the qualifying reservoir pressure) and the resultant potential for leakage, all equalising reservoir piping was removed. This was accomplished by mounting all valves requiring equalising reservoir connections on the same manifold. *With respect to the possibility of leakage, note that some of the QR units were to be fitted with the LOCOTROL distributed power system, which requires a stable and leak-free equalising reservoir circuit.*

A feature of the QR installation is that the usual LOCOTROL push-button brake console has been dispensed with; the P85 brake controllers being used and braking signals being picked up by sensors for supply to the LOCOTROL II radio equipment. Interestingly, this same design feature would again become available some years later with the advanced Integrated Distributed Power version of LOCOTROL that incorporated the NYAB CCBII electronic brake valve into the distributed power system, thus eliminating—in a similar fashion—the need for a separate air brake operating console.

To minimise the amount of piping required, the equipment is arranged in four groups: driver's Automatic and Independent brake controllers and the Automatic and Locomotive brake control units. The manifolds for these latter two units are mounted on a brake equipment frame that incorporates additional brake and auxiliary equipment assemblies.

The driver's Automatic Brake Controllers (one per cab) are of the T-handle design. Because the fore-and-aft movement and desk mounting of these controllers imposed constraints in the degree of angular movement available for handle operation, and to provide optimal spacing between handle positions, the controllers have a two-shaft construction (this feature not being depicted in the brake valve drawing – Figure 251). The top shaft holds the handle-locking mechanism and stop peg, and gear-drives the bottom shaft to provide approximately 160 degrees of rotation from the 90 degrees of movement of the top shaft. The bottom shaft incorporates a horizontal camshaft with one single and three double cams. The principal cam—providing the indexing function—operates the Regulating valve, controlling equalising reservoir and hence brake pipe pressure.

One double cam operates the control pressure dump valve and the vigilance suppression pilot valve, while the other operates the *direct release* and *emergency* brake pilot valves. A full-bore emergency valve operated by the single cam vents the brake pipe directly to atmosphere when the handle is placed to EMERGENCY position. The handle is removable in HANDLE-OFF position. A Changeover Cock incorporated in the valve enables selection of Direct or Graduable Release as well as brake valve cut-out for the purposes of a brake pipe leakage test. For New Zealand operations, the GRAD position is blanked off.

The driver's Independent Brake Controllers (one per cab) also operate in the vertical plane and are of similar design to the Automatic Brake Controller. Handle movement varies the setting of the integral pressure-regulating valve to provide an Independent brake control pressure in the usual manner. A separate, manually-operated push-button pilot valve provides the actuation or 'bail-off' function to release an Automatic brake application (or prevent one occurring) on the locomotive. The handle is removable in the RELEASE position.

The Automatic Brake Control Unit controls air flow into and out of the brake pipe and, therefore, the operation of train brakes. It incorporates the break-in-two protection, the flowmeter, and the double check valves necessary for correct operation from whichever controller is in use. A Lead/Trail selector cock permits isolation of the control unit of a trailing locomotive.

The Locomotive Brake Control Unit controls the flow of air into and out of the locomotive brake cylinders as required either by Automatic or Independent brake operation (with the locomotive set up for Lead operation) and—via the brake cylinder equalising pipe connection—to trailing units. On a trailing locomotive, this unit controls brake operation in accordance with pressure in the brake cylinder equalising pipe (for which the reference is brake cylinder pressure on the middle bogie of the lead locomotive).

The EST4f/g distributor valve controls the operation of three brake cylinder relay valves - one for each bogie. This arrangement minimises the effect of a burst brake cylinder supply hose to one bogie. A timing changeover valve provides a shorter brake cylinder fill time during an Emergency brake application, and a dynamic brake interlock prevents the development of locomotive brake cylinder pressure from an Automatic application when dynamic/regenerative braking is in effect. Under these conditions, Independent brake operation remains enabled. A pressure-regulating valve limits distributor output (and hence maximum possible brake cylinder pressure) during a commanded Emergency brake application.

A control reservoir dump valve prevents the occurrence of dragging brakes on the locomotive, should the Automatic brake controller be re-set to a lower brake pipe pressure or—in a multiple-unit consist—when the locomotive driver changes cabs. As previously mentioned, this valve also provides break-in-two protection, irrespective of whether the locomotive is a Lead or Trail unit.

A Dead-Engine selector cock is also incorporated to provide brake operation on the locomotive when it is being hauled dead (non-energised) or by another locomotive fitted with a brake pipe only.

Mention has previously been made in this book of the enduring use by some Australian railways of earlier models of triple valves historically used with non-maintaining air brake schedules. Since these older triples allowed the auxiliary reservoir to leak off at a similar rate to the brake pipe, the new P85 equipment introduced a problem - as had the 26-L before it. If auxiliary reservoir pressure is maintained while brake pipe leaks off, unintentional brake release may occur. This potential concerned these railway administrations and resulted in an additional requirement on the new system. At the same time, equalising reservoir stability had to be maintained for any operators who might use the P85 with Locotrol distributed power; thus far only QR National (Aurizon). The solution was to provide a means to interrupt the main reservoir feed to the transmission valve on demand, retaining equalising reservoir integrity while permitting brake pipe to reduce at whatever rate was produced by leakage on the train.

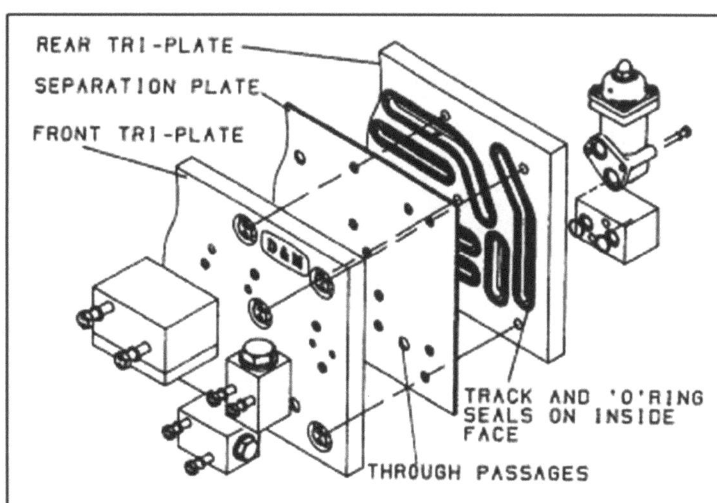

FIGURE 248: THE METCALF P85 BRAKE EQUIPMENT TRI-PLATE BRAKE MANIFOLD

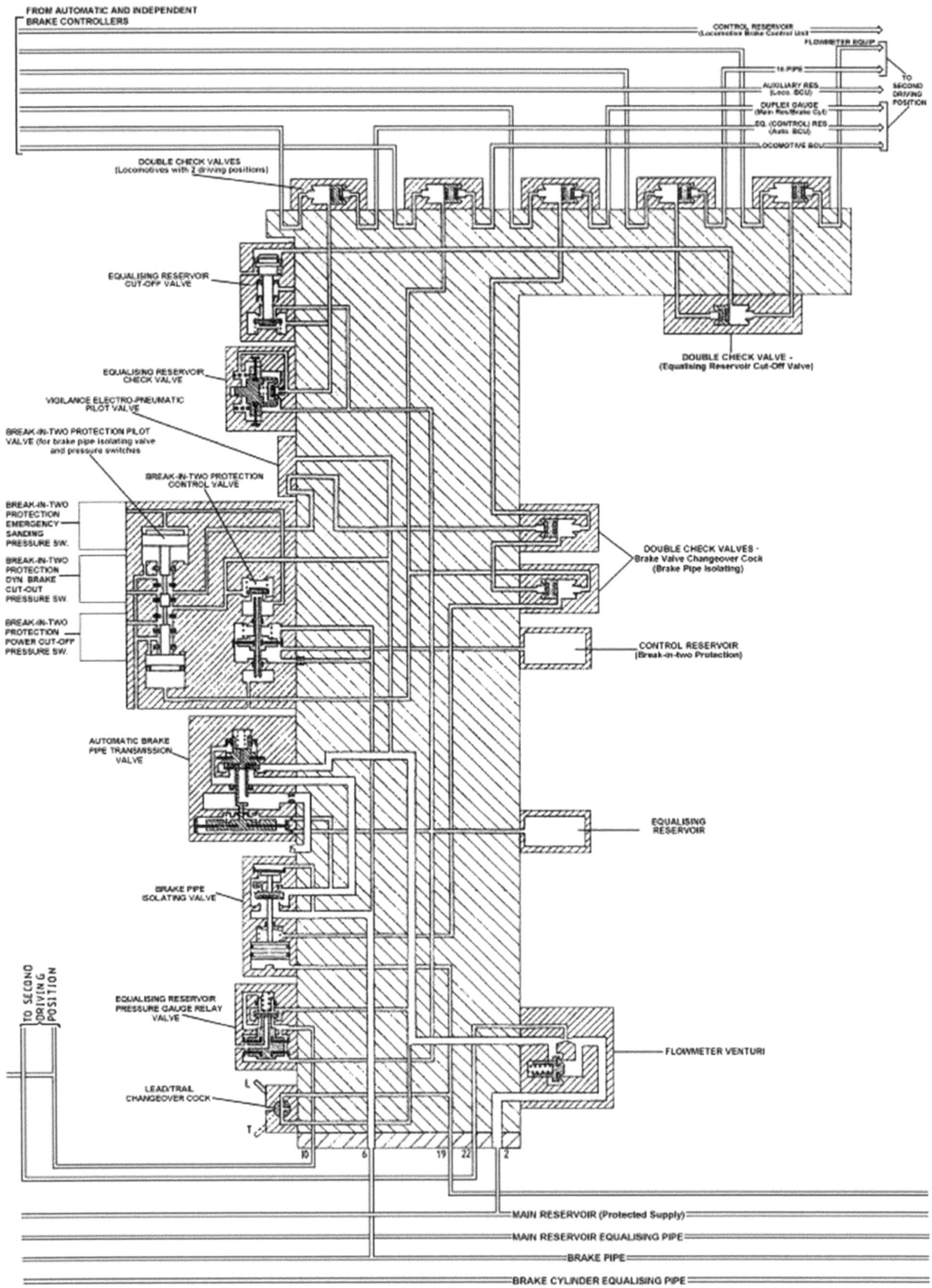

FIGURE 249: THE METCALF P85 BRAKE EQUIPMENT *AUTOMATIC BRAKE CONTROL UNIT* – P85Z VERSION

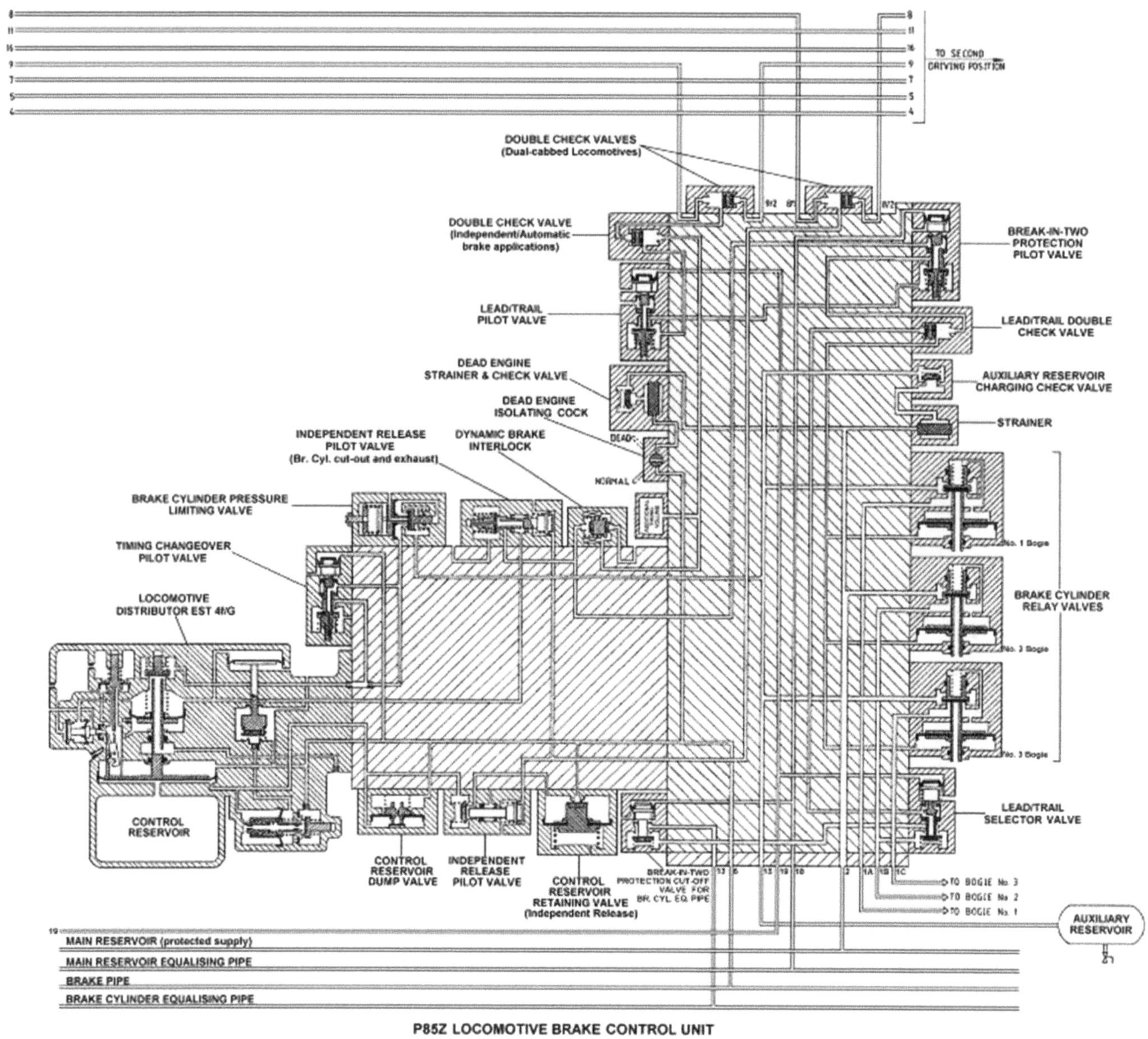

P85Z LOCOMOTIVE BRAKE CONTROL UNIT

FIGURE 250: THE METCALF P85 BRAKE EQUIPMENT *LOCOMOTIVE BRAKE CONTROL UNIT* – P85Z VERSION

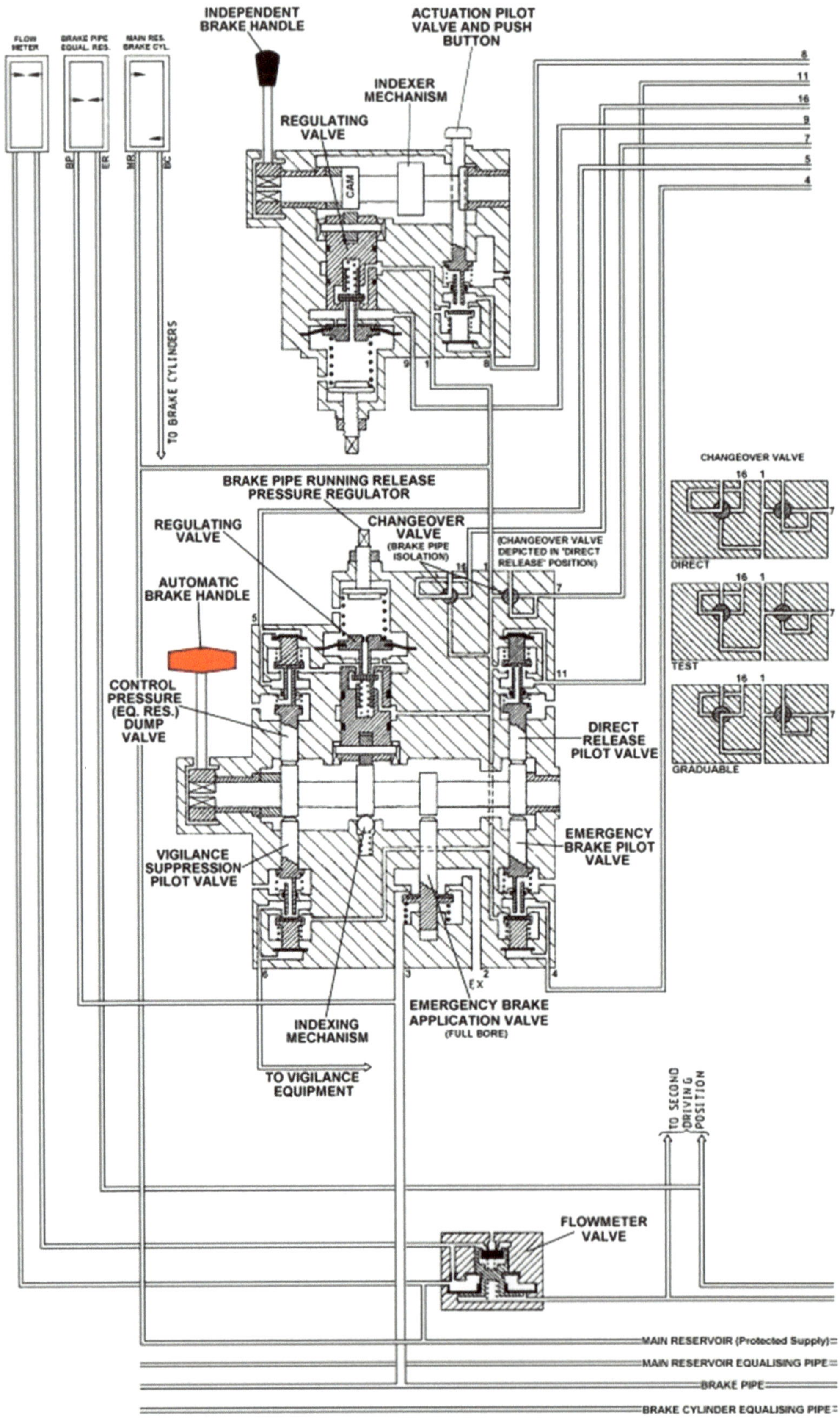

FIGURE 251: DAVIES & METCALF P85Z AUTOMATIC AND INDEPENDENT BRAKE CONTROLLERS

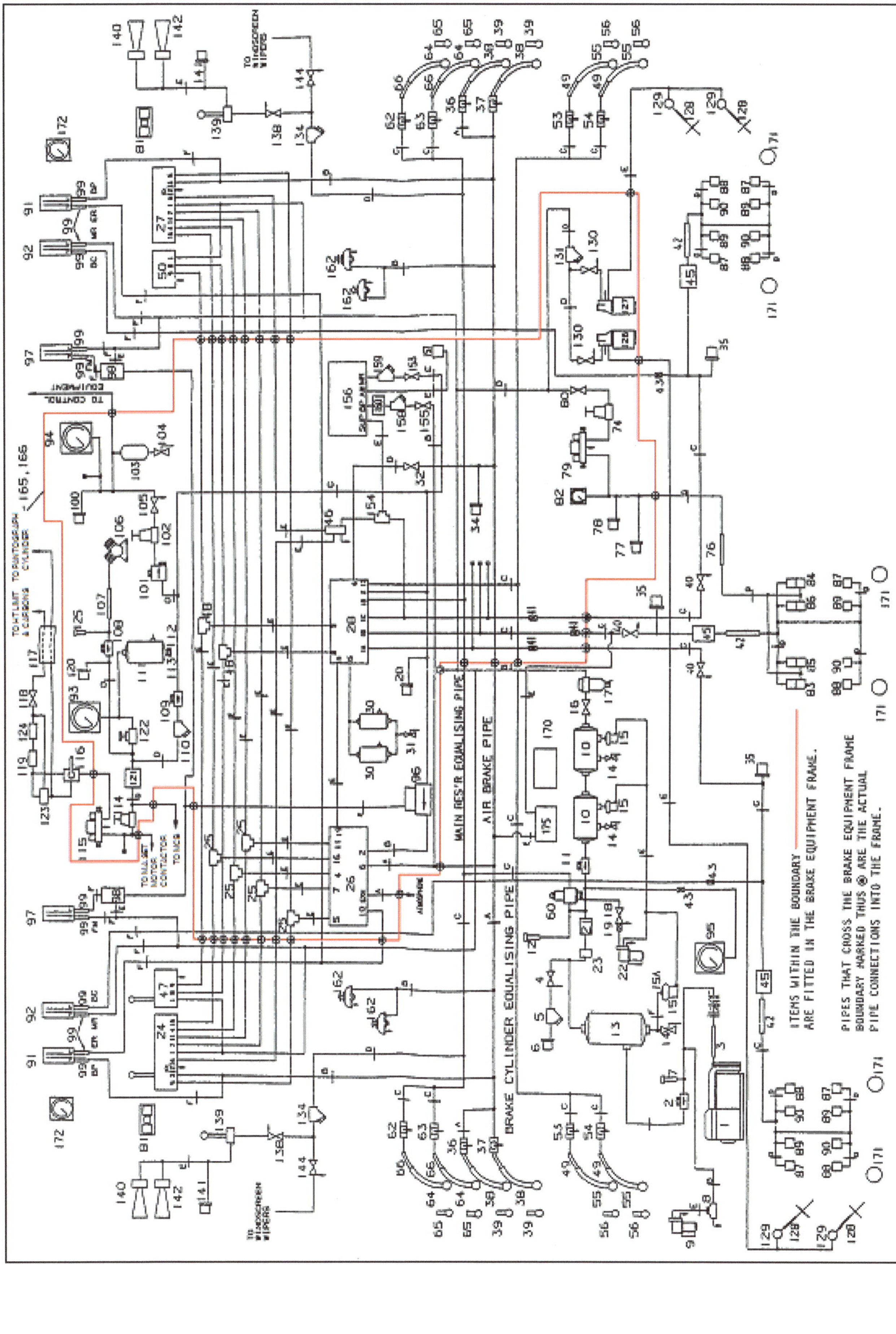

FIGURE 252: NZ RAILWAYS CORPORATION BRUSH EF-CLASS 25 KV ELECTRIC LOCOMOTIVE SCHEMATIC DIAGRAM OF COMPRESSED AIR SYSTEM

1	MAIN AIR COMPRESSOR
2	CHECK VALVE
3	DELIVERY HOSE
4	ISOLATING COCK
5	AIR LINE SIEVE
6	COMPRESSOR GOVERNOR
7	SAFETY VALVE
8	EXHAUST VALVE
9	ELECTRO-PNEUMATIC VALVE (UNLOADER)
10	MAIN RESERVOIR
11	CHECK VALVE
12	SAFETY VALVE
13	MAIN RESERVOIR
14	DRAIN COCK
15	AUTO DRAIN VALVE
15A	MOUNTING BLOCK
16	ISOLATING COCK
17	DRIP CUP
18	ISOLATING COCK
19	CHOKE
*20	PRESSURE SWITCH
21	AIR DRYER
22	ELECTRO-PNEUMATIC VALVE
23	OIL SEPARATOR
24	DRIVER'S AUTO BRAKE CONTROLLER
*25	CHECK VALVE
*26	AUTO and INDEP BRAKE CONTROL UNIT
27	DRIVER'S AUTO BRAKE CONTROLLER
*28	LOCOMOTIVE BRAKE CONTROL UNIT
29	(NOT ALLOCATED)
*30	AUXILIARY RESERVOIR
31	DRAIN COCK
*32	ISOLATING COCK
33	(NOT ALLOCATED)
*34	BRAKE PIPE GOVERNOR
35	BRAKE CYLINDER PRESSURE SWITCH
36	END COCK
37	END COCK
38	HOSE & COUPLING
39	DUMMY COUPLING
40	ISOLATING COCK, VENTED

*41	CHOKE
42	FLEXIBLE HOSE
43	CHOKE
44	(NOT ALLOCATED)
45	DUMP VALVE
*46	POPPET VALVE
47	DRIVER'S INDEP BRAKE CONTROLLER
*48	DOUBLE CHECK VALVE
49	BEND
50	DRIVER'S INDEP BRAKE CONTROLLER
51	(NOT ALLOCATED)
52	(NOT ALLOCATED)
53	END COCK, VENTED, R-H
54	END COCK, VENTED, L-H
55	HOSE & COUPLING
56	DUMMY COUPLING
57-59	(NOT ALLOCATED)
60	DUPLEX CHECK VALVE
61	(NOT ALLOCATED)
62	END COCK, VENTED, R-H
63	END COCK, VENTED, L-H
64	HOSE & COUPLING
65	DUMMY COUPLING
66	BEND
67-73	(NOT ALLOCATED)
*74	PRESSURE REGULATING VALVE
75	DOUBLE CHECK VALVE
76	FLEXIBLE HOSE
*77	PRESSURE SWITCH, N/C
*78	PRESSURE SWITCH, N/O
*79	LATCHED SOLENOID VALVE
*80	ISOLATING COCK
81	PUSHBUTTON INDICATOR PANEL
*82	PARK BRAKE INDICATOR GAUGE
83-86	BRAKE UNIT/PARK BRAKE
87-90	BRAKE UNIT
91	DUPLEX PRESSURE GAUGE, BP-ER
92	DUPLEX PRESSURE GAUGE, MR-BC
*93	SINGLE PRESSURE GAUGE, PANTOGRAPH
*94	SINGLE PRESSURE GAUGE, CONTROL AIR
95	SINGLE PRESSURE GAUGE, MR

*96	FLOWMETER
97	FLOW INDICATOR
98	FLOW METRE VALVE
99	HOSE CONNECTIONS
*100	EQUIPMENT GOVERNOR
*101	STRAINER CHECK VALVE
*102	PRESSURE REGULATING VALVE
*103	CONTROL AIR RESERVOIR
*104	DRAIN COCK
*105	ISOLATING COCK (VENTED)
*106	AUXILIARY COMPRESSOR
*107	DELIVERY HOSE
*108	CHECK VALVE
*109	CHECK VALVE
*110	AIR LINE SIEVE
*111	PANTOGRAPH RESERVOIR
*112	CAP NUT
*113	SEAL
*114	PRESSURE REGULATING VALVE
*115	LATCHED SOLENOID VALVE
116	EARTH SWITCH
117	AIR FEED INSULATION
118	ISOLATING COCK
119	CHECK VALVE
*120	PRESSURE SWITCH
*121	AIR DRYER
*122	DIAPHRAGM COCK
123	PANTOGRAPH AUTO-DROP VALVE
124	FLOW REGULATOR
*125	SAFETY VALVE
*126	ELECTRO-PNEUMATIC VALVE, LEAD SAND
*127	ELECTRO-PNEUMATIC VALVE, TRAIL SAND
128	FLEXIBLE HOSE
129	SAND TRAP & EJECTOR

*130	ISOLATING COCK
*131	AIR LINE SIEVE
132	(NOT ALLOCATED)
133	(NOT ALLOCATED)
134	AIR LINE SIEVE
135-137	(NOT ALLOCATED)
138	ISOLATING COCK
139	HORN VALVE WITH SWITCH
140	PNEUMATIC HORN Gb
141	PRESSURE SWITCH
142	PNEUMATIC HORN Eb
143	(NOT ALLOCATED)
144	ISOLATING COCK
145-152	(NOT ALLOCATED)
*153	ISOLATING COCK
*154	DOUBLE CHECK VALVE
*155	ISOLATING COCK
*156	VIGILANCE UNIT
*157	PRESSURE SWITCH
*158	AIR LINE SIEVE
*159	AIR LINE SIEVE
*160	EXHAUST VALVE
161	(NOT ALLOCATED)
162	EMERGENCY BRAKE VALVE
163-164	(NOT ALLOCATED)
165	BRAKE FRAME
166	No. 1 LOCOMOTIVE MANIFOLD
166-167	(NOT ALLOCATED)
170	WHEELSLIP/SLIDE CONTROLLER
171	AXLE-MOUNTED ALTERNATOR
172	DRIVER'S SPEEDOMETER
173-174	(NOT ALLOCATED)
175	EVENT RECORDER

KEY TO SCHEMATIC DIAGRAM

* INDICATES THESE ITEMS ARE FITTED IN THE BRAKE EQUIPMENT FRAME

FIGURE 253: NZ RAILWAYS CORPORATION EF-CLASS LOCOMOTIVE COMPRESSED AIR SYSTEM SCHEMATIC DIAGRAM - LEGEND

THE KNORR-BREMSE D2B-A BRAKE VALVE

This brake equipment is widely used across Europe and has arrived in Australia on a group of ex-Danish State Railways NoHAB[146]/GM 3900 hp (gross) MZIII Class locomotives imported during 2006 by Lachlan Valley Rail Freight. Expected to have been fitted with flow indicators as well as Independent Release and Control Pipe features for multiple unit operation, these locomotives also introduced to Australasia the eminent Knorr-Bremse KE distributor; in this case, the KE Oc SL (refer to section 3).

Conforming to modern standards, the D2b-A brake valve, a development of the D2, is self-lapping and pressure maintaining and—due to the rubber diaphragm assemblies and seats instead of rubber sealing rings—is relatively leak-free and has an easy operating action. Conforming also to standard European air brake practice, it provides the capacity to graduate the release as well as the application—each position of the driver's brake valve handle corresponding to an Application or Release step—and also affords a Full Release capability (called the CHARGING position and providing the High-Pressure filling stroke) whereby main reservoir pressure is supplied directly into the brake pipe at the commencement of a brake release. This latter facility has been covered in sections describing the American No. 6 and 7 equipments and is intended to assist with triggering triple valve release activity toward the rear of long trains. A period of low-pressure charging automatically follows the filling stroke once the brake valve handle has returned to the RUNNING position, before the rate of recharge again increases to the feed valve capacity. This sequence is designed to prevent brake pipe overcharge and also an unintentional re-application by reducing brake pipe charging at a controlled rate. The driver may choose to initiate a release using low-pressure filling only.

The pressure regulator is situated high in the box-shaped brake valve housing and is adjusted to provide the specified brake pipe pressure by a set-screw that is secured by a lock-screw. Turning the handle of the brake valve within the normal application range reduces the tension on the regulating spring and double-valve **V1** (Figure 254). Since diaphragm **K1** recoils from **V1**, equalising reservoir pressure below **K1** can exhaust via nozzle **1**, as far as a Full Service reduction if necessary. The relay valve (**K2/V2** assembly) that transmits the pressure request to the brake pipe is mounted in the lower part of the housing. Mounted along the same axis of the relay valve is the high-pressure valve (**K3/V3** assembly).

During a filling stroke (recharge), pressure is admitted to diaphragm assembly **K3** by the filling stroke valve. **K3** moves to open large inlet valve **V3**, permitting main reservoir pressure to flow to the brake pipe. After the filling stroke, a small amount of main reservoir pressure remains against diaphragm **K3**, having accumulated in the timing reservoir. This pressure increases the quantity of air delivered by the relay valve, raises the normal pressure, and ensures inlet port **V3** will be opened during the low-pressure filling period.

The brake valve handle is attached to and rotates the control sleeve which, in turn, actuates the filling stroke valve, the rapid-acting (emergency) valve, and the tandem valve. The assimilator is a manually operated valve that fills the chamber between the relay and high-pressure valves when it is desired to momentarily raise brake pipe pressure above the feed valve setting without using a filling stroke (FULL RELEASE position). Brake pipe overcharge can be eliminated in this fashion.

Operation:

The brake valve handle is attached to the control sleeve, which rotates as the handle is turned. The control sleeve is a shroud that covers the spring pipe that, in turn, contains the regulating spring. (To maintain

146. Nydqvist & Holm AB; Swedish manufacturing company based in Trollhättan, originally manufacturing turbines for hydraulic power plants. Between 1865 and 1912, the company built one thousand steam locomotives. In the 1950s, NOHAB started manufacturing diesel locomotives under license from Electro-Motive Division of General Motors. The Danish and Norwegian State Railway systems becoming significant customers.

an appropriate degree of clarity, the spring pipe is not depicted in Figure 254). Handle positions—marked on the brake valve housing in luminous paint on a segment attached to the handle—are tactilely located by means of the catch pin (index mechanism) within the locking device. Around the periphery of the control sleeve are cast cam surfaces that actuate the tappets of the tandem valve, the filling stroke valve, and the rapid-acting brake valve via rollers when the handle is turned. Due to the action of several threaded rings between the control sleeve and the spring pipe, rotation of the control sleeve displaces the spring pipe vertically. Along with the set-screw adjustment, this action varies the tension on the pressure-regulating spring and causes—through the action of the pressure regulator diaphragm, double valve, and regulating spring components—an adjustment in pressure below diaphragm **K1**.

The relay valve then transmits this pressure adjustment from the regulator to the brake pipe. The piston-and-diaphragm assembly is acted-on by the regulated pressure and existing brake pipe pressure to allow either charging of the brake pipe via valve head **V2** or exhaust of the brake pipe via the hollow piston rod and apertures in its head.

The high-pressure valve diaphragm assembly **K3** also includes a collared piston rod and valve head **V3**, which is balanced by a double-grooved sealing ring. This valve allows a large flow of air from main reservoir to brake pipe during filling strokes and in the low-pressure filling period.

The "assimilator" is a lever-operated valve by which main reservoir pressure can be admitted to the piston of the high-pressure valve as desired (via nozzles **5** and **3**).

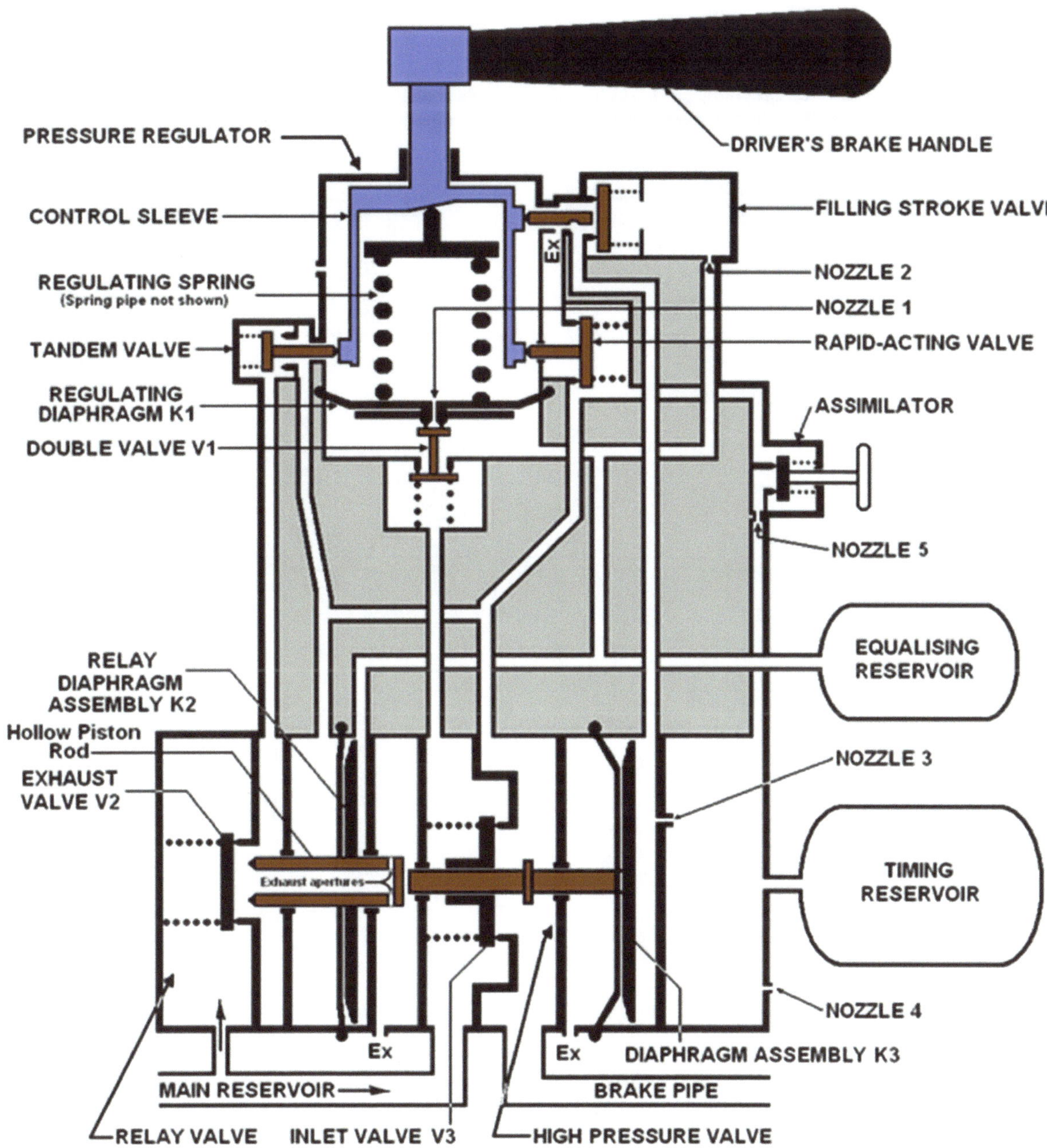

FIGURE 254: KNORR D2B BRAKE VALVE – SIMPLIFIED DIAGRAM *GRADUAL APPLICATION* POSITION

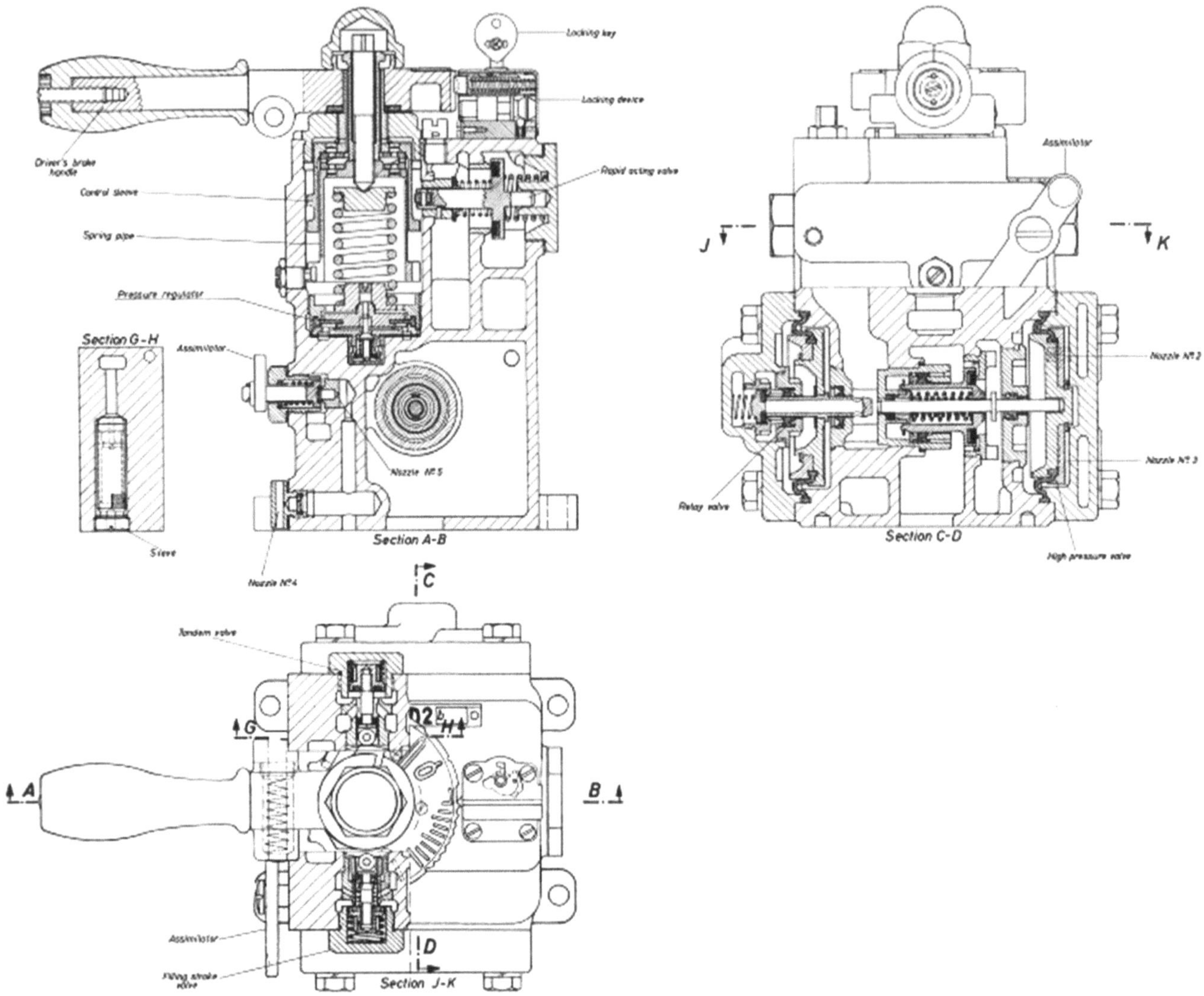

FIGURE 255: KNORR D2B BRAKE VALVE – SECTIONED VIEWS

<u>CHARGING position</u>: The brake valve handle is moved against spring force as far as the stop so that the brake pipe receives air at full main reservoir pressure. The tandem and filling stroke valves are held open, the filling stroke lasting as long as the handle is in this position. Air from the pressure regulator is admitted via nozzle **2** to the chamber to the right of **K3**, moving it and opening high-pressure inlet valve **V3**. Main reservoir pressure now flows freely into the brake pipe, while the piston rod of **K3** lies against the head of the piston rod of **K2** so that exhaust valve **V2** is held closed. Simultaneously, the timing reservoir is partly charged via nozzle **3**. When released, the handle returns to the RUNNING position under spring pressure. In Europe, for the length of a filling stroke after a full application, the following rule of thumb—from which the driver was permitted to depart owing to particular conditions or for a special purpose—was applied: filling stroke (seconds) = 1/10 number of axles. The D2b Automatic brake valve allows for a 50 percent longer filling stroke without causing harmful overcharging.

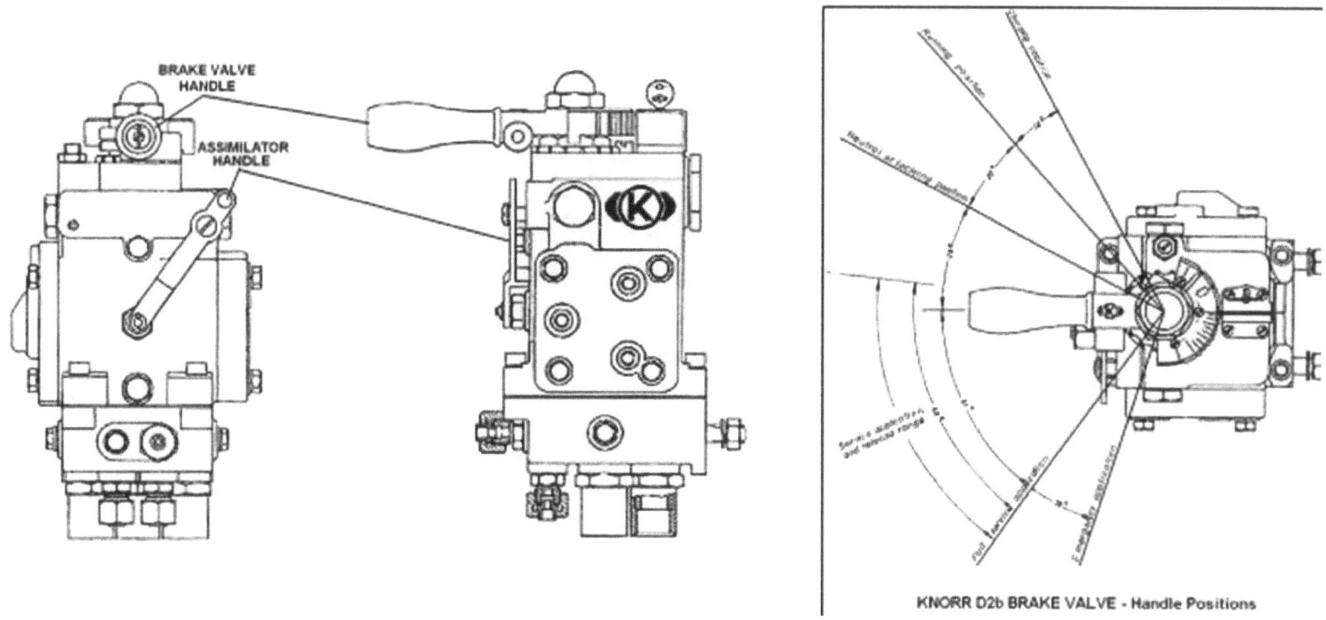

FIGURE 256: KNORR D2B BRAKE VALVE – HANDLE POSITIONS

RUNNING position: The filling stroke valve is closed, and brake pipe pressure should be as regulated (i.e. 500 kPa). The regulator can be accurately adjusted by means of a set-screw and check-nut located beneath the handle cap nut. The low pressure filling period also takes place in this position, either following a filling stroke or when the assimilator has been used to adjust an overcharge (or for the supplementary full release of 'sluggish brakes'). Accumulated pressure in the timing reservoir from any preceding filling stroke continues to act on **K3** and—through the two piston rods—also affects diaphragm assembly **K2** so that variation of pressure in the brake pipe depends on both the regulated pressure and the pressure in the timing reservoir acting against the outer face of **K3**. While assimilation of pressures is proceeding, the pressure in the timing reservoir is constantly decreasing as pressure exhausts through nozzle **4**, which is so adapted to the size of the timing reservoir that the brake pipe overcharge is reduced without causing an application of the train brakes. When the timing reservoir has been completely exhausted, the pressure value in the brake pipe is determined only by the regulator pressure, which is set to 500 kPa.

NEUTRAL position: Marked by an 'O' on the handle segment. The *filling stroke*, *rapid-acting*, and *tandem valves* are all closed, thus isolating the relay valve from the brake pipe. This position is used to 'cut out' the brake valve for the usual reasons - that is, during a leakage test, when changing cabs, for double-heading, and when conditioning the brakes for helper (banking) service. In all cases the process only requires placement of the brake valve handle to this position; no other cock need be opened or closed.

SERVICE APPLICATION range: Marked by nine equally-spaced notches on the handle segment, each providing a discrete value of pressure reduction. Spring tension in the pressure regulator is reduced by a value commensurate with the degree of brake valve handle movement, thus providing—via the relay valve—a brake pipe pressure range for Service applications from approximately 462 kPa down to 338 kPa. The first Service application notch is a wide segment providing a Minimum Reduction of around 40 kPa (slightly less than 6 psi; note that Wabco pamphlets warn against making a minimum reduction of less than 5 psi for safety reasons). Each subsequent notch produces a contiguous brake pipe pressure reduction of 15 kPa (slightly more than 2 psi).

The filling stroke and rapid-acting valves are closed, and the tandem valve is open. The reduced brake pipe pressure is maintained against leakage, and the reduction itself is retarded by the aperture of nozzle **1** and the equalising reservoir volume so that emergency actuation of car operating valves does not result. In performing a graduated release, returning the brake valve handle to the first Service application (MINIMUM REDUCTION) position restores brake pipe pressure to approximately 470 kPa.

RAPID-ACTING position: This is the Emergency position of the brake valve. To access it, the driver's brake valve handle is rotated as far as the backstop. The filling stroke and tandem valves are closed, while the rapid-acting valve remains open to allow brake pipe pressure to exhaust rapidly and completely. The tandem valve isolates the relay valve from the brake pipe.

Assimilation: The assimilator is manually operated by the driver when the brake valve handle is in the RUNNING position. By holding this lever over, the timing reservoir is pressurised such that—via nozzle **3**—the **K3** diaphragm assembly is induced to move so that the solid piston rod contacts the hollow piston of the **K2** diaphragm assembly. By this action after a filling stroke, brake pipe pressure is adjusted to slightly exceed its normal value. The driver releases the lever when brake pipe pressure is judged to have risen to a value necessary to cause a release of the overcharged brakes. Overcharges up to 100 kPa can be released in this manner. When the lever is released, brake pipe pressure rises by another 20 kPa while the timing reservoir is exhausted through nozzle **4** at a controlled rate that is slow enough to allow pressure in the brake pipe and auxiliary reservoirs (or control chambers) to automatically adjust to prevent a re-application. If this process has to be interrupted by a brake application, it is necessary—upon releasing the brakes—to raise brake pipe pressure 20 to 40 kPa above what it was prior to the application. Thus, the assimilation process must be continued at increased pressure; otherwise, the brakes at the rear of long trains will not be fully released.

When normal pressure is restored in the brake pipe, an Emergency application will also register as a pressure reduction on the brake pipe pressure gauge.

It is interesting to note that the Rail Regulator in New South Wales approved this equipment—and its graduated release functionality—for freight train operations, and it is the intention of Lachlan Valley Rail Freight (now known as Independent Rail of Australia) to retain it in service. This operator thus joins Aurizon in Western Australia in a railway air brake operating practice (that is, the operation of conventional Direct Release freight equipment with a locomotive brake that is configured for Graduated Release operations) that is anything but World's Best.

LOCOTROL DISTRIBUTED POWER BRAKE EQUIPMENT

Since the late 1960s, the LOCOTROL™ system has been a well-proven scheme for the radio remote control of distributed power to provide precise control of long freight trains, and its story includes the two major air brake companies of the United States. The idea of using automatic or remote-controlled distributed motive power to relieve drawgear stresses and reduce in-train forces in long freight trains had long been considered by railroad managements as an obvious way to reduce the train operating costs intrinsic to manned helper and pusher operations.

The idea of unmanned distributed power was first demonstrated practically by the experimental Train Multiple Uniter (TMU) system developed by the General Railway Signal Company of Rochester, New York, whose concept used a computer connected to a pair of strain-gauge transducers fitted to the rear coupler of a modified locomotive. This scheme could not cope with the abrupt nature of vigorous in-train slack action and this led to the early demise of this technology.

Originally devised by the Westinghouse Air Brake Company and their affiliate, Union Switch & Signal, another development, the Remote Multiple Uniter (RMU) system, made its debut in the early 1960s. Later, referred-to as Remote Controlled Equipment (RCE-1), this was the first successful system employing a radio link to pass control information between two locomotives not physically coupled together.

Within a short time, the North Electric Company in Ohio, introduced a similar system to RMU based on their nascent SCADA[147] technology.

North Electric—who eventually conferred the *Locotrol* product name—later merged with Radiation Incorporated of Florida, and their locomotive remote control product was developed jointly with New York Air Brake. Radiation was later taken over by the Harris Corporation (Controls & Composition Division) of Melbourne, Florida, and the product was greatly refined and proven in large-scale unit train operation - first on the Southern and later on the Norfolk & Western and Canadian Pacific Railways. Having evolved to incorporate the latest electronics and microprocessor technology, the Locotrol product and name is nowadays owned by General Electric Transportation who has added the following designations: RCL (Remote Control Locomotive) as the designation for their new ground-operated yard switching remote control system, DPU (Distributed Power Unit) to describe a locomotive equipped to provide distributed power, and IDP (Integrated Distributed Power), which is described in following paragraphs.

FIGURE 257: VIEW OF THE DRIVER'S CONTROL DESK OF A BHPIO GENERAL ELECTRIC AC6000 LOCOMOTIVE, SHOWING DISPLAY SCREENS FROM WHICH THE DRIVER CAN CONFIGURE CCB II AND FROM WHICH IDP IS ALSO CONFIGURED AND CONTROLLED

Installed by Harris Corp. on some eighteen railways in ten countries, including on four railway systems within Australia, and with over five thousand systems sold worldwide, Locotrol has now evolved to what might be described as its fourth generation,. While the first and second generations required use of a proprietary push-button air brake console and utilised less-compact on-board electronics packages (with some of this unitised air brake equipment being manufactured under contract by Davies & Metcalf in the United Kingdom) and on-board electronics, the third generation equipment—LEB (Locotrol Electronic Brake), developed in association with New York Air Brake—is nowadays also compatible with both CCB II and EPIC.

On these air brake systems, the driver's control handles are used to operate both Lead and Remote unit air brakes (refer to previous sections on EPIC and CCB), separate push-button operation (Figure 258) having become redundant where an electronic brake system is fitted to the locomotive. With this later generation, GE has developed IDP by integrating Locotrol with their screen-based locomotive control scheme (Figure 257) and the NYAB CCB II air brake system and is developing its further integration with ECP braking. The product has regained its availability for installation on competitor EMD locomotives, and while GE does not appear to have vigorously projected Locotrol internationally until recently, the product will no doubt continue to evolve, and its future might be observed with interest. Indeed, large-scale sales to China a have been responsible for something of a renaissance in its global profile.

As alluded-to in the foregoing description, the Locotrol product has an important air braking functionality since such control within the system is also accomplished via the established radio link between Lead and Remote units. A radio remote control Distributed Power system permits the control commands of the driver in the Lead locomotive to be converted by on-board logic circuitry into radio signals for

147. See Glossary, chapter 8.

synchronous transmission to unmanned locomotives (or consists of locomotives) dispersed at an appropriate position or multiple positions within the length of a train. Thus far, the proprietary LOCOTROL product is the only such radio-based system to be proven in everyday mainline railway operation and to have sold in quantity. The colloquial "Master" and "Slave" terminology originally used with this equipment to designate the *controlling* and *controlled* units, respectively, has nowadays given way to the more politically correct (and perhaps more professional) "Lead" and "Remote" descriptors.

For the purposes of a general description of its air brake functionality, the LOCOTROL II system, now superseded in the Pilbara but still in use in Queensland on Aurizon electrically-hauled coal trains at the time of publication, will be discussed. This remote control scheme includes the management of air brake functions, these being accomplished within the LOCOTROL II system by use of solenoid valves and pneumatic timing and sensing devices that are mounted to a brake rack or 'manifold' properly known as the Brake Control Unit (Figure 259). The Brake Control Unit can be used interchangeably for either a Lead or Remote locomotive by easily adding or deleting certain components as required, and includes a dual cut-out cock to isolate both its main reservoir air supply as well as its emergency brake functionality when the locomotive is in conventional (i.e. non-DP) use. The push-button Air Brake Console (Figure 258), which mounts three illuminated control push-buttons and a large electro-pneumatic Emergency Stop mushroom button, is used by the driver to initiate command signals for operating the Brake Control Unit on both Lead and Remote units.

Figure 258 compares (in the upper picture) LOCOTROL I and LOCOTROL II air brake consoles. Note the LOCOTROL II console has only one button (push/pull) to control Automatic air brake functions. The photos also show a LOCOTROL II air brake console integrated into a conventional EMD control stand.

Momentarily depressing the Automatic brake button produces a Minimum Reduction in the equalising reservoir, and each successive momentary depression produces a further 14 kPa (2 psi) reduction. Sustained depression results in a continuous equalising reservoir reduction at the Service rate to whatever reduced pressure is desired. A programmed safety feature automatically provides an Emergency brake application if brake pipe pressure drops or is reduced below 315 kPa (45 psi). A momentary *pull* on the Automatic brake button will release either a Service or Emergency brake application. (Note: The Emergency button, if used, must first have been reset; see below.) Another programmed safety feature provides an Emergency brake application if—when an Automatic brake application is in effect—an uncommanded increase in brake pipe pressure is detected at either the Lead or Remote unit.

Repeated depression of the Independent Application button will incrementally apply the locomotive air brakes by a value that will vary depending upon the number of locomotives in both Lead and Remote consists, while sustained depression will result in a full Independent application. Momentarily depressing the Independent Release button will cause a locomotive brake release. Pushing the Independent Release button with an Automatic brake application in effect, or pushing it while simultaneously holding the Automatic Apply button depressed, provides the Quick-Release (or 'bail-off') function. The conventional pneumatic locomotive Independent brake valve will operate normally at all times—quite separately from push-button applications—on the Lead consist only; however, if the Independent brake is applied with the Independent Application button, it cannot be bailed off using the pneumatic Independent brake valve handle. The conventional Automatic brake valve handle is latched or (for 30-type valves) wedged in the RELEASE position to prevent inadvertent use when the LOCOTROL system is energised. In this case, any movement of the Automatic brake handle away from RELEASE position will cause an Emergency brake application generated by the LOCOTROL logic.

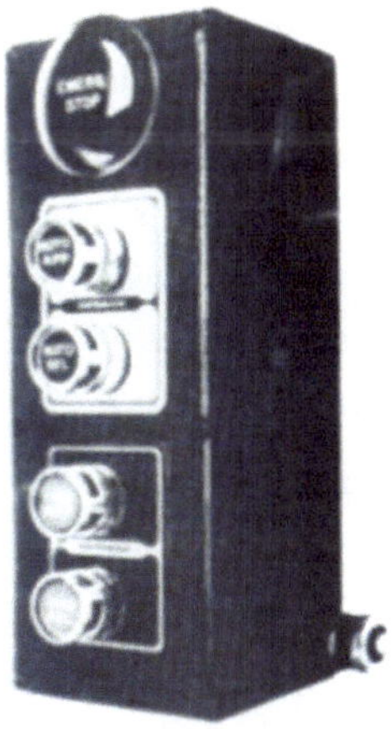

Locotrol 105
air brake console

Locotrol II
air brake console

Comparison between locotrol I and Locotrol II air brake consoles. The locotrol II console has one button for automatic brake functions (Push-APPLY/Pull-RELEASE). The image above shows a Locotrol II console as a standard installation in an EMD locomotive control stand.

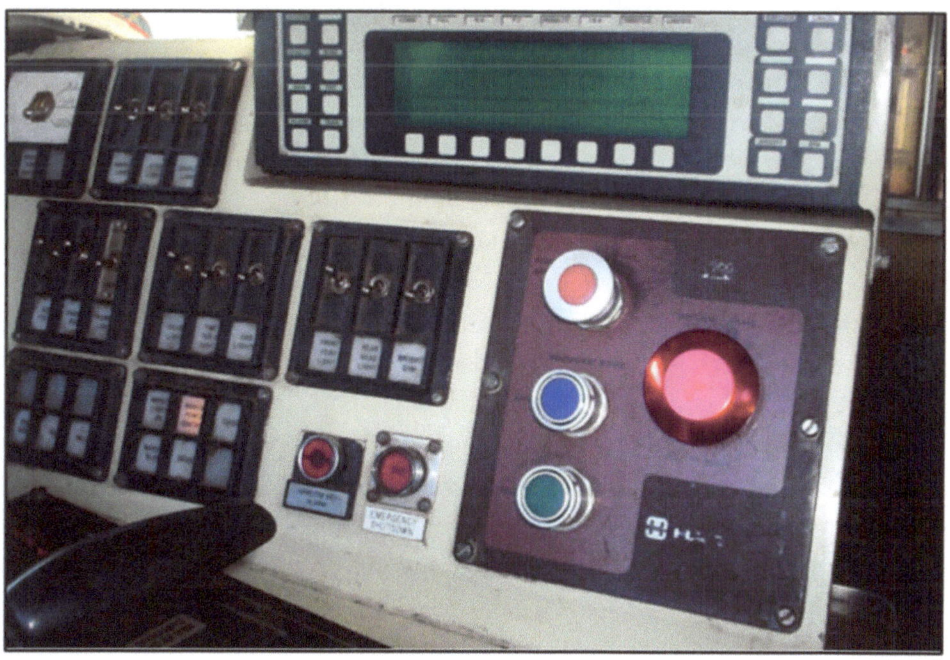

This image depicts the 'space-saver' option of the Locotrol II air brake console, in this case installed on a Western Australian narrow-gauge Locotrol III-equipped 2500-class Goninan-GE Dash 8 locomotive.

With the modern EPIC/FastBrake and CCB equipments, distributed power air brake control can be incorporated into the locomotive braking controller, thus negating the requirement for a separate air brake console.

FIGURE 258: LOCOTROL AIR BRAKE CONSOLES

Depressing the Emergency brake button provides an Emergency brake application—both pneumatically and electrically—on the Lead unit, while the Remote unit simultaneously receives a radio command to also initiate an Emergency application. This telemetry is backed up by the pneumatic Emergency signal propagated through the brake pipe from the Lead unit. Note that even when a Lead unit is operated in 'conventional' mode (i.e. Locotrol system inoperative), an Emergency brake application is always available from the Air Brake Console Emergency push-button due to its mechanical functionality. Once pressed, the Emergency button latches in and must be manually reset by being physically pulled out.

The Central Processing Unit (CPU) microprocessor controls the air brake functions on both the Lead and Remote units. It continuously monitors the Automatic brake button on the Lead unit Air Brake Console, while a pressure transducer in the Brake Control Unit continuously monitors equalising reservoir pressure. The electrical output of this latter pressure transducer—an analogue voltage signal proportional to pressure—is converted to digital data through an Analogue-to-Digital (A/D) converter and is always available to the CPU.

If the air brake is in Release and the Automatic brake button is depressed momentarily, the microprocessor senses the digital equivalent of the equalising reservoir pressure value and subtracts from it a digital equivalent of 14 kPa (7 psi) with the output then being sent to a Digital-to-Analogue (D/A) converter. This analogue value, which represents the desired equalising reservoir pressure, is used to control three magnet valves—Decrease/Increase (DEC/INC), Fast (FST), and Slow (SLO)—that, in turn, control the actual equalising reservoir pressure. The equalising reservoir transducer is also connected to the magnet valve control circuitry, by which it provides direct feedback to maintain the desired equalising reservoir pressure against leakage or unintended increase.

If the Automatic brake button is again pressed momentarily to increase the brake application with the brake already applied, the microprocessor senses the digital equivalent of the existing equalising reservoir pressure value and subtracts from it a digital equivalent of 14 kPa (2 psi), with the result being output to the D/A converter that emerges as an analogue value. This is the new desired equalising reservoir pressure that again controls the three magnet valves as previously described. The equalising reservoir pressure transducer continues to provide direct feedback to maintain the desired ER pressure.

The logic circuitry of the CPU controls the three magnet valves. The DEC/INC magnet valve responds to brake application or release commands (i.e. equalising reservoir pressure decrease or increase). The FST magnet valve is energised and opens when the difference between actual equalising reservoir pressure and the pressure called-for is greater than 7 kPa (1 psi). In this case, once the pressure difference is reduced to within 7 kPa, the SLO magnet valve operates to control the final pressure difference to within ±1.4 kPa (0.2 psi).

One significant difference between the Lead and Remote locomotive Brake Control Units is the airflow sensor module incorporated on the latter. Consisting of two differential pressure transducers connected across the A-19 flow adapter, this module performs an essentially identical function to the normal brake pipe flow indicator, except that it also outputs an electrical signal proportional to the pressure difference; this information being supplied to the Remote unit CPU microprocessor to provide:

- a digital airflow indication presented on the Lead unit driver's Control Console display;
- a warning to the Lead unit when any unexpected or uncommanded airflow is detected (also initiates a communications check); and
- initiation of a Remote unit *Idle-down* sequence (also cuts out the Remote unit Feed Valve in the event of a radio communications interruption – "Comm Int").

This last point is significant in that both brake pipe as well as radio continuity are required for operation of the Locotrol system.

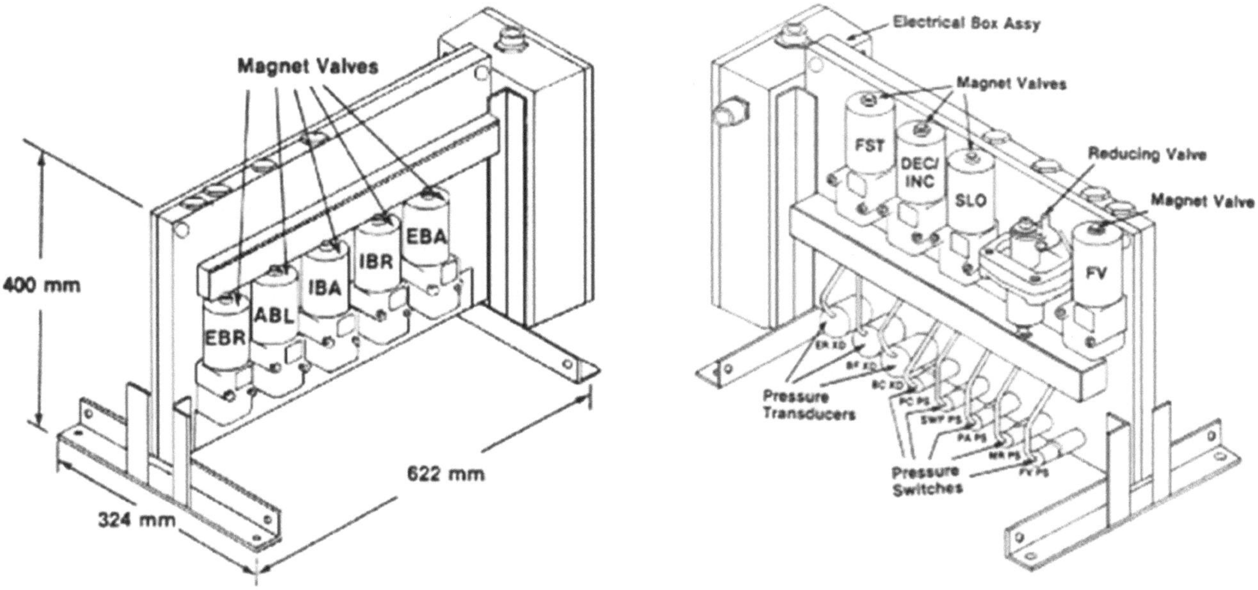

FIGURE 259: LOCOTROL II – AIR BRAKE CONTROL UNIT (MANIFOLD)

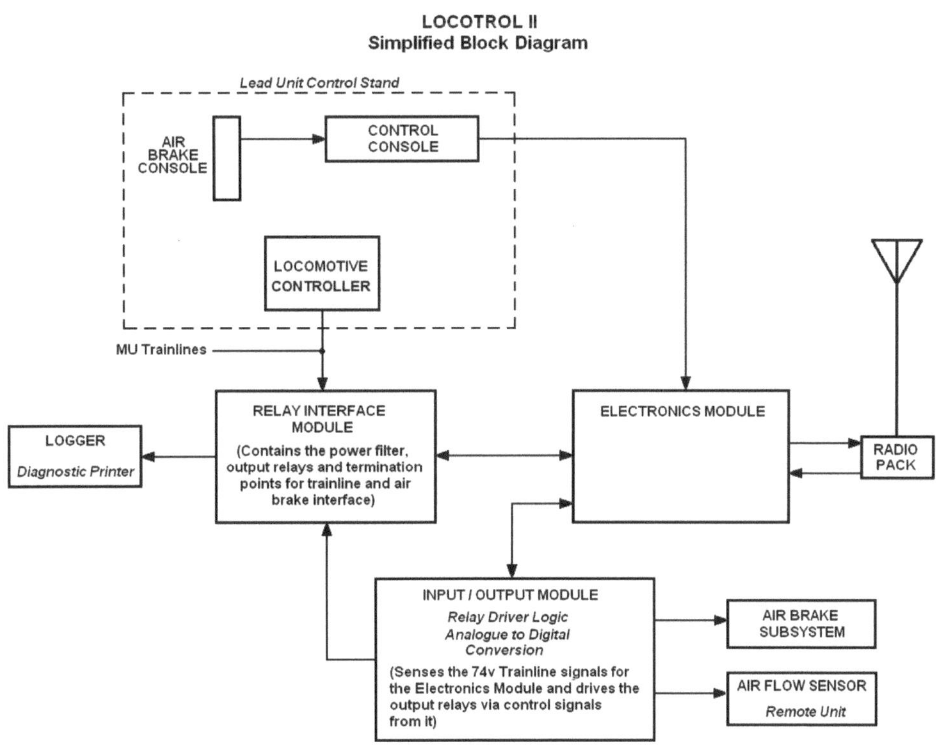

FIGURE 260: LOCOTROL II – SIMPLIFIED BLOCK DIAGRAM, LEAD UNIT

The OzECP Equipment

The genesis of ECP brake development in Australia might possibly be traced back to 1998–99 when a prototype ECP control valve was produced at Westinghouse Brakes Australia in Concord, Sydney, as an exercise to develop and demonstrate a homegrown ECP system. This was done because Wabtec had withdrawn TMX, EPIC, and their ECP system from the available product range for WBA. At the time, Wabtec had apparently mooted the acquisition of WBA, a decision believed to have been a strategic move by Wabtec to force the sale. Ultimately, the buyout did not proceed, and further development of a local ECP product stalled for the usual associated reasons of cost and available market.

At about this time the Central Queensland University—in association with (then) Queensland Rail—were studying aspects of train braking and new systems. CQU had acquired the 75-wagon Research & Development brake rack from WBA in Sydney, and while this equipment was being dismantled and prepared for shipment to Rockhampton, the WBA "brake room" and its contents were openly accessible. These contents included prototype equipment and componentry remaining from the early ECP development work undertaken by WBA; now surplus to their requirements and—to all intents and purposes—forgotten by them, at least at the corporate level. WBA personnel from that period have been unable to confirm the eventual disposition of this early componentry – it is assumed the equipment was removed to Queensland with the rest of the apparatus.

The product that became OzECP and went on to be developed by Faiveley Transport was the brainchild of Barry Payne. It became a collaborative project between Rail CRC members Monash University, Central Queensland University, and rail industry partners who had expressed support for the concept of an 'Australian' ECP system designed for relay braking[148] and intended for easy adaptation to European and American brake systems.

During the late 1980s and early 1990s, Barry Payne was Queensland Rail's Assistant Brake Engineer and later, Principal Brake Engineer. Through his relationship with Pat Wilson, manager of the Train Driving Simulators at QR's Rockhampton Training Centre, Payne was regularly asked for advice on ideas about QR's possible future needs with regard to maintaining simulator functionality at the latest technological level. At the time, Wilson was studying at CQU, to whom he would sometimes direct requests for help with various simulator problems. Through his connections with both the university and QR, Wilson became a member of the Cooperative Research Centre for Railway Engineering (Rail CRC) upon its eventual formation[149]. Thus, Payne became a source of advice to Rail CRC regarding possible future projects, and he recalls having proposed about ten different ideas, one of which was to build an Australian ECP train brake system.

Payne had made two trips to the United States to examine the development and use of ECP braking, during which he had ridden an ECP-equipped taconite train operated by the Duluth, Missabe & Iron Range Railroad and visited TSM (see footnote 68) who were developing and had trialled an ECP system destined, eventually, for AAR endorsement. He considered that this technology was specific to the North American AAR air brake scheme of arrangement and that it would not be entirely applicable to the distinctive Australian braking system. He believed that ECP as detailed in US specification S-4200-00 tended to be overly engineered; in his words, *"a sledgehammer for a tack-hammer's job."* Payne felt that in its use of

148. Refer to Glossary, chapter 8.

149. The Rail Cooperative Research Centre (CRC), based at Rockhampton's Central Queensland University, was established in June 2001 and is supported under the Australian Government's Cooperative Research Centres program. Its members include Monash University, Central Queensland University, the University of Queensland and the Queensland University of Technology, University of Wollongong, and University of South Australia as well as QR, RailCorp (as constituted until July 2013), Pacific National, ARTC, and (then) TMG Rail Technology.

AAR reservoir volumes and devices, the system—for example, as applied in the Hunter Valley on Wabtec-equipped Pacific National coal trains—provided for excessively high air flows and was unnecessarily heavy.

This view would later be reinforced during testing at Callemondah depot of the Wabtec ECP-4200 system intended for installation on narrow-gauge coal trains, and Payne freely shared his views with the American suppliers. As a result of his static and over-the-road testing[150], Payne informed the Americans that in his view their system was inherently unsafe. He and other QR brake engineering staff met with Wabtec and together they devised a number of improvements; enhancements that Payne believes were destined to be accepted by the Americans and to be added to the AAR performance standards. He states, though, that his suggestions regarding the use of a lower power requirement, reduced mass, and simplicity of installation fell on deaf ears.

Soon after the Rail CRC was formed, Payne was invited to Melbourne to discuss his suggestion to develop an Australian ECP brake system capable of being applied to all types of passenger and freight rollingstock hauled by American locomotives or those of American parentage built in Australia; in other words, EMD and GE[151]. There was no intention at that time to develop a specific ECP control system for locomotives.

Having convinced the committee that his proposal was worth proceeding with, Payne was soon meeting with learned staff from Monash University. Since these people had no experience with or knowledge of train air braking, Payne and other QR brake section staff commenced to inform them and to begin the joint development of the basics of an ECP system built specifically for safe and reliable operation on Australian rollingstock using a relayed brake system. Early challenges involved the development of reliable software and the sourcing of appropriate and dependable valving applied in a failsafe and practical layout.

The project had been dubbed MECP (***M**onash*ECP), but Payne soon queried this moniker. He was pushing for a quintessentially "Aussie" ECP product and felt strongly that the project was in fact pursuing the development of <u>his</u> ideas from his involvement with QR rather than from a Monash concept. This led to the adoption of the product name "OzECP."

Since Monash did not possess the capability to undertake static train testing to ascertain whether OzECP would work on rollingstock hauled by an American locomotive with ECP control functionality, CQU was called-upon to help. QR Brake Section personnel set up a rake of old coal wagons and a compressor at the Rockhampton rail yard, and using an electro-pneumatic valve bank (built by Monash University with assistance from QR) on each wagon, connected through a laptop computer control system and cabling from CQU, the system was proved to work effectively. At this point, OzECP was a crude but marketable product requiring further development.

Payne suggested to CQU that the combined system as applied to two coal wagons be installed on a car trailer so that when QR had a defective Wabtec ECP-equipped locomotive or train in the Gladstone/Callemondah area, they could drive to Gladstone depot, couple this to the unit, and carry out testing and troubleshooting. This initiative worked well and resulted in a number of CQU students becoming skilled in the software characteristics necessary to permit OzECP to satisfy ARA and QR operational requirements with a US-licensed, ECP-equipped locomotive. Payne felt that this had provided a very efficient way to develop and prove the on-train software and pneumatic performance of OzECP and that it supported his enthusiasm for an Australian university subsidiary company to further develop the product to a commercially viable degree. It eventually became apparent, however, that the universities were disinclined to maintain trained staff in such a specialised role.

---

150. See section 2, *The Wabtec ECP-4200 equipment*.

151. Referring to locomotives designed and built in Australian under EMD and GE licensing arrangements.

Intent on maintaining the momentum, Payne—who believed there was no ECP in their product line-up and that they might be interested—suggested that European brake manufacturer Faiveley Transport be approached to gauge their interest in collaborating. Headquartered in Italy, Faiveley (who incorporate SAB-WABCO) has an office in Sydney and are active rail air brake players in the Australasian region. After a series of preliminary meetings regarding commercial and financial aspects, agreement was reached that Faiveley would equip a number of QR coal train consists with OzECP once it was fully developed, and the company "came on board." Two CQU engineers travelled to Italy to brief Faiveley Transport on the project, and later, a student from CQU also travelled to Italy to assist with further development.

Payne and his QR Brake Section staff assisted in the preparation of the product design for submission by Faiveley to QR, and during this time they decided they would like to appraise one of the multifunctional European proportional valves for inclusion in the OzECP scheme. If effective, such a device would improve both the simplicity of OzECP as well as its performance under all conditions and, accordingly, such a valve was acquired from Faiveley. None of the American ECP systems have this style of proportional valve; multi-functional because it can be used for ECP Emergency as well as for conventional pneumatic-only (Emulation) operation. At the air brake test rack at Redbank Railway Workshops near Ipswich, west of Brisbane, Payne and a colleague proved that even with extremely slow variations in pneumatic pressure, the valve could not be induced to malfunction. The device is now an integral part of OzECP, as is *quick-service* functionality.

Barry Payne has since been confirmed by the European Patent Office as the acknowledged "inventor/co-inventor" (along with four others from Faiveley Transport Australia) of this product. After Faiveley became involved, discussions revealed that they had, in fact, previously developed an ECP device to UIC standards, but none of the potential customers could agree on the required functionality, so the system was never built. Payne believes that Faiveley views OzECP as a way forward in Europe; otherwise, they would not have become involved.

Obviously, the trade name "OzECP" might not be overly useful to a European air brake corporation marketing the system within Europe, so the product no longer carries this name. In addition, it might be expected that the American manufacturers would take a dim view of an upstart newcomer with a competitive product attempting to horn in on a huge potential market, and corporate relations between Faiveley Transport and at least one of the US brake companies are currently awkward, to say the least. Faiveley Transport Australia was not prepared to provide information about OzECP or its current status or product name.

To maximise its market potential, OzECP was always intended as a retro-fit to existing pneumatic relayed brake systems; design criteria requiring that it be state of the art, reliable, lightweight, affordable, and functionally adaptable. The system is intended to fulfil local requirements for equipment that can be mixed in one train with non-ECP wagons ("single-system interoperability"), thus facilitating its staged retro-fit to existing conventional rollingstock at a greatly reduced total cost. The distributor will operate either in 'direct' or 'graduated release' mode, although it can only do so after the brake pipe recharge occurs at each individual wagon. In this case, the current Australian W-type air brake equipment fitted to rollingstock affords the dual function of providing both an Emergency portion and functional air brake backup. The software is being designed to AAR Standards and for interoperability with current Australian ECP-equipped locomotives utilising North American (AAR) systems.

Possible evolutionary progress with the system might see the inclusion of basic "get-home" functionality (although on this, Payne is doubtful; the AAR get-home feature requires the train to be operated at a greatly reduced speed with the resultant risk of losing Service braking ability as the wagon batteries are slowly depleted).

The system is expected to evolve such that this air brake "backup" element will eventually be replaced with basic get-home functionality and a purpose-built ECP Empty/Load relay valve. Ultimately, it is also intended to develop a stand-alone locomotive brake controller.

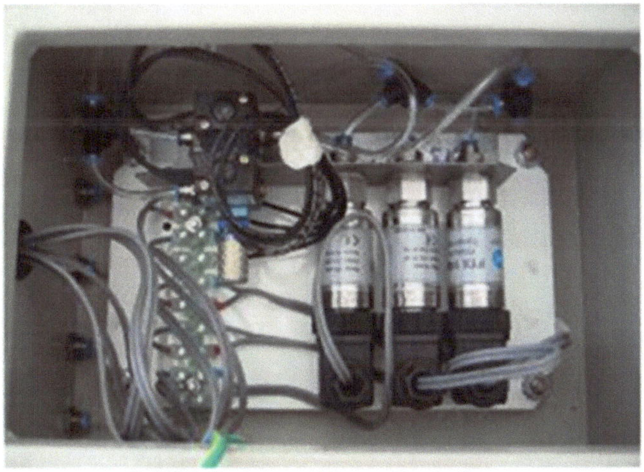

OzECP Prototype Wagon Controller
Rail CRC

Oz-ECP Application Valve
Rail CRC

FIGURE 261: PROTOTYPE OzECP COMPONENTRY

The original Rail CRC claim of benefits for their low-cost retro-fit OzECP system included the following:

- Versatility and flexibility. While US systems require conversion of the whole train before it can be replaced into revenue service, OzECP can be progressively added to rollingstock until all wagons are equipped and then "switched on";
- Modularity of design to provide ease of installation and repair;
- It is safer and more reliable, as the conventional brake system on the wagons is retained as a backup;
- The electronic controller can also be packaged with a Rail CRC Health Card[152], as the powerline network is already there for OzECP;
- The system can be integrated with the locomotive dynamic brake.

I believe there are two ways to view OzECP. On the one hand, the partners can be congratulated for "having a go" at providing an ECP system manufactured in and tailored for the unique requirements of Australasia, and perhaps the greater Asia-Pacific region. If, as Rail CRC has claimed, the system can be packaged with rollingstock and track fault diagnostic tools also being developed by them, then perhaps the total challenge facing railway operators in the region in re-equipping with ECP can be meaningfully reduced.

On the other hand, though, there may be grounds for caution. The North American ECP freight brake systems have been designed primarily for new wagon equipment. While upgrading existing rollingstock with an ECP overlay system might be less expensive in the immediate term than the capital acquisition of new fleets of freight wagons, such configurations might prove to be problematic from the viewpoint of their cost and complexity (i.e. the requirement to maintain two control systems on each wagon). For these reasons also, overlay systems would likely not be competitive for new wagons and thus might be expected to have a limited manufacturing life, meaning higher cost.

152. The Train and Track Health and Speed Setting System (TTSS) or "Health Card" is envisaged as a train-based system that automatically determines the safe operating speed and records any observable defect on the train or track for maintenance purposes. The technology incorporates a low-cost sensor device placed on rollingstock to detect vehicle and rail track faults and capture vehicle responses. The health cards are located on multiple train wagons and simultaneously communicate with an on-board computer system to monitor and report on train and track health and defects and assist to reduce the risk of train derailments (*Rail Innovation Australia Pty Ltd.*).

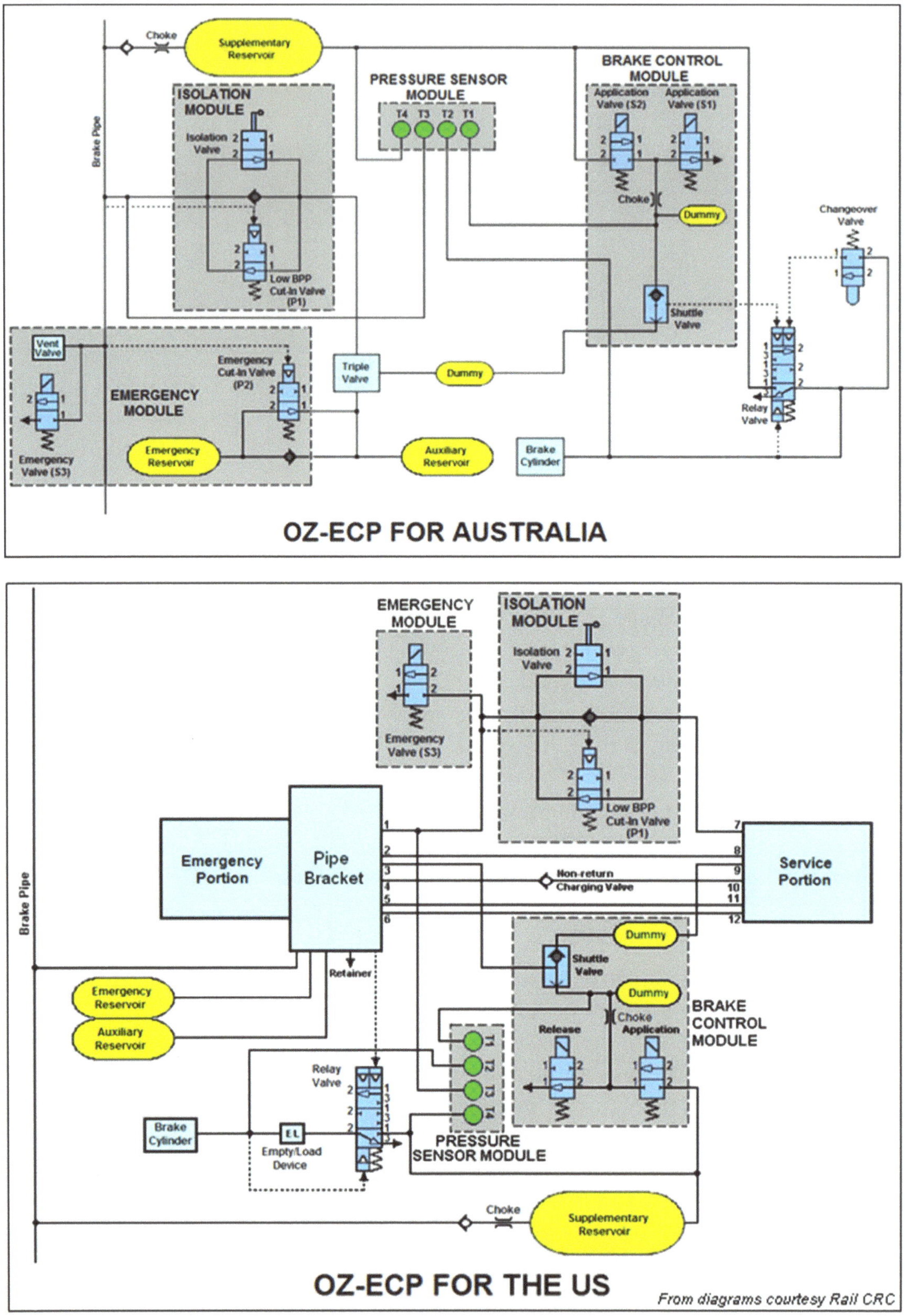

FIGURE 262: OzECP PROTOTYPE SCHEMATIC DIAGRAM

Australian operators who might introduce an overlay ECP system to current rollingstock would be wise to ensure that it was fully AAR-compatible (as was agreed by the ARA Working Party on Air Brake Systems and Equipment for Freight Wagons, intent on preventing a repeat of the previous local experience that resulted in Australian automatic freight brake equipment not being fully compatible—and thus interoperable—with overseas control equipment)[153]. Prospective purchasers will be looking to the longer-term design intentions for OzECP to be fulfilled.

OzECP global contribution

Barry Payne advises that when the Wabtec ECP-4200 system was originally tested on a train in Queensland, it became apparent that air reservoirs were slightly undersized and could deliver a brake cylinder pressure of only 335 kPa during an Emergency application. During the recharge, it was found the CCD Emergency portion incorporated an *accelerated release* feature, directing reservoir pressure from each wagon into the brake pipe. When the CCDs on each wagon then detected this depletion in local reservoir pressure, they treated it as an anomaly and responded by cutting themselves out. As the reservoirs were recharged, the CCDs would cut back in; accelerated release would again activate; reservoir pressure would again reduce; and the CCDs would again respond by cutting out. This continuous cycling of CCDs cutting in and out occurred to almost all wagons in the train, and to recover the situation it became necessary to isolate individual wagons one at a time using the driver's screen display and wait for the reservoir(s) to recharge sufficiently, whereby accelerated release was rendered ineffective. Wabtec altered the software to eliminate the effects of accelerated release, although this was not adopted by the AAR.

1. Car Control Device
2. Distributor
3. Faiveley 2-stage relay valve
4. Single-stage relay (conveys braking signal to the E/L relay on the coupled Slave wagon
5. CCD isolating cock

FIGURE 263: OzECP PROTOTYPE APPLICATION ON QR FREIGHT WAGON

153. Prior to 2000, the Australasian Railway Association (formerly Railways of Australia) contracted a consulting firm to compile codes of practice covering all aspects of the Australian rail industry. From this initiative, an industry working party was formed to cover Air Brake Systems and Equipment for Freight Wagons. The then-current ROA Standard was used as a basis to compile the new air brake Code of Practice, resulting eventually in RCP-2105: *Train Brake Systems* in January 2002. The final codes were never completed because they were to be adopted as Australian Standards and available at the usual fee. Rail industry members—disputing the notion that ownership of this information should revert to the ARA and federal government—baulked at the prospect of having to eventually pay for what it considered was its own intellectual property. The ARA (through its wholly-owned subsidiary, the Rail Industry Safety and Standards Board) is currently creating a suite of industry standards that are based on existing Queensland, national, and international standards.

5.

THE FUTURE

AS IS SO often the case with rapidly advancing technology, the future is already here. Throughout Australia both Diesel Multiple Unit and Electrical Multiple Unit railcar sets are operating with advanced electro-pneumatic brake systems and, insofar as multiple unit passenger train sets might be considered to be both "locomotive" and "train" combined, modern brake equipments currently being used will be briefly described in this section.

The Metcalf E-P Brake System

This equipment is fitted to three-car QR suburban EMU trains in Brisbane. An Auxiliary Braking Unit controls brake pipe pressure by means of a relay valve that responds to a reference pressure supplied by the driver's brake controller. The Service brake is electrically-controlled with pneumatic control being automatically locked out. Movement of the brake controller to apply the brakes causes a reduction in the regulating valve output pressure but the control reservoir is maintained at full charge, and therefore the relay valve maintains the brake pipe at full charge. An electrical failure, however, will result in the control reservoir (and thus brake pipe) reducing to the brake controller regulating valve output pressure, thus producing an equivalent pneumatic Automatic brake application without any additional action on the part of the driver. In both cases, the brake is graduable in Release as well as Application.

Load compensation of both tractive and braking effort is provided in conjunction with the braking distributor valve by use of a Load Weight Shaft and potentiometer plus a Variable Pressure Limiting Valve that regulates the reference pressure input signal to the Distributor application valve in proportion to the vehicle weight.

The Metcalf EBC/5 E-P Brake System

This is an analogue ***E***lectronic ***B***rake ***C***ontrol system utilising Knorr-Bremse wheelslip/slide protection. Versions of this equipment are utilised on the WA *Australind* and NSW *Explorer* Diesel Multiple Unit trains as well as the Transperth A- and B-Series Electric Multiple Units and certain variants of the Queensland Rail suburban and outer suburban EMU stock (e.g. ICE/SMU/IMU sets). The version applied to QR electric passenger rollingstock—***I***nter***C***ity ***E***xpress trains (now reassigned since the introduction of the electric Tilt Trains), ***S***uburban ***M***ultiple ***U***nits, and ***I***nterurban ***M***ultiple ***U***nits—includes an electric code demand transmission system that controls the rates of acceleration and deceleration during motoring and braking, respectively. The pneumatic Automatic brake is controlled conventionally by the variation of brake pipe pressure, the operation consisting of two distinct stages - the control of brake pipe pressure and the control of brake cylinder pressure in relation to the brake pipe. The arrangement allows blended braking (i.e. combining air and regenerative braking) with preference given to the latter on the power or "motor" cars. In this case, the system is a graduable application and release Automatic air brake incorporating E-P control with programmed reversion to Automatic Brake Mode in the case of electrical failure. The EBC/5 Brake Control Unit is used for Service braking and a "mini distributor" for Automatic braking, all components being mounted to slab manifolds.

Load compensation of both tractive and braking effort is also provided; this system—simpler than that for the EMU trains—using a Variable Pressure Limiting Valve in concert with a Pressure Transducer.

The SAB-Wabco/Davies & Metcalf CAB 2000 E-P Brake System

This ***E***lectronic/***P***neumatic brake control system (***C***ontrol ***A***ir ***B***rake) represents an incremental advance from EBC/5. The Automatic brake backup function is identical to the previous equipment, but this system incorporates digital electronics and an integrated, proprietary wheelslip/slide component. An early version is fitted to some QR SMU and IMU stock. The demand signals for Power and Brake are generated by on-board electronics mounted on the driver's controller. These Pulse Width Modulated control signals are processed by the Brake Control Unit to generate power, dynamic brake, and air brake signals. The power and dynamic brake signals are transmitted to the Traction Control Equipment on each motor car, while the air brake signals are handled by the Brake Control Unit located on each car. Being a *dynamic brake priority* system, each motor car is permitted to achieve a dynamic braking effort "in excess" of its own requirement, so reducing the air (friction) brake demand on the trailer cars.

A later version of this equipment is fitted to the NSW 4G (4th Generation) *Millennium* (or M-Class) double-deck EMUs operated by Sydney Trains.

The Knorr-Bremse KBGM-P Brake system

The QR Diesel and Electric Tilt Trains and some SMU and IMU sets utilise this new-generation equipment, the Electric Tilt Train being equipped with KBGM-P (***K****norr-****B****remse* ***G****leitschutzsteuerung* ***M****ikroprozessor:* ***P****neumatischen*; or 'K-B microprocessor-controlled anti-wheelslide braking control system: Pneumatic'), while the Diesel Tilt Train has KBGM-P plus ESRA (***E***lectronic ***S***ystems for ***R***ailway ***A***pplications).

The new Transperth B-Series EMUs have what is effectively an "electric brake pipe"- in other words, a wire rather than a pneumatic brake pipe, while QR—maintaining a degree of conservatism—continues to opt for a conventional pneumatic brake pipe with their KBGM-P systems. The new *Prospector* and *AvonLink* diesel railcars are also equipped with the Knorr-Bremse KBGM-P equipment (pneumatic portion used for backup), as are the Siemens EMUs operated in Melbourne by a franchise operator and the VLocity DMU sets operated by V/Line in regional Victoria.

Although the driver's control action to 'command' the brake operation is increasingly becoming an electrical function, pneumatic power currently remains the best method of providing the train braking 'muscle.' However, since the on-board production of electrical energy on a rail vehicle is now a simpler proposition than is production of pressurised air (and requires less equipment), this will no doubt change. The universal use of fully electric braking systems may denote the next generational evolution. Perhaps the word "air" in relation to railway braking is becoming redundant.

6.

Handling the Air

"The human element is involved with every phase of train air brake control. Whilst the brake is termed an 'automatic' brake, that does not mean that human skill and intelligence is not of prime importance in its use. Safety, the most important function that the brake performs, is not automatic. The brake's effectiveness is attained in proportion to the proper train brake control knowledge of the person operating it." [154]

An Exercise in Judgment

ALTHOUGH—IN THE MOST fundamental terms—the task for a locomotive driver is to move his or her locomotive from one place to another as required, as a vital role-player in the pursuit of an appropriate commercial outcome there is more to achieving this than would be obvious to the casual observer. In most cases, there is a train attached to the locomotive, and this is where the competency requirements of this particular land transport craft begin to diverge from those of others. A train is a flexible entity and—generally-speaking—the longer it is, the more flexible it is. Indeed, the railroad train—in particular, the freight train—has been described as the world's largest machine (although operators of naval aircraft carriers might beg to differ).

The competent operation of any locomotive-hauled train of any reasonable size (mass and length) requires of its driver a quite significant cerebral and intuitive contribution, and part of the reason for this mission may be best explained by an analogy. Imagine a thin elastic cord as representing the length of a train, with a heavy lead weight—representing the total mass of the train—tied to one end. Imagine your thumb and forefinger—representing the locomotive (the motive power)—gripping the free end of the elastic. The

154. Master carman and US railroad air brake instructor, the late C. M. Drennan, *ABC's of Air Brakes*, Chalk Talks, vol. 1 (Simmons-Boardman Publishing Corporation, 1954/1971).

lead weight at the nether end is very heavy; if you wrench the elastic, the weight will scarcely move, and the elastic might break at a random point in its length. However, if you stretch the elastic gently until it is taut (representing the take-up of coupler slack) and continue to apply a moderate and increasing force, the weight will begin to move. If the lead weight is carried on roller bearings running in a bearing race (wheels on rails), it will eventually move freely and travel at the same linear rate at which you are pulling the elastic. If, however, you dramatically slow the rate at which you are pulling the elastic, the weight will tend to continue at its original velocity until it moves up and impacts your finger (the locomotive). This exemplifies coupler and drawgear slack action - or the length-wise rate-of-change of velocity of one part of the train in relation to another. This fore-and-aft action is the "flexibility" mentioned above. It is progressive, and the forces generated are usually referred-to as draft and buff, although railway operating people often refer to them colloquially as "run-out" and "run-in", respectively. A large—perhaps the greater—part of a driver's efforts in operating a locomotive-hauled train is directed toward controlling this slack action; that is to say, minimising its impact so as to run the train smoothly, and the more cars or wagons comprising the train, the greater is the degree of intellectual effort required.

Of all the imperative functional capacities required of a locomotive driver, that of well-developed perceptual/motor coordination is possibly the most vital. This capacity provides the ability to perceive information that affects the safe operation of the train and to apply it for safe and effective control via the power and braking systems. This means the development and use of "judgment"; that undefined, seat-of-the-pants skill by which a locomotive driver demonstrates *anticipation* or the intuitive ability to make control actions—using throttle and brake—sufficiently in advance of such territorial and infrastructure features as curves, grades, crests, dips, speed restrictions, signals, and station stops so as to control the train appropriately.

It is the author's opinion that good judgment derives from experience, and the gaining of experience includes making mistakes. Although many of us look back, with relief that it is an era past, upon a time when "firemen" (or driver's assistants) spent a sometimes inordinately lengthy period—even after having been certified by examination—waiting to be advanced to "driver", it was this supervised apprenticeship that gave one an opportunity to make the mistakes that provided the experience that led, in turn, to the promotion of good judgment.

Throttle manipulation or *power control* is critical but is also relatively simple - and on most diesel-electric and straight electric locomotives provides instantaneous or near-instantaneous response and results. Air brake manipulation, on the other hand, has traditionally been a different story. Good air brake operation requires anticipation (the finely tuned "judgment" previously mentioned), some knowledge of the air brake system being used and—in my view—a reasonable grasp of basic, generic air braking theory because—with the exception, perhaps, of ECP braking—there is a time lag between manipulation of the control device (the air brake valve or braking controller) and achievement of the desired result. There is also the added complication that if the driver's control input is neither well-judged nor well-made, then the result—when obtained—may not be as expected, and this can be critical since the flexibility-of-use of the air brake system is limited.

In other words, once a locomotive driver has initiated an air brake action, that driver is usually then committed to an outcome that may or may not develop as desired in terms of slack action or correct speed management. Train air brake applications and releases cannot be thrown "on-and-off" with alacrity nor instantly varied in intensity to suit the developing conditions as is possible with manipulation of the brake pedal of a motor vehicle. This is where the functionality of locomotive air brake controls and rollingstock air brake operating valves (especially the latter) become critical, and the need for the locomotive driver to be better-equipped to smoothly handle the train has been the crucial influence in the technical development of both devices and their associated systems. This evolutionary progress accounts for the unique operating characteristics of these devices as described in earlier sections of this book.

Consider a train—for the purposes of this hypothetical, running on level track—approaching a speed restriction (e.g. due to track maintenance, a curve speed, or turnout requirement). Braking requires the dissipation of energy and this, in turn, requires time. Since, in this context, "time" relates to *distance*, the driver considers the speed and mass of the train versus the distance to run to the commencement of the speed restriction (the *target location*) with consideration to the restricted speed ahead (the *target speed*). Any control action (e.g. reducing or closing the throttle or establishing dynamic braking) will require a judgment call, as previously explained, as to **where** (how far in advance) to initiate the action and to **what degree** (*how much* speed retardation will be required); however, since we are discussing air braking, we shall look specifically at such requirements.

From the considerations listed in this paragraph, the driver initiates a service air brake application by making an appropriate **equalising reservoir** pressure reduction, most likely a "Minimum Reduction" (sometimes colloquially referred to as "first service" or "initial application"). This must occur a sufficient distance from the speed restriction such as to permit speed to decay to that required at the commencement of the restriction (the *target location*). The driver does not directly control the reduction in brake pipe pressure. As has been explained earlier in this book, this automatically follows the equalising reservoir reduction via the internal workings of the driver's air brake valve or other on-board equipment. The driver may then decide to increase this reduction (thus producing what is often referred to as a "split reduction," which is usually desirable). But in any case, the *total* eventual brake pipe pressure reduction—which should ideally be the *least possible* consistent with good train-handling practice—needs to be of sufficient magnitude and to be completed suitably in advance of the target location such that it is *effective* prior to brake release. That is, the exhaust of air from the brake pipe should have ceased and the reduced pressure have settled along the length of the brake pipe, a condition referred to as *service equalisation* (sometimes also "brake pipe equalisation").

Effective brake pipe equalisation is a highly desirable goal since a good, positive brake release (i.e. individual cars or wagons not left with "sticking" or dragging brakes) is more likely from an *effective* brake pipe reduction than from an *ineffective* one. An *ineffective* brake pipe reduction occurs when the release is initiated prior to the brake pipe exhaust having terminated, thus brake pipe equalisation has not yet occurred, and Service application activity at the triple valves (in particular, toward the rear of the train) is incomplete. Proficient locomotive drivers will monitor the progress of their brake pipe reduction by listening to the sound of the brake pipe exhaust; that is, its *length* and *strength*. This provides an indication of progress toward that desired outcome of **equalisation** and must be considered in relation to distance-to-run and rate of deceleration.

To be able to hear the brake pipe exhaust occurring has traditionally been fundamental to achieving good air braking outcomes. However, with the advent of electronic air brake control on newer locomotives, the brake pipe exhaust may be much less audible; indeed, on double-ended units, it may be completely absent from one cab! In this case, the locomotive driver is required to monitor the digital equalising reservoir and brake pipe displays (and, if one is provided, the End-of-Train monitor display) when making an air brake application to ascertain when brake pipe equalisation has been achieved, or the rate of progress toward it. On the other hand, modern screen-based driver displays are likely to include a computer-generated speed *trend* or "rate-of-change-of-velocity" indication. Properly used, this display can be a powerful tool for a locomotive driver in making judgments regarding speed control, and it might be relevant to mention here that experienced locomotive drivers have for many years been using the almost infinitesimal flicker of the Hasler speedo needle in either direction as a reliable early indication of the speed trend.

A further important element for deliberation in air brake releases is *train speed* at the point of release. This factor—which must be considered in relation to the state of slack within the train (i.e. stretched or bunched) and whether or not equalisation has been achieved—is discussed in following paragraphs. Meanwhile, having attained equalisation, the driver must then decide when to release the brake, since—once this is initiated—there is again a time lag as the triple valves respond to the restoration of brake pipe pressure (usually sequentially from front to rear, as explored in the early sections of this book). During this time, speed will continue to decrease, and the driver's objective is to arrive at the speed restriction (the target location) at the target speed, with the brakes released and with a power setting sufficient to maintain that speed. Although this optimum result may not always be realised in practice, it is an important goal, and a competent driver will always strive to achieve it. If the train arrives at the target location with the brakes released but speed still too high, the driver is faced with the challenge of taking belated recovery control actions. On the other hand, should the train arrive at the target location traveling significantly *below* the target speed, then extra fuel and time are going to be expended in the recovery.

Again, the in-cab audible cue is important - no less so for the brake Release than for the Application. Traditionally, while recharging the brake pipe, the driver listens to the sound produced by the brake pipe flow meter (in addition to the corresponding visual signal of flow indicator needle deflection) to be able to judge the comparative degree of brake pipe recharge. However, since this prompt has been effectively engineered out of the newer electronic locomotive brake systems, the driver is now required to scan the digital flow-rate display for a computer-generated numerical indication of the measure and magnitude of brake pipe recharge. This aspect of brake pipe "charging air flow" merits further consideration and will be examined later in this section.

The considerations for a planned stop—especially in passenger train service—are similar; perhaps the greatest difference with passenger train operations being the need to bring the train to a stand with minimal or nil car brake cylinder pressure in order to prevent *brake snatch* (Americans sometimes refer to it as *stonewalling*). This phenomenon is the insidious, upsetting "jerk" that sometimes accompanies the instant of stopping and invariably throws passengers off balance as they stand in the aisle or reach for their luggage (or that sloshes the soup or wine on the table linen in the dining car!). However, in all train-handling scenarios, there will always be variables such as gradient, train brake performance on the day (i.e. the physical state of the equipment and possible assortment of wagon types), or weather influences to complicate the task. Additionally, the locomotive driver may desire or feel obliged to augment the slowdown by assisting it with dynamic braking, an effective adjunct but often an added complication.

So, in achieving the desired air brake slowdown, the locomotive driver must be able to make instantaneous and intuitive decisions as to **when**, **where**, and by **how much** to apply the brake, **how long** to leave it applied, and **when** to release it. With a *graduated release* system (and with regard to locomotive-hauled trains, this is not something that Australian locomotive drivers of air-braked trains have had the opportunity to experience), one could add, "and by **how much** to release it" to this list. Effective decision-making of this sort is critical for competent train handling; however, the outcomes rely on the degree of *expert judgment* exercised by the locomotive driver and this judgment is only developed over time; in other words, from constant practice.

It is therefore a matter of professional concern to me that one of the results of the contemporary obsession of some railway operating managers against use of the air brake (on the simplistic premise that the resultant saving on brake shoes and fuel is of overriding significance) is contributing inordinately to a stealthy decrease in basic technical knowledge and expertise within the ranks of both locomotive drivers and their supervisors. The myth—perpetuated widely within Australia—that 'it costs the company $500 every time a driver makes an air brake application' is also ultimately counterproductive (and would hardly survive engineering scrutiny). I believe there is a sensible compromise to be found between the maintenance, generally, of good train-handling performance and the valid pursuit of commercial financial prudence. However, such endeavour is hardly promoted by proscriptive operating instructions that actively discourage air braking and peculiar—even bizarre—rulings such as that 'dynamic braking should be considered the primary braking system on a train.' From a safety perspective, such dogmatic directives are unwise, counterproductive, and extremely ill-advised.

The critical need for competent training and instruction of locomotive personnel starting out in their careers and the vital requirement to provide a practical and legitimate experiential path toward proficiency are utterly fundamental, yet I believe there is a dangerous tendency by present-day railway management in Australia to overlook this in the rush to provide legally qualified staff in an industry environment characterised by chronic shortage. It would not be unreasonable to suggest that the current ownership structures of some rail operators—with their business background in logistics and road transport—militate against a clear understanding and appreciation of this philosophy. With some transport companies who view locomotive drivers (and their competency requirements) as differing little from truck drivers, and with start-up operators who are keen to secure and retain business, there is potential for a dangerous dilution of the "judgment" gene pool.

Perhaps there is an archetypal story able to be related of operators in any rail operations sector, but for me—resident in Western Australia at the time—the abiding lesson was provided one sad night in January 1996 at a single-line crossing location on the standard-gauge main line across that state. As an eastbound intrastate freight train pulled into a crossing loop known as Hines Hill, a westbound interstate freighter entered the location on the main line. This was the classic scenario requiring the exercise—by the drivers of both trains—of my definition of "judgment." But those qualities were lacking in one of the drivers whom I believe was not sufficiently qualified by experience (although he was by legal definition) to have been given the responsibility with which he was charged that night. Operating in an advanced category of train service for which—in my view—his brief locomotive career had not thus far adequately prepared him, this young driver misjudged his approach to and overran his stopping point at the Departure signal (all colour-light signals look closer on a dark night than they actually are) and ploughed into the tail-end of the opposing train that had not yet cleared into the loop. The wagons he impacted were tank cars conveying fuel, and the resultant conflagration took the lives of this driver and a supernumerary (a child) who was riding with the crew, wrecked the two locomotives of the interstate freighter and much of the train, and led to an engineering safety action that forever reduced the degree of incisiveness in judgment required to be exhibited by train drivers in arriving and stopping at crossing locations.

The outcome of this tragedy has been to cement in the collective mind of contemporary railway management the view that the less critical the judgment skills you need to require of your drivers, the less total training they will need in order to be qualified for their role. It remains to be seen whether this *lowest common denominator* approach will, in the long run, contribute effectively toward improving rail safety. Although, prima facie, the incident resulted in a safer environment that now prohibits the "close meet" of two opposing trains, it has also served to forever provide an impediment—in an extremely competitive transport environment—to the activity of train crossings on the single-track interstate standard-gauge network. It remains, also, to be seen whether this initiative will prove over time to have served to improve train operating safety so long as engine service personnel might continue to be advanced—by their employers—beyond their capabilities.

Air Flow Indication

The air **flow indicator** is basically a differential pressure gauge and is designed to provide the locomotive driver with an indication of the quantity of air being delivered by the brake valve to the brake pipe. This provides an indication of the integrity of the brake pipe as well as its relative state of charge or recharge. Brake pipe flow indicator gauges have been associated mostly with electric and diesel-electric locomotives, possibly because it was assumed the operation of longer trains would inevitably follow the introduction of this new motive power.

In New South Wales, a flow indicating device had been invented locally in the 1950s (see chapter 4 on the SL equipment) although in its earliest incarnation it provided an audible cue only. Also, in 1956, the in-house journal *Victorian Railways* reported that two Melbourne railwaymen—Wally Minchinton (Assistant Metropolitan Rollingstock Superintendent) and Arthur Richards (locomotive design section engineer in the Rollingstock Drawing Office)—had been credited with the invention of *"an ingenious device that proved successful after months of research."*[155]. The device was known as a Brake Pipe Indicator; there was no allusion in the VR Westinghouse Air Brake Book of Instructions to "flow" and nothing in the description to the effect that it displayed a brake pipe charging airflow.

The case of the Brake Pipe Indicator is a story in itself. Wally Minchinton had progressed through the ranks of Victorian Railways locomotive engineman to a senior operational position. Unlike most other Australasian and North American railway organisations, the Victorian Railways did not—at least at this time—view the niche operational area of air brake policy and equipment management as being worthy of a specialist technical officer - a supervisory role into which most other railway organisations placed either a professional engineer or at least a specialist technician[156]. Perhaps VR considered the new generation of motive power that sported the diesel version of the tried-and-tested A6-ET brake schedule as basically "more of the same" and not deserving of any special attention. In any case, Mr Minchinton was apparently acknowledged as a competent air brake instructor and something of a specialist on the subject of A6-ET/A7-EL air brake equipment. Perhaps also, a senior person well-versed in what was at the time the contemporary apparatus, might be reticent about having to acquire expertise on an emerging technology such as the 26-L equipment. Could this have been at least one reason for the Victorian Railways' late conversion to the 26-L schedule?

Whatever the background, Minchinton and Richards were either subject to immaculate inspiration, or they had become aware of such devices elsewhere. Both men would have been involved with the introduction of early diesel types to VR and must have known, through Clyde Engineering and their EMD connection, of such brake pipe metering devices, even though—by the absence of these on the early VR diesels—one must assume that VR had originally decided not to accept the flow indicator as an option. Perhaps the two men sought to develop a version they felt was uniquely appropriate for VR.

The *Victorian Railways Magazine* described the device thus (and, in doing so, introduced a reference to air "flow", about which the enginemen's manual is silent):

Their brake pipe indicator is simple, effective, and comparatively inexpensive to install and maintain. It consists of an illuminated gauge without calibrations, incorporating the usual type of Bourdon tube used in pressure gauges. The indicator is connected to the brake system in such a way that any flow of air into the brake pipe is immediately registered...

155. *Victorian Railways Magazine*, January 1956, page 3.

156. Until the eventual appointment by VicRail of a professional engineer as 'air brake inspector', what might be generally thought-of as the air brake 'engineering' function was undertaken by Design Office staff.

The VR Westinghouse Air Brake Book of Instructions described the device as providing locomotive drivers with an immediate indication that an irregular brake pipe pressure reduction had occurred. The gauge, lacking any calibrated scale, was connected to the M3 feed valve on the brake pedestal and presented a rotating mask for adjustment as required by the engineman (via the knob marked **D** in Figure 264).

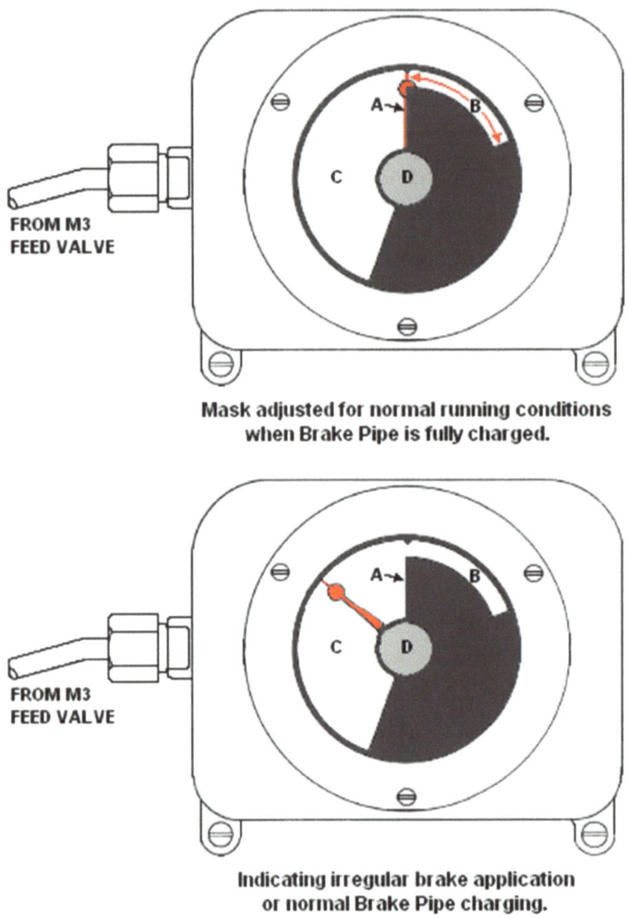

FIGURE 264: VICTORIAN RAILWAYS BRAKE PIPE INDICATOR

While the brake pipe was being charged, the needle would be somewhere within the face of the gauge marked **C**. As the brake pipe charge continued, the needle would rise clockwise to a point that indicated the brake pipe was fully charged. At this point, the Regulating portion of the M3 Feed Valve would be causing the needle to indicate Main Reservoir pressure. This reading would fluctuate on the dial in the cutaway section of the mask labelled **B** in accordance with the normal variation of MR pressure as controlled by the compressor governor. The engineman rotated the mask, adjusting it so that its edge marked **A** just covered the needle when the latter was indicating its lowest normal MR value, just prior to the compressor loading.

When the main reservoir pressure was at its minimum value and the M3 Feed Valve opened to replenish any reduction in brake pipe pressure arising from normal leakage, the needle might move slightly to the left away from edge **A** of the mask. In this case the locomotive driver would now rotate the mask slightly to the left to just cover the needle (at its lowest "normal" position).

With the mask thus correctly adjusted, should any irregular brake pipe reduction be initiated (for example, from a passenger emergency, conductor's, or guard's application, or brake pipe separation or rupture), the needle would drop to the left away from edge **A** of the mask. The warning cue was visual only, and the driver was then responsible for ascertaining the reason for the irregular brake application.

By way of comparison, an early (pre-26-L) EMD brake pipe flow indicator is depicted. This gauge provides a moving scale incorporating a fixed (painted arrow) scale indicator and a needle that is free to move. The moving scale fluctuates with variations in main reservoir pressure, while the movable needle indicates brake pipe pressure. In this way, the needle moving across the fixed scale on the moving indicator depicts a relativity between brake pipe and main reservoir pressures, the latter fluctuating according to the degree of brake pipe depletion and air compressor activity.

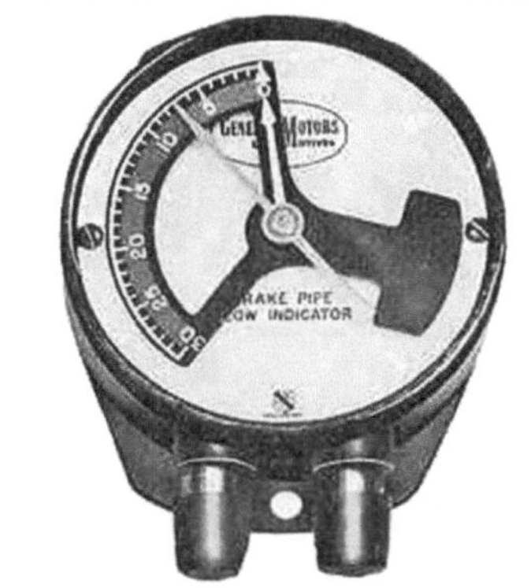

Partially Charged Train, Or Reduction Made From Rear End Of Train

FIGURE 265: EARLY EMD BRAKE PIPE FLOW INDICATOR

Locomotive drivers in North America have not traditionally been supplied with quite the same assistance as our Aussie and Kiwi drivers; they were not always given the audible cue. Their arrangement (assuming 26 or later equipment, since earlier brake systems used slightly different pipe connections) utilises a small check-valve-and-choke device called a "flow indicator adapter" located in the main reservoir pipe leading to the locomotive brake valve and connected directly to the flow indicator. Consisting of a check valve and parallel orifice, this adapter operates with the characteristics of a venturi such that the velocity through the venturi tube when a charging airflow is present reduces pressure in the second side branch by a proportional amount, producing a pressure differential across the choke orifice. Using two pipes, this discrepancy—a relatively weak signal—is fed to and displayed on the flow indicator.

In North America, a brake pipe flow indicator is a distinctive, box-shaped device that could, perhaps, best be defined as a *manometer* since its display scale provides for indications of relatively small degrees of magnitude. Regardless, the American version of this device is entirely adequate for its task. The display includes a movable pointer that can be set by the locomotive driver to any significant flow value that, if attained, will trigger the illumination of an amber warning lamp. This flow indicator is visual only, providing no audible cue and is connected as shown in Figure 266.

The non-provision of an audible cue is, perhaps, understandable given that freight car control valves in North America generally provided the full Emergency feature and any major disruption to the continuity of the brake pipe would become directly apparent as the air "dumped." Within Australasia, train operating practices and thus brake pipe flow display requirements differed. Since the full Emergency capability has only been provided on some rollingstock in recent decades, Australian and New Zealand locomotive drivers were traditionally provided not only with a flow indicator that afforded a more robust visual cue (greater flow-needle deflection utilising a full gauge dial), but also—courtesy of the **flow meter**—with that all-important audible cue that has become so recognisable to generations of enginemen as the familiar *chuff-chuff-chuff* noise signifying the degree of brake pipe charge or recharge in its secondary and final stages.

Westinghouse Australia had long produced a proprietary version of a flow indicator adapter, but these days the A-19 is used almost universally. As mentioned above, though, this device can only produce a relatively weak pneumatic signal, thus the flow meter is provided to amplify the adapter's output such that a more dynamic visual signal is available to the driver and (perhaps more importantly) to provide the "noise."

The flow meter (Figure 268) utilises a flexible diaphragm having main reservoir pressure on one side via passage FM1 and venturi side branch pressure on the other via passage FM2. Any pressure differential across the diaphragm resulting from a main reservoir charging flow through the adapter as previously described will result in an output signal via FM3. This latter pressure will induce movement of the red "flow" needle, while the flow meter exhaust produces the characteristic noise. The main reservoir signal from the first side branch of the adapter (FM1) also supplies the black MR needle. The flow indicator scale, graduated but unmarked, is used as a comparative reference only.

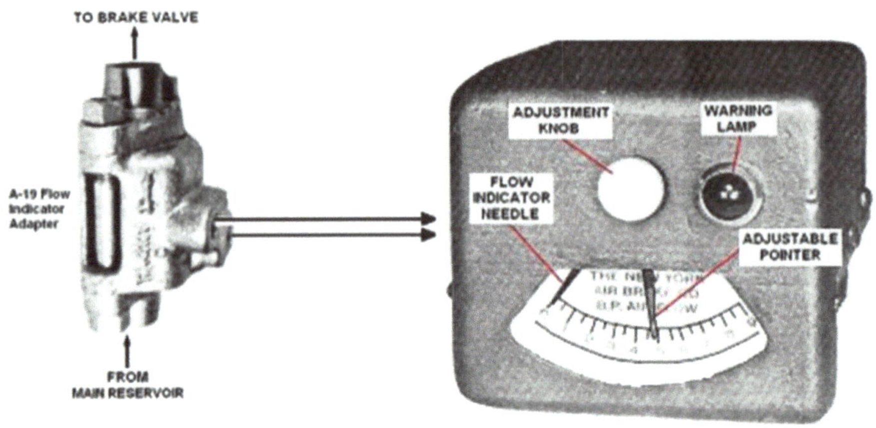

FIGURE 266: CONTEMPORARY NORTH AMERICAN ANALOGUE BRAKE PIPE FLOW INDICATOR

FIGURE 267: NORTH AMERICAN BRAKE PIPE FLOW INDICATOR. THIS PHOTO SHOWS A CANADIAN NATIONAL RAILWAY EMD SD60 WITH THE ORIGINAL DESK-TYPE 'COMFORT CAB' CONTROL ARRANGEMENT INCLUDING 30-CDW BRAKE VALVES. THE FLOW INDICATOR GAUGE IS MARKED

Considering that the Main Reservoir/Brake Pipe duplex pressure gauge is typically located (on the control stand or console) adjacent to the flow indicator, one could argue that the main reservoir pressure indication on the flow-meter gauge is somewhat redundant.

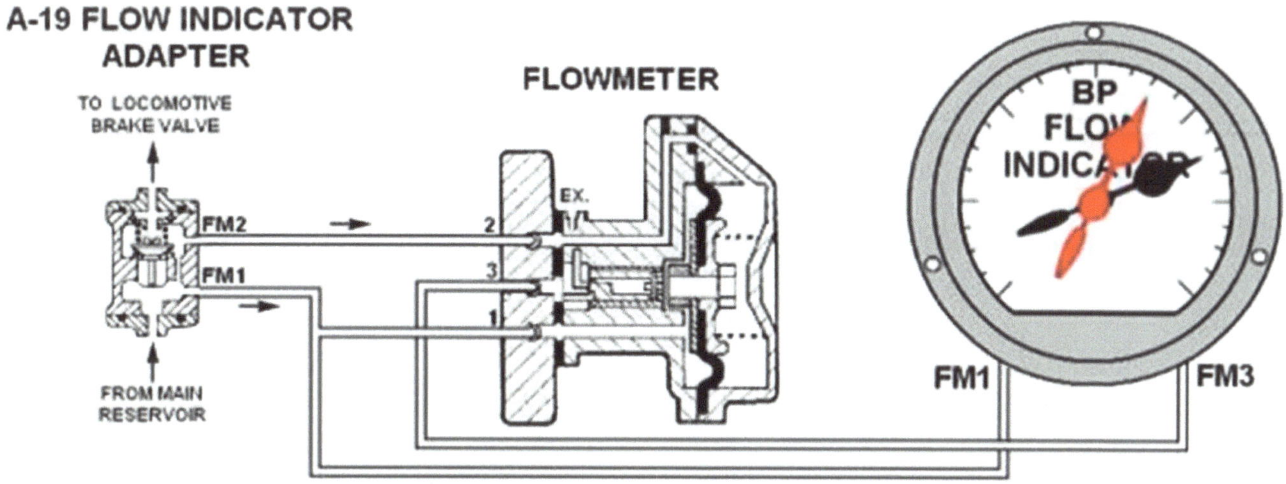

FIGURE 268: AUSTRALASIAN BRAKE PIPE FLOW INDICATOR ARRANGEMENT - TYPICAL

The Long and Short of It

It is an acceptable generalisation—I contend—that short trains are easier to run than long trains; the aspects of train handling discussed in the preceding chapter being simpler to manage on a train with a shorter rather than a longer brake pipe. A major exception to the above simplification, though, occurs when the short train is particularly heavy. For the purposes of this discussion, I am going to boldly declare a personal view: trains in present-day Australasia only "begin-to-become-long" when they have more than 700 metres (say 2 000 feet) of brake pipe. Of course, every train-operating environment is comprised of its own unique characteristics, and I recall running freight trains of between 500 and 700 metres in length in New Zealand during the busy 1970s—not long trains by some standards—that sometimes, nonetheless, consisted of up to 70 wagons. That was 70 mixed AF, early WF, and possibly a few lingering ITVs to contend with during air braking. I did not again encounter trains as "flexible" as these until I arrived in the Pilbara! With (in those far-off days) a Guard riding in the van at the rear behind 69 relatively lightweight chopper couplers, any mishandling by the driver was transparently obvious and was going to be common knowledge by the time one made it to

FIGURE 269: A NEW ZEALAND RAILWAYS FREIGHT WITH MIXED 4-WHEEL AND BOGIE WAGONS ROLLS AROUND THE FOREST LAKES CURVES NORTH OF WELLINGTON IN MARCH 1983. THE CONTAINER FLAT WAGONS ARE EQUIPPED WITH HEAVY-DUTY CHOPPER COUPLERS. AHEAD LIES 650 KM OF MOSTLY UNDULATING COUNTRY, MUCH OF IT HEAVILY GRADED AND TIGHTLY CURVED. THIS WAS A CLASSIC WF TRIPLE VALVE OPERATING ENVIRONMENT AND THIS IMAGE IS A CLASSIC DEPICTION OF THAT AND REVEALS WHY NZR WERE AT THE FOREFRONT OF THE DEVELOPMENT OF THE WF TRIPLE VALVE. *D.M. Cole*

the depot office and booked off duty! To some shaken Guards (and, in such instances, their contrite drivers), *these* were long trains! So, train length is relative, and when we consider it in the broad context of air braking and train handling, an interesting paradox arises: longer may sometimes be shorter!

For the locomotive driver, the major considerations of "long train" handling with respect to air braking are:

- The challenge of achieving an *effective* brake application and concomitant release without "coming to a stand" (i.e. making an undesired stop). In such cases, an enforced stop will be required if the length of brake pipe and number of wagons in the train combine to result in a brake pipe exhaust that is sufficiently lengthy that the brake application stops the train before a state of "brake pipe equalisation" is reached; and
- To ensure that train speed is still high enough, and the train is stretched, to be able to make a safe release.

This last point is important with long conventional trains (i.e. head-end power only), since reaction of the operating valves to the brake pipe "Release" signal is somewhat sequential from front to rear, and the danger always exists, no matter how rapid the propagation, that—with the forward portion of the train more advanced in its release than is the rear portion—bunched slack can be extended violently, resulting in a broken coupler or a "pulled" drawgear package. A slow-speed brake release with a long train is generally something to be avoided, and historically this phenomenon (or at least its possibility of occurring) has had a substantial effect upon the manner in which train operations are conducted on some railways. Some US railroads that operated over relatively undemanding territory ran extremely long, relatively underpowered trains known as "drag freights." These operators accepted the inevitable fact that any air brake application could not safely be released "on the run" and was going to result in the train coming to a stand. Under these circumstances, use of the "air" while running over the road was generally prohibited unless a stop was intended or could otherwise safely and practicably be accomplished before a positive brake release was initiated.

A similar situation existed in Western Australia in the late 1960s and during the 1970s. In this latter case, the standard-gauge rail connection to the eastern states was new and the West Australian Government Railways' experience with air braking likewise adolescent. An unfortunate result of the dogged perseverance of some eastern railways with the prehistoric Improved Triple Valves and, to a lesser extent, the slightly less-outmoded AF triples (not to mention some infamous maintenance standards previously discussed) was that interstate freight trains often arrived in WA sporting a rather antediluvian assemblage of triple valves. Westrail operating officials were cautious as to the performance standard that might be expected from such a conglomeration of equipment on the one train. Given their lack of corporate experience with air braking, their instruction to locomotive drivers, perhaps understandably, was pragmatic and similar to the American drag freight operators: "Don't use the air brake if you can avoid it - and if you can't avoid it, then stop and release."

The West Australians were not alone in this position as the trans-Nullarbor Commonwealth Railway held a similar view; indeed, it was the CR who received the (typically) 75-wagon mixed interstate freight trains of that era from the New South Wales and Victorian Railways, forwarded them across the Nullarbor, and handed them over to Westrail at Kalgoorlie for onward dispatch to Perth.

The above advice to drivers may well have been prudent; however, the instruction does not appear to have been accompanied by a coherent explanation, as the end result of this culture is an unfortunate and pervasive view—still prevalent across Australia—that good triple valve performance on long trains, even under appropriate circumstances, cannot be assumed. In my view, this and other similarly prosaic notions concerning the applicability of other valid but sometimes "unfashionable" train-handling techniques (e.g.

appropriately used power or "stretch" braking, to be discussed later) are incompatible with the general need for rail operators to permit their drivers an acceptable degree of flexibility and discretion in terms of train-handling practices. Unnecessarily prescriptive managerial viewpoints toward train-handling practices can be counterproductive and will only be dispelled by a greater appreciation among train operating supervisors of contemporary air braking philosophy and equipment functionality, plus a more detailed comprehension of train-handling theory. Such a degree of awareness, I am certain, is a fundamental responsibility of railway operating managers and trainers. That said, it has to be acknowledged that the new ECP technology—as a transformational innovation—will change some rules. One of these, arising from an expected increase in brake block consumption, is likely to be that limitations on the use of air braking (such as a doctrine that air braking always be accompanied by dynamic braking as far as possible) will gain credibility.

The above sentiments notwithstanding, there is at least one other train operation in Australia that is currently notable for its requirement of limited air brake use. Since the standard-gauge line to Leigh Creek in remote northern South Australia was completed in 1957, coal trains servicing the Port Augusta power stations from this mine have increased in length to 161 wagons. This is often claimed to be the longest coal train in the world, and the claim may be true. Certainly, in the Australasian context, a 161-wagon consist is an extremely large and heavy head-end power train, this example being about 2.84 km long and over 14 000 gross tonnes.

FIGURE 270: AN EMPTY COAL TRAIN ROLLS NORTH AWAY FROM THE PORT AUGUSTA POWER STATION AND TOWARD THE MINE AT LEIGH CREEK. *Pacific National*

Unlike the lengthy Pilbara iron ore trains with head-end power, the modern Leigh Creek "coalies" have historically been equipped with Australian Westinghouse WF triple valves; indeed, the challenges of this particular operation have been directly responsible for many of the incremental improvements to the WF triple valve over the years, most particularly the introduction of the piloted reduction-ensuring feature developed by Jack Washbourn. This and other operational augmentation features enabled the operation of longer coal trains but still did not provide complete flexibility of control. If the air brake had to be used, Commonwealth Railways (later to be Australian National Railways) would only allow a release to be made from a Full Service application. The situation did not improve when the NSW agency FreightCorp (later to become a part of Pacific National) took over operation of the Leigh Creek coal trains. Use of the ANR gondola wagons with their contemporary WF-type triple valves ceased, and into this coal train service came NSW hopper-type wagons with triple valves of lesser sophistication.

During testing, FreightCorp found that on their coal train, a brake application of any magnitude less than Full Service did not result in a complete response of all triple valves throughout the train. To overcome this disadvantage, FreightCorp introduced an active End-of-Train Monitor attached at the rear of the last wagon. This device, utilising radio telemetry, reports the state of the brake pipe at the rear of the train as well as last-car "movement" and responds to brake operation on the lead unit to make Service brake applications at the rear of the train. Although—for the purposes of air braking—this practice has the effect of halving the length of the train, the lengthy brake pipe exhaust for any brake application continues to ensure that running releases still cannot be made. Additionally, of course, the EOTM is unable to assist with recharging the brake pipe and, therefore, cannot contribute to brake releases.

On this point, the last word might be given to three Australian engineers with considerable air brake expertise. One of them has concluded—in a comprehensive discourse on some differences in operation between US and Australian operating valves[157]—that, since 1989, there has been a dearth of research into improvements to brake application delay times, especially in Australia, as wagon and total train brake pipe lengths have increased. The author of this thesis contends that further research into the assistance of the propagation rate of brake pipe reductions on Australian brake systems is needed and that he has demonstrated that with modifications to the brake system, improvements to the performance of brake application delays are achievable. Taking up his challenge, two others have shown that it is possible that a relatively simple modification can be made to the wagon brake system to ensure the transmission of the brake signal throughout the train, regardless of its length[158]. It remains to be seen how soon or widely these initiatives will be applied.

The last twenty years of railway operations in Australia have changed the meaning of "train length" in the context of air braking and related train-handling issues. Except for the giant iron ore trains in the Pilbara (and even there we shall gain some insight, as shall be seen), the number of wagons no longer necessarily defines the length of a train. Instead, for example, with an intermodal freight train, one now often refers to "platforms" or—for crossing purposes—a suitable linear value. A long (let's say 1 500 m/4 500 ft) interstate freight train may at one time have consisted of, say, 75 individual wagons; each with a triple valve, an auxiliary reservoir, and a brake cylinder. An air brake application on this train resulted in the operation of 75 triple valves that responded by directing part of the contents of 75 auxiliary reservoirs into at least 75 brake cylinders. To effect a release and to recharge the brake pipe, this volume of air (and perhaps a little more to top up any Emergency or Accelerated Release reservoirs) had to be replaced, and this took time - and fuel! An effective re-application, should it be required, might not be available until the brake pipe recharge was at least well-advanced.

By comparison, a modern 1 800 metres-long intermodal or coal train may consist only of some thirty wagons due to the increasing use of articulated multi-platform and semi-permanently coupled, multi-unit wagons (e.g. a five-unit container well-wagon has five individual "platforms"—in this case, *wells*—each capable of holding two ISO containers stacked one atop the other, while a single five-unit coal wagon consists of five separate hoppers). Other multiplatform wagons include *2-*, *3-*, and *5-Packs*, each providing 2, 3, and 5 platforms, respectively - either wells or flat-tops (and sometimes a combination). But overall, due to fewer brake pipe hose couplings and other convolutions to the run-of-pipe, *the total brake pipe is materially shorter*. In the latter case, an air brake application may only result in the activation of 30 triple valves

157. Ian Ripley, *An Investigation of Brake Application Delays in Australian Train Brake Systems* (Central Queensland University, 2004), pp. 46, 97, and 99.

158. M. McClanachan (Central Queensland University) and B. Payne (QR), *Improving Brake Propagation in Long Freight Trains* (CORE, 2008).

(perhaps more, since some of these long, multiplatform wagons require either more than one triple valve or the addition of an auxiliary brake pipe reduction device). So, although the train may be physically the same length, there are only 30 (or 35 or 40) auxiliary reservoirs feeding air to the various brake cylinders as against 75 in our first hypothetical train. In addition to a shorter elapsed time for the brake pipe exhaust, since the volume of air to be replaced during brake release is much less, the release will occur quicker and the recharge completed sooner, making the driver's task easier.

Conventional (head-end power) "long train" operations in Australia are currently typified by the SCT Logistics freight services—usually around 75 wagons—running between Melbourne (and Parkes) and Perth; the 161-wagon coal trains operating between Leigh Creek and Port Augusta; and some of the coal operations in the Hunter Valley, around 80 wagons. Another significant long-train operation that requires a high level of train-handling finesse due to the extreme roller-coaster gradient profile of the route and the standard train mass, is the ARG (Aurizon) iron ore haul for Portman Mining over that portion of its route between West Kalgoorlie and the port of Esperance. These trains convey 120 Pilbara-style ore cars over some exceptionally challenging countryside, including a permanent way that was—when I operated these trains over it during the period 2002–2004—barely fit for purpose. Significantly, QR National's operation in the Hunter Valley is the setting for Australia's first ECP braking (apart from the BHP Billiton railroad in the Pilbara), currently 74-wagon consists.

FIGURE 271: PHOTGRAPHED FROM THE LEAD LOCOMOTIVE, SCT LOGISTICS TRAIN 2MP9 ROLLS WESTWARD INTO THE SETTING SUN OVER THE REMOTE AND UNDULATING TERRITORY BETWEEN PORT AUGUSTA AND TARCOOLA. THE TRAIN IS 74 WAGONS/1 739 METRES/5 593 TONNES, AND IS EQUIPPED WITH WF TRIPLE VALVES. *F. Moffat*

The Pilbara iron ore railroads provide another example and the opportunity for a classic comparison. Prior to changing from conventional operation to a full LOCOTROL distributed power schedule during the mid-1980s, Mount Newman Mining (now BHP Billiton Iron Ore) had standardised on 192-car trains (about 2 100 metres long) while, at the same time, another major Pilbara producer, Hamersley Iron (now part of Pilbara Iron and most latterly named Rio Tinto), was running 210-car trains (about 2 200 metres long). This longer HI train, though, would have been somewhat different to handle from an air braking perspective because HI operates their ore car fleet in double-unit "married pairs", whereby each two semi-permanently coupled ore cars share a single air brake operating set (control valve/auxiliary reservoir, etc.), although there is a brake cylinder supply reservoir and separate relay valve on each car. Therefore, a Hamersley Iron driver making an air brake application on a full relay-equipped consist—reducing pressure in a brake pipe of similar total volume to the MNM train—was depleting 105 auxiliary and 210 supply reservoirs (i.e. a total of 315 separate volumes) that required to be recharged when the brake was released. By comparison the MNM driver—actually operating a physically shorter train (also a full relay-equipped consist)—depleted a similar-length brake pipe but was then required to recharge 192 auxiliary reservoirs plus 192 brake cylinder supply reservoirs - or a total of 384 separate volumes. Thus, the MNM train used more air for the same outcome resulting in greater time lags and elapsed times during air braking events.

In the train-handling context, it could be argued that this, in turn, required greater effort in planning and executing these events. Note that during the brake application, for both trains the elapsed time for brake pipe exhaust would have been similar due to the action of the car control valves' *accelerated application* feature whereby during the Minimum Reduction, only a relatively modest amount of air is required to exhaust through the driver's brake valve, the balance occurring locally at the individual control valves.

The boot, though, went onto the other foot when, during the mid-1980s, MNM moved to a full distributed power schedule. The standard train length initially became 270 cars (almost 3 000 m/9 850 ft) before operational considerations saw it reduced to 240 cars (2 700 m/8 860 ft). However, these trains now had a *Remote consist* of locomotives situated roughly midway in the body of the train, and these units would react synchronously during air braking events to provide a second point within the train for brake pipe supply and exhaust. Thus, the MNM locomotive driver, while now running the longer train in a physical sense, enjoyed the coordinated operation of what were, effectively, two quite shorter trains coupled together. Contemporary Locotrol equipment—providing for *multiple* Remote unit locations—as well as ECP braking (which provides almost infinitely variable graduated release and which has also been tested in revenue operation at BHPBIO) are two technologies that have only multiplied this potential train-handling benefit, to which some modern-day BHBIO and Aurizon drivers will no doubt enthusiastically attest.

Less can, indeed, sometimes be more!

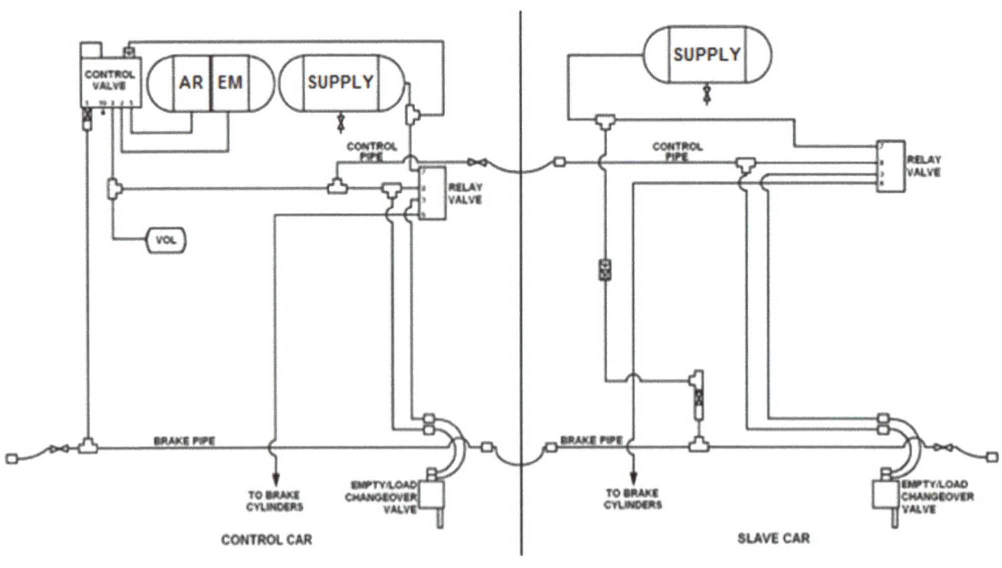

FIGURE 272: HAMERSLEY IRON (PILBARA RAIL) BRAKE CYLINDER RELAYING SYSTEM APPLIED TO "MARRIED PAIR" ORE CAR-SETS

Figure 272 depicts the AAR brake cylinder relaying system as applied to Pilbara Iron double-unit ore cars. These cars are semi-permanently coupled using solid drawbars, with one control valve operating the brakes on both cars. This is an efficient operating practice that has long served Hamersley Iron well;

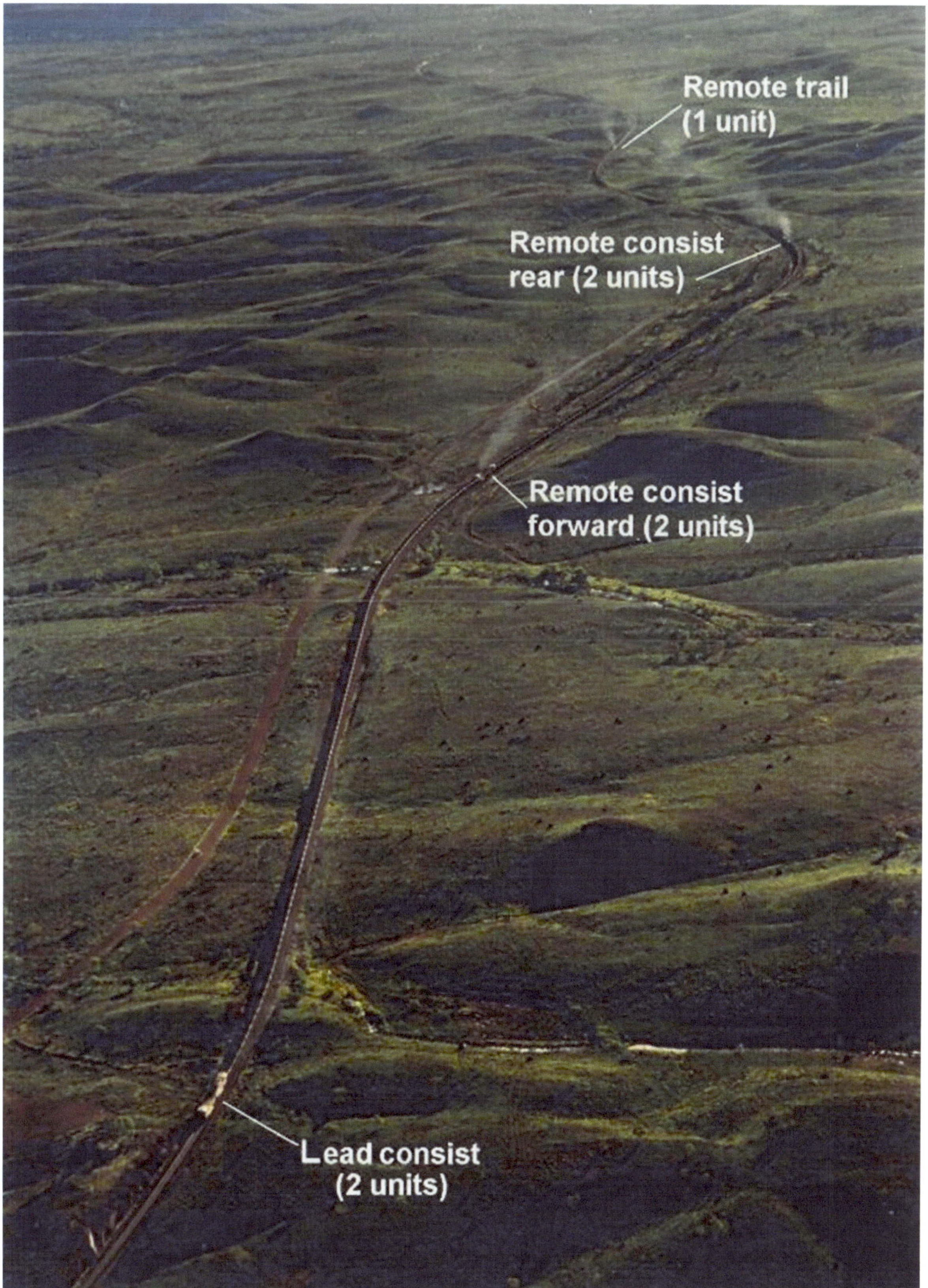

FIGURE 273: A 350-CAR LOADED ROBE RIVER IRON ASSOCIATES (NOW PILBARA IRON) Locotrol III TEST TRAIN CLIMBS OUT OF THE ROBE RIVER VALLEY EN-ROUTE TO THE PORT AT CAPE LAMBERT

however, from a brief look at the diagram, the interesting question arises as to how a Slave car, if running as the last car on a train, might be controlled were it to break away from its Master car (e.g. in the event of a drawbar or draw package failure). In such a circumstance, although the train's brakes might be in "Emergency", the loss of control pipe pressure on the Slave car would mean that it could not develop brake cylinder pressure. Should the train be moving upgrade, the separated Slave car would be free to run backward, uncontrolled.

Figure 273 depicts an exceptional distributed power train; around 3 800 m/12 500 ft long, and with a mass of 45 500 tonnes. With 7 x GE Dash 8 units, the train is configured as 2 units + 120 cars + 2 units + 115 cars + 2 units + 115 cars + one unit. Therefore, with regard to air braking, this entire train will handle as if it were about 58 cars in length (half the average brake pipe length between locomotive consists).

A further fundamental aspect of air brake handling is consideration of the "state-of-charge" of the brake pipe in its entirety, since the expected performance of operating valves is dependent upon a good brake pipe pressure signal, and reservoir recharge is dependent upon the supply of an appropriate volume of air. Leakage in car and wagon piping and devices—often to a subtle degree in individual vehicles—and mechanical friction can inhibit airflow within the brake pipe, creating a pressure differential between the front (close to the point of pressure supply) and the rear of a standard-configuration train. This condition is referred-to as *brake pipe gradient* (known colloquially as *taper*), and it has historically been crucial to considerations of the effectiveness of operating valve performance between the front and rear of longer trains, since its existence results in a lower air pressure toward the rear.

The operating characteristics of modern-day air brake operating valves as well as the increasing use of multi-platform and multi-unit cars are such that these considerations are less important than they once were. (Note: There is a difference; multi-platform cars are articulated with bogies being shared at the intermediate couplings, whereas with multi-unit cars the constituent portions are connected by solid drawbars and do not share bogies.) Indeed, brake pipe gradient is nowadays a less prevalent, or at least a less significant, phenomenon than has historically been the case. Nevertheless, it can be considered good air brake practice to allow for it when making brake pipe reductions. This is yet another factor for the locomotive driver to consider when initiating an air brake application, and the following diagrams—with the scale exaggerated for effect—are provided by way of illustration.

The first diagram shows that if the brake pipe is not fully charged throughout, then a reduction in pressure at the head-end of the train may not be replicated on a rearward portion (even taking account of Quick-Service activity), with the result that a certain portion of rear-end operating valves may not receive an "application signal." In this example, if the driver makes a Service brake pipe reduction of, say, 70 kPa, the rearmost wagons or cars will not apply. As a result, the driver, will not receive the degree of total train braking intended and will probably incur unwanted and potentially severe "buff" slack action as the rear wagons run into and bunch up on the braked wagons ahead of them.

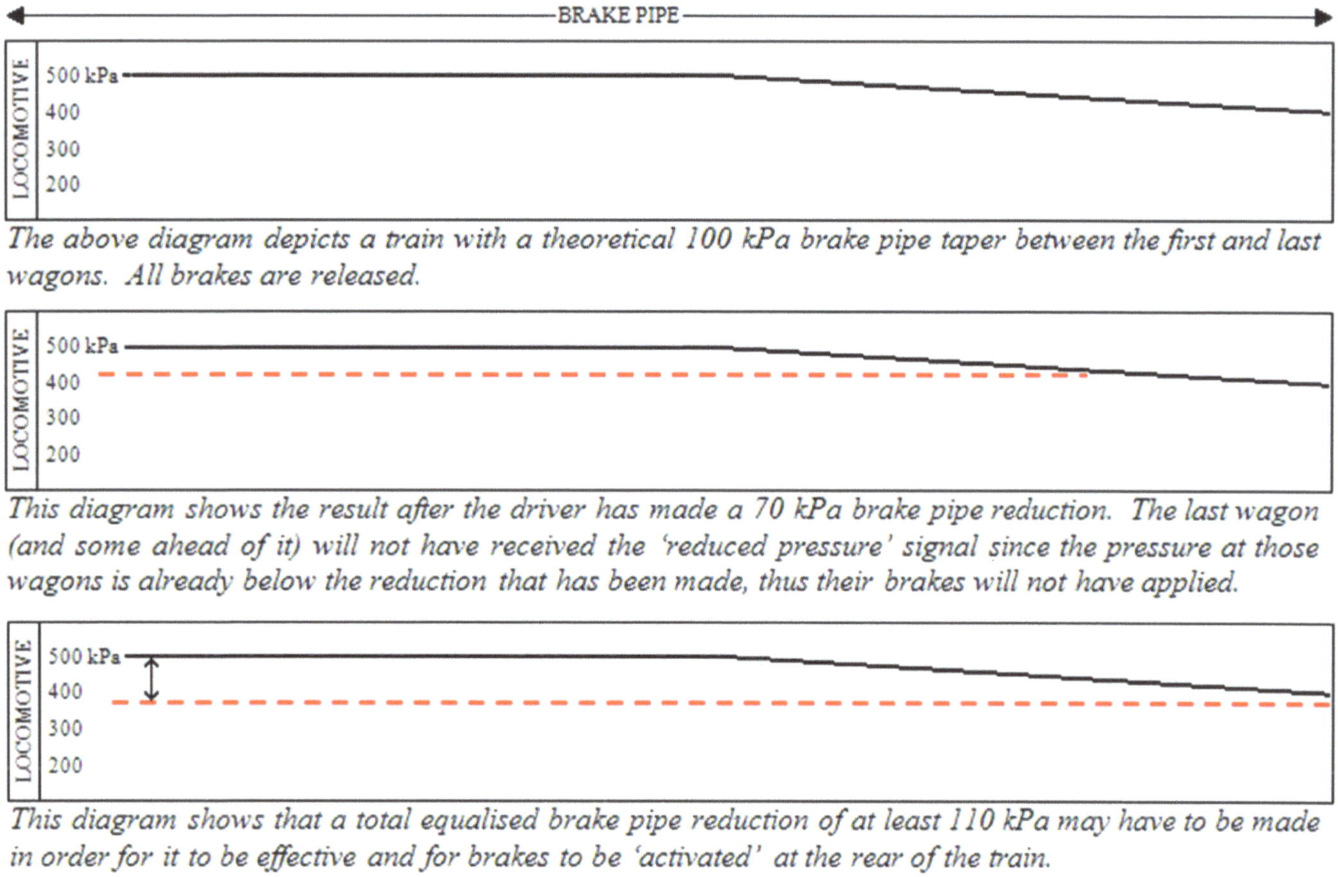

FIGURE 274: THE EFFECTS OF 'BRAKE PIPE GRADIENT' WITHOUT *ACCELERATED APPLICATION*

Train Operating Practices

It is often stated, when locomotive drivers are discussing aspects of train handling, "If there are ten individual drivers, there are ten different ways to run the section." The drivers' position, inevitably, is that each of them is equal in ability, that each is entitled to their opinion, and that each has an equal capacity to operate successfully while pursuing his or her own train handling strategy. I beg to differ.

There are parallels between train handling and commercial aviation, insofar as—in almost all things—there is one best way to achieve any given outcome even allowing for the vagaries of "human nature" and personal preference. Aviators will call this "flying the numbers", which means operating to parameters based in scientific law rather than the pilots' own opinion of what he or she should do at any given point in a handling sequence. Since there are laws of mechanics and fluid dynamics as well as considerations of the management of energy that are pertinent to train handling generally—and its most critical aspect, air braking, specifically—locomotive drivers are no less bound by the laws of science and physics in their endeavours at running trains.

There may indeed be more than one way to run a train through a particular territory. However, depending on such variables as permissible track speed, the presence or not of a speed restriction, and train length and mass on the day, this choice will likely consist of no more than two options where speed retardation is required; whether to **stretch brake** or **bunch brake**. Whichever option is more appropriate, there is but one set of laws by which the desired outcome can be satisfactorily achieved in any given set of circumstances. Usually, this "one best way" can be prescribed and taught easily (assuming there are competent trainers) and measured and monitored effectively (assuming there are proficient supervisors). It then becomes incumbent upon locomotive drivers to do what is required within the limits permitted by—and subject to the provision of a necessary degree of latitude arising from—the disparity in individual performance and competency standards. It may be this latter measure of variety that gives rise to the view among drivers that there are diverse methods by which to run a train over any given section of track. In fact there are not, but credible and appropriate train-handling strategies need to be disseminated by knowledgeable and capable operations managers, training staff, and supervisors.

Stretch braking

Strategically, all train braking falls into one or the other of the two categories mentioned in the preceding paragraph. Of these categories, *stretch braking* is by far the most contentious and arguably the most misrepresented and misunderstood. Also known as "power braking", it is guaranteed to provoke much emotion and impassioned comment. The very words seem to conjure up threatening imagery of trains being operated with the brake applied and the throttle wide open, brake shoes smoking, engines notched out in a thunderous cacophony, and couplers straining at their tensile limits. This agitation, of course, stems from concerns regarding fuel use, brake shoe wear, and possible excessive coupler tension. While these are valid concerns wherever this train-handling method might be used at an inappropriate intensity, their contemplation in the extreme should not be permitted to override the rational consideration of power braking as a legitimate and reliable train-handling method.

However, one can perhaps understand that confusion exists given the span of expert opinion on the matter, some of it divergent. Consider the following four passages, each with its individual context, each from a credible industry source:

Slow-downs can be made using either the slack-stretched or slack-bunched procedure. The choice of method depends largely on the state of the slack prior to initiating the slowdown and the slack is in turn influenced mainly by the grade characteristics of the territory over which the train is moving, and to some extent by the existence of sharp curvature, particularly reverse curves.

Stretch braking has the advantage that it can allow a train to enter or exit from a particular slowdown location more rapidly than if bunch braking is employed. It is disadvantageous, of course, in that by working additional power in the braking operation there is a corresponding increase in fuel consumption and brake shoe wear.

—From **The Requirements of Good Train Handling for Heavy, Long Trains**.
A technical paper by the late Charles W. Parker, (then) Chief Motive Power Engineer, Canadian Pacific Rail, consulting to Mount Newman Mining Co. Pty Ltd (BHP Billiton Iron Ore Railroad Library, TR 00 01522)

There are situations where braking against power can be used to obtain satisfactory speed reductions and train handling. However it must be recognised that it is not always possible to keep every train stretched at all times. Severe strains on couplers and draft gears can result if proper attention is not given to the load indicator as speed reduces... Throttle should be reduced as speed decreases to prevent tractive effort from building up too high...

—From **Management of Train Operation and Train Handling.**
The Air Brake Association, reprint 1993, p. 153

Working power with the locomotive consist against the action of brakes and resistance can provide the net retarding force needed to hold speed steady or make minor speed reductions on the downgrade while running over undulating terrain. This type of train handling is called "power braking" or "stretch braking" and is widely practiced all over North America. Obviously, extreme caution must be used in applying power to protect against pulling the train apart. This requirement to use varying amounts of power and braking at the same time to suit a wide range of [track] *profile conditions is the reason why single-handle power and brake control does not seem to be practical on road locomotives. Train slack is a potential problem with the centre-of-car automatic coupler and draft gear system...*

Change in train slack may be appreciably snubbed by the action of the brakes. The train is kept in a generally stretched condition as it comes over the crest, usually with full power on... The serial action of the train brake application from front to rear is desirable as the train progressively passes down through the vertical curve to the basic downgrade.

Obviously, timing and location of the start of the application and release are very important. The engineman must have a detailed knowledge of the railroad profile [and] *must also be able to relate his train length, tonnage and rollability to the specific profile. Response of the train to power and braking is important. The experienced engineman will find the opportunity to make a running test of the train brakes as soon as practical after leaving the initial terminal.*

—From **a paper presented to the 1978 conference of the Air Brake Association** by the late David G. Blaine, Senior Marketing Representative, WABCO[159]

Keeping the throttle open while braking to keep the train stretched is believed unnecessary and foolish for braking a train of high-class passenger equipment. It is a waste of fuel and an abuse of equipment to attempt to accelerate and decelerate at the same time. If excessive slack exists in the train it may be desirable to use the throttle while braking to keep the train stretched but care must be taken to observe the ammeter and prevent traction motor current from increasing while brakes are applied by reducing the throttle position.

—From **Management of Train Operation and Train Handling**
The Air Brake Association, reprint 1993, p. 186

This last passage illustrates my earlier point that discussion of power braking can sometimes lapse into the extreme. The passage opens with two sentences of emotional prose against the use of power braking but then moderates to sagacity in offering a scenario in which the method is acceptable and provides some prudent advice on its employment. The fact remains that power braking is a legitimate train-handling tactic that, as in all things, has its place and its limits and limitations. These need to be competently taught and credibly enforced, since air braking competency—in all its permutations—remains a fundamental skill requirement for locomotive drivers. Despite what may be spoken or written by misguided individuals at times, the **air brake** must still be acknowledged as the primary braking system on the train, and locomotive drivers should be permitted and expected to maintain proficiency in *all* aspects of its use.

In one of the passages quoted above, David Blaine states, "*...extreme caution must be used in applying power to protect against pulling the train apart.*" In this case, the sentiments are sound, the language itself somewhat extreme. Perhaps just an instructive "caution" rather than an emotive "extreme caution" might have sufficed; common sense, after all, is what is required. Let's allow one locomotive operator's manual (for an EMD model) to provide a final word. Under the section headed **Air Braking with Power** are these wise words for the driver:

The method of handling the air brake equipment is left to the discretion of the individual railroad. However, when braking with power it must be remembered that for any given throttle position the draw-bar pull rapidly increases as the train speed decreases. This pull might become great enough to part the train unless the throttle is reduced as the train speed drops. Since the pull of the locomotive is indicated by the amperage on the load meter, the engineman can maintain a constant pull on the train during a slow-down, by keeping a steady amperage on the load meter. This is accomplished by reducing the throttle a notch-at-a-time whenever the amperage starts to increase. It is recommended that the Independent brakes be kept fully released during power braking. The throttle MUST be in Idle before the locomotive comes to a stop.

159. Dave Blaine had a long involvement in the design and development of Wabco control valves and with railroad air brake operations. Refer to chapter 7 for a biography.

I recall learning these skills at an early stage in my career. They have stood me in good stead throughout the great variety of my forty-plus years in locomotive and train operations, including thirty years in a manifest variety of locomotive engine service environments.

Bunch braking

Bunch braking is train braking that involves gathering the slack and keeping the train in a state of "buff" as opposed to the "draft" condition discussed in the preceding section. The most common form of bunch braking is *dynamic braking,* by which the locomotive traction motors are converted into generators and forced to rotate—by the kinetic energy of the train—within a strong magnetic field. This is an extremely effective and efficient method of speed control that provides a retarding force against train velocity and, under suitable conditions, can be used to bring a train to a stand.

But bunch braking is also an air brake function by which the brake application is made such as to cause or permit the locomotive to "back into" the train so that slack is gathered (bunched), and the train is slowed or stopped maintaining this state. A running release of the brakes with the train bunched should be carried out with great care (and a full appreciation of the relevant aspects of train dynamics and air brake response), since the release occurs from front to rear, providing the potential for a run-out of slack which, if violent, could part the train. I have seen a crew member catapulted out of his seat and thrown against the front of the cab by violent run-out. A glass water bottle strapped into a refrigeration cooler was likewise launched across the cab and shattered. On another occasion, a large metal toolbox fixed to the cab floor behind the crew member's seat was dislodged to slide across the cab floor. Finally, the most spectacular outcome from violent run-out that I have seen is "stringlining" of the train[160]. Uncontrolled tensile slack action can be dramatic and damaging - and expensive to clean up!

Some rail authorities prefer to promote bunch braking at the expense of power braking, even almost to the extent of banning the latter. I restate my view that locomotive drivers should be expected to maintain proficiency in all aspects of air braking and should be thus encouraged to practice power braking to an appropriate measure.

"Maintaining" braking

Not to be confused with the various pressure-maintenance strategies and sometimes dubious practices used by locomotive drivers in a previous era to air-brake down long grades[161], this is a downhill train-handling method that involves setting a light brake application and maintaining it for an extended time in pursuit of a steady speed. It is possible only with a modern maintaining air brake system (e.g. the 26-L schedule) and is often used in conjunction with dynamic braking, in which case the air set will probably be a Minimum brake pipe reduction with around three-quarters dynamic effort used - this strategy leaving plenty of braking authority in reserve. In this case, the dynamic braking effort can be varied as required to control speed through curvature, grade easements, and speed restrictions. This is an efficient and effective way to bring a heavy train down a long grade when dynamic braking alone is insufficient, since energy dissipation and thus brake shoe heat and wear remains within acceptable limits. *Maintaining* braking is regularly used for descents of all Australian Great Divide and many other gradients and in numerous locations throughout the mountainous New Zealand rail

160. See Glossary, chapter 8.

161. See "The Development of Maintaining Braking" in this section for further information regarding an earlier era of non-maintaining brake valves, when locomotive drivers sought to bring trains down long grades with a minimum number of brake applications and releases.

network. When dynamic braking is unavailable, the air brake application may have to be slightly heavier, and maintaining braking may only be possible with some power applied, that is, by power braking.

FIGURE 275: "MAKE A MINIMUM REDUCTION AT PRECISELY THE RIGHT LOCATION AND SPEED AS THE TRAIN CRESTS THE GRADE." IN THE EARLY 1980'S, A 144-CAR MOUNT NEWMAN MINING ORE TRAIN BEGINS THE 1-IN-50 DESCENT OF THE CHICHESTER RANGE, EN-ROUTE TO PORT HEDLAND. *F. Moffat*

Cycle (or Serial) braking

The knowledgeable reader may have seen this referred-to in North American literature as *short-cycle braking*. It is another train-handling method for downgrade operation and is the principal technique used with older non-maintaining equipment such as the No. 4, 6, and 7 brake systems. Since AH-7 brake valves with the A- or B7-EL equipment are still in regular use throughout Australia, this method is still contemporary. Because these driver's brake valves are unable to maintain the brake pipe against leakage when the brake is applied, any brake pipe reduction—if held long enough—will slowly increase, all the time increasing brake cylinder pressure until the equivalent of a Full Service application is attained. Since this will stop the train and require it to be held—perhaps requiring the application of hand brakes—while the brake is released and the brake pipe recharged, it is problematic and undesirable.

Therefore, to maintain a steady progress down the grade while keeping the train under control, the brake application is only held long enough for it to become effective (i.e., brake pipe becoming settled at Service equalisation) and for train speed to be reduced sufficiently that a release can be safely made. This "release point" has to be finely judged by the driver, since the train now begins to accelerate and the release speed must have been low enough that the brake pipe can be recharged or at least almost recharged to enable the next brake application before the train speed becomes excessive. This method derives its name from this cyclic apply-release-apply sequence. Partial or incomplete recharges must not be made serially, or the train may run away due to insufficient auxiliary reservoir and thus available brake cylinder pressure.

Cycle braking, though, was one instance where brake pipe leakage could be of assistance. I well remember running trains down some of New Zealand's many grades with AH-7-equipped locomotives and just enough leakage that if the brake valve handle was placed directly from Running to Lap position at precisely the right location and speed as the train crested the grade, the application gradually induced by the leakage and concomitant increase in brake cylinder pressure as the brake pipe reduction intensified could balance the speed as the train settled onto the downgrade. In this state, the train could be operated for a considerable distance before release, recharge, and the commencement of the next braking cycle. Thus, brake valve manipulation could be minimised. From many discussions with drivers throughout Australia, it is obvious that this technique is relatively well known.

Brake checking and testing - and the "Full-Service" Obsession

Several test procedures are routinely employed to validate the integrity of air brake systems and correct functioning of equipment. These range from brake system checks made prior to placing a locomotive into service or changing control stations on a locomotive, to testing brake pipe continuity and the performance of equipment on individual wagons. This regime of testing is what validates the air brake system of a locomotive or train far beyond any confirmation of the reliability of a dynamic braking system. It is why the air brake system of the locomotive (and the train) must be considered the *primary* braking system and dynamic braking relegated to a distant *secondary* place. Any railway official who sought to prioritise these schemes otherwise would, at the very least, be an extremely bold individual.

In the global context, there are some interesting anomalies in this brake-testing regime. It is customary for either a train crew or specialist staff to conduct a terminal air brake test and, in certain prescribed cases, a train examination prior to departure. This involves one person walking from front to rear of the train with its brakes applied (typically with a minimum 100 kPa brake pipe reduction), observing certain critical elements on each wagon. Once at the rear of the train, this member then checks the rear-end pressure to verify brake pipe continuity and the brake pipe pressure on the last vehicle (confirmation and measurement of any existing brake pipe gradient) before calling the driver for a brake release. The member then observes brake pipe pressure restoration, and returns to the head-end of the train along its opposite side, similarly inspecting each vehicle now in the "release" condition.

During this time the locomotive driver remains in the cab and in radio communication with the inspecting member and normally conducts a brake pipe leakage test. This involves "cutting out" the brake valve (isolating it from the brake pipe so that pressure cannot be automatically maintained therein, thus revealing any existing leakage) and, by reference to the brake pipe gauge, timing the pressure drop over a period of one minute. Any reduction in brake pipe pressure represents leakage from the brake pipe and/or connected equipment—for example, reservoirs, branch piping, or operating valves—and is not permitted to exceed 35 kPa over that minute. Upon completion of this test, the equalising reservoir pressure should be reduced to fractionally less than the brake pipe (assuming the existence of leakage) and the brake valve "cut in" so that it can maintain the brake pipe reduction at this latest reduced value. The driver is now in a position to effect a brake release when requested by the inspecting member.

At this point, we encounter a curious phenomenon that has always puzzled me; one that I refer to as the "Full Service obsession." Once the inspecting member at the rear-end requests the brake release, local procedure often mandates that the driver increase the brake pipe reduction to Full Service (the maximum Service brake application available) before initiating the release. One ponders why this is considered either necessary <u>or</u> desirable. An effective brake application has existed for the period of the train inspection to a value that itself will promote a good, positive release. The driver needs to know that the train brakes will function correctly in the manner in which he expects to use them throughout the trip ahead. For this reason, the brake test should ideally be released from an effective Service application only, not from Full Service, which the driver does <u>not</u> expect to use (and should have no reason to use) in the normal conduct of the journey. Additionally, since all air pressure used has to be made up at the cost of extra fuel use (even for locomotives with electric compressors), there would appear to be a case for reviewing this requirement for increasing the brake test application to Full Service prior to release. Apart from the unnecessary use of air, one could legitimately pose the question; *What is achieved, during brake testing, by insisting upon increasing the brake application to Full Service before making the release?*

Another example of the "Full Service obsession" is the requirement to hold a Full Service brake application while standing in a crossing loop. Obviously, the quite proper intent is to secure the sidelined train against unintended movement, especially in the face of an opposing train. One may question, though, whether this extreme degree of brake application is warranted. For one thing, the perception that a train with its Automatic brake fully applied is somehow less likely to suffer release or otherwise respond in some way to a release signal than one with the brakes "partly" applied is a furphy. For whatever reason that it may occur, the tiny rise in brake pipe pressure that will trigger the car operating valves to go to Release will have exactly the same result no matter if the application is at Full Service or Minimum Reduction. The only difference will be a slightly longer brake cylinder release time, in this case, an irrelevancy.

For another thing, given that a Full Service brake application from the 500 kPa brake pipe operating pressure commonly used throughout Australia provides approximately 350 kPa of brake cylinder pressure, there is a total brake force developed on each wagon of around 32,000 lb or 14,500 kg. Expressed another way, a brake shoe force against each wheel of almost 1.8 tonnes! One has to query why any railroad—unless the crossing loop is located on a Swiss mountainside—feels that this magnitude of static holding power is necessary to temporarily secure an attended train.

Why does it matter? Again, because all pressure exhausted from the brake pipe has to be replenished when the brake is released - at the penalty of fuel cost and mechanical wear. Why increase both unnecessarily? One might also add that a train sitting on a Full Service brake application may have very little braking capacity in reserve should an uncommanded release occur, requiring a rapid re-application. Should the driver be forced to go to Emergency to stop the train movement, at the very least the result will be a significant delay. I believe the preoccupation with Full Service brake applications when a train is stationary and/or being tested is unwarranted and unwise.

Finally, we might consider the Holding Test. This is a safety check that is made to test the ability of a vehicle or rake of vehicles to maintain brake cylinder pressure in isolation from the brake pipe, or—in other words—if cut off from a maintaining supply of air pressure. The test validates the capacity of the wagon or wagons to be able to stop and be held by their brakes following the resultant Emergency brake application should they become detached from the train. As a safety procedure, this test is unique to Pacific National operations and appears to have originated from long-established South Australian railway custom and its influence on the development of operating practices at the nascent National Rail Corporation. Indeed, the Holding Test would appear to have been a cultural legacy of traditional South Australian rollingstock maintenance practice. The Victorians were not to blame for the introduction of the holding test (they had not traditionally had even a brake pipe leakage test – a state of affairs that came to an end following national railway brake engineer's conferences held between 1978 and 1982). By comparison, it is standard practice elsewhere in the world for freight and passenger vehicles to be subject to periodic, prescriptive maintenance schedules that are designed to validate the functionality of air brake equipment on a wagon or car for a set time interval—perhaps a period of months—within which correct operation of the equipment is able to be expected and is assumed. The unique Australian *Holding Test* exists in lieu of (or perhaps nowadays in addition to) such conventional maintenance regimes and, while acknowledged as an extra layer of safety certification, is extremely expensive in time. Certainly, the Holding test is not a 'World's Best Practice'... it is not even a World Practice.

The development of Maintaining Braking

This section is intended to provide some technical background to the evolution of brake pipe pressure-maintenance, possibly the greatest railway air braking advance prior to the advent of ECP brakes. For this narrative I acknowledge the Locomotive Engineer's Journal[162] *and information contained within numerous issues from the 1950's.*

As far back as the era of the H-6 brake valve (6-ET equipment), enginemen had used forms of what they thought was 'trick' braking to prevent the gradual reduction in brake pipe pressure during the time that the Automatic brake valve was being held applied with the brake valve handle in the LAP position (see page 341, Cycle/Serial Braking).

This gradual drop in brake pipe pressure was due to brake pipe leakage and had the effect of increasing the brake application until the full (Service) equalisation pressure was reached. Another effect of brake pipe leakage was a gradual drop in pressure within the brake pipe as the distance away from the locomotive increased. This difference in pressure between the locomotive and the last wagon of the train was called 'brake pipe gradient'. When the Automatic brake valve handle was moved to a position that did not supply air to the brake pipe (e.g. LAP position), the pressure throughout its length tended to equalise over a period of time. Because the pressure at the front end of the brake pipe tended to be higher, it would drop by a greater value than the pressure at the rear, reducing the gradient but causing the brakes on head-end wagons to apply harder than those closer to the rear.

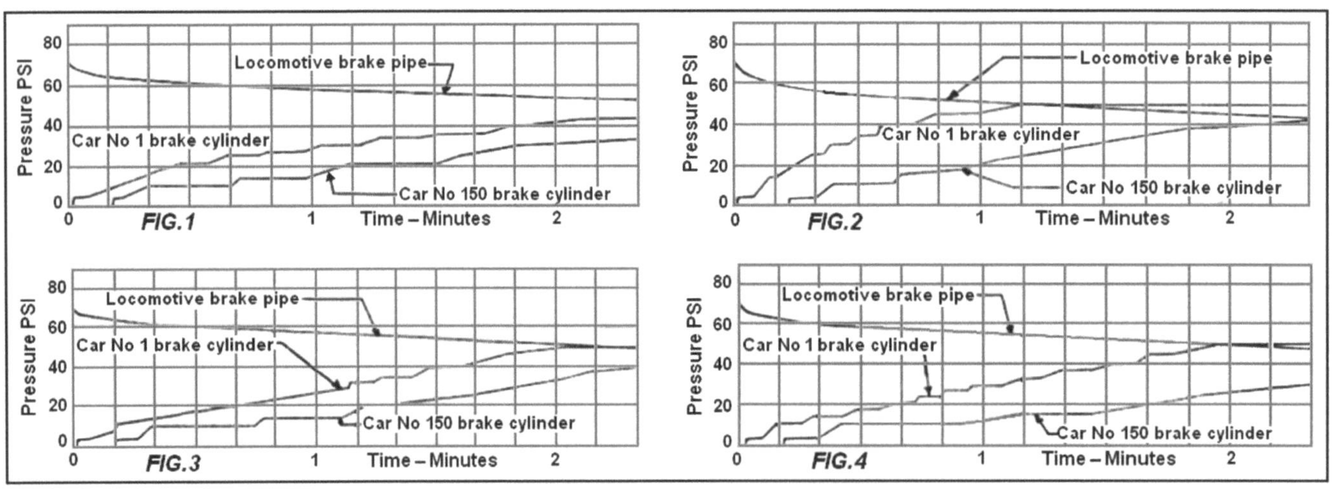

FIGURE 276: THE DEVELOPMENT OF MAINTAINING BRAKING. THE EFFECTS OF BRAKE PIPE LEAKAGE

Figure 276 consists of a series of diagrams used to illustrate the content of a Letter to the Editor in a 1950's issue of the Locomotive Engineer's Journal. The diagrams are not referenced or credited and I assume they have been drawn by the author of the letter, a representative of the Westinghouse Air Brake Company. Within Figure 276, FIGS 1 and 2 show the effect of two different rates of brake pipe leakage upon the brake pipe at the locomotive, and brake cylinder pressures on wagons 1 and 150. The data from which these plotted curves were derived was obtained by making a 6 psi (40 kPa) Service reduction—First Service or Minimum Reduction—and then placing and leaving the Automatic brake valve handle in LAP position. The slow reduction in brake pipe pressure due to brake pipe leakage is apparent in FIG 1.

162. The official journal of The Brotherhood Of Locomotive Engineers (now the Brotherhood of Locomotive Engineers and Trainmen), Cleveland, Ohio, USA.

Brake cylinder pressure on the first wagon gradually increased as brake pipe pressure reduced. Brake cylinder pressure on the 150th wagon also built up gradually but was always less than that for the first wagon due to brake pipe gradient causing a small reduction in brake pipe pressure at the rear of the train. In FIG 2, the more rapid reduction in brake pipe pressure due to greater leakage can be seen. This resulted in a faster build-up of brake cylinder pressure on both the 1st and 150th wagons. However, the greater difference in brake cylinder pressure between these should be noted. This difference was especially important during the first half-minute of the brake application because it was during this time that the slack would be adjusted. With the greater amount of brake pipe leakage, more severe slack action could be expected to result.

The use of so-called 'trick' braking methods was an attempt by locomotive drivers using this equipment to minimise these two effects of brake pipe leakage while braking down a grade (in other words, to attempt to obtain a more steady and stabilised brake cylinder pressure rather than one that incrementally increased after the application was Lapped). Probably the two most common methods used were *bridge braking* (a term often misused in Australia) and *feed valve braking*.

To 'bridge-brake', the engineman—after making the desired brake pipe reduction—would move the brake valve handle back from the Lap détente onto the 'bridge'[163] adjacent to the RUNNING (or HOLDING) détente. This action—requiring some precision in manipulation of the handle—provided a feed valve supply of main reservoir air to brake pipe just enough to keep the brake pipe pressure from dropping further. With 'feed valve braking', the feed valve was manipulated[164] (by the engineman) to provide a desired brake pipe reduction.

Although the use of these techniques enabled the engineman to get down the grade with fewer brake applications than did the short-cycle method, improper manipulation or poorly-maintained equipment could lead to several undesirable conditions. Equipment leakage caused brakes to leak off; a light re-application made without fully recharging the system sometimes caused brakes to release at the rear of the train; and use of feed valve braking could result in stuck brakes.

Except with poorly-maintained equipment, the arrival of the AB control valve practically eliminated the effects of equipment leakage. Use of WABCOTITE fittings minimised auxiliary reservoir leakage so that the brake was less liable to release. Use of rubber packing reduced brake cylinder leakage, and the limiting valve maintained at least 10 psi (70 kPa) pressure in the brake cylinder to counteract any remaining brake cylinder leakage.

In developing the higher-capacity № 8-ET locomotive equipment, the manufacturers not only recognised the desirability of maintaining against brake pipe leakage, but also recognised the disadvantage of undesired release of train brakes due to improper manipulation. The result was the First Service position[165] which—from a 70 psi (500 kPa) brake pipe pressure—gave a 6 psi (40 kPa) reduction at the standard Service rate followed by a slow continuous decrease or "blow-down" of equalising reservoir pressure to effect a 20 psi (140 kPa) reduction within two minutes. This slow blowdown was called *fading*. Throughout the reduction, locomotive brake pipe pressure was prevented from dropping below equalising reservoir pressure by the maintaining feature. This was the first application of the concept of pressure-maintenance that would eventually mature with the development of the 26-L equipment.

A comparison of the locomotive brake pipe pressures as shown in FIGs 3 and 4 of Figure 276 shows that the reduction in brake pipe pressure was virtually the same with either rate of leakage. The data from which

163. A portion of the brake valve quadrant that is raised in 'relief' so as to provide a tactile and physical barrier for the brake valve handle locating latch between the 'Application/Lap' and 'Release/Recharge' sectors of the quadrant.

164. Sets the pressure value carried in the brake pipe. This should be per company regulations, and feed valve adjustment (except to maintain the company-prescribed BP pressure) was not generally permitted.

165. Nowadays called 'Minimum Reduction'.

these curves were derived was obtained by placing the brake valve handle in FIRST SERVICE position and leaving it there.

In spite of this improvement incorporated into the № 8-ET equipment, certain railroads with long, steep grades continued to request a closer approach to pressure-maintaining. *Pressure-maintaining* is the maintenance of **locomotive** brake pipe pressure (and thus, theoretically, the **train** brake pipe) at any desired value while the brake is applied. The common request was to increase the blow-down time—usually to five minutes—by using a smaller choke or to replace the choke with a plug to eliminate the blow-down altogether. The manufacturers were reluctant to make these changes because tests showed that increasing the blow-down time to greater than two minutes resulted in an undesired release of train brakes and established conditions of *false gradient*.

As has been explained, the brake pipe pressure at the last wagon was typically lower than the brake pipe pressure at the locomotive (and thus the first wagon) when the train was (nominally) fully charged. This pressure differential was the true brake pipe gradient provided that there was no equipment leakage, and any maintaining feature would endeavour to reproduce this. During the time a train was being charged or recharged, the pressure on the last vehicle was lower than when it was fully charged although the locomotive brake pipe would reach normal pressure in just a few seconds. This pressure difference was the 'false gradient'.

If a light brake pipe reduction were made while a false gradient existed, Quick Service would carry the application through the whole train. However, if an attempt were made to maintain brake pipe leakage, the true brake pipe gradient would be established. The possibility existed that this would actually increase the brake pipe pressure at the rear of the train and release those brakes.

In order to improve train handling on one particular grade, one railroad modified the First Service feature of the № 8-ET equipment. It was this railroad's requirement that pressure-maintaining be used on only this one grade with all retainers[166] turned up to prevent undesired release of train brakes and that the operation be adequately supervised[167] to prevent improper manipulation of the brake valve. The modification was made by replacing the conventional fading choke with a cock-controlled fading choke; the cock to be open at all times except when 'maintaining'. The FIRST SERVICE position was used for braking on the specified grade. When the desired brake pipe reduction had been made, the cock was closed. Further reductions were made by opening the cock. Although the operation was satisfactory in the hands of skilled enginemen, the air brake companies felt that the degree of skill required to obtain the desired results prevented its general adoption.

Another railroad encountered considerable trouble with conventional methods of cycle-braking on an important hill. Most numerous were complaints of severe slack action and low brake pipe pressure at the caboose, braking concentrated at the front of the train, and long delays in recharging the brake pipe and cooling the wheels. By 1949 a scheme had been developed whereby the fading choke was replaced by two chokes and a plug controlled by a three-position cock. These were arranged so that following the initial 6 psi reduction, (1) no further reduction; (2) a further reduction at the normal Service rate; or (3) a further reduction at the slow 'First Service' rate could be made with the Automatic brake valve in FIRST SERVICE position. This modification provided a FIRST SERVICE, a SERVICE and a LAP position all incorporating the

166. The retainer valve is a manually-operated device mounted on the exhaust port of the wagon brake cylinder. When set to an operating position it will restrict or close off that exhaust. This restriction retains pressure in the brake cylinder, thus keeping the brake applied or only allowing it to release slowly, even though the triple or control valve might be in Release and the wagon reservoirs being recharged.

167. Meaning that a manager of operating practices (e.g. a Trainmaster or Road Foreman of Engines) accompanied the locomotive crew on the grade to monitor and instruct as necessary.

maintaining feature - therefore pressure-maintaining was available for grade operation. Road tests showed that this scheme eliminated the complaints found with the short-cycle method of braking.

An improvement to the effort required in manipulation of the brake valve to provide pressure-maintaining was made during tests on this particular railroad. Instead of measuring the equalising reservoir reduction from the pressure of the feed valve setting, the reduction was measured from the pressure at which the equalising piston raised. This eliminated false gradient and appeared to prevent undesired releases resulting from it. However, the schemes used on both railroads introduced an additional operating lever—the extra cock-handle—into the locomotive cab.

Introduction of the *Service application* portion into the brake valves of the 24-RL and the 8-EL locomotive equipments posed a new problem because the fading choke was installed in the Service application slide valve and—unlike the 8-ET equipment—was inaccessible for modification.

A third railroad developed the simplest scheme of achieving pressure-maintaining with this equipment. Their scheme plugged the rotary valve seat passage leading to the *reduction limiting reservoir* and the fading choke. This changed FIRST SERVICE position into a MAINTAINING LAP position. All reductions had to be made in SERVICE position, but since most enginemen on this railroad had not been using FIRST SERVICE there was no disadvantage in its elimination. In addition to elimination of the First Service feature, no maintaining was provided during movement of the brake valve handle from FIRST SERVICE to SERVICE position and back to FIRST SERVICE, to prevent degradation of head-end brake pipe pressure such as might occur in making additional, light reductions with a moderate-to-large brake pipe gradient.

The second scheme provided maintaining in SERVICE and LAP positions of the brake valve as well as in FIRST SERVICE. This was done by plugging the *maintaining supply passage* below the rotary valve seat to cut off feed valve supply and by adding an external pipe to supply main reservoir pressure to the maintaining valve; this to ensure an adequate supply of maintaining air. In addition, the rotary valve was modified to drain equalising reservoir and thereby prevent 'maintaining' when the brake valve was in EMERGENCY position.

The provision of a main reservoir supply to the maintaining valve, though, was undesirable because of the adverse conditions that could result if it should stick open. Service brake applications would be hampered, overcharge might occur, and if the valve should then close, stuck brakes would result. Under some conditions the scheme was undesirable in passenger service. If an Emergency application was made when operating with the electro-pneumatic brake, the rotary valve was positioned to cut off feed valve supply to the brake pipe but the equalising reservoir would not be exhausted, so the maintaining valve would be open and able to feed air to the brake pipe.

The practice of trying to prevent leakage from causing a further reduction in brake pipe pressure during Service braking was not new but it was not until the introduction of the diaphragm-type equalising piston that the practice was considered entirely acceptable. To ensure safe and effective operational pressure-maintaining, the equalising piston needed to function with extremely small changes in pressure and be entirely free from ring leakage.

The decade of the 1950s was an exceptionally productive one in air brake development. The improvements in basic research, new materials and new techniques resulted in outstanding progress in the air brake art. One of the common elements long used in air brake control valves had been the ring-fitted piston. However, a well-fitted piston ring, free from leakage, had always been difficult to achieve and maintain under service conditions. It followed that a positive method of eliminating ring leakage around the piston and, at the same time eliminating the friction variables of the ring-fitted piston, had become a much-desired goal. One of the foremost improvements in air brake progress at the time had been the development of the diaphragm-operated piston to take the place of the ring-fitted piston.

The diaphragm had long been recognised as having possibilities for the solution to this problem, however it was not until the era of the late 1940's and early 50's that it was possible to produce diaphragms that had all of the characteristics required for application to air brake service. Diaphragms for this critical use needed to be highly sensitive to pressure changes, be efficient (so that space was not wasted), to operate with little or no change in performance over a wide temperature range, to withstand any adverse effects of oil, and to perform reliably over long periods of service. Constant and intense research eventually made it possible to produce diaphragms that could unhesitatingly be used in places where formerly ring-fitted pistons were thought to be essential. Brake devices could now be designed using these diaphragms in which the variable effects of ring leakage were completely eliminated and the reliability of the operating devices materially improved.

Next in importance to the diaphragm—and not to be forgotten—has been the application of the 'O' ring and spool-type valve to replace, wherever practical, lap-fitted type valves. The 'O' ring—previous to its application in the air brake—was used to some extent by a number of other industries. Before it could be applied to the railway air brake it was necessary to determine the special design requirements for air brake services. This was successfully accomplished through an extensive series of tests and limited service applications so that today it is practical to use the 'O' ring design for the realisation of many structural and economic improvements.

The DS-24-M brake valve

With an increasing demand from industry for a brake valve incorporating pressure-maintaining, the manufacturers sought to develop a scheme without the previous functional disadvantages. In addition, technological advances since the development of the D-24 brake valve (24-RL schedule) made improvements to several details of the brake valve possible. This new brake valve was designated DS-24-M.

Foremost was the addition of the "***M***aintaining" feature that was available in FIRST SERVICE, LAP, and SERVICE positions. This feature could be readily cut out to permit making the brake pipe leakage tests or at any other time it was not desired. The conventional First Service function with its many advantages was retained and was independent of the new maintaining feature. Air for the maintaining feature was supplied at feed valve pressure through a strainer. The capacity of the maintaining supply was controlled by a choke to maintain against maximum permissible AAR brake pipe leakage on a 150-car train. The maintaining supply choke was also beneficial in assisting propagation of brake applications initiated from the rear of the train. The maintaining air supply was automatically cut off during a Safety Control[168] or automatic Train Control[169] brake application by movement of the *Service application* slide valve or by a reduction in brake pipe pressure below a predetermined value such as would occur with a locomotive pull-apart. This latter would prevent the undesired release—due to 'maintaining' being cut IN—of an Emergency application.

The addition of the new maintaining feature necessitated a new equalising portion. This was achieved by designing a diaphragm-actuated piston valve structure. The diaphragm eliminated any trouble that might have resulted from air leaking from the brake pipe past the equalising piston rings to the equalising reservoir, and the piston valve arrangement together with the strainer in the maintaining supply passage minimised trouble with dirty air. The design was such as to eliminate 'pumping' and thus prevent fluctuations in brake pipe pressure during 'maintaining'.

Several other features were included in the design of this brake valve:

168. Penalty brake application from activation of alerter/vigilance control.

169. Penalty Service brake application initiated due to overspeed or lack of driver response to a signal indication.

- A *brake pipe cut-out* valve of piston-valve type with a manually-operated pilot valve replaced the customary brake pipe cut-out cock. This was designed to decrease maintenance and could be removed without dismantling the brake valve;
- The rotary valve was redesigned to increase the capacity of its porting in order to eliminate the effect of manufacturing tolerances that tended to produce variations in the brake pipe charging rate in RUNNING position;
- The diaphragm-actuated Service application valve was included in the design;
- Port changes and a piston-valve type First Service were designed to prevent an illegal suppression of Train Control when the First Service feature was cut out;
- Changes in passage capacities were made to properly time the rate of brake pipe exhaust and to improve operational Safety Control or Train Control applications.

Additionally, a new filling piece was offered to convert the contemporary D-24 brake valve for 'maintaining' operation. This filling piece incorporated the new equalising portion which was necessary for proper 'maintaining' operation.

Ironically, grade operation—where the greatest benefit of pressure-maintaining was to be obtained—was also where the greatest danger due to its potential improper use existed. The one effect of pressure-maintaining operation most liable to get an engineman into trouble at this time was undesired release of train brakes, and the most likely cause of undesired releases was false brake pipe gradient arising from partially charged equipment. Another cause was making an initial reduction of a lesser amount than the drop of brake pipe pressure that had occurred due to Quick Service activity. To prevent such an undesired release required a new technique in Automatic brake valve manipulation. It was essential that the equalising reservoir pressure be reduced through the brake valve by not less than 6 to 10 psi (40-70 kPa) below the point where the equalising piston lifted for all initial applications or re-applications following a release. When the equalising reservoir reduction was continued by this amount beyond the point at which the equalising piston lifted, the brake pipe pressure at the rear of the train would not increase above the pressure obtained there following Quick Service propagation. The 6 psi further reduction was sufficient for a fully-charged or nearly-fully-charged train. The 10 psi further reduction was required when a reduction was made immediately following the release of a heavy reduction or following several smaller reductions and releases. The amount of leakage present in the brake pipe had very little bearing on this problem.

The brake pipe flow indicator could be used to give a useful measure of the amount of false gradient present. If the red hand of the indicator were set—following a long charging period—to indicate a fully-charged train, the difference between the red and black hands indicated—subject to a degree of interpretation—the amount of false gradient.

If some brakes leaked off or released, the train speed would increase. Further reductions would then need to be made to offset this increase in speed. A good understanding was required in order to not continue the reductions and reduce the brake pipe and auxiliary reservoir pressures to a dangerously low value.

No attempt was ever to be made to kick off a few brakes with 'maintaining' cut IN. During ordinary braking some enginemen had become adept at releasing a few brakes when holding the brake applied by moving the brake valve handle momentarily from LAP to RUNNING position. This manipulation sent a pressure wave through the brake pipe that was sufficient to release a few of the front brakes when too heavy an application had been made. With 'maintaining' cut IN, such an action would increase equalising reservoir so that the entire brake pipe would be maintained at a higher pressure, and all brakes would release. If too heavy an application were made, it would be necessary to release all brakes, recharge the train, and then make another (lighter) application.

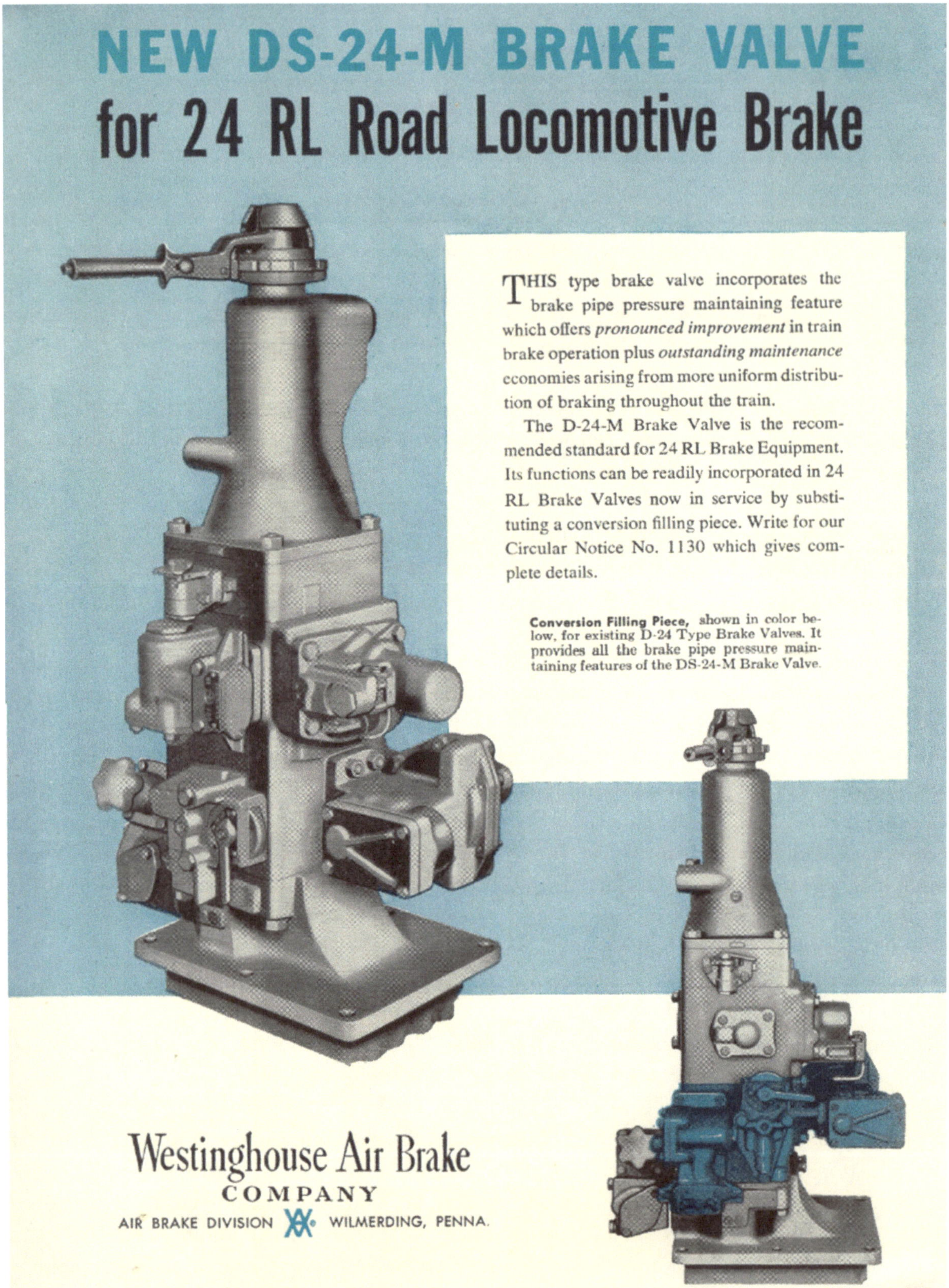

FIGURE 277: D-24-M MAINTAINING BRAKE VALVE – 24RL EQUIPMENT. *Railway Age*

It must be emphasised that although the 'maintaining' feature would minimise some of the adverse effects of brake pipe leakage, it was not intended to be a substitute for thorough equipment maintenance procedures designed to reduce brake pipe and equipment leakage to a minimum. Excess leakage remained as undesirable as ever - in fact more so, because use of 'maintaining' tended to hide from the engineman its effect on brake pipe pressure. 'Maintaining' was designed to care for that margin of difference between thorough inspection and maintenance and the ultimate of an absolutely leak-free air brake system.

Following the introduction of the DS-24-M brake valve, a number of railroads experienced greatly-improved train handling as well as considerable savings on wheel maintenance and shoe wear. Because the variable effects of brake pipe leakage were eliminated, pressure-maintaining gained particular favour with the trainmen (that is to say, those riding the caboose). Braking was more uniform which resulted in smoother operation without the objectionable (and dangerous) slack action.

The DS-24-M brake valve incorporated a number of structural and maintenance improvements in addition to the pressure-maintaining feature. The *service application* portion as well as the *equalising* portion now employed the use of the diaphragm piston. The *First-Service cut-out valve, brake pipe cut-out valve, maintaining cut-out valve and brake pipe cut-off valve* all employed the 'O' ring valve design. A number of other modifications were also made to provide improved brake valve operation. Both the brake pipe angle cock and the combined dirt collector and cut-out cock were redesigned to help reduce brake pipe leakage; the keys of these devices were provided with the new 'O' rings to prevent leakage to the atmosphere above the waterway; and, on the angle cock, a WABCOSEAL fitting was provided to facilitate installation and to eliminate leakage at the pipe joint (the new angle cock with these features was designated the 'WABCOSEAL angle cock'). The standard angle cock for passenger service employed the same leakage-prevention measures with the additional feature of a spring-loaded locking handle. These, together with other improvements, are considered to have provided a modern and effective brake valve for road service that would be the contemporary standard until the advent of the 26-L equipment.

In time, considerable emphasis was placed on reducing the weight of valve portions to facilitate maintenance operations. In some evolutionary developments at this time, aluminium came to be used extensively for this purpose. Improved manufacturing and heat-treating processes enabled aluminium castings to be made with similar pressure-tightness and strength characteristics as existed with cast-iron. Newly-available processes ensured that aluminium castings—when subjected to proper treatment—would have the corrosion-resistant characteristics necessary for use in air brake service.

Contemporary advancements both in freight and passenger brake control made extensive use of these three and other modern developments. Application of these improved methods and materials enabled the brake designers to develop valves having many functional improvements, simplified structures, and more maintenance advantages than were previously possible.

AFTERWORD

This book has described the evolution in design of air brake operating valves and drivers' air brake valves and has discussed the actions required of the locomotive driver with regard to achieving effective air braking. These requirements have resulted in successive advancements in operating valve design to improve response times and sensitivity to the brake pipe pressure signal; however, locomotive-hauled train air brake systems have probably now reached their operational limits without electrical augmentation of triple valve operation. ECP braking has made its introductory appearance in Western Australia's Pilbara region, the New South Wales Hunter Valley, and in Queensland, although it can be expected that some time will transpire before this technology sees more extensive use within Australia (and it might never be required in New Zealand).

Even in 2009, when these advances were just around the corner, Australian railways—struggling in the newly-emergent commercial reality—continued to sometimes fit compromise equipment packages to new and refurbished rollingstock. Australasia is a minor market, and there may be insufficient economy of scale to pare equipment capital costs back to an acceptable degree. The locally designed-and-made product cannot compete in price with that from the United States, and even this latter option is often viewed by Australian operators (Pilbara iron ore roads excepted) as being too expensive. Even the realities of the relentless global economy intrude, as Australian air brake equipment manufacturers and agents find that they are no more immune from the "Made in China" label than were local producers of footwear, apparel, consumer whitegoods, and electronics. Drivers of locomotive-hauled trains in Australia and New Zealand will continue to be required, for the foreseeable future, to apply expert anticipatory judgment in air brake manipulation, so long as management permits them to gain and practice the requisite skills.

However one chooses to look at it, the exigencies of good air brake management demand much of the locomotive driver - the unique competencies requiring a high degree of sophisticated information processing and intuitive action in order to control what is a very complex man–machine system. These competencies can be developed only with knowledge and experience and, in this regard, there is a crucial necessity for our rail industry in Australia and New Zealand to understand and recognise the consequences of the massive loss of corporate technical knowledge that has occurred over the past thirty years. Those current railway managers whose background is in rail operations are the custodians of essential railway lore. Upon them lies the responsibility of assisting and guiding colleagues who have joined the industry from other environments, unaware of *essential* custom and practice, and who may believe that their atypical background is more meaningful than is merited. This responsibility includes ensuring the next generation of locomotive engine service personnel—currently being sought with palpable desperation after three decades of denial by our industry of the need to prepare for the future—receives appropriate education and direction. No longer does the rail industry have the historical "government railway" luxury of bringing new entrants slowly up through the ranks at taxpayer expense while they incrementally acquire their knowledge and experience. Rail operators need to find a way by which their workforce can be replenished and then trained with expedition, all the while ensuring that overall competency standards do not regress. Specifically, new entrants must be afforded the opportunity to acquire practical experience and theoretical instruction to an acceptable standard.

As this author—a proud and unapologetic product of an era fast fading into history—casts a jaundiced eye over the current state of the rail industry with regard to engine service recruitment, selection, and lower-level promotion as well as the diminished focus on technical proficiency (especially in the field of training and instruction) he cannot suppress some nagging doubts. One thing, though, seems certain. As we contemplate the inexorable march of technological development, we can expect the evolution of electronics and automation to render the practice of locomotive and train operation physically easier and perhaps intellectually less demanding. And, consequent to that, one surmises that the rail industry will eventually require of their engine service personnel no greater degree of technical awareness for adequate performance than is currently required of any householder to operate a TV remote control, a cell phone, or a digital set-top box, and the remorseless advance of train braking technology will undoubtedly expedite this attenuation of technical responsibility. Indeed, as these very words are written, Rio Tinto is testing the operation of unmanned ore trains in the Pilbara; one cannot but ponder how long the *air will be there*. We can, however, expect that as long as there <u>is</u> a pneumatic brake pipe and a locomotive driver, there will be a need for that driver to have some appreciation of air brake theory and to know how to "work the air."

7.

Biographies

The history of arresting the speed of railway trains by the application of compressed air is one (to which the records of the patent office bear frequent witness) of a gradual progress from rude and imperfect beginnings, step by step, to a final consummation, which, in respect to this invention, had not been reached when the patent in suit was taken out, and which, it is quite possible, has not been reached to this day. It is not disputed that the most important steps in this direction have been taken by Westinghouse himself.[170]

THE STORY OF railway air braking is so inextricably linked to the personalities of the engineers, technicians, and thinkers who delivered us the technology, that it seems entirely appropriate that some historical notes of the lives and personalities of a few who are a part of the story be included in this narrative. The earliest engineers, whose concepts first resulted in the use of steam as a motive force and the parallel advancement of civil engineering that made possible the lines of railway, were often the original thinkers

170. Justice Henry Billings Brown - US Supreme Court; Boyden Power-Brake Company v. Westinghouse (et al): May, 1898. Brown—who had earned a BA but never a law degree—had practiced Admiralty Law and was known as a competent "legal technician". He was appointed by President William Harrison, and earned fame (or notoriety) for a landmark ruling in 1896—the 'Separate but Equal' provision—that upheld the legality of racial segregation in public facilities and remained standard doctrine in U.S. law until its repudiation in 1954.
George Albert Boyden, of Mount Washington, MD, was born in 1856. Having attended public schools and then college in Altoona, PA, he served an apprenticeship in pattern making, and worked in the Pennsylvania Railroad foundry and machine shops in Altoona, later studying mechanical engineering and architecture. He became interested in patents, and studied Patent Law - practicing before the U.S. Patent Office for several years. Later, in failing health, he turned his attention to the invention of a railroad air brake arrangement and became "...the inventor of the well-known and highly meritorious brake which bears his name, and the rival of the Westinghouse air brake. Mr. Boyden is the president of the Boyden Power Brake Co., Baltimore, MD, and is actively engaged in the manufacture of the same, and has thus far successfully rendered void all the efforts of the Westinghouse Air Brake Co. to show that his device is an infringement upon theirs. The issue involves millions of dollars. Recently the last and highest court has sustained Mr. Boyden's claims."
The *American Railroad & Engineering Journal*, Vol. LXIII, 1889 said of the Boyden brake, "...in its general character, resembles the Westinghouse, although the mechanical appliances used differ considerably from what may be called the organs of its prototype."
It seems that the Westinghouse Air Brake Co., unable to prevail against Boyden in court, set out to secure his patent. "After long negotiation he finally accepted a settlement on terms that afforded him the means of living in ease and retirement the rest of his life." (*Maryland Historical Magazine* - The Maryland Historical Society, Vol. XLIII, Baltimore 1948).

of these disciplines or were "pupil-trained" novices who learned the practical skills of their nascent crafts as apprentices and associates of close relations or mentors. These individuals—as well as others who pondered these questions even earlier—included such people as Babbage, Brunel, Cugnot, Newcomen, Norris, Stephenson, Trevithick, and Watt. The formal tertiary education process with which we are now familiar as a part of the making of an engineer of any discipline did not exist when these men began to conceive, calculate, and construct. The earliest of these engineers gained their expertise from thinking, listening, observing, and experimenting. Many traveled to other countries, and so the various disciplines of railway engineering were exported.

The time would inevitably arrive when the progress of university learning and course development enabled engineers to be "professionally" educated. In turn, this would create tensions between those whose education had been "pupillage" and those who considered themselves "professional" (see the comment on W. C. Kernot included in the biography of Thomas Higinbotham). Although this book does not specifically explore the development of vacuum braking—itself an "atmospheric brake" technology—in many respects, this story is as much a part of the development of air braking as are the early attempts to produce some sort of mechanical patent braking system. For that reason, details of one of the vacuum brake pioneers whose products and efforts had particular relevance to the evolution of air braking are included, as is the life story of a modern-day air brake luminary.

I have compiled the following biographical accounts from a variety of sources. I hope these stories will entertain and inform the reader and—by revealing something of the personalities and their lives and achievements—provide insight into the evolution of railway air braking and how it was promoted within Australia and New Zealand.

George Westinghouse

"Well gentlemen, this only compels me to do something else."
— GW, as the financial panic of 1907 began to affect him.

The various Westinghouse companies were the product of the mechanical inventiveness and the business acumen of George Westinghouse - inventor, manufacturer, and entrepreneur. This prolific discoverer influenced the course of history by enabling the growth of the railroads through his inventions and by promoting the use of electricity for power and transportation. As an industrial manager, his influence on industrial history is considerable, having formed and directed more than sixty companies to market his and others' inventions during his lifetime. His electric company became one of the greatest electric manufacturing organisations in the United States, and his influence abroad was evidenced by the many companies he founded in other countries.

In the ninth century, the Wistinghausen family was prominent in Westphalia, Germany, and in the fourteenth century a branch of the family emigrated to England and later the United States. George Westinghouse's great-grandfather, John Hendrik Wistinghausen (also recorded as "Westinghausen") arrived in America with his widowed mother in 1755, aged fifteen years. They settled in the locality of Pownal in what would become Bennington County in the state of Vermont—acquiring and clearing land and building a log cabin. John (by now, "Westinghouse") raised a large family and died in 1802. One of John's sons, John Ferdinand, spent his life in Pownal and had twelve children, the fifth being George senior who had ten children. The eighth of these was the inventor of the continuous air brake.

From his father, George junior inherited his physical stature and a knack for the mechanical. From his ancestors he inherited a nobility of bearing and an advanced mental faculty (one ancestor, George, had been a captain in the Russian navy and distinguished himself in battle in command of a frigate; several of the family were ennobled in the Russian service). Both sides of George junior's family had a farming and mechanical background and were financially comfortable, self-respecting, self-reliant, and competent. From his mother, Emeline Vedder, George is said to have inherited imagination, good taste, vivacity, alertness, and a cheerful altruism.

In the early part of the nineteenth century, George Westinghouse senior moved the family from Vermont, eventually settling in Central Bridge, in upper New York State, where he built a machine shop and manufactured farming and other machinery. Here his sons developed their taste for mechanics and received their first lessons in manufacturing. And it was here, on October 6, 1846, that the young George was born. When George was ten years old, his family moved to Schenectady, New York, where his father established G. Westinghouse & Company to manufacture farm implements. While working with his father, young George acquired a realistic sense of tools, materials, machinery, and structures.

George and his two elder brothers served in the Union army and navy during the American Civil War and won commissions. Albert was captured at the Battle of Gaines' Mill (June 27, 1862) and was in the infamous Libby Prison in Richmond, Virginia, for a short time before being exchanged and receiving his commission as 2nd lieutenant in the 2nd NY Veteran Volunteer Cavalry. Biographer Prout states that he served with distinction, being mentioned in dispatches, but was killed leading a cavalry charge in a raid, reportedly on the 'Leaf River Railroad'[171] in Louisiana in December 1864. John Westinghouse served as an engineer officer in the navy, returning after the war to his father's business in Schenectady. Here he worked indefatigably in the interests of the less fortunate in the community before dying at a comparatively young age. George himself was halfway through his fifteenth year when the Civil War broke out. He promptly ran away to join up, which small adventure was, in turn, promptly ended by his father. However, two years later—at aged 16½—he was permitted to enlist. After a short service in the infantry and cavalry, he became an engineer officer in the navy, possibly as a result of a dawning awareness that a boy so young in a veteran army had little chance of distinction. Certainly, it is recorded that military tedium of caring for the army's livestock did not enthuse the young Westinghouse.

After three years of largely noncombatant military service, he returned to Schenectady and, in September 1865, enrolled as a sophomore at Union College. Within three months, however, he convinced himself and his teachers the college curriculum had little to offer someone with his mechanical leanings. He dropped out upon his Christmas vacation, apparently with little argument from his instructors, and returned to his father's factory. One important thing did happen during his short stay in college; on October 31, 1865, at age nineteen, he obtained his first patent - for a stationary rotary steam engine (which would later be used to power a dynamo).

171. Prout's details are inaccurate. In fact, in *The Story of a Cavalry Regiment: "Scott's 900" Eleventh New York Cavalry, from the St. Lawrence River to the Gulf of Mexico, 1861-1865* (Gale Cengage Learning/HardPress 2012), author Thomas West Smith recounts that 1st Lieut Westinghouse was fatally wounded leading a charge on a Confederate force encountered during an expedition to destroy a Mobile & Ohio RR bridge near the town of State Line on the Mississippi-Alabama border. In the course of this march they encountered the New Orleans, Jackson & Great Northern Railroad and crossed the Leaf River (among others), but there was no railroad in this vicinity with 'Leaf River' in its name until incorporation of the Pearl & Leaf Rivers logging railroad in November 1897. The force, of which Lt Westinghouse was a part (two mounted divisions under Gen. John Davidson), had set out from Baton Rouge on 27 November 1864 on a planned action that has become known as the 'Pascagoula Raid', with the goal of destroying Mobile & Ohio RR infrastructure, including the burning of bridges. Lt. Westinghouse's group encountered a Confederate force before they could reach the railroad, and were ordered to charge. Some accounts describe the action as having occurred near a grist mill belonging to Roderick McLeod, near the Chickasawhay River a few miles south of Leakesville, MS. Other accounts mention Robert's Ferry (probably at Robert's Bluff), some way further south on the same river. The Westinghouse family received the news on Christmas Day, 1864. George Westinghouse was greatly affected by his brother's death, not least because he himself—in comparison—had served out the war in relative safety.

In 1866, Westinghouse took a train ride that changed his life. While travelling for his father's business, his train was brought to an abrupt halt to avoid hitting a train derailed on the tracks ahead. From his contemplation of this experience he invented a re-railing ramp for placing derailed trains back on the rails with greater ease and in shorter time - and more importantly, he began to consider ideas for improved braking systems for trains. And so began his long and illustrious career. He formed a business in Schenectady with two men who put up $5 000 each to finance the manufacture of the "car replacer." That same year, he married Marguerite Erskine in Brooklyn, New York. The couple would have one child, George Westinghouse III.

His liking for making new things was stabilized by routine work in his father's factory for manufacturing agricultural machinery. Hands and head worked together. The engineering of his day was empirical. He developed his own ideas of engines and mechanisms and made them himself before the days of schools with engineering laboratories. Air pressure and its control by valves became matters of intuition. He was alert, ingenious, capable, and ambitious. In his advance from a boy working at the bench to the head of great industries, he acquired experience in every stage of industry, and he knew the workmen as his fellow-workers.

When he entered young manhood, the environment also was propitious as the stage was then set for a stirring act in the drama of transportation and power. The expansion following the Civil war created new demands with alluring opportunities; the railroads with forty years of pioneering experience and century-old steam power, now giving new impetus to industry, were the beginnings upon which to build.

Westinghouse moved to Pittsburgh and approached the Anderson & Cook Foundry to negotiate production of the railroad car replacer. They hired him as a traveling salesman and, through these arrangements, he has been credited with—or at least being involved with—the first successful steel castings produced in America. During visits to Pittsburgh, he made the acquaintance of people who shared his interest in railroads and his work on inventions and manufacturing for the industry, and who would eventually help him with his Pittsburgh companies.

The air brake

Westinghouse must have been aware of the growing fervour within the rail industry, both in America and Great Britain, for a safe and reliable power braking system for trains and locomotives. He saw that railroads could never live up to their potential until trains had a safe and effective method of brake control, and for three years he worked at improving train brakes, failing at numerous attempts. But Westinghouse finally hit on the idea that worked; he would place an air compressor on the locomotive, and pipes would carry the air to the brakes on each of the cars. The locomotive engineer could admit compressed air into the system to stop the train and release the air when he wanted to move. Previously, train accidents were frequent since brakes had to be applied manually on each car by a crew of brakemen reacting to an engine whistle signal from the engineer.

The first form of power brake that had occurred to Westinghouse was a buffer brake; the brakes on each individual car being automatically applied by impact when the brakes were set on the locomotive. After some shop experiments, this idea was abandoned for a design that contemplated a coupled chain running through the length of the train, by which the car brakes could be applied through the manipulation of some power device on the locomotive. Visiting Chicago with this in mind, he found a chain brake installed on the *Aurora Accommodation* of the Chicago, Burlington & Quincy Railroad which, in a large measure, anticipated his own idea and at the same time demonstrated its inherent weakness. This brake, patented in 1862, was the invention of Augustine I. Ambler[172], of Milwaukee. His Chicago experience led Westinghouse to design

172. See footnote 182.

a chain brake in which a steam cylinder under the locomotive displaced Ambler's clumsy friction-drive windlass for tightening the brake chain. This idea was superseded by a more practical one that, in order to avoid excessive slack, each car must have its independent brake cylinder, supplied with steam from the locomotive by a continuous pipe with the necessary flexible couplings between the cars.

At this stage of his study, when he was wrestling with the problem of condensation, occurred the interesting and almost romantic incident of the magazine subscription which he made while at work in the Westinghouse shop at Schenectady at the solicitation of a young woman, evidently for personal reasons and not because of any particular interest in the magazine. This purchase turned out to be one of the most important he ever made, although he apparently never saw the young lady again. The first or second number of the magazine received brought to his eager attention an illustrated article on the Mont Cenis tunnel, then under construction. Both headings of this tunnel, then 3 000 or more feet from the entrances, were being driven by rock drills worked by compressed air. This gave him his cue, and thereafter his efforts were centered on brake designs in which the operating force was compressed air as the medium for transmitting power from the locomotive to the brake mechanism on each car.

WHO INVENTED THE AIR BRAKE? As we take up a description of the apparatus and an account of its introduction, let us note that the first air-brake patent of George Westinghouse was 88,929, dated April 13, 1869. This patent was reissued July 29, 1873, as 5504.

The essential parts of the air brake as first assembled were:

- *An air pump driven by a steam engine receiving its supply from the boiler of the locomotive;*
- *A main reservoir on the locomotive into which air was compressed to about sixty or seventy pounds per square inch;*
- *A pipe leading from the reservoir to a valve mechanism convenient to the engineer;*
- *Brake cylinders for the tender and each car;*
- *A line of pipe from the engineer's brake valve passing under the tender and all of the cars, with a connection to each brake cylinder. Flexible hose connections between the cars provided with couplings having valves which were automatically opened when the two parts of the couplings were joined and automatically closed when the couplings were separated.*[173]

173. Colonel Henry Goslee Prout: *A Life of George Westinghouse*, written for The American Society of Mechanical Engineers (New York: The Scribner Press, 1921). Later editions are available.

FIGURE 278: THIS WELL-KNOWN IMAGE DEPICTS A PENNSYLVANIA RAILROAD CLASS D1 LOCOMOTIVE FITTED WITH EXPERIMENTAL WESTINGHOUSE AIR BRAKE EQUIPMENT DURING THE 1869 TRIALS ON THE PRR THAT LED TO THE ROAD'S ADOPTION OF IT IN 1870. SOME OF THE AIR BRAKE ARRANGEMENT CAN BE SEEN BETWEEN THE DRIVING WHEELS.

After having obtained approval for this first air brake patent, Westinghouse had a difficult time securing financial backing in order to commence production. It was not until W.W. Card, Superintendent of the Steubenville Division of the Pan-Handle Railroad inspected the device and agreed to allow Westinghouse to install the braking system on a test train that the invention began to establish some credibility.

Almost immediately upon arriving in Pittsburgh Westinghouse met and formed a friendship with a local industrialist Ralph Bagaley. With Superintendent Card's endorsement, four passenger cars of the Pan-Handle Railroad were equipped with the new straight-air brake at Ralph Bagaley's North Side plant, with Westinghouse doing most of the work himself. With the assistance of Bagaley, Westinghouse had a successful trial run from Pittsburgh to Steubenville and back. Minutes after departing on the trip—from Pittsburgh's Union Station—as the train was exiting Grant's Hill tunnel, the engineer saw a horse-drawn wagon blocking the tracks ahead. The brake was fully-applied and the train came to a stop just feet from the wagon driver.

The last car on the train would carry some special guests including Card and some of the Panhandle directors. Daniel Tate[174] *was the engineer and Westinghouse had trained him personally with how to use the brake. He also gave Tate fifty dollars as the train left Panhandle* [Union] *Station on its fifty-mile trip to Steubenville. Westinghouse went to sit in the locomotive cab. Normally, 'Accommodation' trains went about a quarter-mile through the Grant Hill tunnel*[175] *and stopped at Fourth Street* [outside the south portal] *to pick up passengers. This reserved train, however, was not scheduled to stop in Pittsburgh that day.* [Its route] *crossed Second Avenue*[176] *and the Monongahela Bridge and then* [ran to Steubenville on] *open track following the Ohio River. The train left* [the south portal of] *the tunnel at thirty miles per hour only to see a merchant and his cart crossing the track at Second Street* [sic]. *It is not clear where the planned stops were that day, but this was an unexpected emergency requiring Tate to use the brake for the first time. He pulled the hand lever, quickly bringing the train to a jarring stop. The guests in the last car were thrown from their seats in confusion. This first successful trial is immortalized in a painting today*[177] *and verified in personal letters of the engineer* [Tate] *and Superintendent Card.*[178]

In July 1869, aged 22, George Westinghouse organised the Westinghouse Air Brake Company. He did not forget those who helped him; Bagaley was elected vice-president, W.W. Card was appointed General Agent and Robert Pitcairn treasurer. The first factory, on Liberty Avenue in Pittsburgh, began production in 1870. This was soon outgrown and the company moved across the river to a larger plant in Allegheny City, on Lacock and General Robinson streets (now situated beneath I-279, Parkway North). In 1889, Westinghouse moved his operation to the Turtle Creek Valley, about 12 miles east of Pittsburgh. The 500 acres of land he purchased became the borough of Wilmerding, named after Joanna Wilmerding Bruce, the wife of a friend of company secretary Pitcairn.

174. In other references the spelling is 'Tait'.

175. See author's note inside the front cover.

176. Now Boulevarde of the Allies.

177. See cover of this book.

178. Quentin R. Skrabec - *The World's Richest Neighborhood: How Pittsburgh's East Enders Forged American Industry*: Algora Publishing, 2010 (available as a Google eBook).

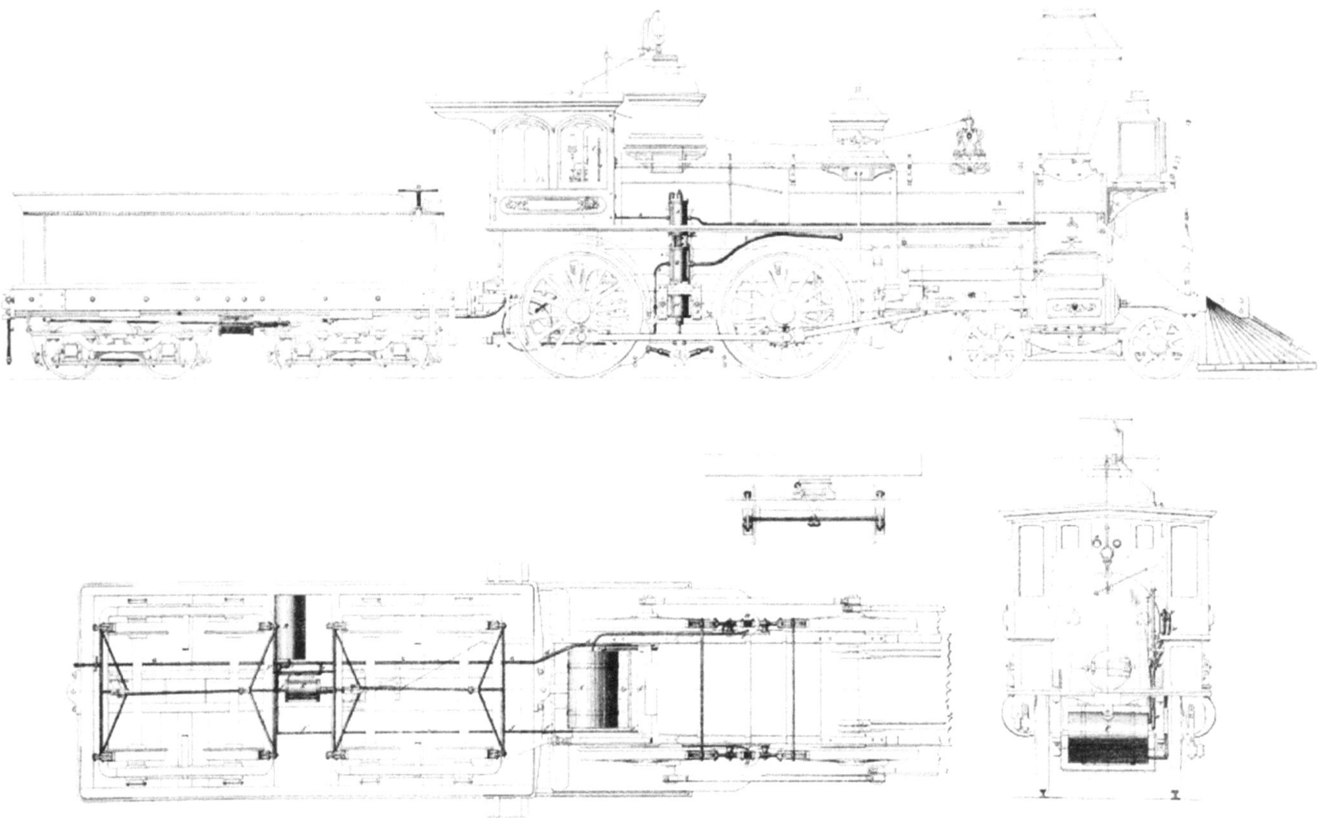

FIGURE 279: THE FIRST AIR BRAKE INSTALLATION - LOCOMOTIVE & TENDER EQUIPMENT

Westinghouse was an enlightened employer; Wilmerding was a factory town but the stores were privately run. He had pleasant houses built for the employees which they could rent or buy at a low cost, and he instituted the half-day off on Saturdays, paid vacations, pensions, and sickness and accident benefits. The company office building, referred to as 'The Castle', was built in 1890 on Marguerite Avenue in Wilmerding. It had a swimming pool, bowling alleys, restaurant and a library. Westinghouse's office, along with that of his staff, was on the third floor. It also had a tower that contained steel vaults on each floor. Westinghouse had valuable papers stored in the tower every day in case of fire including documents, inventions and drawings. On April 8, 1896, the general office did in fact burn. But the foundation, tower, and all of his valuable papers were saved. Reconstruction began immediately—the office building reopening early in 1897—but this time the building was to accommodate only offices since a new YMCA had been built in town, and this provided for staff recreational needs.

The company, with Westinghouse's inventions for braking and signalling systems, helped to transform the railroads. He continued to evolve and modify his air brake design and soon developed the Automatic air brake and its operating device, the triple valve. In 1893 the United States Railroad Safety Appliance Act mandated that all American trains must be equipped with the air brake, and the invention was eventually adopted by most of the world's railroads.

In fact, though, despite his work and unarguable achievements, George Westinghouse was not the first to conceive the idea of operating a railway brake using air pressure, and as he incrementally established his profile with successive ideas and patents, others sought to challenge his emerging primacy. Indeed, with the obvious exception of the triple valve, he was not even *...the inventor of the larger part of the devices employed for such purposes...*[although] *such fact does not detract at all from his merit or rights as a successful inventor.*[179]

Justice Swayne and Judge Welker, though, added:

The organisms covered by [his] *patent reissue 5504 seem to have been entirely new with him, and the incorporation of these elements, together with that of graduating the air pressure in the brake cylinders—also shown to be new and of the highest importance and utility—...with other substantial and material differences not necessary to enumerate, fully substantiate his pretentions as an original and meritorious inventor...*

Westinghouse's contribution is further acknowledged and summarised (in typically protracted legalese) from the above worthy benchers:

So far as appears from the testimony in this case, none of the alleged prior inventions of air-brake apparatus have ever been successfully applied to practical use, and when we consider the immense importance of the introduction of the air brake on railroads, and the incalculable benefit which it has conferred on the public in the readiness and certainty with which trains can thereby be controlled, and the comparative immunity from accidents thus secured, and also the number of devices which have been patented for this purpose, in connection with the fact that Westinghouse was the first, so far as appears in the records and proofs, to put an air brake into successful actual use—such considerations only strengthen and confirm the soundness of the conclusions to which a careful examination of these prior patents has led us—that there are substantial and essential differences between these prior patents and the Westinghouse apparatus, and that to these differences we may justly attribute the successful and extensive introduction of the Westinghouse air brake.

In the context of any discussion about what Westinghouse may or may not have been singularly responsible for "inventing," the following words from a biographer[180] are apposite:

All of this goes to show that there is very little new under the sun in the sense of complete and unanticipated inventions, as every student of the history of patents knows. A famous inventor says... "Our ancestors were very dishonest; they stole all our best inventions." In 1868 Westinghouse was only 22 years old. His experience in taking out patents

179. Proceedings and decision of the US Circuit Court, Northern District, OH: G. Westinghouse Jr v The Gardner & Ranson Air-Brake Co., April, 1875. Judicial determination on the question of Westinghouse's claim to be the original inventor of the first form of air brake. The case was tried at Cleveland, Ohio, from April to November 1874 before Justice Noah Haynes Swayne and Judge Martin Welker. Prout's biographical description of this legal action contains significant errors in detail.
Thomas W. Ranson was born in England in 1846 and had some early experience in the Royal Navy. He came to Cleveland when he was about twenty years of age and was for many years prominent in railroad work. At one time he was superintendent of motive power for the old C.C.C & I. (the Cleveland, Columbus, Cincinnati & Indianapolis Railway) before it was acquired by the Big Four System (the Cleveland, Cincinnati, Chicago & St. Louis Railway). He was also, for some time, an employee of the Erie Railroad. In 1872 he co-patented an air brake arrangement, and with John W. Gardner (of whom little is known) founded the Gardner & Ranson Air Brake Company to market it. The rights to the new air brake invention were contested in an infringement suit brought by the Westinghouse Company, and being defeated in a lower court, Ranson and his partner never carried the matter to a higher tribunal. After leaving railroading Ranson was engaged in the manufacture of artificial ice machinery, as a member of the Arctic Ice Machinery Company at Cleveland and Canton. He also had charge of the American Ice Company's plant at Baltimore and Washington. He retired from business about eight years before his death on March 26, 1916.

180. Prout: *A Life of George Westinghouse*.

was limited, and his knowledge of the prior art of power braking was confined to the steam brake of Goodale[181] *and the mechanical brakes of Ambler*[182] *and Loughridge.*[183]

[Note though, that Westinghouse was bright and well-read and despite Prout's loyal support we might ponder the extent of his knowledge of experimentation and research then being carried out in Britain; and we might surmise that it was considerable.]:

Therefore he was an independent but not original discoverer of the fundamental idea of applying brakes to the wheels of all the vehicles in a train through the instrumentality of air, compressed by an independent steam-driven air pump, stored in a tank or reservoir and conveyed at will by pipes to brake cylinders under the cars. From 1852 onward, various patentees in England and the United States filed applications disclosing this general purpose, and partly or wholly providing for and describing in more or less detail the principle devices necessary to accomplish it. None of these patents, however, nor all of them together, covered a complete and workable combination, as did Westinghouse's original patent which, in addition to all the essential devices enumerated and described in the previous patents, provided at least two additional novel devices equally important in actual operation. The first of these was the three-way cock which served as the first form of engineer's brake valve, and the other was the hose coupling for connecting the air pipes between the cars.

With these two elements providing the basis for the success of his invention, Westinghouse with his vision, his sheer will, and his business acumen, created a *"beautiful new art."*

In July 1871, Westinghouse and his wife made their first trip to Europe to introduce his invention, remaining abroad for a year. Despite being able to demonstrate the rapidity with which American railroads were equipping with his straight-air brake, *"...he found it exceedingly difficult to make any impression on the railway managers of Great Britain."*[184] Seeking the assistance of the prominent journal ***Engineering***, he was interviewed at length. Later, one of the journal's editors produced for Westinghouse's review a draft editorial that he proposed to publish, arguing the case for better brakes on British railways. In it, the editor discussed the qualities he considered to be essential for the satisfactory operation of continuous brakes. While a number of these specifications were fully met by the Westinghouse straight-air brake, at least two were not. The editorial stated these as being (1) that if the train should be parted, the brake must apply automatically; and (2) a failure of the equipment on one or more cars should not interfere with the brake response on the remainder of the train.

Biographer Prout writes:

The precise statement of these two requirements may have suggested new ideas to Westinghouse or may have crystallised ideas already on his mind. At any rate it was not very long thereafter that evidence of his serious study along these lines appeared in his patents.

181. Samuel N. Goodale of St. Louis, Missouri; The Goodale Steam Car-Brake Manufacturing Co.

182. In 1862, Augustine Irel Ambler of Milwaukee patented a chain brake that was operated by the revolving of a friction-driven windlass when engaged with a driving wheel on the locomotive, the chain thus being drawn taut to operate the car brake levers. Westinghouse claimed to have thought of this idea himself but saw the principle applied in service for the first time on the *Aurora Accommodation*, a local train on the CB&Q Railroad in Chicago. Seeing the equipment, he had the idea of substituting a steam cylinder for the windlass, the piston of which would have to be of sufficient stroke to be able to draw in the chain, providing more accurate control of the brakes than was possible with the windlass device. Tests quickly showed, however, that an insufficient piston stroke was possible to operate the chain on more than five or six cars. Westinghouse overcame this problem by placing a cylinder beneath each car with a flexible steam pipe connecting each one to the locomotive. The concept, of course, was a failure due to condensation, even in warm weather.

183. Developed by William Loughridge in 1855, this was an earlier chain-operated brake (it must have been improved-upon by Ambler in relatively minor ways). When the inventor succeeded in getting the Pennsylvania Railroad to purchase it in 1859, he was so overwhelmed he reproduced the railroad's letter of acceptance in his advertising. The device was not successful because in service the brakes locked and slid the wheels of the front car, and there was almost no braking pressure at all after the third car. In addition, the wheel-flange drive to the windlass wore out, and in winter the chain was prone to freeze around the numerous pulleys. By the 1860s, Loughridge was experimenting with steam and air pressure.

184. Prout, 1921.

On March 5, 1872, the first public disclosure of Westinghouse's industry in this regard was evident with the issue of two patents addressing these issues, among other things. Patent 124,404 specified in part that:

In the air-brake apparatus heretofore in use a single line of pipe conveys the compressed air from the main reservoir on the locomotive to each brake cylinder. If this pipe becomes accidentally broken at any point it is, of course, useless for braking purposes from that point to the rear end of the train. For this and other reasons I have devised an apparatus consisting in part of a double line of brake pipes, which may be cooperative or independently operative in braking, at the pleasure of the engineer...

The improvement herein described consists in the features of construction and combination by which... when a car becomes disconnected from the train by accident or otherwise... air from auxiliary reservoir will be admitted freely to the brake cylinder so as to automatically apply the brakes...

According to Prout, four significant things were mentioned for the first time in this patent:

1. The double line of brake pipe (being an "operating" and a "reservoir" pipe);
2. The installation of separate or "auxiliary" reservoirs on each car;
3. The employment of the brake system as a means of communication by signals; and,
4. The installation of an independent compressing apparatus on each vehicle, including a suggestion of axle-driven compressors.

"Priority of invention" was claimed for the first three of these. The patent described the usual method of operation of the straight-air system using *either* of the brake pipes. The other brake pipe would be available to connect the main reservoir with the auxiliary reservoirs and at all times be charged to a pre-determined pressure. At the end of each car, "cross-pipes" would establish communication between the two brake pipes via *...three-way valves that were operated automatically to charge the brake cylinder with auxiliary reservoir air in case of a break-in-two or in the event of a car leaving the rails* [refer to the piping system diagram in the lower portion of Figure 8, page 14]. *The various devices provided to accomplish this purpose are crude and their availability in actual service somewhat questionable, but as the first step in the development of an automatic brake, the idea disclosed by this patent is interesting.* [Note the designation, "three-way valves". There is as yet no reference to the famed "triple valve".]

The new automatic emergency feature, then (for that is what this device was), functioned as an independent "secondary" brake system not controllable by the locomotive driver. Normal service stops would continue to be made with the straight-air system (Figure 282, page 370).

Another air brake historian has suggested an additional reason that Westinghouse felt a revision of the straight-air brake was necessary:

The straight-air system was hardly patentable; competitors could easily point to prior inventions. A growing business such as Westinghouse's would attract rivals, and he needed a specialised design to protect his valuable monopoly.[185]

The communicating signals mentioned in point 3 above were achieved by admitting air from the charged reservoir pipe into the operating pipe. In each passenger car and on the locomotive were installed pressure gauges scaled with eight equidistant marks upon their face. Each of these points conveyed a distinct order or message (such as "flag station," "stop for orders," etc.). The position assumed by the gauge-indicating needle was determined by the amount of pressure admitted into the operating pipe through the controlling valve on each car as manipulated by the operator (conductor, brakeman, engineer, etc.). An alarm whistle

185. John H. White Jr.: *The American Railroad Passenger Car, Part 2* (The Johns Hopkins University Press, 1985).

to engage the attention of the engineer was provided. Westinghouse soon found that audible signals alone were able to meet every practical requirement, and the index gauges were eventually abandoned. This idea would evolve into the well-known and simple air signal later installed for conductor-to-engineer communication on American passenger trains.

The devices necessary to render the brake operation "cooperative" as mentioned in the quote from Westinghouse's patent on the previous page were covered by another patent (124,405) issued concurrently. So well does the specification for this second patent describe the operation of an automatic brake controlled by the locomotive driver through the variation of air pressure in the brake pipe(s) that it is worthy of inclusion here:

I now propose further to improve this system of railway brakes by bringing this continuous reservoir pipe into communication with the brake cylinders at pleasure, through the agency of compressed air admitted from the main reservoir into the other brake pipe; by such construction and arrangement of intermediate devices that, by simply discharging compressed air from the auxiliary reservoir to the brake cylinder, whereby the brakes will be applied; by a system of valves and ports which shall effectuate all these results by their automatic action, except as their action is governed by the engineer at the main reservoir.

Thus do we learn of the intellectual leap by Westinghouse from the concept of a straight-air to an automatic pressure brake. The vaguely referred-to "intermediate devices" and "system of valves and ports" by which the functions described by Westinghouse were accomplished would eventually evolve—through his diligence and ingenuity—to become the "simple and beautiful plain triple valve in its final form" (patent 220,556), providing the means whereby the controlled discharge of air pressure from the "reservoir pipe" would apply the brakes "at will," or the rupture of the brake pipe would apply them in "emergency." The triple valve of 1879 can be considered the legitimate successor to the collection of valves previously conceived. From this great leap, suggests Prout:

The student of the air-brake art...will get a new conception of the mechanical ability of the man who worked out this combination, and a still higher appreciation of the later developments by which even more remarkable results were obtained.

Westinghouse had, in fact, added complexity to his original concept by using two pipes, neither necessarily used for the purpose designated by their respective names. But the inventor originally considered it essential that either pipe be used as an "operating pipe" with the inevitable result that an "*...extravagant number of independent valve devices was required to accomplish this purpose.*" The structure embodied in patent 124,405 was apparently successfully rack-tested but never placed in service, as less complicated devices soon superseded it.

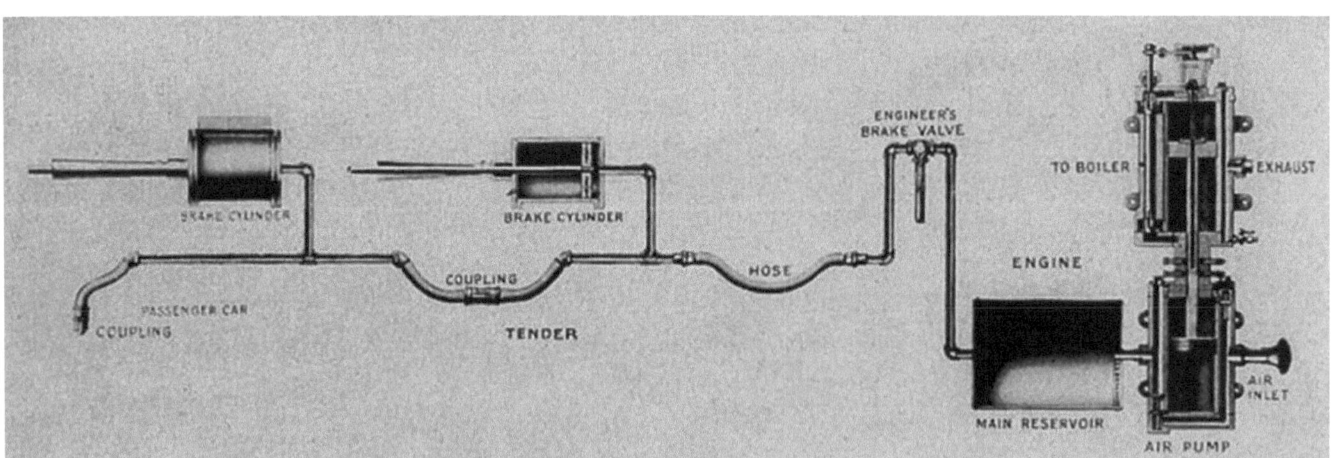

FIGURE 280: THE WESTINGHOUSE STRAIGHT-AIR BRAKE - SECTIONED VIEW

FIGURE 281: THE ORIGINAL WABCO PLANT ON LIBERTY AVENUE IN PITTSBURGH

The triple valve first assumed its general form so well-known—and to be maintained until the emergence of the control valve—in patent 138,827 issued May 13, 1873. In this progenitor valve, the brake pipe and auxiliary pressures were separated by a rubber diaphragm (Westinghouse's "flexible annulus").

This triple valve was a triumph, being both sensitive and efficient, but the leak-prone poppet valves and perishable rubber diaphragm were a problem. Westinghouse soon developed the piston-and-slide-valve triple, the experimental version gaining patent 168,359 of October 5, 1875 (Figure 9, page 15) and the first commercial version in standard use until 1879, patent 172,064 of January 11, 1876. This device embodied a unique improvement, the feature of "lost motion" in the slide valve, the great value of which was in eliminating the hindrance of friction in the slide valve and enabling the feed port to open and close in response to only slight variations in brake pipe pressure, thus increasing its sensitivity still further.

The final step in the development of the plain triple valve was the introduction of the *graduating valve*, which provided for even more delicate regulation of the brake operation; this discovery being documented in patent 220,556 of October 14, 1879. With minor modifications, this version of the plain triple valve remained the standard for passenger car application until the invention of *quick-action* in 1887 and, for locomotives (and tenders), until the introduction of the ET equipment in 1908.

It may have seemed that Westinghouse had the answer to every railroad executive's dream, but the take-up of his original straight-air system was not without resistance. To begin with, the equipment was not cheap; the compressor and other locomotive-mounted appliances cost almost $400 (about USD$9 500 in 2013 terms) and the car devices another $50 ($1 200). And once the company decided to thus equip, they had to do so completely because any car lacking this equipment would interrupt the brake pipe continuity. One contemporary writer[186] records that:

...the air brake created a public sensation even before the tragedy at Revere.[187] Just two months after the initial trial, the inventor negotiated his first sale to the Pennsylvania Railroad. In seeking to persuade Pennsylvania President J. Edgar

186. Steven W. Usselman, *Regulating Railroad Innovation: Business, Technology and Politics in America, 1840–1920* (Cambridge University Press, 2002).

187. On the evening of August 26, 1871, the Eastern Railroad's *Portland Express* collided with the rear of a stopped local train at Revere, Massachusetts. Under hand braking, the express managed to slow from 30 to about 10 mph, but the collision telescoped the rear cars, and coal-oil lamps ignited the wreckage. Twenty-nine people died (some sources say thirty-two) and others were injured. The railroad was condemned by the public for its conservative management, old-fashioned equipment, not having the Westinghouse air brakes on its locomotives, not using the telegraph, and its insistence on using the old time-interval safeworking system. The company was sued and almost went bankrupt, leading eventually to the Eastern being merged with its rival, the Boston & Maine Railroad, in 1884.

Thomson to try his device, Westinghouse informed him of two major railroads that had already adopted it, and that other western roads would probably do the same, with a view to featuring its use in their advertising. The Chicago, Burlington & Quincy became an enthusiastic supporter in the west but in other parts CB&Q executives were skeptical. The railroad's western devotees, though, were persistent; "I have no doubt..." said one western manager, "...that [his region's use of the air brake] *will be made a subject of reference in advertisements and that whether the travelling public would really be more safe or not, they would think so.*

Railroads were uncomfortable with Westinghouse's monopoly and tried to diminish it. The Pennsylvania, for one, purchased some equipment and then asked Westinghouse for a license to manufacture it. Westinghouse, of course, rejected the notion, explaining that he wished to maintain uniformity in the apparatus; something—he noted—that would be essential in order for railroads to be able to interchange air braked rollingstock. Remember, too, that some railroads were choosing the vacuum brake ostensibly because it was cheaper. Westinghouse considered that a major reason for the existence of this irritating vacuum brake market was to encourage competition to weaken his position. However, despite all efforts, says Steven Usselman:

Railroads ultimately foundered in their attempts to undermine Westinghouse. In a little more than another year's time, the vacuum brake was a dead letter in the United States and virtually all passenger trains had air brakes. Westinghouse accomplished this reversal primarily through a relentless patent fight. The inventor had built up a formidable array of patents, including a basic one covering continuous brakes operated by compressed air, which he could plausibly argue encompassed the vacuum principle. Whenever railroads expressed interest in vacuum brakes, Westinghouse appeared with threats of infringement suits. When railroad executives persisted in obtaining vacuum brakes, Westinghouse offered to supply them cheaper himself...

Westinghouse met enough resistance [from some railroads] *that in early 1875 he purchased the Smith company and its patents. Meanwhile, Westinghouse conducted his own* [vacuum versus air brake] *comparison, declared the air brake superior, and in 1876 discontinued production of vacuum brakes. He had for all intents and purposes secured a monopoly over a technology most Americans had come to consider indispensable for railroad passenger service.*

The global Westinghouse company

Westinghouse's plans for commercial expansion included the world; these vast plans apparently having become defined early in his career. Three years after the air brake was started in America, he was in England, and within another year a Pennsylvania company was formed to handle export business. Nine years later, the first English company was organised. This was soon followed by German, French, and Russian brake companies, and later by the brake companies of Italy and Australasia. As the electric field opened up, another group of Westinghouse companies spread over Europe. Some twenty-two foreign companies are listed, including a Hungarian automobile company.

Westinghouse wrote:

I have never had much difficulty in working out plans, but my greatest difficulty seems to have been in finding enough men to carry out such plans. Indeed, young Napoleon could not find a man in Italy, and the Pharaohs had to import ability from Asia. This difficulty in finding men is common to the human race. Lord Fisher said, "...the secret of efficiency is favouritism..." which can be interpreted as "...when you find capacity grab it and push it forward, regardless of age or rank."

This principle appealed to Westinghouse's practical common sense, and he acted on it often. Yet, he was loyal to his men and had a punctilious sense of justice. He kept some men too long in high places, and he injected into the organisations some men who had later to be ejected.

In 1896 Westinghouse wrote a letter to an officer of the Westinghouse Electric Company Limited (called the 'London Company'), setting forth in some detail his idea of a vast scheme of foreign companies. At that time there existed—besides the London Company (registered in 1889)—the Westinghouse Brake

Company Ltd. in England and the Westinghouse Bremsen Gesellschaft in Germany, all having broad territorial rights. The (American) Westinghouse Electric Company transferred to it all of its patent rights for the whole world outside of North and South America. This was a trading and constructing company; it did not manufacture but did install machinery on contract.

The three brake companies, American, English, and German, covered the world, trading within their defined territories. The 1896 plan contemplated the formation of new companies; British, French, Belgian, German, Russian,[188] and Austrian (to include the Balkan states). Norway, Sweden, and Switzerland, he said: *...should probably be reserved for the Westinghouse Company (British) as part of its territory.* Italy is not mentioned in this letter, but an Italian brake company was formed in 1906 and an electric company in 1907. These European companies and the parent companies in America were to take care of the habitable globe.

Canada was not included in the plan of 1896. In October of that year, the Westinghouse Air Brake Company (United States) began the purchase of land, buildings, and machinery in Hamilton, Ontario, with a view to the manufacture of brake equipment, and in the following January the Westinghouse Manufacturing Company Ltd., was incorporated under the laws of Canada. In July 1903, the Canadian Westinghouse Company Ltd., was chartered and on October 8 was formally organised with a capital of $2 500 000. *The new company is a consolidation of all the Westinghouse interests in Canada, which heretofore have been conducted individually.*[189] These interests and operations included electrical, steam engineering, and air brake. This company still exists.

The British Westinghouse Electric & Manufacturing Company was organised in 1899. It built great and famous shops at Manchester and became an important producer but is counted among Westinghouse's mistakes. It was never successful financially and was a heavy burden upon the home company and upon Westinghouse personally. Nevertheless, it was a striking example of his prescience.

The plan of 1896 was always in mind and was pretty well executed, but often with disappointing financial results. Writing in 1909 to the directors of the Electric Company, Westinghouse said:

The extraordinary growth of your business in America required the whole time of your best officials, so that neither they nor I had adequate time to effectively carry out the plans outlined. Nevertheless, there was established a strongly favorable financial situation which existed for several years, and a manufacturing record was created in Europe which has made the name a real power in business.

These words were written two years after the panic of 1907 and the receiverships of several Westinghouse Companies. The triumphant and audacious note of the past years is not heard, but the Old Man was still struggling to save for his stockholders what might be saved. Large equities were saved; all of the important companies except the Russian Electric Company are still operating.

The results of the world plan cannot be estimated; they will not be developed for years. It is not at all probable that they will ever develop—as Westinghouse had hoped—into a great system of allied companies cooperating closely under one central management; indeed, it is not certain that such a development was ever practicable or desirable. But those bold and varied enterprises carried the air brake into Europe and Australasia and so promoted the evolution of transportation. They made familiar to British and Continental engineers and financiers the distribution of energy by the alternating-current, and that spread abroad the fundamental idea of the central power station. Thus, they stimulated and advanced the manufacture of power and pushed along the industrial revolution, the most important phenomenon of the nineteenth and twentieth centuries.

---

188. A gentleman who saw the St. Petersburg works in July 1921 wrote that about two hundred men were at work. "They produce some air brakes, carts for the army, and farm implements which they use to exchange with the peasants for their wheat, etc. The manager received 300 000 rubles a month, plus rations of bread, butter, and a little sugar, etc. He was bartering away his clothing and other things to live."

189. *The New York Times*, Friday, October 8, 1903.

Westinghouse's industry expanded as he opened companies in Europe and Canada. In the United States, he expanded into the railroad signaling industry by announcing in 1881 that he had purchased the rights to William Robinson's[190] signaling patents, including the one covering the fundamental fail-safe principle of the closed circuit. Through this acquisition he was able to organise the Union Switch & Signal Company, where devices based on his inventions and the patents of others were designed to control the increased speed and flexibility of train operations made possible by the invention of the air brake.

Apart from mechanical modifications of greater or lesser significance, the invention of an improved engineer's brake valve in 1896 and of a quick-service triple in 1907, Westinghouse's direct personal contributions to the air-brake art ended with the perfection of his quick-action triple valve in 1887. When, in order to cope with changing conditions within the railroad environment, the next step forward became necessary, he was so deeply engaged in other work—largely, but by no means exclusively—of an administrative character, that the Westinghouse Air Brake Company was able to rely upon the engineering talent that in the meantime had been developed within its own ranks or acquired by additions thereto.

Natural gas caught Westinghouse's attention after a well drilled on his own property produced a large flow of gas, and in 1885 he purchased the charter of a holding ironically called the Philadelphia Company. He went on to overcome issues of logistics and safety to supply gas to thousands of private houses in Pittsburgh through many miles of reticulation.

During his development of the braking and signaling systems in the mid-1880s, Westinghouse became interested in electricity. His interest was piqued by the obvious disadvantages of Edison's direct-current system. He began pursuing the technology of alternating-current, and he associated with those who were developing AC devices. He obtained the US rights to Gaulard and Gibbs' system of distributing AC and hired William Stanley to redesign and improve the Gaulard-Gibbs "secondary converter", or transformer, as the device was later called.

Westinghouse organised the Westinghouse Electric Company to manufacture and promote the use of alternating-current equipment and became a spirited competitor of Edison and his DC system. He acquired exclusive rights to Nikola Tesla's patent for the polyphase system in 1888 and lured Tesla to join the electric company and continue his work on the AC motor he had been developing.

In 1892 Westinghouse won the contract to light the 1893 Columbian Exposition at Chicago, manufacturing over two hundred thousand lamps for lighting and replacements. The Westinghouse exhibit also included a complete working model of a polyphase system, including step-up and step-down transformers, a short-length transmission line, and a switchboard. At about the same time, Westinghouse was negotiating with the Cataract Construction Co. of Buffalo, New York, to supply AC generators to harness the energy of Niagara Falls. He succeeded over six other American companies that had been asked to bid. The generator specifications called for three 5 000 hp, two-phase generators (2 200 V, 250 rpm) complete with switchboard and auxiliaries. The system was placed in commercial service in late 1895. There followed many successes for Westinghouse's company in the fields of power generation and the application of electricity to industry, rail, marine transportation, the military, and to the home.

In 1895 he also began the development of gas engines and built high-speed steam engines designed by his brother, Herman. Westinghouse made further industrial history by acquiring, in 1896, exclusive rights to manufacture the Parsons steam turbine in America.

---

190. William Robinson was born November 12, 1840, in Brooklyn, NY, and invented the automatic electric and electro-pneumatic signal systems known as the Robinson Electric Railway Signal System. He also patented the Robinson Radial Railway Truck and a Repeating Telephone. He was a high school principal, taught in the Classical and Commercial School, Stamford, Connecticut, and was principal and teacher of Natural Science in Spring Valley Academy, New York. From 1875–81, he was engaged in business in Boston, Massachusetts, and in 1878, he organised the Union Electric Signal Company. Robinson is credited in 1872 with the first successful application of the railway track circuit and became president and general manager of the Robinson Electric Railway Signal Company. He also did considerable work in developing a safe third-rail electrification solution. He died in 1921.

At the turn of the century, the various Westinghouse companies were worth about $120 million and employed approximately fifty thousand workers. By 1904, there were nine Westinghouse manufacturing companies in the United States, one in Canada, and five in Europe.

The first major application of alternating-current to railway systems was in the Manhattan Elevated railways in New York, and later the New York subway system. The first single-phase railway locomotive was demonstrated in the East Pittsburgh railway yards in 1905, and, soon after, the Westinghouse Company began the task of electrifying the New York, New Haven & Hartford Railroad with the single-phase system between Woodlawn, New York, and Stamford, Connecticut.

Then came Westinghouse's personal tragedy. The great financial panic of 1907 caused the failure and receivership of the Electric Company, the Machine Company, and some minor companies although it did not affect the Air Brake Company or Union Switch & Signal. A reorganisation was eventually engineered, based upon a brilliant project devised by Westinghouse, but the actual control of the Electric Company passed out of his hands, and in less than four years he ceased to have any official relations with the company.

Although these circumstances affected Westinghouse greatly, there remained some fragment of his indomitable spirit that was not beaten. Spending much of his later life in public service, the short years that remained after the calamity were filled with the same unceasing activity and the same undying hope, warding off the depression and despair that might otherwise have engulfed him. The affairs of the Machine Company, the development of the steam turbine and of the reduction gear, the invention in 1910 and subsequent development of a compressed air spring for automobiles, and various minor interests completely filled the busy hours of the long days.

But Westinghouse showed signs of a heart ailment by 1913 and was ordered by doctors to rest. After deteriorating health and illness confined him to a wheelchair, he passed away on March 12, 1914; close at hand were drawings for an electric wheel chair. As a war veteran, Westinghouse received the privilege of a plot in Arlington Cemetery in Washington, D.C. Marguerite died three months later and is buried with him. With a total of 361 patents to his credit, his last was received in 1918 - four years after his death.

Many honors accrued to Westinghouse, and although the Union College, where he had spent only three months as a youth, conferred upon him the degree of Doctor of Philosophy, he declined honorary degrees from several colleges. He was awarded the John Fritz Medal and the Franklin Institute's John Scott Legacy Medal and Premium.[191] He was one of two honorary members of the American Society for the Advancement of Science. Westinghouse also received the AIEE Edison Medal—named for his strongest opponent—in 1911 for "meritorious achievement in connection with the development of the alternating current system for light and power." Abroad, he was made a member of France's Légion d'Honneur; King Humbart of Italy decorated him with the Order of the Crown; and King Leopold II of Belgium conferred the Order of Leopold. In Germany, he was the first American to receive the Grashof Medal, the highest honour bestowed by that country on an engineer.

George Westinghouse was a hero in an age of heroes. Even in a period exploding with scientific creativity, he was hailed as the world's leading inventor-engineer. Patenting useful devices before he was old enough to vote, he worked throughout his career at the cutting edge of technology. His legacy has maintained its momentum for well over a century, and the Westinghouse name continues to denote technical innovation and practical application.

He left no written record except in the files of his numerous companies. He wrote almost no private letters. He kept no journals or even note-books [the author doubts this]. He made but few addresses and wrote few papers. Some record of his work might be made after a laborious search of office files, so far as the files of forty-eight years still exist, but the result would be formal and without color. It would not be a life, and George Westinghouse was a very human being. Furthermore no one man has the range of knowledge and the comprehensive judgment of the relative importance of the things done to fit him to write an

191. The origin of these awards is well-documented in Wikipedia.

adequate life of George Westinghouse. Lord Rosebery said that it would take a syndicate to write the life of Gladstone; perhaps this is quite as true of a life of Westinghouse.

"If someday they say of me that in my work I have contributed something to the welfare and happiness of my fellow men, I shall be satisfied."— GW

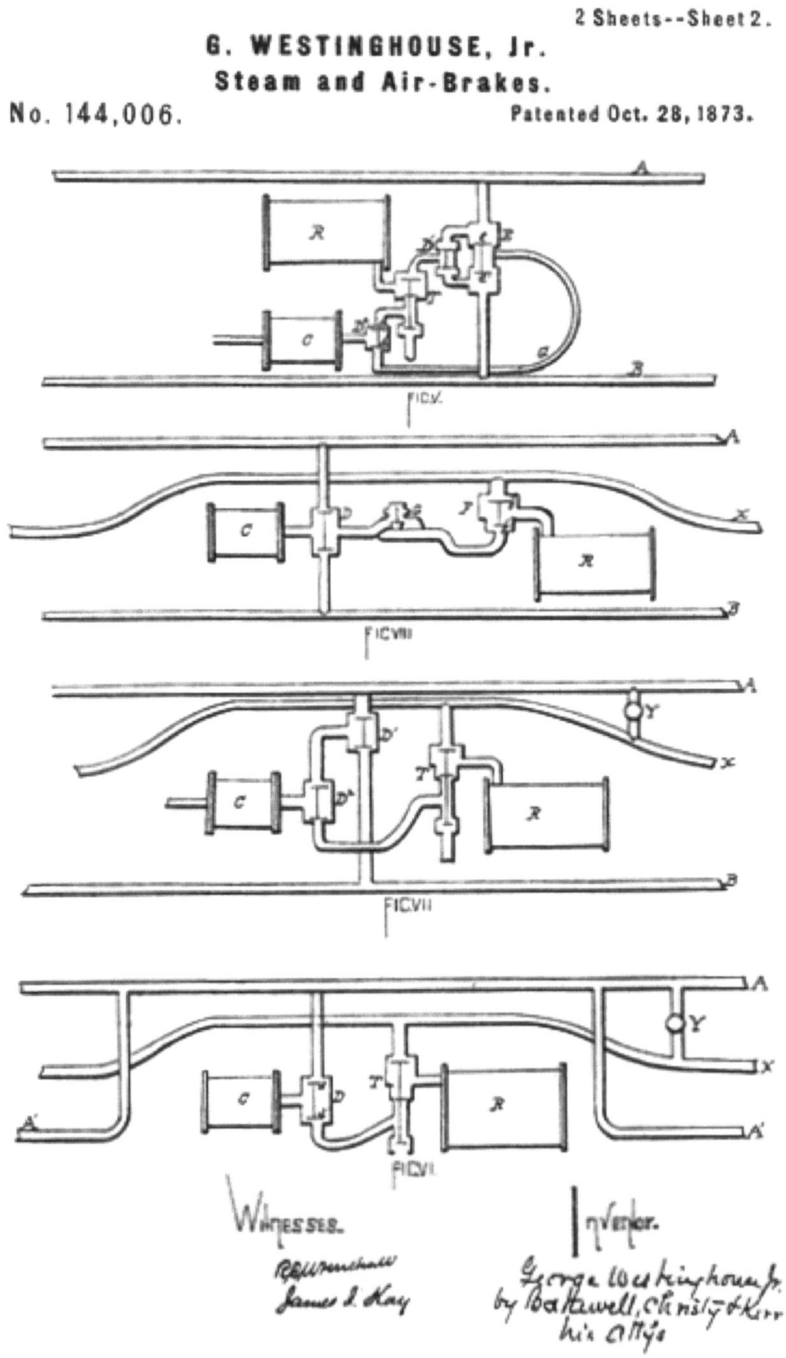

FIGURE 282: I DO NOT HAVE AN ILLUSTRATION OF THE DEVICE AS RENDERED IN PATENT № 124,405; HOWEVER, THE 'EXTRAVAGANT NUMBER OF INDEPENDENT VALVE DEVICES' MENTIONED PREVIOUSLY ARE SHOWN IN THIS ILLUSTRATION AS PART OF THE RECORD OF PATENT № 144,006

The first air brake

Some historical commentary of the original Straight-air brake, including reminiscences of its first locomotive engineer; February 1900.[192]

The first train controlled by air brakes in America, and doubtless in the world, was a local passenger train on the Pan-Handle Railroad[193] in April 1869. Early in the spring of that year arrangements were made by George Westinghouse with W.W. Card, superintendent of the Pittsburgh, Cincinnati & St. Louis Railway, to test his 'straight-air' brake. The train was the *Steubenville Accommodation*[194] making a round trip of 86 miles each day between Steubenville, Ohio, and Pittsburgh, PA. The engineer of this train was Daniel P. Tait, afterwards road foreman of engines, and at the time of this interview General Foreman Shops at Steubenville, Ohio. Johnson Mooney, of Steubenville, Ohio, was the conductor.

After the air brake had been experimented-with for several months on the Steubenville Accommodation, the officers of the Pennsylvania Central Railroad—having taken great interest in the experiments, and ever watchful of the safety, welfare and comforts of its patrons—arranged to have the brake applied to the *Wall's Accommodation*[195], which was accordingly equipped in September of the same year, with a view to giving the brake a thorough practical test and adopting it on all of their trains should the tests prove satisfactory. So eminently satisfactory were the subsequent trials, that shortly afterward the Pennsylvania began the application of brakes to their entire passenger equipment, being the first railroad to adopt and use the air brake as its standard.

To practically demonstrate the value and utility of the new device, the Westinghouse Air Brake Company, which had meanwhile been formed to manufacture air brakes, equipped an engine and passenger train of six cars with the brake, and made an exhibition tour of the principal railroad cities between Pittsburgh and Chicago. At Chicago, on the Galena division of the Chicago & North Western Railway, and before the most prominent and notable gathering of railway officials of the times, a successful and memorable series of tests were made. The Chicago Evening Post of November 26, 1869, commented upon the tests as follows:

A large party, composed mainly of prominent railway officials, took an excursion yesterday afternoon on the Galena division of the Chicago & North Western Railway, for the purpose of testing an atmospheric brake, invented by George Westinghouse, Jr., of Pittsburgh. The train, consisting of a powerful locomotive and six fine passenger cars of the Pennsylvania Railroad, was brought here at the suggestion of the superintendent of the North Western road, expressly to exhibit the improvement to our railway men. The brake, which was subjected to the most thorough tests, worked to the most perfect satisfaction of all present, who declared that its equal had not been seen. The heavy train was put under headway at the rate of 30 miles an hour, and stopped within a distance of 80 feet more than its length, in 19 seconds after the brake was applied. Another trial was even more successful, the train making nearly 40 miles an hour, and being stopped within the length of about seven cars, or some 350 feet, in 18 seconds, without jerking, the track, too, being rather slippery.

The Pennsylvania was now vigorously pushing to finish the application of straight-air brakes to all passenger equipment, cars and engines. Upon the assurance of Mr Card that Dan Tait was the engineer who really ran the first air-brake train, [Railway & Locomotive Engineering] immediately called upon Mr Tait in

---

192. Compiled from *RAILWAY AND LOCOMOTIVE ENGINEERING: A Practical Journal of Railway Motive Power and Rolling Stock*: John Alexander Hill, 1858-1916; Angus Sinclair, 1841-1919; (New York: A. Sinclair, J.A. Hill [etc.]).

193. Spelling is per original Pennsylvania Railroad timetables.

194. Steubenville is a city in the county seat of Jefferson County, Ohio, and is located along the Ohio River. An 'accommodation' train was an early railroad term for a local service that stopped at all or nearly all stops and conveyed both passengers and freight.

195. Wall is a borough in Allegheny County, Pennsylvania. It is located to the east of Pittsburgh, adjacent to Wilmerding. It was originally the site of a farm purchased by James Walls in 1829.

his suburban home near Steubenville, where he related interestingly his experience with the early struggles of the first air brake as follows:

"*Yes, my boy,*" he said, "*I ran the first air brake* [train]. *Moreover, I—along with Nick Johnson and Harry Frazer—helped George Westinghouse put his first air pump on an engine. We put it on the '23', a little eight-wheel Dickie Norris engine*[196]. *The air pump was made-over from an old Worthington water pump and laid flat on its back on the running board outside of my front cab window. I'll never forget, it had a leather packing in the air cylinder. After a while Mr Westinghouse took this pump off and put on one that stood upright on the side of the boiler, and we used to take turns pushin' and pullin' at the trigger valve motion to keep it going when it wanted to lay down.*"

"*What did people think of the Westinghouse brake in those days?*" Dan was asked. "*Well,*" replied the veteran, stroking his long, flowing beard and running his fingers through his silver-grey hair, "*...the brake had its friends, and its enemies, too. Why, my boy, I think its best friends were the brakemen and myself. They liked it because it did their work for them. I liked it because it did the work quicker, surer and better than any human beings could. It gave me the control of the train, which seemed the right thing to do, for I was in the best place to see what was going on ahead, had the dose of right kind of medicine right under my thumb, and didn't have to rely on help from anybody to apply it.*

"*The president of the road didn't like it, for it seemed to be always getting out of order. Hanged if it didn't seem to go wrong every time he rode on the train! But there was never a trip but what it worked part of the time. When it got out of order and we had to go back to hand brakes, we would work all night but what we got it in shape for the next day's trip. I was only too glad to work on it nights, for I saw it was a 'comer' and told Mr Westinghouse so.*"

"*Did Mr Westinghouse seem to think the brake would ever amount to anything?*" "*Did he? Well, I should say he did!*" replied Dan emphatically. "*He was always enthusiastic over it; never got stumped, and was the quickest man to find a remedy I ever saw. 'Keep a stiff upper lip, Dan,' he would sometimes say as we wearily pulled the old trigger valve or when something broke. 'There'll come a day when we get this brake in shape that no railroad can do without it.'*"

"*And what did you think of it, Dan?*" "*Why, my boy, from the very first stop I ever made with it—the stop at Fourth Avenue the first trip*[197]*—the stop where the Westinghouse air brake drew its first breath of life—I saw it was a wonderful thing and was bound to become a great success. After that stop I turned around to Mr Westinghouse, who shared my seat box, and said: 'If you get this thing going all right it'll be worth millions. Mark my word!'*"

"*Those were prophetic words, Dan.*" The old patriarch sat silent and continued to stroke his silver-grey beard. His eyes wandered through the bay window of his room, where he was convalescing from a sharp illness, to the orchard trees standing forth black and bare from the heavy snow-carpeted orchard below. "*Yes,*" he resumed, "*I told Mr Westinghouse that there would come a time when every passenger car in the country would have an air brake. He had a way of quietly smiling when I got enthusiastic over the doings of the brake, but said nothing.*"

Here Uncle Dan broke off his narrative and gave his attention to a weeping child at his knee. "*Fix it, grandpa,*" she wailed, holding up a broken doll. "*I can't fix it, dear,*" he said, soothingly. "*Its neck is broken square off. There, don't cry.*"

"*Broken necks remind me,*" continued Uncle Dan, "*that the air brake couldn't have been introduced on a better road; for nowhere in the land was there a pike where the engineer and fireman had to so quickly, regularly and frequently unload their*

196. Richard Norris & Sons, locomotive builders, Philadelphia, 1853-1867. The Norris Locomotive Works, which had begun manufacturing locomotives earlier under another brother, held a leading position for a number of years regarding the magnitude of their operations, the speed capabilities of their locomotives, and the company's readiness to innovate. They built more than 1 000 locomotives between 1834 and 1867. A Norris design is thought to be the genesis of the 'American' locomotive (4-4-0); credited by some archeologists as being the first thoroughly successful locomotive in the United States. The Norris companies exported 170 locomotives to Britain, France, Germany, Prussia, Austria, Belgium, Italy, Saxony, Canada, Cuba, and South America.

197. Also referred-to as Fourth Street.

baggage out the cab window, gang-way or off the tender as on the Pan-Handle in those days[198]*. One of the first things we learned was the soft spots along the right-of-way and how to strike them. About that time you were likely to see most anybody pop out on you from around a curve. I soon saw that the air brake was indispensable for the Pan-Handle road and was bound to succeed."*

"Did it ever save your neck, Uncle Dan? "*Indeed it did - and the necks of others too,"* he replied. He sat quietly gazing into the big grate fire that threw out hospitable warmth. He said nothing, but was evidently growing reminiscent. *"Yes,"* he resumed, *"it did save necks - a woman's neck once, and I'll never forget the incident if I live a hundred years. One day, coming west, just as I came 'round the curve out of the cut east of Holliday's Cove, I caught sight of two women running on the track, trying to reach the station ahead of me in time to take the train. The skirt of one of the women caught on a splinter of the old iron rail, tripped her up and threw her heavily to the ground, right across the rail, not more than 150 yards ahead of me. I was running 25 miles an hour when I banged on the air, for no one can understand how horrible it is to see a human being ground to pieces under your wheels - most of all a woman. Ten men at each brake wheel couldn't have done the work the air brake did on that stop. The train pulled back like she was dragging a heavy anchor from her rear end. When I stopped the pilot was one side of the narrow road crossing east of the old station, and the nearly crazed woman lay stunned and fainting at the other side. For the hundredth time I thanked heaven that we had been given a 'George Westinghouse' to make us an air brake."*

"Another time I had nearly reached the end of the long straight line on the up-grade above Hanlon's, going east, when I saw several cars rounding the curve ahead, coming down. In an instant I had the air on and perceived that the cars were run-aways from Dinsmore tunnel above, that was being newly-arched. I reversed the '23' on sand, stopped in a surprisingly short distance, and then began to back up. I never saw a train run downhill so slow as we did. It seemed hours, and every moment those nine cars of sand and brick came rapidly closer. Finally they caught me, but I was going pretty fast then, too, back toward Hanlon's, and no damage was done. I dropped the cars into Hanlon's siding, and again thanked my lucky stars that the air brake put the train under my individual control; for had I called for handbrakes, the brakemen would have set them, then looked ahead and seen the cars coming. Then they would have jumped off and left the passengers and me to get smashed up.

"Once I had orders to meet No. 1 at Collier's. I was a little late, and came around the curve into the east end of the yard pretty lively. No. 1 was standing on the main track and hadn't opened the switch for me. Mr Durand, the general manager, standing alongside No. 1 with other passengers, saw me coming in lively and supposed I had forgotten my orders. He ran up the track towards me, waving his hat excitedly. I stopped in regular exhibition style before reaching him, stuck my head out the window and enquired, "Is there anything you want, Mr Durand?" He stuck his hat back on his head, and says, rather sheepishly, 'No. I forgot you had the air brake.'"

"The first emergency stop was made at Fourth Avenue on the second trip[199]*. A coal train, with no flag out, as usual, tried to cross over ahead of me. I didn't see her until she came out from behind the high coal-yard fence, and she was then not more than 40 feet away. I banged on the air, and things looked bad for all hands. The waiting passengers on the station platform scattered and ran in a panic. When the train stopped, the Twenty-three's pilot stood off just four feet from the coal train. That was a fine exhibition stop."*

"Well, my boy, I could spin you yarns all day about the marvellous stops and close shaves had with the first air-braked train. And you can say that during the whole time I believed, as Mr Westinghouse did, that the brake was destined to someday become a great thing. I am the only one of the crew of that first air-braked train that is now alive. I have rail-roaded nearly fifty years—went firing when I was fourteen—and feel happy to know that it was I who helped this big giant—the Westinghouse air brake—draw its first infant breath more than thirty years ago."

And so the Westinghouse air brake came into life.

198. The author assumes Tait is referring to the frequency of derailments caused by the soft track condition and the encountering of track obstructions and unauthorized movements, at which engine crews often immediately jumped since stopping in time was not an option.

199. See footnote 197.

The first automatic air brake

The first automatic brake was experimentally introduced in October 1872, and was applied to the train known as the *Wall's Accommodation*, running between Walls Station and Pittsburgh. Although the straight-air brake had been giving excellent satisfaction and wonderful service, Mr Westinghouse early discovered the importance of having a safety brake which, while doing its ordinary work, would give notice and stop the train should anything happen to the brake to put it out-of-order. It was on this basis, then, that Mr Westinghouse designed the automatic brake.

While the automatic brake on the *Wall's Accommodation* was being tested, the new automatic attachment was being discussed much by the railroads, many of whom took steps to place it upon their cars. A series of interesting tests of the improved attachment was held on the Eastern Railroad in Boston in April, 1873, and made such a good impression that many of the roads started in immediately to equip their trains with the new brake. About this time, the vacuum brake was struggling hard in its efforts to become the leading power brake, and some interesting and exciting competitive trials were made between the vacuum brake and the Westinghouse automatic. In the summer of 1873 the railroads of New England were equally divided as to the superiority of the vacuum and compressed air brake. Some had equipped with one and some with the other. Unsuccessful efforts had been made to bring about a series of tests between the two forms of brakes to demonstrate superiority, and in June of that year, while the Master Car Builders were in session at Boston, arrangements were made to hold these.

The 14-mile stretch between Reading and Lawrence, on the Boston & Maine Railroad, was selected as the running ground, and twelve coaches were used in each train. The trains ran side-by-side on double track. At 27 miles per hour the vacuum brake stopped its train in 20 seconds and 804 feet. The compressed-air brake [straight-air] required 26 seconds and 825 feet, the vacuum winning by 6 seconds and 21 feet. In the next run of 32 miles per hour, the vacuum won again by 5 seconds and 25 feet.

At this stage of the tests, a train of nine cars arrived from the Eastern Railroad, fitted with the automatic air brake. The Westinghouse representatives had unsuccessfully endeavoured to get this train to the scene in time to enter it instead of the straight-air train, against the vacuum train. The vacuum train was now reduced to nine cars, and similar tests were made as with the straight-air brake.

At 28 miles per hour the Westinghouse automatic stopped in 13 seconds, and the vacuum in 22 seconds; the automatic winning by 9 seconds and 315 feet. At 32 miles per hour the Westinghouse automatic stopped in 6 seconds less time and 265 feet less space. At this stage of the proceedings the victory of the compressed-air brake was so pronounced that the vacuum train was entirely withdrawn from the contest. Thus the superiority of the Westinghouse automatic brake over the vacuum was so satisfactorily demonstrated to this body of railroad men that the equipping of cars with the automatic brake went rapidly forward.

The engineer who ran the first automatic-braked train was Levi Close, with John Rauth the conductor. Mr Close ran the experimental train for nearly six months before the above-mentioned trials were made, and went with his train to Philadelphia for exhibition trials in May 1873. The train consisted of seven passenger cars and engine No. 45, Altoona-build, Class C. Here trials were held before a large committee of the Franklin Institute and a large gathering of railway officials - nearly 100 persons altogether.

The tests consisted of getting the train up to speed, when a flag was thrown from the cab window and the brakes applied; therewith, simultaneously, the time of stop and distance being taken. In the first test, the train was severed while running at a speed of 32 miles an hour and the stop was made in 11 seconds and 367 feet. Next, the engine was severed from the train and given steam while running at 45 miles an hour, and a remarkable stop of the cars was made in 10 seconds and 318 feet. So wonderful did the committee think this stop that they requested Mr Westinghouse to repeat the experiment, which he gladly did. From this point on, the success of the automatic brake was assured, and finally displaced the straight-air brake and vacuum forms.

In 1876 there were in use in the United States 15 569 locomotives and 14 055 passenger cars, of which 2 645 locomotives and 8 508 passenger cars, or 37.7 percent of the whole, were equipped with Westinghouse straight-air brakes.

At this time the Pennsylvania Railroad operated trains consisting of a locomotive, one express car, one baggage car, three coaches and six Pullmans. The average weight of a passenger locomotive and tender with fuel and water was fifty tons; baggage and express cars, ten tons; passenger coaches, sixteen tons, and Pullman coaches, twenty-seven tons. On that basis, the Pennsylvania trains above described would weigh 580 000 pounds. With the straight-air brake the average length of stop from thirty miles an hour was 500 feet.

The engineer's operating valve of 1870 was improved in 1879, by what was called the "excess-pressure" valve which was designed to hold in the main reservoir a pressure greater than the standard brake-pipe pressure. This was important in providing for emergencies and permitting prompt release of the brakes and quick recharge of the auxiliary reservoirs. This seems to have met the requirements of passenger service, but when the air brake eventually came to be applied to very long freight trains the need of further improvement was disclosed by an interesting experience. An experimental train of fifty stock cars fitted with automatic brakes was being run over the Pittsburgh Division of the Pennsylvania Railroad. In those days, very few engineers had handled so long a train with the air brake. The Pennsylvania Railroad engineer on this train was a good air-brake man in passenger service, but he had no knowledge—either experiential or theoretical—of the difference between braking a ten-car and a fifty-car train. When he came to let his train down the eastern slope of the Alleghenies, he got into trouble. The grade is long and often well over 100 feet to the mile, and there are sharp curves - including the famous "Horseshoe Curve". As he started down the grade, the engineer did what he would have done with the trains he was used to; he put on brakes and quickly "lapped" his brake valve; that is, he put it in the proper position for running with brakes set. The result was to set the brakes on the leading cars and to cause a surge of air from the rear cars that at once released them. This operation was repeated three or four times; air pressure was getting low, and it looked as if there must be a call for hand brakes [to prevent] a runaway. Westinghouse was in the cab, and it was suggested that he should take the brakes, a suggestion promptly and gratefully accepted by the engineer. By delicate handling, Westinghouse let the train smoothly down the grade, and at the same time built up the air pressure. From this experience quickly came the invention of the engineer's equalising valve - a most important improvement in air-brake equipment.

FIGURE 284: BURLINGTON BRAKE TRIALS

The equalizing valve was designed so to control the discharge of air as to [permit equalisation of] the pressure through the whole length of the brake pipe, and secure uniform application of the brakes. This prevents the things that happened on the Altoona grade setting brakes on the head cars, failure to set them on the rear cars, and the release of the forward brakes by a surge of air from the rear. Probably no man could have reasoned out such conduct by compressed air; it had to be developed and discovered by experiment. The history of the art is full of such situations. This valve was the joint invention of George Westinghouse and his nephew, Frank Moore. In its first form it was used in the classical Burlington brake trials of 1886 and 1887. It was improved in 1889, and then no important change was made in it for twenty years.

Early railroad industry perceptions of the Westinghouse brake

Report of the Proceedings of the Annual Convention of the American Railway Master Mechanics' Association, 1874. London: Forgotten Books. (Original work published 1874)

There is but little difference of opinion, as to the requirements of a train brake, in the minds of well-informed persons having experience in the management of trains running at high speed. The requirements are, namely:

1. Certainty of instant application to all or as many of the wheels of the train as practicable.

2. The retarding power should be capable of being graduated at the will of the person applying it, under all ordinary circumstances.

3. When applied, from any cause, it should so remain until the train is brought to a stop, unless persons in charge of it desire otherwise, in which case it should be capable of instant, partial, or total release.

4. The brake power should be under the control of the engineer of the train. It should apply itself automatically in case of derailment of any part of the train, or in case the train should part accidentally, and should be capable of application from any car in the train.

The above being the main points in the requirement of a theoretically perfect brake, the question for consideration is, how near are these requirements fulfilled by the best continuous train brakes yet brought to public notice?

The system of running trains in this country, the character of the country through which they run, and the causes so frequently occurring, requiring quick stops to be made at unlooked-for times and places to avert danger, imperatively demands the adoption on all trains conveying passengers, running at high speed, some good system of continuous train brake, embracing in the main these features. A long list of the different kinds of continuous brakes, which have been tried on the different roads in this country during the past twenty years could be given, but as the greater number have been abandoned, reference to them is unnecessary. At the present time two classes only seem to give promise of fulfilling the requirements of a brake approaching perfection, namely, the one using compressed air as a means of applying the brakes; the other, a vacuum. Of the former class, that known as the Westinghouse Air Brake is the only one that has so far come into general use.

The Gardiner-Ranson Brake, in some respects a duplicate of the Westinghouse, has been tried on several roads, but it has not, as yet, met with general favour. The Loughridge Air Brake, in some respects also similar to the Westinghouse, is now being applied to the passenger equipment of the Baltimore & Ohio Railroad and the lines under its control, and up to the present time about forty-five locomotives and one hundred and fifty cars have been equipped with this style of brake.

The Westinghouse Air Brake, up to the present time, has been applied to 2,232 locomotives, and about 6,900 passenger, baggage, and express cars, on 141 different railroads (not counting branches) in the United States and Canada; and on 66 locomotives and 448 cars on sixteen different railroads in England, Scotland, Wales, Belgium, and in South America, Cuba, and Mexico.

Of the roads that reported to your Committee, two have adopted the Smith Vacuum Brake, viz: the Central Railroad Company of New Jersey, and the Hartford, Providence & Fishkill Railroad and the Old Colony Railroad of Massachusetts, have it on one engine and four cars, but use the Westinghouse Brake on the balance of their passenger equipment.

The Vacuum Brake has been introduced on a number of other Railroads, principally in the Eastern States, but which have made no reports to your Committee of its operation…

…As quickness of action in a brake is of vital importance, it is then a matter of interest to know in which system the force will be transmitted the quickest through the pipes from end to end of the train, whether by one cubic foot of air compressed to forty-five pounds flowing into the pipes and cylinders of the train, or by three cubic feet at the atmospheric pressure flowing out of the pipes and cylinders in producing the vacuum, distance remaining the same in both cases.

Your Committee have been unable to get correct data as to the difference in time in the transmission of power by these two systems, other than that given by those forming an opinion from their own observation of the working of the two systems of brakes, and their testimony is that compressed air acts quickest in the transmission of power in the proportions given above.

The ingoing force of compressed air is a fixed and positive force, capable of being instantly applied, and can be graduated to any force desired on the brake shoes of the train, between nothing and the maximum, due to the pressure of air in the reservoir. Of the stem of brakes operated by compressed air, that known as the Westinghouse Air Brake has so far met with favour from the greatest number of railroad managers; in fact, with the exception of the Loughridge Air Brake, and the Gardiner-Ranson, it is the only one in general use, and at the present time it is in use on about fifty-seven per cent of the miles of railroad in operation in the United States and Canada.

The Westinghouse Brake is not considered perfect as yet, although it has been in general use for the past five or six years, the chief defect complained of being the valve motion to the steam cylinder and the length of time required to release the brake after coming to a stop, especially with a long train of cars. The latter is, doubtless, due somewhat to want of a better system of springs for effecting the release than is used on the cars complained of, and the former to a want of better proportions and arrangement of the valve motion, and intelligent attention on the part of the persons having charge of the brake. The want of a better or more reliable valve motion to the air pump has been the most serious objection urged against this brake, as reported to your Committee. Improvements have been made in the valve motion within the past two years, yet owing to its delicate proportions objections are urged, and a desire expressed for a more reliable valve arrangement than that heretofore used.

The distinguishing features of this brake are: the independent pump for compressing air; the automatic valves in the couplings between the cars; the automatic "take-up" arrangement for taking up the lost motion, or "slack" in the levers and rods under the cars, keeping the brake shoes close to the wheels, and thus preventing loss of air in allowing a greater motion of piston in the cylinder than necessary; the "triangular lever" by which change of leverage is made to compensate for a decrease in the pressure of air incident to an application of the brakes; and uniformity in all the various parts of the brake. One important feature claimed is the automatic valve in the couplings between the cars which, in case the train should be severed, closes the ends of the couplings so that the brake on the cars next to the locomotive can operate as well as before without any attention whatever, and in case the train should become severed after the brake is applied the brake will remain set on the detached part of the train.

An instance of this kind occurred on the Louisville, Nashville & Great Southern Railroad on December 9, 1873, to a passenger train approaching a trestle at a speed of twenty miles per hour, upon which an obstruction had been placed. The facts were furnished by Mr. Thomas Walsh, Master Mechanic of the Memphis Branch, and are as follows. It being dark, the obstruction was not perceived until the locomotive was within sixty yards, when the brake was applied. The locomotive and tender, baggage car, and second-class car were thrown down into the creek, a distance of twenty-five feet, taking the trestle down with them, but the remaining three cars, all full of passengers came to a stop, and with the brakes tightly set remained standing on the main track. The forward pair of trucks alone were pulled off the track by the previous car as it went down before the coupling pin broke. We believe that this automatic arrangement by which their pipes or hose are closed, in case the train is severed, either intentionally or otherwise, is not found in any of the other brakes referred to. With the Loughridge, Gardiner-Ranson, and Smith Vacuum Brakes, in case the train is severed while the brake is on, it is released from all the cars in the train, and cannot be applied to any part of it until the hose, where parted, is in some way closed again. Next to the Westinghouse Brake, the vacuum brake has, perhaps, come into use, so far as road and rollingstock is concerned. This system of brake, like all the others, has its defects. One is, that to a certain extent it lacks the positively certain, unvariable element of the compressed air brake, as the quickness of action and effectiveness depends upon the steam pressure in the boiler, and its power is liable to the same fluctuations. Mr Loughridge in his answer to the American Railway Master Mechanics' Association Circular on the subject of brakes, in alluding to the vacuum brake, says in reference to the ejector:

"The laws governing it are singular. The vacuum varies much when the steam pressure in the boiler varies, and the nozzle is very uncertain in its action, something like the injector. The force changes with the collapse of the sack or diaphragm, and the liability of the rubber to rot, or be cut by boys or discharged employees, and freeze in the winter, are objections." Of his brake Mr Loughridge further states: "I canvassed this mode of applying the force, fully, up to 1870, and after a full and careful review of it, I preferred paying twenty thousand dollars for my patents of 1864 to enable me to use compressed air and produce better effects."

So far as the facts have yet been demonstrated, and from our experience and observation, and the opinions of others elicited in the foregoing investigation, your Committee are of the opinion that under all ordinary circumstances, and the conditions likely to arise in the use of a continuous train brake, compressed air supplied to a reservoir by an independent pump, and thus storing the power for applying the brakes whenever needed will give the best results, all things considered. The Westinghouse Air Brake seems to be, in the opinion of a majority of Master Mechanics, on the whole, nearer perfect than any other that has yet been brought into general use.

The above allusions to the Westinghouse Brake refer to the brake as heretofore used. As the Westinghouse Improved Automatic Brake is being introduced on some of the leading railroad lines in this country and in Europe, and as there may be persons interested in this subject who are not familiar with its peculiarities, we consider a description of its distinguishing features wherein it differs from other brakes necessary. The air pump and main reservoir are situated on the locomotive, and are the same as used for the common [Straight-Air] Westinghouse Brake; the pipes and hose of the train and the brake cylinders are also the same; but under each car, near the brake cylinder and connected with it by a pipe, is an auxiliary reservoir from which the brake cylinder receives the air when the brake is applied. In the pipe between the auxiliary reservoir and the brake cylinder is an automatic triple valve, with an elastic diaphragm so adjusted and arranged that when the air fills the pipes of the train it flows into the auxiliary cylinder through a small opening, filling it at the same pressure as that in the main reservoir and pipes of the train; but no air is admitted to the brake cylinders while the air pressure in the pipes of the train is greater or remains as great as that in the auxiliary reservoirs. To apply the brake it is only necessary to turn the three-way cock on the locomotive, cutting off the air from the main reservoir and allowing that in the pipes and hose of the train to escape, [whereupon] the pressure of air from the auxiliary reservoir, acting upon the elastic diaphragm of the triple valve, lifts a valve admitting the air from the auxiliary reservoir on the car to the brake cylinder, and thus applying the brake; while to release the brake the air is again allowed to fill the hose and pipes of the train, which, acting upon the opposite side of the diaphragm, the movement of the triple valve is reversed, cutting off communication between the brake cylinder and the auxiliary reservoir, and at the same time opening a relief or exhaust valve which allows the air in the brake cylinder to be exhausted direct into the open air, releasing the brake on the instant.

To apply the brake with its full force, all the air is allowed to escape from the pipes; to graduate this brake so as to admit only a part of the full reservoir pressure to the brake cylinders, only a part of the air is allowed to escape from the pipes and hose of the train, the operation of the triple valve being such that the pressure admitted to the brake cylinder is proportional to the amount of pressure released.

The old [Straight-Air] brake being applied and graduated by admitting the full or part of the pressure, while the automatic is applied by the reverse taking place, so far as the pipes and hose are concerned. The automatic brake has the disadvantage of a constant pressure inside of the pipes and hose the entire length of the train at all times, except when the brake is on, thus rendering it more difficult to prevent leakage of the joints of the pipes and hose. Another objection is that it is liable to be applied by accident or designedly when it should not be done, and which might be detrimental in its results. The bursting of a hose in any part of the train would apply the brakes with their full force, and they could not be released until such hose was removed and another one substituted and the air pressure restored again from the main reservoir, or except a cock was opened letting the air out of each auxiliary cylinder in the train. These are some of the objections that may be urged against the automatic brake aside from the additional care and attention requisite to keep the triple valve arrangement and other parts, not found in the old brake in thorough and perfect working order.

The advantages of the automatic brake are, that in case the train is parted it applies itself on the instant to each car that is attached to the locomotive, as well as the detached part. Another is, that should a car leave the rails from any cause, the [derailment cross-bars][200] coming in contact with the ground or obstruction, the brake is applied at once with its full force to the entire train. Another, and perhaps the most important advantage, is its quickness of action. Experiments have demonstrated that from one to two seconds only, depending upon the length of train, is required in applying it to the wheels with its maximum force. The force required being stored in the auxiliary reservoirs, close to the brake cylinders, and when the automatic valves open the brake is applied with the full force due to the pressure of air in the reservoirs. Quickness of action gives to this brake its chief value, and if it supersedes the old one, or other brakes, it will be largely due to that fact. Comparing the old [Straight-Air] brake with the improved automatic, it is probable that on a large majority of the passenger trains run on the roads in

200. An invention (patent № 156,322, October 27, 1874 - Discharge Valves for Fluid Brakes) originally included in complete sets of automatic brake equipment for passenger cars, and subsequently abandoned, this was a 'trip' device for automatically setting the brakes in the event of derailment. A valve controlling the discharge of air from the brake pipe was actuated by a stem extending downward beneath the car, carrying a cross-bar in such proximity to the rail that if the car were to derail the cross-bar would strike the track, lift the valve from its seat, and—by discharging air from the brake pipe—apply the brakes. The arrangement worked effectively; however, experience showed that a brake application sometimes occurred when the cross-bar was struck by loose objects on the right-of-way—including livestock—resulting in an emergency stop under inconvenient or even dangerous conditions. The use of the 'trip' device was discontinued, and although numerous variations have been patented, the objection has never been overcome.

this country, the Westinghouse Brake as now in general use [i.e. the Straight-Air brake], will give, on the whole, better satisfaction than the improved automatic brake, on account of its simplicity, compared with the other. But there are fast trains run on many of the leading roads where, in our opinion, the automatic brake would be the means of insuring greater safety in certain [emergency] cases. In such cases the automatic brake has the advantage, and in these respects is superior to the old brake.

But little information was elicited on the subject of driving-wheel brakes; comparatively few of them have been used up to the present time. With but one or two exceptions, those who have had experience with such brakes speak favourably of them. Brakes upon the driving wheels of passenger locomotives would, doubtless, give good results, and the time may not be far distant when such brakes will be applied to all locomotives hauling fast trains, for the reason that a retarding force equal to the propelling power of the locomotive can be utilized in that manner to bring the train to a stop, and there is no reason why a retarding force applied to the tires of the driving wheels, should be more injurious in its effects on a locomotive than an equal propelling force applied to them from the cylinders, and we think that it will be less.

It will be shown in practice, we think, that good brakes applied to the driving wheels of the locomotive will very materially add to the efficiency of a train brake in making quick stops, being in effect the same as reversing the steam but without the injurious effect on the machinery of the locomotive that results from reversing while running. A train equipped with good brakes applied to all the wheels, except perhaps the engine truck wheels, can be brought to a stop in the shortest possible distance run, after the brake is applied, because to each wheel of the train is applied a retarding force about equal to that of the adhesion of the wheels on the rails, and when that point is reached nothing more in the way of a retarding force can be obtained from the use of a brake applied to the wheels. In making the ordinary stops of a train the driving wheel brake will be found to be advantageous, at least so far as a proper distribution of the retarding force applied to the train is concerned, as it gives to each wheel in contact with the rails its proportion of the retarding power necessary to bring the train to a stop. If a train equipped with a brake to the driving wheels, in addition to the train brake, at a speed of thirty miles per hour, is brought to a stop in a distance run of six hundred feet, it is evident that the train cannot be brought to a stop within that distance when the driving wheel brake is not used. All other conditions being equal, in other words, if no retarding work is done by the driving wheels, more must be done by the other wheels of the train in a given time to effect the same result.

Brakes to the driving wheels of locomotives drawing freight trains have not heretofore been used to any considerable extent, and where their use involves the cost and maintenance of an independent or other style of air pump, it is not probable such brakes will be generally used, unless the brake can be applied to a part or all of the train also. There may be exceptional cases, however, where the advantages derived from the use of the brake on the driving wheels and tender only will compensate for the cost of the same, but further experience and investigation will be necessary before a correct conclusion can be arrived at on this point. Under all ordinary circumstances the hand brakes on freight trains are sufficient in descending grades and for ordinary stops; and in order to avoid accident the locomotive can be reversed, and in that way exert a force equivalent to a brake on the drivers. Brakes upon the driving wheels of locomotives used for switching purposes are in use on some of the roads in the United States, and, so far as reported, give good results. It is not probable that the benefits to be derived would compensate for the cost of such brakes in a majority of cases, yet there may be others where the advantages will warrant the cost.

Excerpt from a Westinghouse Air Brake Company 'Reference Booklet'

The Westinghouse Air Brake Co.

Office, Cnr, Liberty Avenue & Twenty-fifth St's

OFFICERS

GEORGE WESTINGHOUSE, Jr. ______ President
RALPH BAGALEY ______ Secretary & Treasurer
T. W. WELSH ______ Superintendent

PITTSBURGH, PENN'A
1876.

THIS COMPANY DESIRES to so conduct its business that none of its customers shall have cause for complaint, and we believe that care should be taken not to encourage the infringement of our patents by any one. It is our intention to adopt all improvements that may tend to perfect the brake, and to purchase those not originated at our own place if necessary for this purpose. This Company has purchased all the patents owned by the Vacuum Brake Co., which, with those previously owned, gives us the exclusive right to manufacture vacuum brakes of whatever kind or nature, and it is our intention to protect ourselves in this exclusive control of both air and vacuum brakes. We call especial attention to the great superiority of the automatic brake, and believe that a careful study of the manner in which so many desirable results are produced, will not be a loss of time to Railway Officers.

The Westinghouse Air Brake

Now in use upon two-thirds of the railways of the United States, [the Straight-Air brake] consists of an air pump driven by steam taken from the locomotive boiler, by which air is forced to a pressure of about eighty pounds per square inch into a reservoir fastened under the foot-plate of the engine. From this reservoir a pipe leads to one opening of a three-way cock, while from a second opening the brake-pipe is extended down beneath the tender and the entire length of the train. Under each car, near the center, a cylinder, fitted with a piston, is firmly bolted, having a connection from the piston to the ordinary brake lever, and a branch leads from the main brake-pipe to one end of this cylinder.

The pipes from car to car are joined by flexible tubes, having malleable iron couplings, which are provided with valves. When these couplings are united the valves are forced open, so that there is a free passage from the three-way cock the entire length of the train to the last coupling, where the valve not being unseated prevents the escape of air. The handle of the three-way cock before alluded to, is under the immediate control of the engineer, so that by a simple movement of the hand he can bring the pipe from the reservoir in communication with the brake-pipe, whereby air is forced through the pipe and into each of the brake cylinders, setting the brakes almost instantly, with any force desired up to the maximum, which is only limited by the amount of pressure carried in the reservoir.

A return of the handle of the three-way cock to its former position first closes the communication between the brake-pipe and reservoir, and then opens a communication from the brake-pipe to the atmosphere, whereby the air is discharged and the brakes released by strong springs, which force the pistons back.

The Westinghouse Automatic Brake

The introduction of the brake just described has demonstrated the fact that other features required to make a brake meet all emergencies. Continued experimenting has led to the perfection of a brake which, we believe, meets every contingency that can possibly arise. The Automatic Brake differs from the [Straight-] Air Brake by having, in addition to the main reservoir on the locomotive, an auxiliary reservoir located upon each car in the train. The branch from the main brake-pipe, instead of leading directly to the brake-cylinder, leads to one opening of a small triple valve device, passing through which it enters this reservoir; through a third opening the pipe leads to the brake-cylinder. The construction of this valve device is such that, if air be turned from the main reservoir into the brake-pipe, the auxiliary reservoir will be charged with the same pressure, while at the same time an opening will be made from the brake-cylinder to the atmosphere. After the reservoir has become charged, if the air be discharged from the brake-pipe by the engineer (or otherwise), it will cause this triple valve to shift so as to close the opening from the auxiliary reservoir to the brake-pipe, and from the brake-cylinder to the atmosphere, at the same time opening a direct passage from the auxiliary reservoir to the brake-cylinders, applying the brakes to the full force of the pressure in the reservoir, less the reduction caused by expansion in a larger space.

It will be seen by the above description that the brakes are off when the pipe is charged with compressed air, and that the brakes are applied by the reduction of pressure in this pipe. A valve arranged in the brake-pipe on each car, with a cord extending through the inside of the car, enables the conductor or brakemen to discharge the air from the pipe, thus setting the brakes on the entire train. The separation of the train, and consequently of the hose couplings between the cars, discharges the air and applies all of the brakes. At both ends of each car a cock is arranged in the pipe, and the cock at the rear of the train is closed to retain the pressure. When the train is separated intentionally, the cocks in the pipe are closed where the separation is to take place before the couplings are disconnected, preventing the application of the brakes. The force with which the brake is applied is entirely under the control of the engineer, who governs the amount of the reduction of pressure in the brake-pipe. A slight reduction, owing to leakage in the brake-pipe, does not apply the brakes, whether the locomotive is connected to the train or not. Both halves of the hose coupling are alike, which permits of the use of a single line of pipe, and these couplings are so constructed that they are perfectly air tight when united, but can be separated forcibly by the breaking of a train without injury.

The advantages possessed by the Automatic Brake are so great that we believe it will be generally introduced on all roads throughout the world. It was, by far, the most efficient and reliable tested at the recent great brake trials in England, and fulfilled all the requirements of the engineers appointed by the Royal Commission on Railway Accidents. It will stop a train in less distance than any other brake; it will not fail if a hose or pipe bursts; it is automatically applied to the whole train by the separation of the same, and can be applied from any portion of the train. It is instantaneous in its application and release, and costs less for repairs than any other brake ever offered railway companies. It has been adopted by a large number of railroads, has been in use over three years with most satisfactory results, and is considered by those who are familiar with all kinds to be as much superior to all other continuous brakes as they are superior to the hand brakes. The [Straight-] Air Brake, as already introduced, by a slight addition of parts, can be changed to the Automatic Brake. By the use of the same couplings on both kinds of brakes, they can be run in the same train without difficulty, as each of the triple valve devices is provided with a four-way cock, by which it can be changed into the [Straight-] Air Brake at pleasure.

Whereas, The Westinghouse Air Brake Co.

...of the City of Pittsburgh, Pennsylvania, operating under sundry letters patent of the United States, some of which are owned or controlled by the said Company, and others of which are owned by George Westinghouse, Jr., of said city, as patentee, is engaged in making and selling apparatus for operating railway car brakes by compressed air, which apparatus as adapted for use on the different parts of a railway train, or as employed on trains in different modes of operation, is for convenience classified as per the following schedules:

Schedule A (constituting the equipment of a locomotive in the system of brake apparatus heretofore in use as the Westinghouse Air Brake): One Steam Engine and Pump, One Main Air Reservoir, One Air Gauge, One Three-Way Cock, One 1¾-in. Lubricator, One Oil Cup for Air Pump, One Steam Cock, One Drain Cock, One Check Valve, One Safety Valve, One Air Pipe Strainer, Hose and Coupling with Valves, One Three-Foot Hose Connection, one hundred and twenty-five feet of Pipe, assorted sizes, thirty-three Pipe Fittings, assorted.

Schedule B (constituting the equipment of a tender in the same system): One 8-in. Air Brake Cylinder complete, with sleeve.

Schedule C (constituting the equipment of a passenger car in the same system): One 10-in. Air Brake Cylinder complete, with Releasing Springs, Hose and Couplings with Valves complete, one hundred and thirty feet of Pipe, assorted sizes, thirteen Pipe Fittings, assorted.

Schedule D (being the additional apparatus provided for equipping a locomotive driver brake in the same or in AUTOMATIC system): Two 8-in. Air Brake Cylinders, Thirty feet ½ in. Pipe, Six Pipe Fittings, assorted, One Check Valve, One Safety Valve.

Schedule E (constituting the equipment of a locomotive in the Westinghouse AUTOMATIC system): Same as Schedule A, less thirty feet of Pipe, and for "Hose and Couplings with Valves", substituting "Hose and Couplings without Valves".

Schedule F (constituting the equipment for a tender in the AUTOMATIC system): One 8-in. Air Brake Cylinder, complete, with sleeve, One Triple Valve, One Leakage Valve, One Auxiliary Air Reservoir with Drain Cock, One Drain Cup and Cock, One ¾-in. Stop Cock, ten feet ½-in. Pipe, Six Pipe Fittings, assorted.

Schedule G (constituting the equipment for a passenger car in the AUTOMATIC system): One 10-in. Cylinder as in Schedule C, One Auxiliary Air Reservoir with Drain Cock, One Drain Cup and Cock, One Triple Valve, One Conductor's Valve, One Leakage Valve, Two ¾-in. Cocks, Hose and Couplings without Valves, seventy-five feet ¾-in. Pipe, Twelve Pipe Fittings, assorted.

Schedule V (being the apparatus provided as additional to that mentioned in Schedules A and B, for equipping a locomotive and tender to work both the "old" and "automatic" systems): One Auxiliary Air Reservoir with Drain Cock, One Triple Valve, One Leakage Valve, Hose and Coupling without Valves, One —-in. Cock, ten feet ½-in Pipe, thirty feet ¾-in. Pipe, Five Pipe Fittings, assorted.

Schedule W (being the apparatus provided as additional to that mentioned in Schedule C, for a passenger car under both the "old" and "automatic" systems): One Auxiliary Air Reservoir with Drain Cock, One Triple Valve, One Leakage Valve, One Conductor's Valve, One Double Check Valve, One Drain Cup and Cock, Two ¾-in. Cocks, Hose and Couplings without Valves, eighty feet of Pipe, assorted sizes, Twelve Pipe Fittings, assorted.

Schedule X (being apparatus required to change a passenger car from the "old" to the "automatic" system): One Auxiliary Air Reservoir with Drain Cock, One Triple Valve, One Leakage Valve, One Conductor's Valve, One Drain Cup and Cock, Two ¾-in. Cocks, Hose and Couplings without Valves (Pipe of old brake can be utilized.).

Schedule Y (constituting a brake equipment complete for a locomotive and tender in the "old" system): The articles named in Schedules A, B and D when ordered together.

Schedule Z (constituting a brake equipment complete for a locomotive and tender in the "automatic" system): The articles named in Schedules E, F and D when ordered together.

Schedule:		Schedule:	
A —	Three Hundred Dollars.	G —	One Hundred and Thirty-Eight Dollars.
B —	Twenty-Five Dollars.	V —	Thirty-Five Dollars.
C —	One Hundred Dollars.	W —	Eighty-Eight Dollars
D —	Fifty Dollars.	X —	Eighty Dollars.
E —	Three Hundred Dollars.	Y —	Three Hundred and Twenty-Five Dollars
F —	Sixty Dollars.	Z —	Three Hundred and Sixty Dollars.

Steam Engine and Pump

Sheets 3 and 4 represent the engine and pump for compressing air for the operation of the brakes. A differential piston valve movement is used, in which the difference in area between the two ends is such that when steam is admitted between them, the tendency of the valve is to move upward, which gives a downward stroke to the main piston. When the main piston reaches the bottom of its stroke it operates upon the reversing valve rod, causing the valve 13 to uncover a port by which steam is admitted above the piston 20 in the cylinder 19, which, by excess of pressure, causes the main valve to descend, exhausting from the upper part of the steam cylinder and admitting steam below the main piston. As the main piston completes its upward stroke, the valve 13 is again moved so as to exhaust the steam from the reversing cylinder, whereby the reversing piston is moved upward, together with the main valve, by the difference of pressure between the two valve pistons.

The operation of the valves of the air pump is clearly shown; the lift of each discharge valve should not exceed one-sixteenth of an inch, while for the receiving valves one-eighth of an inch is sufficient. These valves are easily removed and examined, proper wrenches being sent with each set of apparatus for removing the caps 29 and 30, and the valve seat 31 in which the upper discharge valve works.

Care has been taken in the arrangement of this valve motion to enable us to utilize and alter the principal parts of pumps furnished with the valve motions shown in our former catalogue. For this alteration the pumps should be shipped to our works when they will be made as good as new, at an expense of eighty dollars and freight charges. This charge includes the replacing of defective parts, re-jacketing and thorough testing. Our arrangements for the license under which we make these alterations are such that it is necessary that they should be changed at our works. Very large numbers of pumps are being sent to us for this purpose, as the decreased cost of maintenance and the increased efficiency much more than compensate for the small expense.

Driver Brakes

Sheet 11 shows one mode of arranging the driver brakes which we have so strongly urged Railway Managers to use, both upon freight and passenger locomotives. The retarding force of brakes upon the driving wheels and tender of an engine is double that of the engine reversed with the tender brakes set by hand. An ordinary freight train can be stopped in less time and distance with the engine and tender alone equipped with this apparatus, than by the usual number of brakemen when at their posts; an ordinary passenger train can be stopped in nearly one-third less distance when the engine is fitted with these brakes in addition to the others, than when without the driver brakes. The saving effected each year by the prevention

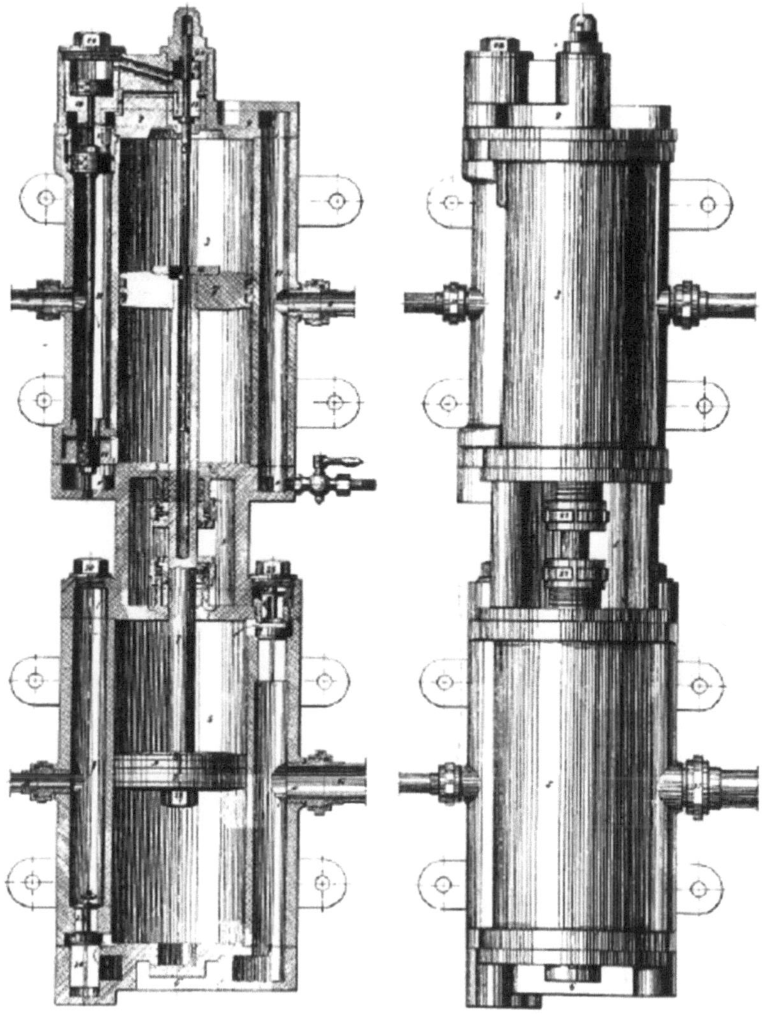

FIGURE 285: AIR PUMP

of accidents will more than pay for the first cost of equipping freight locomotives, while the large decrease in the number of wheels destroyed by the injudicious use of the hand brakes would also effect a large saving. These brakes are arranged so that the wheels cannot be caused to slide, by the use of a safety valve, which blows off the pressure in the brake cylinder at any desired point which can be regulated by those in charge to do the most efficient work. A check valve, No. 20, is also used to prevent a too sudden application of the brakes, by causing the air from the three-way cock to seat the valve, through which is drilled a small hole for feeding the brake cylinders properly. When the brakes are released this valve unseats and permits a quick escape of the air. This valve should be inserted in the pipe before it is branched to lead to the driver brake cylinders, and care must be taken to arrange it so that it will be seated by the pressure passing to the brake cylinder. The safety valve 21 is put in the pipe between the check valve 20 and the brake cylinders; the opening from this safety valve is large enough to blow off the pressure at the desired point as fast as it passes through the small hole in the check valve. These check and safety valves are formed in one piece when desired as shown No. 28, Sheet A.

In furnishing driver brake equipment we provide only the cylinders with pistons and crossheads, which fixtures are included when desired with the regular locomotive equipment without extra charge, no deduction, however, being made if they are not taken. To insure the successful working of these brakes we will furnish drawings without charge, showing their application to any class of locomotives, and to make these drawings it is necessary that a sketch of the engine be sent us, showing the parts between the driving wheels, the position of the connecting rod, and such other information as will make it impossible for us to interfere with existing devices; the weight of the driving wheels upon the rails should also be given, that the levers may be properly proportioned.

When desired we furnish the brake shoe castings 17 and eccentric levers, with nuts and screws 15, 18 and 19, charging therefore a fair manufacturing rate, which can only be determined after the drawings have been prepared. The brake shoe faces 16 are made of wrought iron, we having found that preferable to any other material. In adjusting the brakes at least one-half inch clearance should be allowed between the brake shoe faces and the wheels, and these faces can be kept an equal distance from the wheels throughout their length by the small set screw in the extension of the part 19. Large numbers of locomotives are being fitted with driver brakes, some of which have been in use over three years, and full particulars of the roads using them, the number and time, will be furnished upon application.

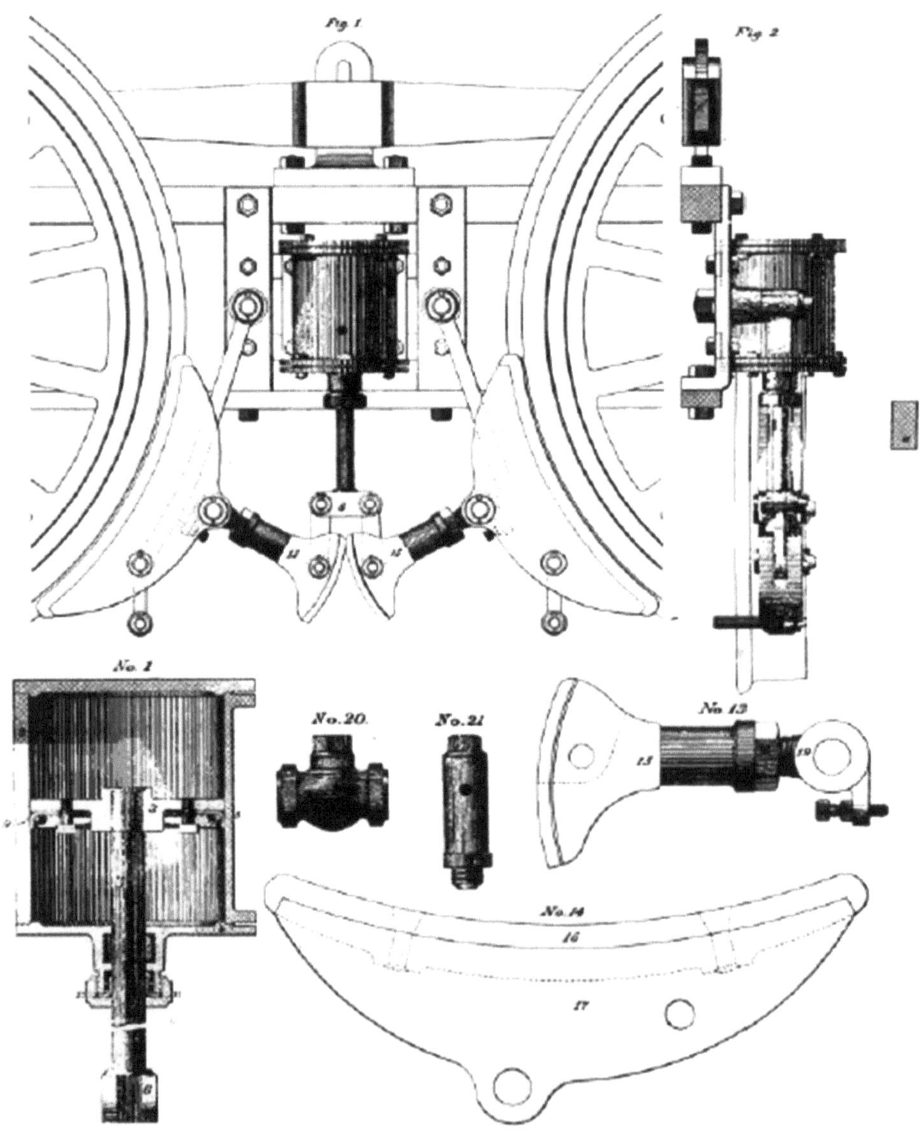

FIGURE 286: DRIVING WHEEL BRAKE

Triple and Leakage Valves

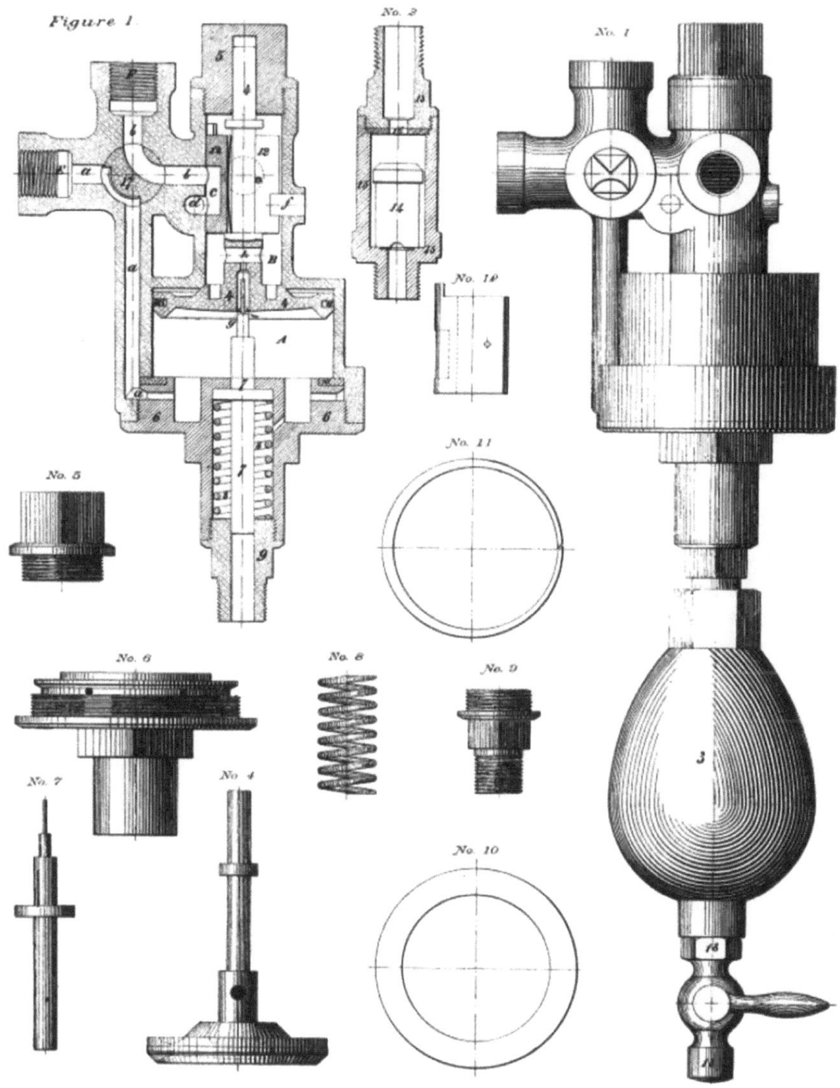

FIGURE 287: TRIPLE VALVE

Clutch Coupling

Sheet 6 shows our improved clutch coupling with details, the construction of which is such that the tendency of the air pressure is to more firmly hold it together. The joints formed by rubber rings 8 are also made tight by internal pressure, while these same rings form seats for the valves 5 when the couplings are separated. In renewing the rubber packing rings, it is necessary to remove the cap 5, which should be carefully replaced. These couplings are provided with valves when used with the air brake, but when used for the automatic brake the valves are removed and the piece 10 inserted to expand the packing.

Both halves of this coupling being exactly alike, and there being no tendency whatever for the air pressure to blow them apart, a single line of pipe and couplings of this kind between the cars answers the purpose of the double couplings heretofore sent out. This, with the fact that these couplings are used for the automatic brake, warrants us in recommending that all railway companies not already provided, should substitute this clutch coupling for their present style. All of the lines controlled by the Pennsylvania Rail

Road Co., have already made this change on their equipment of about two thousand cars and four hundred locomotives. The decreased cost of having only one-half the number of couplings to keep in repair, will more than compensate for the outlay. We only furnish new couplings at our regular price to replace the kind shown in Sheet 7, on condition that all of the old ones are returned to us, for which we will allow two dollars and fifty cents per car with hose and nipples, or two dollars if hose and nipples are not returned. This price is more than double what they are worth, as we are obliged to remelt them.

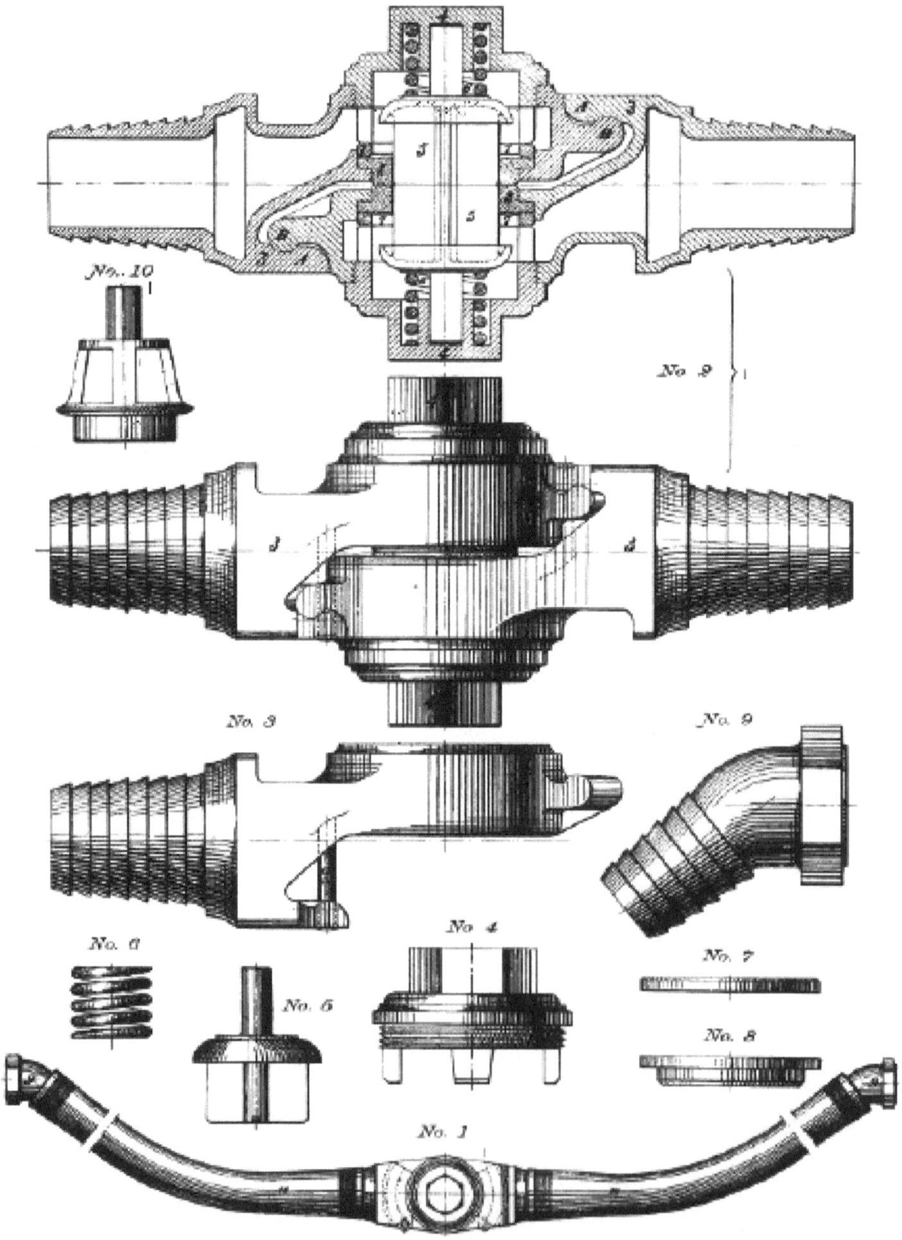

FIGURE 288: WABCO BRAKE HOSE CLUTCH COUPLING

The following illustration depicts the earlier Westinghouse air hose coupling.

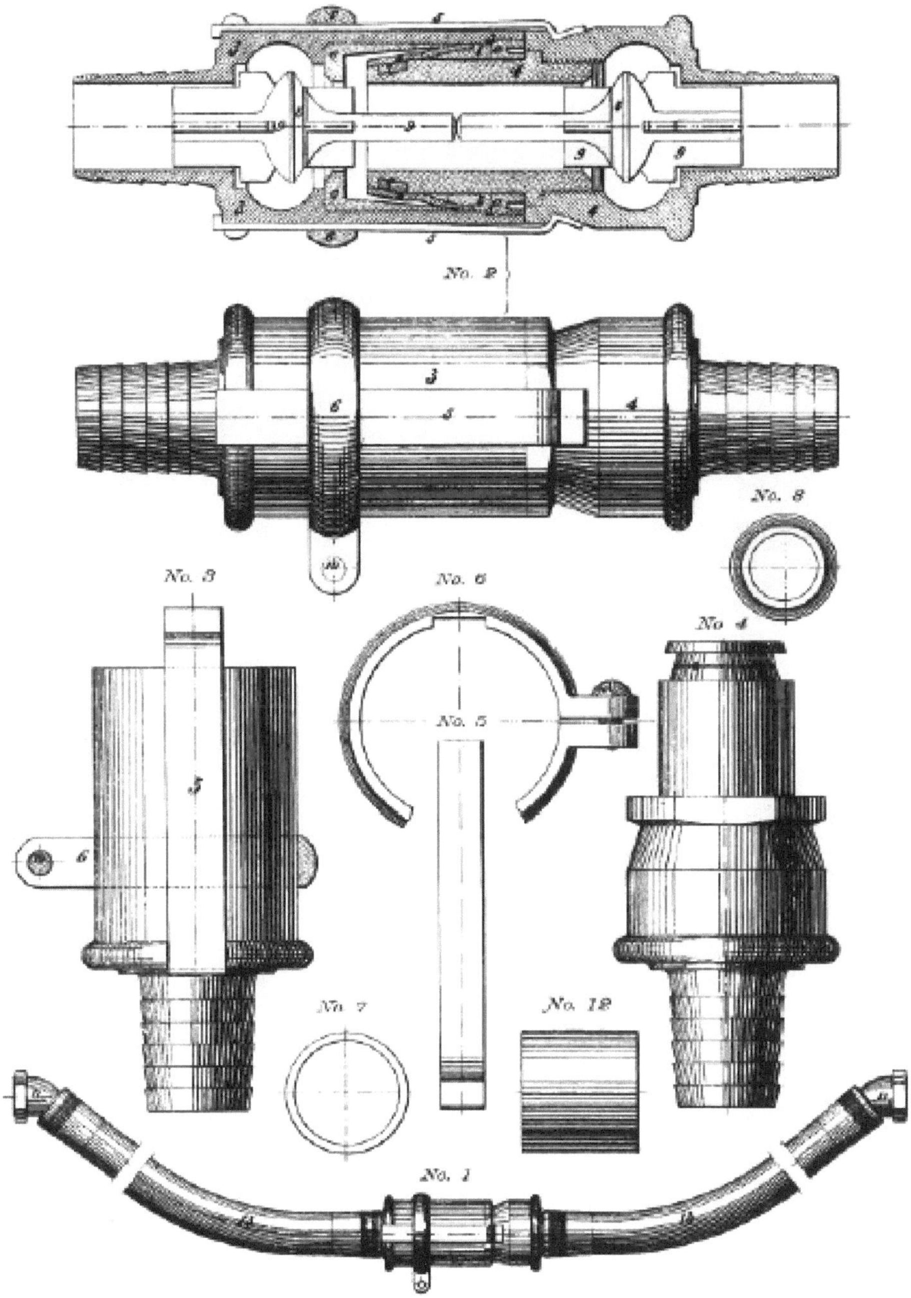

FIGURE 289: ORIGINAL WABCO MALE & FEMALE COUPLING

The following illustration depicts the Gardner & Ranson air hose coupling as patented. The reader will note the physical similarity to the Westinghouse Male & Female coupling.

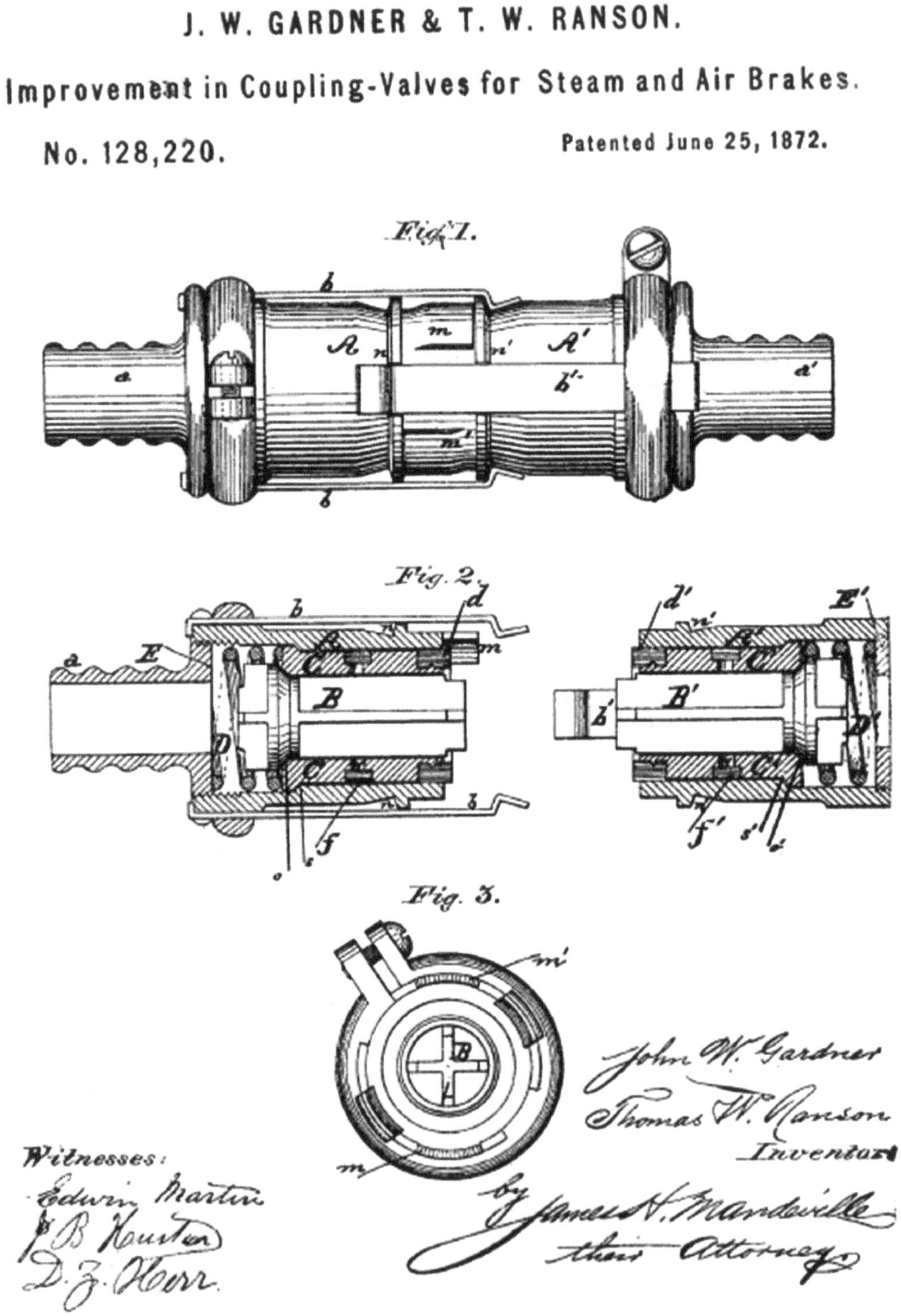

FIGURE 290: THE GARDNER & RANSON MALE/FEMALE-TYPE COUPLING

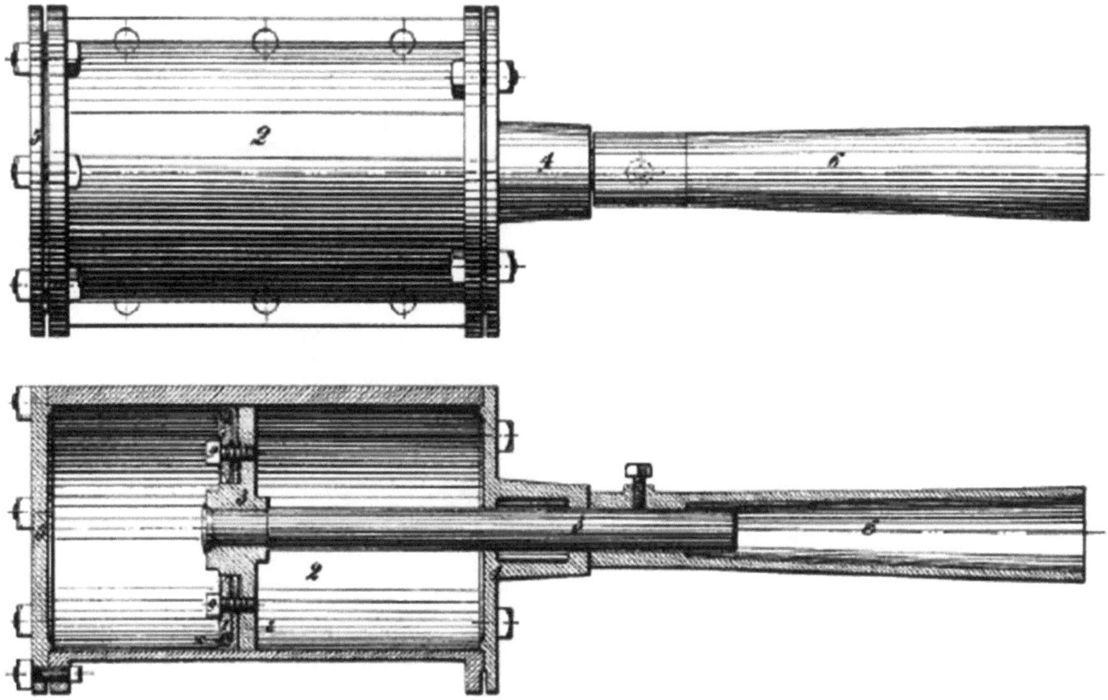

FIGURE 291: WABCO 8-IN. TENDER BRAKE CYLINDER

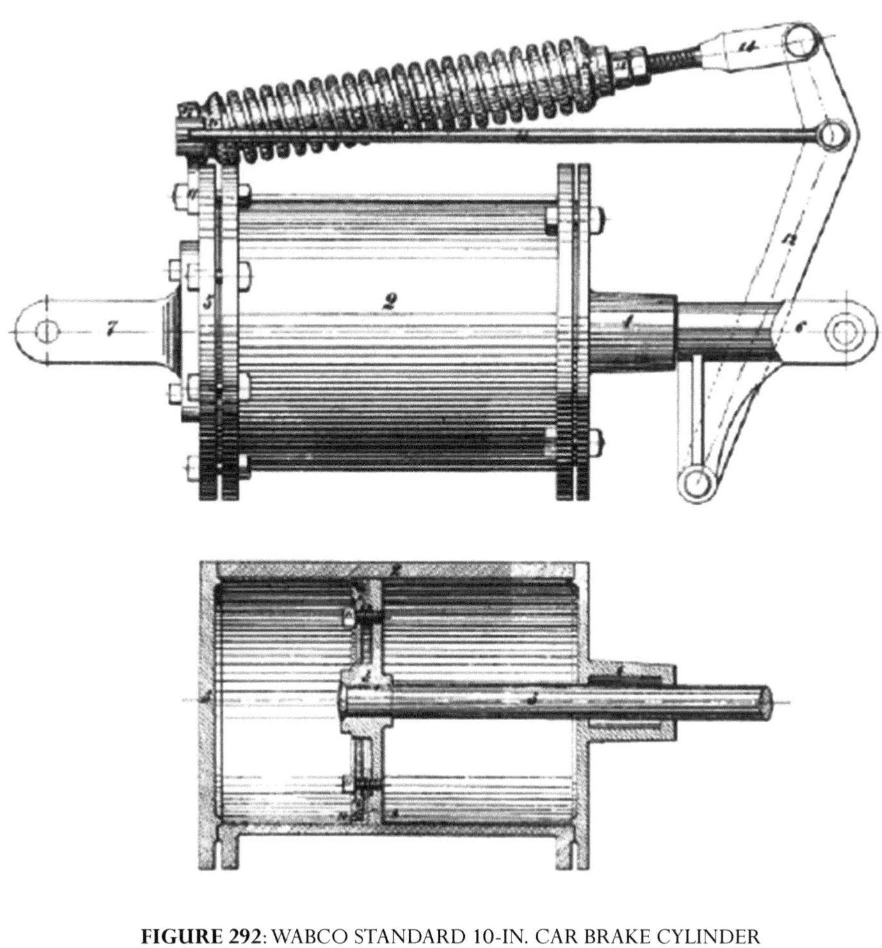

FIGURE 292: WABCO STANDARD 10-IN. CAR BRAKE CYLINDER

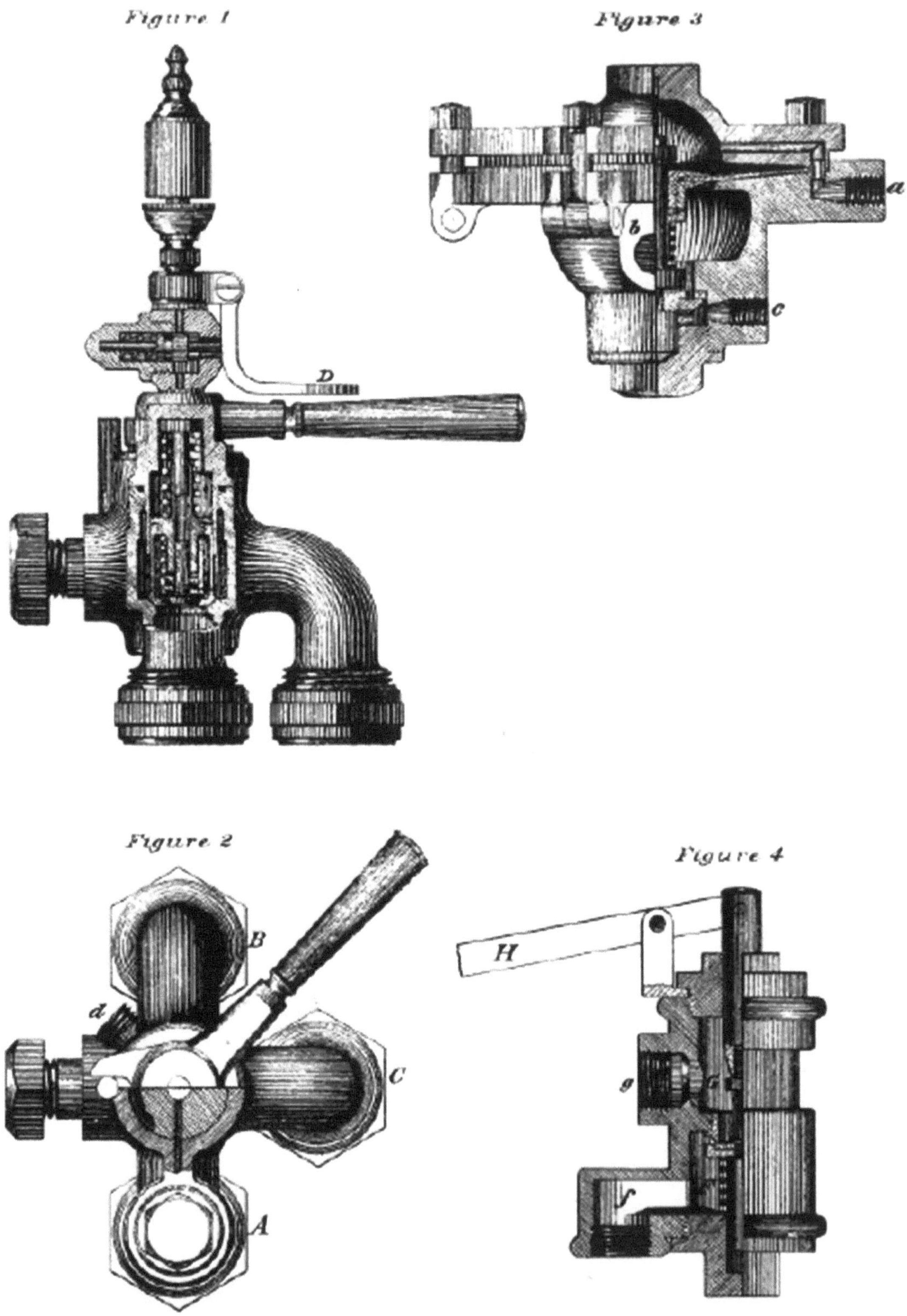

FIGURE 293: WABCO TRAIN SIGNALLING APPARATUS

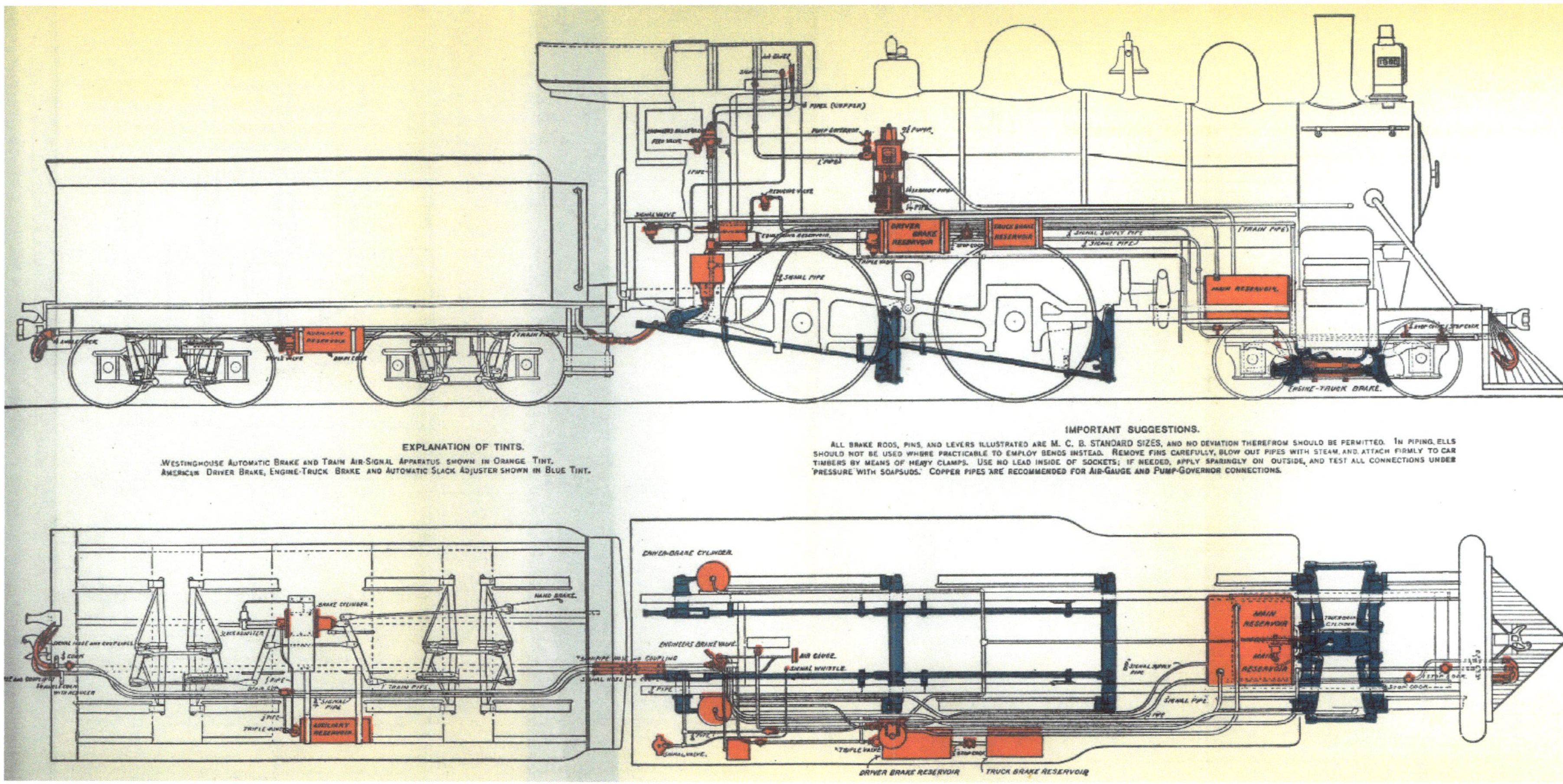

FIGURE 294: THE WABCO LOCOMOTIVE EQUIPMENT INSTALLATION

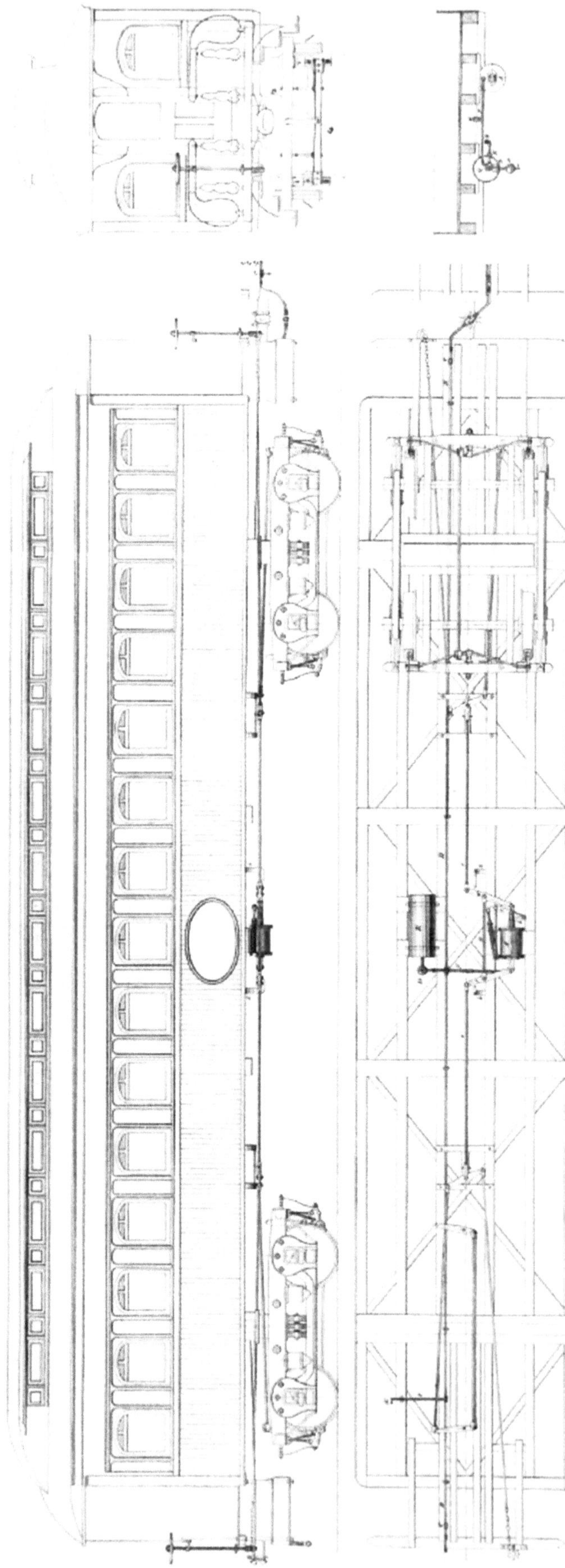

FIGURE 295: THE WABCO CAR EQUIPMENT INSTALLATION

Westinghouse Air Brake patents:

Following is a list—reproduced from Prout's biography—of some of the more significant patents awarded to George Westinghouse in the development of the air brake. Readers should understand that the date of the granting of a patent might bear no relationship to the date in which that invention was applied or placed into service.

№ 88,929, April 13, 1869. Steam-Power Brake [Straight-Air]:

This was the first patent issued to Westinghouse for [a railroad power] brake. It formed a firm foundation for the air-brake structure that was built upon it, and its chief characteristics will be found stated in the opinion of Justice Swayne and Judge Welker, of the United States Court, in litigation between the Westinghouse Air Brake Company and the Gardner & Ranson Brake Company. [The conclusions of the court are recorded elsewhere in this book.]

№ 115,667, June 6, 1871. Steam-Power Brake Apparatus:

This patent proposed a device to produce a vacuum on the non-pressure side of the brake cylinder piston for the purpose of quickening the release of brakes to remedy a defect of the straight-air system. It was not used in practice, but was one of the earliest disclosures of the vacuum-brake system, as by reissue 5506, July 29, 1873, a claim was allowed for the operation of power brakes by atmospheric pressure.

№ 117,841, August 8, 1871. Steam-Power Air-Brake Devices:

The purpose of this invention is described in the following quotation from the specification. "In applying car brakes it is desirable that the movement of the brake shoes at first be rapid, so that they shall engage the wheels as quickly as possible, and after they have engaged the wheels that they be pressed against them with great force. Before they touch the wheels they offer no great resistance. After they engage the wheels their motion is little, but the resistance is great." The form in which the invention was patented was used to a limited extent in the early days of the application of air brakes, but was abandoned as not satisfactorily accomplishing the desired result. In a modified and improved form it is now an essential feature of what is technically called the "empty and load brake", one of the latest air-brake developments.

№ 122,544, January 9, 1872. Improvement in Exhaust Valves for Steam and Air Engines:

This patent is for a valve device, to be used in connection with the straight-air system to provide an escape of pressure directly from the brake cylinder to atmosphere when it is desired to release brakes so that the time of release will be reduced as compared with that required to permit the air to escape through the train pipe and out of the engineer's valve on the locomotive. Very few were put in service.

№'S 124,404 and 124,405, March 5, 1872. Improvement in Steam-Power Air Brakes and Signals:

These patents for the first time reveal the basic invention of the automatic brake and also the system of train-signalling that subsequently became standard on passenger trains of this country.

№ 134,178, December 24, 1872. Steam and Air Brakes:

In this patent means are proposed for automatically compensating for the wearing-away of brake shoes, which must otherwise be done by hand adjustment. In improved forms, which embody the basic idea exhibited in this patent, many thousands of these devices are employed, and are practically standard in passenger service.

№ 136,631, March 11, 1873. Steam-Power Brake Couplings:

The purpose of this invention was to remove the necessity for a double line of pipes under the cars due to the type of coupling theretofore used. The following quotation from the specification describes the condition to be remedied and the method proposed in the patent for doing it: "In the patent granted to me, August 8, 1871, No. 117,841, provision is made for the reversal of a car without changing the relative arrangement of the couplings. This is done by branching the air-brake pipe at or near each end of the car, and attaching a male coupling to one branch and a female coupling to the other, as therein described. In my present improvement I accomplish the same useful result by making a coupling wherein each half shall have a male and female part to couple into or with the female and male parts of the next coupling. With couplings so made there will be no occasion to branch the pipes, and the half coupling on either end of either car will couple on to any other half coupling on the train." Couplings of this form were experimentally used but did not become a part of standard apparatus.

№ 138,827, May 13, 1873. Valve Devices for Steam and Air Brakes:

Describes the first form of triple valve experimentally tried in road service, but as it could not graduate the brake pressure it was not introduced into general service.

№ 141,685, August 12, 1873. Valve Devices for Fluid Brakes:

Describes a triple valve capable of graduating the brake cylinder pressure, and was the first form supplied for service use.

№ 144,005, October 28, 1873. Locomotive Air Brakes:

This is the first of Westinghouse's inventions describing the application of power brakes to driving wheels of locomotives. The form here shown was applied to a limited extent.

№ 147,212, February 3, 1874. Locomotive Brakes:

An important improvement in details of construction on Patent No. 144,005, for limited space between driving wheels.

№ 149,901, April 21, 1874. Valves for Fluid-Brake Pipes:

Improvement on Patent No. 141,685. Valves constructed of the design shown in this patent succeeded those of Patent No. 141,685 and were largely used in service.

№ 156,322, October 27, 1874. Discharge Valves for Fluid Brakes:

The device shown in this patent was intended to provide for the automatic application of the brakes in case of a derailment of the car. It was included among the devices furnished with the automatic brake when it was first introduced, but as a result of experience its use was discontinued because of its undesired operation, due to its being operated by flying missiles [and livestock that had been run over] when the train was in motion.

№ 168,359, October 5, 1875. Air Valve for Power Brakes:

An important improvement in the automatic brake, in which a slide valve and piston is substituted for poppet valves and diaphragms used in preceding structures. It was also the first triple valve with a normally open exhaust port.

№ 172,064, January 11, 1876. Air-Brake Valve:

An improvement in triple-valve construction in which a limited amount of lost motion between the valve stem and the slide valve is the important feature. This particular feature [the graduating valve] is an important element in all subsequent triple-valve constructions.

№ 214,336, April 15, 1879. Coupling Valve:

This patent is the first to describe the combination of an air hose coupling with a valve arrangement controlling the flow of air through the train pipe in which the valves are automatically opened when the couplings are united, and closed when they are manually separated by partially rotating the two halves of the coupling with reference to each other. The valves, [should the train become parted], provide for the escape of air from the train pipe, causing the automatic application of the brakes. The successful employment of this device would dispense with the train pipe cocks that are otherwise required at each end of the car. Many variations embodying the basic idea have been proposed, but no satisfactory substitute for a train-pipe cock has been found.

№ 214,602, April 22, 1879. Cocks for Fluid-Pressure Brake:

This invention was an important contribution to the improvement of air brakes, as it is the first engineers' valve arranged to store pressure in the main reservoir in excess of the brake-pipe pressure, to facilitate the release of brakes. It is of particular importance in trains of considerable length and the principle of excess pressure has ever since been employed in all operative air brake systems.

№ 235,922, December 28, 1880. Fluid-Pressure Brake:

The first patent to describe the combination of the brake-cylinder auxiliary reservoir and triple valve in a single structure; created the general type of freight-car brake that has, since its invention, been employed almost exclusively in freight-car service.

№ 243,417, June 28, 1881. Fluid-Pressure Brake:

This is believed to be the first patent issued for an air brake in which the air valves are electrically actuated. Improved and expanded by other additions, it is now largely used in some classes of railway service. The general principles revealed in this patent were largely used in electro-pneumatic switching and signalling.

№ 360,070, March 29, 1887, and No. 376,837, January 24, 1888. Fluid-Pressure Automatic-Brake Mechanism:

These patents disclose the invention of the quick-acting brake. The specification of Patent No. 360,070 clearly states the difficulties to be overcome and the general principles of the method employed to do it. The detailed construction shown in Patent No. 376,837 was embodied in the standard triple valve thereafter for both freight and passenger service. Next to the original invention of the automatic brake, the development and introduction of the quick-acting triple valve is the most important event in the history of power braking, for it resulted in the general use of power brakes in freight service on long trains.

№ 448,827, March 24, 1891. Air Brake:

A quick-acting brake in which the train-pipe vent valve is not combined with the triple valve; a variation of the original quick-acting triple valve; not put into practical service.

№ 538,001, April 23, 1895:

This patent describes a type of quick-acting triple valve in which the quick-acting feature is improved. It differed from previous patents in the fact that its operation now depended upon a relatively quick movement of the triple-valve piston that was sufficient to move the slide valve to a position that provided a large opening for a supply of auxiliary reservoir pressure to an Emergency piston as well as the brake cylinder. Activation of the Emergency piston unseated the Emergency valve, permitting brake pipe pressure to also flow into the brake cylinder. [refer to The Introduction of Quick-Action – 1889, p23]

Thomas Higinbotham

FIGURE 297: THOMAS HIGINBOTHAM

Thomas Higinbotham was born in Dublin in 1820, the third son of Anglican parents; his father a merchant. He received a general education at Castle Dawson School near Blackrock and afterward attended the drawing and mathematical classes and lectures at the Royal Dublin Society House, Kildare Street, Dublin, where he acquired the rudiments of his professional education. Around 1839 or 1840, he left Ireland and entered the London office of Sir William Cubitt, former president of the Institute of Civil Engineers, then in the front rank as a civil engineer. He was elected a member of the Institution of Civil Engineers on February 7, 1854.

Shortly afterward, the railwaymania in Britain reached its height. Sir William Cubitt's office was full to overflowing with work of all kinds, and Mr. Higinbotham had opportunities—of which he apparently diligently availed himself—to gain a complete and accurate knowledge of the theory and the "indoor practice" of railway construction, and of the proceedings before railway committees of the House of Commons. He was subsequently appointed assistant engineer on the Ashford and Canterbury branch of the South Eastern Railway, and he later filled a similar position on another branch line in Lancashire.

Having won the approval and confidence of his superiors in these works of construction, he was next promoted to the office of resident engineer on the Huntingdon section of the Great Northern Railway, of which Sir William Cubitt was the consulting engineer. After the completion of this line, Mr. Higinbotham practiced in London for several years.

At the end of 1857—following his younger brother George—he emigrated to Melbourne, Australia. He joined his brother's household first at Emerald Hill and after 1860 near the beach at Brighton in a villa (which Thomas was chiefly responsible for designing). He never married and lived with his brother, sister-in-law, nephews, and nieces till 1880 in a relationship characterised by remarkable tolerance, friendship, and respect, despite strong differences in political opinion.

After a short time in private practice in Melbourne, Higinbotham was invited by the government to accept the office of Chief Engineer of Roads and Bridges of the colony. Three years later he was appointed engineer-in-chief of Victorian Railways, succeeding George Christian Darbyshire (himself an English civil engineer with impeccable connections with George and Robert Stevenson). Higinbotham supervised the surveying and construction of all new Victorian lines and also guided the settlement of such railway questions as city stations and facilities and the lighting of trains. He fearlessly contested proposals that he considered unsound, such as cheap narrow-gauge lines and showed great vision in advocating a railway renewals fund, construction of Melbourne's outer-circle railway and adaptations to permit unbroken rail traffic between Sydney and Melbourne.

In March 1874, he traveled overseas at the government's request to investigate and report upon the latest developments in railway construction and management in Europe, North America, and India. Among a host of other things, this provided him with exposure to the Westinghouse brake and the opportunity to consider the inadequacies of the present Flinders Street Station as a potential "main" station for Melbourne (comparing it with Boston, Massachusetts, a city of comparable size and population). His observations in Canada brought Higinbotham to an understanding of the folly of building narrow-gauge lines on the cheap, which he saw as false economy. He also formed the view that Victoria had nothing to gain by adopting American methods of car and bogie construction, and the comfort (seats with headrests) and convenience (multiple doors) of the English-style passenger car accommodation, he thought, was superior to the American. He did, however, champion the quality of materials, workmanship, and style of American cars over the Victorian product to the extent that he expressed the opinion that no further purchases of railway stock be made from "home makers"!

In Great Britain Higinbotham noticed the move to dispense with the inconvenient broad-to-standard break-of-gauge and compared the Westinghouse and Smith vacuum brakes. Having previously spoken in high terms of the Westinghouse brake, so generally in use in America, Higinbotham delved further into the subject. The Smith vacuum brake, he observed, was in direct competition with the Westinghouse (air), and though both were excellent, he perceived the preference to be toward the former. The vacuum system:

...is more cheaply maintained (on account of its simplicity of construction), its release is instantaneous, and tight couplings between the carriages are not required, because the last car in the train is the first to feel the tension of the action of the brake. With the Westinghouse brake, if the couplings are slack, the slack is taken up by the cars following one another, and when they come together an unpleasant jerking and tremor are felt. There is not much difference in the cost of applying these excellent brakes. (Higinbotham was obviously not overly impressed with the Westinghouse system pre-Quick-Action.)

In France he was impressed by his first sight of bi-level suburban passenger cars, these operating on a ring-line around Paris that connected with the main-line spokes. Higinbotham also visited and documented the railways of Austria, Italy, Hungary, Norway, Sweden, and Prussia.

He returned to Victoria in 1875, the results of his inquiries—having extended over a period of a year and a half—being embodied in an exhaustive parliamentary report. He continued to hold his office until January 1878, when, as a result of the conflict at that time occurring between the two Houses of the Legislature (and the consequent denial by the opposition of "supply"), he—together with county court judges, police magistrates, and other officers of high rank in the public service (including G. C. Darbyshire who was now surveyor-general)—was removed from office by the radicalised Berry government. During the two succeeding years, he consulted to the governments of the adjoining colonies of South Australia, Tasmania, and New Zealand.

In March 1880, at the request of the Premier, the Hon. James Service, he consented to resuming the position of engineer-in-chief of Victorian railways and returned to that office; however, the ministry soon fell, and he was unhappy under its successor. He decided to resign, his long standing and status in the department being rewarded by a sumptuous banquet and presentation. Higinbotham died quietly in his sleep on September 5, 1880, and is buried at St. Andrews Church of England, Brighton.

Higinbotham was one of that select band of English railway engineers who emigrated—perhaps as a result of the contraction of available construction positions following the abatement of railwaymania in the British Isles—and would exercise a profound influence on the development of Australian communications in the second half of the nineteenth century. These men provided practically the only mark of distinction in the Australian colonies' railway departments of the day, but their efforts were not enough to provide firm foundations for sound management as political pressures developed. Though Higinbotham did not live to see the change, his own Victorian service became the first candidate for management by public corporation when the system of political control was formally discredited in 1883.

The railways of Victoria were—at the beginning of 1880—1 182 miles in length. The bulk of these lines were designed and constructed under the superintendence of Higinbotham, and they were considered thoroughly sound and substantial in their general character. With the exception of the trunk lines from Melbourne to Sandhurst (Bendigo), from Footscray to Williamstown and to Geelong, and from Geelong to Ballarat (which were designed and partly constructed before Higinbotham became engineer-in-chief), the average cost per mile of the Victorian railways has been £57,212.

The Government of Victoria at one time had proposed that a narrower gauge than 5'3" should be adopted for the new and cheaper lines, which were then about to be constructed. To this proposal, which was powerfully supported by the newspaper press, Higinbotham "felt it to be his duty to offer the most uncompromising opposition." His official remonstrations and his evidence given at the Bar of the Legislative Council delayed the carrying-out of the proposal and, ultimately—after the lapse of a parliamentary recess—Higinbotham had the satisfaction of finding the government and the whole community converted to his views on a question that he regarded to be of vital importance.

An obituary would say of him:

As the permanent head for nearly twenty years of a large department, and chiefly responsible for the prudent expenditure of many millions of pounds borrowed on public credit for the construction of railways, Mr. Higinbotham earned and generally obtained—to a more-than-usual degree—the trust and confidence of the Government of Victoria. He was sensitive in regard to his rights and public obligations as a member of the profession of civil engineers, and he never shrank from what he believed to be his duty of clearly expressing and recording his professional opinion—however inconvenient or unpalatable—upon any matter coming within the scope of his official functions as a professional adviser to government.

But in the administration of a public department, he never forgot the duty of submission to the advisers of the Crown under whom he served; and it was his constant endeavour in practice to observe the duty, which he conceived that he owed to his profession, and to himself as a member of a profession, as well as that which his official superiors were entitled to claim from him as an administrator. He was strict and vigilant and, at the same time, just, and amongst the very many to whom his personal qualities endeared him, there were few who regretted his death more sincerely and deeply than those who served the Crown under him and with him in the public service of Victoria.

Higinbotham was for many years a member of the Royal Society of Victoria. His loss was greatly lamented by a society in which public officials of such widely acknowledged integrity were all too rare.

Some interesting details on the engineering ethos of the time:

William Charles Kernot was a graduate of the University of Melbourne, having been admitted at age 15, and became its first professor of engineering in 1883. In the following year, the first graduates in engineering—as distinct from holders of the "certificate"—appeared. He was to hold this professorship until 1909. Achieving his BA in 1864 and MA plus the Certificate of Civil Engineering in 1866, he became the first qualified engineer to be produced by the university, though he thought little of the course. He also gained a Master of Civil Engineering in 1898.

Having to battle deep-seated prejudice against the academically trained engineer, Kernot was well-versed in the perennial argument between the "practical man" and the "theorist" and was probably best known as what a later generation would call a "stirrer." Indeed, his character and influence cannot be assessed without examining the stiff sense of professional dignity and ethics, which led him into many disputes. The early public service experiences—the disputes mentioned above—he regarded as humiliations, and they made him a relentless scourge of bureaucratic incompetence, nourishing within him a sense of self-righteousness in his many campaigns for safety, economy, and the application of scientific principles in public works. He lectured government and municipal engineers sternly when he found faults in their designs, especially when these menaced public safety, and referred to the "ordinary ignorant empiric who calls himself an engineer."

Among many other battles, he dedicated himself to a lengthy campaign against the anti-academic bias of the prestigious Institution of Civil Engineers in Britain (to which many of the emigrant engineers to Australia and New Zealand belonged), and to a similar fight against the scientific and anti-theoretical approaches of government departments in appointing and promoting engineers and surveyors.

A paper published by Professor Kernot in 1893 provides insights into the organisation and practice of engineering during his early career in the 1860s and 1870s and of his experiences with pupil-trained engineers. In the Water Supply Office, while he experienced no antipathy to his university education among the staff, and he found experienced engineers to be knowledgeable about practical details of engineering, he observed that even the most senior men were almost entirely without knowledge of elementary theoretical principles. He found that he was not differentiated from those whose only qualifications were the handicraft skills in handling drawing instruments and whose knowledge of routine came from years of ordinary office work. When this office was disbanded, he moved to the Department of Railways for six months in the early 1870s, finding there that the engineer-in-chief, Thomas Higinbotham: *never missed an opportunity of impressing upon me the uselessness and undesirability of university training for engineers.*

He continued [relating Higinbotham's views that]: *It was far better to go straight from school to the office or workshop, rather than wasting 3 or 4 years at the university. Boys straight from school could do work of a mechanical or even menial character that we would hesitate to employ an educated gentleman upon, and would not think they knew more than their superiors, as does the university man on the grounds of his useless theoretical training. There were no real advantages and many disadvantages* [Higinbotham had apparently insisted] *in university education for engineers.*

Prof. Kernot later returned to a reconstituted Water Supply Office and finally resigned from government service in 1875.

John Woods[201]

John Woods, engineer, politician, and inventor, was born on November 5, 1822, at Liverpool, England; son of a railwayman. He was educated locally and trained as a locomotive engineer in Liverpool and then on the Leipzig & Dresden railway. He worked in North America and held positions in railway and iron works in Staffordshire and Lancashire, where he was active in anti-corn law agitation. He won first prize for railway axles at the 1851 Great Exhibition. At Liverpool, aged 21, he married Sarah Gibbons.

FIGURE 298: JOHN WOODS. *By an unknown engraver, 1892, courtesy of La Trobe Picture Collection, State Library of Victoria. IAN02/05/92/16*

With his wife, Woods arrived in Victoria in 1852 and went to the goldfields. He first "mounted the stump" at the Goulburn diggings to lead passive resistance to the miners license fee. He became president of the Ararat branch of the Land Convention and was active in the registration of electors and in obtaining sufficient ballot papers for the 1858 election. At Ararat he was elected to the local court and then to the Mining Board and was a delegate to the 1859 Mining Board Conference. Partnership in a deep wet claim left him penniless, and he took employment at Stawell, building the St. George's crushing plant.

From October 1859 until his defeat in 1864, Woods was a member for the Legislative Assembly seat of Crowlands. On October 30, 1865, he was appointed engineer and surveyor for the Victorian Water Supply at a salary of £300. He worked on the Malmesbury reservoir until summarily dismissed by (Sir) James McCulloch's government on a charge of knowingly accepting faulty pipes and tarring over the cracks, the accusation and dismissal being later privately acknowledged as unfair.

Woods represented Crowlands again in 1871–77, and in 1877–92 the new electorate of Stawell. He was commissioner of railways and roads and vice president of the Board of Lands and Works in the (Sir Graham) Berry governments of 1875 and 1877–80. He sat on many select committees, mainly about railways, and was also a member of several royal commissions, including those on the tariff (1881–83), employees in shops (1882–84), and coal (1889–91). He was a commissioner for the 1880 Melbourne International Exhibition and an honorary commissioner to the 1883–84 Calcutta Exhibition. From 1890 he was a member of the Parliamentary Standing Committee on Railways.

A consistent, active proponent of democratic and radical principles and a staunch local member, Woods worked hard for the interests of miners and small settlers. He advocated a non-legalistic approach to mining on private property and a sliding scale of gold royalties to replace rents for mining land. He steadfastly put forward the Land Convention program, always opposing permanent alienation of crown land and sale by auction and proposing, in 1859, 1864, and 1873, a progressive land tax to "burst up the great estates," publishing pamphlets on the subject in 1873 and 1880. As minister, his overriding interest was to reduce costs to small farmers, a political priority opposed by senior officials. He initiated great railway activity, a major branch line along the Goulburn Valley, the Melbourne-Oakleigh connection and two northwestern extensions into the Wimmera. He reformed the goods tariff to reduce freight charges and leveled and reorganised the Spencer Street yards, but his plans to build a block of grain and wool sheds and a new dock to reduce farmers' storage and cartage costs were canceled by the next government.

A strong protectionist, Woods championed local industry (thereby opposing Thomas Higinbotham) in the use of Victorian-built locomotives. Among his several inventions were a stone-breaking machine (patented 1860) and a hydraulic railway brake (1882), which did not prevail when the Westinghouse air brake was

201. http://adb.anu.edu.au/biography/woods-john-4884 (retrieved September 2013).

eventually chosen for Victorian Railways. He had set up a company in December 1877 to register the patent and manufacture his invention, but this came to nothing when air-powered brakes eventually dominated.

Although a member of the Chamber of Manufacturers, Woods was always proud of his artisan background and claimed to be a representative of labour; "*I graduated in a hard school…I had to work my ten hours a day in England for very little money, and I know exactly what it means; and I know the advantages which have been obtained in this colony by labour in combination.*" However, he opposed political representation of trade unionists as such. He fought for the eight-hour principle and argued for unemployment relief and old-age pensions. In a major article, "Wages," in the *Victorian Review* (1880), he raised the problem of ensuring for a worker a fair share in the profits of industry, not merely a day's pay. After 1880 he became a liberal oppositionist to the James Service/Graham Berry coalition and subsequent ministries.

Alfred Deakin first met Woods in 1879, *"his rotund form, snub nose, glistening eyes and spiky hair rendering him a rather Socrates-Silenus appearance."* Deakin described Woods as a serious reader and a man of original mind and initiative. However, Woods never attained high parliamentary stature. His radical ideas, unorthodox opinions, and abrasive approach offended the community's respectable conformism. Impetuous and unconventional in his actions, he sometimes had to apologise for *"taking intoxicants too freely"* and wrote amorous doggerel, which *Punch* would not publish. He was a vigorous speaker with a gift for phrase-making. Deakin, however, claimed that he lacked industry and self-control. He lent his name to some of James Munro's[202] more questionable promotions and with L. L. Smith[203] "puffed" a tin mine on the London market. Attacked by the conservative press, the nickname "Tarbrush" followed him everywhere.

Woods' first wife died, aged 74, on January 12, 1888, and on September 23, 1891, he married a divorcee, Jessica Muir, née Whitley. He died of heart disease, dropsy (oedema), and gangrene on April 2, 1892, at Brighton. Survived by his second wife and two of his four daughters, and predeceased by his two sons, he was buried in the Boroondara cemetery according to the rites of the Anglican Church.

Norman Selfe, MICE, MIME

FIGURE 299: NORMAN SELFE

Engineer and educationist, Norman Selfe was born on December 9, 1839, at Teddington, Middlesex, England, the son of a plumber. Norman was educated at Kingston-upon-Thames. He arrived in Sydney with his parents in January 1855 and, at age 17, he was apprenticed to P. N. Russell & Co., reputedly becoming chief draftsman before completing his articles. While with the firm, he designed gunboats for service in the New Zealand Maori wars. From 1865–68, Selfe was in partnership with James Dunlop and designed and built major installations for the Australasian Mineral Oil Co., the Western Kerosene Oil Co., and the Australian Gaslight Co. In 1869, he became chief engineer and draftsman for Mort's Dock and Engineering Co., designing the machinery for the *S. S. Governor Blackall*. (The first iron vessel of any size to be built in Australia, 487 gross tons, 2 masts schooner rigged, built in Sydney at a cost of £30,000, and launched in 1871. She was originally ordered by the Queensland Government for a coastal mail-run; however; the contract was canceled, and the vessel was purchased in 1872 by Australasian Steam Navigation Company.

202. http://adb.anu.edu.au/biography/munro-james-4271 (retrieved September 2013).

203. http://adbonline.anu.edu.au/biogs/A060172b.htm (retrieved September 2013).

She became a coal hulk in Sydney in late 1910 and was scuttled July 1931 off Sydney Heads. The Leslie Administration of Queensland was accused of extravagance and eventually brought down over this order.

FIGURE 300: STEAM SHIP *GOVERNOR BLACKALL*. THE FIRST LARGE IRON-HULLED VESSEL BUILT IN AUSTRALIA. MACHINERY DESIGNED BY N. SELFE

Selfe set up in private practice in 1876 and soon became known as a versatile and original engineer. He designed over fifty steamships, including two torpedo boats for the New South Wales government and the first double-ended screw ferry in Australia, *S. S. Wallaby* (164 gross tons, for the North Shore Ferry Co. Ltd., the 2nd double-ended screw ferry to be built in the world, to the order of Thomas Summerbell and built by W. Dunn of Berrys Bay, Sydney, launched in 1879. The vessel was probably meant to be called the *Willoughby*, after the Sydney suburb but was believed to have been mispronounced by the ship's sponsor. She was used as a cargo-lighter before being broken up in 1926.)

Selfe was reputed to have introduced the compound engine in Australian vessels. An expert in dock design, he built the first concrete quay wall in Sydney Harbour and most of the wharves for deep-sea vessels. A skilled refrigeration engineer, he designed the first ice-making machines in New South Wales and introduced the first lifts, patented an improved system of wool-pressing, and carried out many other hydraulic and electric light installations in Sydney. Selfe worked and published on compressed air and its application to railway brakes, designed many friction winches and the acetylene gas storage plant for the New South Wales railways, and planned mills, waterworks, and pumping stations. He is nowadays considered the leading representative of the 'pupil-trained' generation of "the all-round engineer".

In 1884–85, Selfe visited seventy cities in North America and Europe, inspecting engineering works and notable bridges. A strong advocate of a "Circular City Railway" linking Sydney's outlying suburbs to the ferries, he was largely responsible for a design, in 1903 for a cantilever harbour bridge from Dawes Point to McMahon's Point. The plans were entered in an international competition to find a Sydney Harbour Bridge design and were eventually accepted but, to his immense chagrin, not proceeded with, owing to a change of government. It was probably on this trip that Selfe met William Hanscom and became enthused by his proposed air brake system (see section 3, Westinghouse Down Under).

In 1870 he had been a founder of the Engineering Association of New South Wales and president in 1877–79. Like his contemporary in Melbourne, W. C. Kernot (see the biography for Thomas Higinbotham), Selfe had a strong sense of natural justice, and his spirited paper, "Sydney and its institutions as they are, and might be," published in the Proceedings of the Engineering Association in 1900, gives a clear picture of his forthright personality and his professional and social interests. An active member of the Sydney Mechanics' School of Arts and of the Royal Society of New South Wales, in 1901 he was founding vice president of the Australian Historical Society, president in 1902, and a committee-man until 1911. He was also a prominent Anglican and a freemason.

As early as 1865, Selfe had been teaching mechanical drawing at the School of Arts and, in 1883, became a member of the Board of Technical Education. He was acting president for much of 1887–89 when the board confronted the minister of public instruction, James Inglis, and his departmental officers. Selfe wanted an independent system of technical education, with the task of building a trained workforce. The board was abolished in 1889, and Selfe lost a contest in which he had probably gone—tactically at least—too far in expressing his contempt for the civil service and for traditional educational values. But he did see the concept of workshop training introduced into Australia.

Selfe died of heart failure at Normanhurst—the suburb named after him—on October 15, 1911, and was buried in the Gore Hill cemetery. He was survived by two daughters of his first wife Emily Anne (Booth), whom he had married in Sydney on October 10, 1872, and by his second wife Barbara Marion (Bolton) BA, Sydney, whom he had married on May 12, 1906.

(Further information on this outstanding and little-acknowledged engineer is available in Wikipedia)

D.H. Neale

Deodatus Hilin Neale, eldest son of William Johnson Neale, barrister-at-law and Recorder of Walsall, and at one time a naval officer, was born in London on 5 September 1849, and spent his early childhood at Kingswear, Dartmouth. After being educated at King's College, London, and in Germany, he became a pupil in the office of William Dredge—well-known at that time as an engineer and bridge designer—of Buckingham Street, Adelphi.

In October 1868, he was articled to William Adams, locomotive superintendent of the North London Railway, and served his time in the workshops and drawing office at Bow until August 1871. On the appointment of Adams as locomotive superintendent of the Great Eastern Railway (later merged into the LNER), Neale followed that gentleman to Stratford as chief draughtsman, under Works Manager (and later, Locomotive Superintendent) Massey Bromley. After the death of Mr Bromley in the Penistone railway accident in 1884[204], Neale left Stratford, and joined one of the South Wales mineral railways—the Monmouthshire Railway—where he was chief draughtsman in the Locomotive Works at Newport.

After holding that post for about four years, he joined—in the spring of 1876—the Hyde Park Locomotive Works of Neilson & Co at Glasgow and in the following year conducted some elaborate brake trials at the Cowlairs Works of the North British Railway Company between Westinghouse's automatic air brake and Smith's vacuum brake (supplied by the Vacuum Brake Co. of London) which demonstrated the advantages of the air-brake.

Neale then worked for a few months for the Westinghouse Brake Company and, from 1878 to 1881, was again in the Locomotive Works of the Great Eastern Railway Company at Stratford. He was elected an associate of the Institution of Engineers on February 4, 1873, and was placed in the associate Member class on its creation in 1878.

One journal (*The NSW Railway Budget 1892–93, vol. 1*), cites Neale working as inspector of locomotives for the South Indian Railway under Sir Charles Douglas Fox and being well-involved in rollingstock design. Early in 1882, Neale secured the position of assistant locomotive superintendent to the Cape Government Railways (one of three competing rail systems in the British South Africa of the time) but resigned the following year and returned to Britain prior to sailing for America in order to represent the journal *The Engineer* at the Chicago Exposition of Railway Appliances.

204. On 16 July 1884, a few miles to the west of Penistone, near Bullhouse Colliery where the road to Huddersfield passes below the Manchester, Sheffield & Lincolnshire Railway line, the 1230 pm express from Manchester London Road to London King's Cross, with through carriages for Grimsby Docks for the evening steamer sailing, had left Woodhead Tunnel and was accelerating on the downhill gradient towards Penistone. As the train entered the curve at Bullhouse, the driver felt the engine develop an uneasy roll, but before he could apply the brakes he heard a 'crack'. A driving wheel crank axle on the locomotive had snapped, and the resulting spread of the driving wheels distorted the track. The axle fracture was probably caused by metal fatigue. A horsebox coupled behind the engine was derailed but remained upright. The coupling between the horsebox and the following carriages failed, and they ran off the rails and down the embankment on the outside of the curve. Nineteen passengers—a high proportion being women—including Bromley were killed instantly, and sixty injured, five of whom later succumbed to their injuries.

An immigrant list published in *The NewYork Times* on May 8, 1883, lists a 'Neale' as having arrived on the *S. S. Alaska*[205] from Liverpool (berthing on May 5), and by November he had settled in NewYork as an associate editor of the *Railroad Gazette,* one of the principal rail industry journals in the United States. Almost the first work he undertook in this capacity was to compile—under the title of *Recent Locomotives*—an illustrated series of descriptions, with specifications and details, of recent American and European locomotives, reprinted from the *Railroad Gazette*. An enlarged edition with his name attached to the preface was issued in 1886. During this time he was involved, among other things, with the notable Burlington Brake Trials (see Chapter 2) and as an associate member of the Master Car Builders Association was also immersed in the questions of safe passenger car heating and the adoption of an automatic car coupler.

Mr E Eddy, Assistant General Manager of the Caledonian Railway, had taken up an appointment as Chairman of the NSW Railways Commission late in 1888[206], and having met Neale in NewYork, asked him to join him at Sydney, as adviser on matters connected with rollingstock. After five years with the *Railroad Gazette*, Neale resigned in November 1888. In Australia, Neale performed valuable service in bringing into order the much-neglected rollingstock of the colony and—as one of the commissioners appointed to select a brake for all the Australian railways (member South Australian Brake Board 1890 and the New South Wales Brake Board 1891)—was instrumental in the choice of the Westinghouse system. In fact, Neale has been credited with membership of no less than nine brake boards who have conducted various trials resulting in recommendations that have been adopted by 88 railway companies and four brake companies.

During an investigation before a Parliamentary Committee, of accusations against Mr Eddy of mismanagement, Neale's evidence was among the chief causes of the complete acquittal of the railway commissioner, and of the almost universal public verdict in Eddy's favour. When this case was over Neale visited America, and on the invitation of the proprietors of the *Railroad Gazette*, accepted the post of chief editor of that journal. Returning to the United States in the autumn of 1892, Neale rejoined the *Railroad Gazette.* In addition to his work for that journal, he was occupied during the last few months of his life in editing a new edition of *The Car Builders Dictionary*. His career was, however, destined to be prematurely cut short. Overwork brought on an attack of cerebral meningitis, which proved fatal, and Neale passed away at his home at 615 St. Mark's Ave., Brooklyn, NewYork, on the evening of Wednesday, April 5, 1893.

The *Railroad Gazette* eulogised Neale in the following terms:

He was a man of studious habits and of unusual intellectual gifts. He had a keen mechanical faculty, good powers of analysis, fair mathematical attainments, and a mind that was instinctively and habitually active. Added to these gifts he had a remarkable faculty for using the English language. Few mechanical writers were so well-endowed in this respect as Mr. Neale. He died in the maturity of his powers and his death is a real loss not only to us but to the world.

T.F. Rotheram

Thomas Forth Rotheram was born at York in England on June 28, 1850. He obtained his early engineering training in the Manchester, Sheffield & Lincolnshire Railway works at Gorton and subsequently in the shops of the North British Railway at Cowlairs, Glasgow.

With the approval of the officials of the latter line, he took a position with the London and Glasgow Shipbuilding Company to gain experience in the fitting and completion of the engines and boilers of large

205. This vessel was a fast liner and immigrant ship that—at the time of Neale's passage—held the Blue Riband for her six-day crossing of the Atlantic.

206. Edward Miller Gard Eddy almost went to New Zealand. NZ Agent-General in London, Sir Dillon Bell, was anxious to have Mr Eddy for the NZ Railways Commission, but Sir Daniel Cooper, Australia's acting A-G stepped in and—offering Eddy £500 more—secured his service for £3 000pa.

ocean steamers. He then returned to the North British Railway and was appointed to take charge of the erection of indoor and outdoor machinery and plant. He was subsequently engaged for three years with Messrs. Ransomes and Rapier in the design and manufacture of fixed plant and equipment for railways in various parts of the world.

Early in 1875, Rotheram emigrated to New Zealand and entered the service of the New Zealand Government Railways, occupying the following positions:

- General manager of the Picton and Blenheim Section (1875–78);
- General manager of the Wanganui, Foxton and New Plymouth Section (1878–86) - Elected a Member of the Institution of Engineers on January 12, 1886;
- Locomotive superintendent of the Hurunui-Bluff Section (1885–88). In 1887, while still connected with the NZGR, he was specially appointed by the Government of New South Wales to report on the railways and tramways of that colony;
- Locomotive superintendent NZGR (1888 to April 1900). This was the job title of those with ultimate technical responsibility for the provision and maintenance of locomotive power in the original British railway companies. When some of these men emigrated, the designation was exported to the colonies, eventually evolving to become 'Chief Mechanical Engineer'.

While in the service of the NZGR, Rotheram introduced the complete manufacture in the railway workshops of many locomotives and other rollingstock. In 1891 he was appointed by the New South Wales Railway Commissioners to inquire into the merits of the Westinghouse and vacuum brakes as applied to goods trains. In 1893, he was sent to America and Europe to examine the working of compound locomotives and to inquire into the feasibility of introducing electric 'motors' to the New Zealand Railways and, generally, questions regarding electric welding, locomotives, machinery, and lighting. NZGR would later become noticeably 'American' in its selection of steam locomotive designs and equipment as well as many operational practices.

In 1900, Rotheram moved to Western Australia[207] as the first chief mechanical engineer to the Western Australian Government Railways to replace the incumbent officer, locomotive engineer R. B. Campbell (who had fallen out of favour with general manager, John Davies). Rotheram assumed Campbell's responsibilities in planning for the new railway workshops at East Guildford (later called 'Midland Junction', then 'Midland') and was largely responsible for selection of the final construction plan.

Rotheram died suddenly from angina pectoris at his residence in Mount Street, Perth, on September 11, 1903.

[Author's note: Rotheram was not the only such engineer to move from New Zealand to Australia. In 1885, while T. F. Rotheram was still locomotive superintendent of the Hurunui-Bluff Section, Allison D. Smith—the incumbent locomotive superintendent of the NZGR—moved to Victoria to assume the same position. Smith had been born in Jamaica in January 1854 and, like Rotheram, educated as a mechanical engineer at the Cowlairs Works of the North British Railway of Scotland. He had family ties across the Atlantic Ocean, thus becoming acquainted with American practice. Smith had been appointed locomotive superintendent in Christchurch in 1877 and saw American practice as suitable for local conditions. Smith placed the first New Zealand order for US-built locomotives;

207. In *When Steam Was King*, W.W. Stewart recounts that Rotheram—at the time NZGR's Chief Mechanical Engineer—had a violent row with a Member of Parliament over possession of a lady's seat in a "birdcage" carriage on the New Plymouth Express. His position must have become untenable as a result, and Rotheram resigned and immediately obtained a position as CME of the WAGR – taking one of his staff members with him. Stewart also states that—had the NZGR not increased his salary—H.H. Jackson, who later rose to the position of CME and oversaw the final stages of design and introduction into service of New Zealand's famous AB Class locomotives, would have soon followed Rotheram to WA.

with the Rogers Locomotive & Machine Works of Paterson, New Jersey, for eight highly-ornamented 2-4-2 locomotives as the NZGR K Class and later the T Class obtained from the Baldwin works in Philadelphia. In 1892, while Smith was locomotive superintendent at Newport Workshops in Victoria, he was retained as a consultant by the Western Australian government and proposed relocation of the WAGR railway workshops from Fremantle to a new site at East Guildford (Midland), the very project that Rotheram would—ten years later—become involved with. He is reported to have been retrenched in 1893 whereupon he retired.]

F.W. Eames

Frederick William Eames was born at Kalamazoo, Michigan, in November 1843, the son of an inventive farmer-mechanic, Lovett Eames. His mother, Lucy C. Morgan, was the daughter of Rev. Elisha Morgan, *"a man of marked ability, unusual balance of judgment, strong convictions, and conscientious devotion."* His daughter inherited these characteristics and intensified them by increased mental activity. Miss Morgan married Lovett Eames about 1835 and removed to Kalamazoo, where her husband had already taken up residence. There, Frederick W. was born, passing his early years in an intelligent Christian home and enjoying the superior educational advantages of his native village.

He had scarcely entered college when the Rebellion began, and President Lincoln called for volunteers to protect the threatened life of the nation. Then, but 18 years of age, he was the second man to enlist in the first company raised in his own town for defence of the "old flag," the 2nd Michigan Infantry. Captain Charles S. May, who commanded this company, writes of him:

He was one of the gallant and patriotic band of young students in Kalamazoo College who enlisted in my company at the very first call to arms in 1861, and during the time I served as his captain I remember him as a good and brave soldier, bearing himself gallantly and unflinchingly in the early battles in which we were engaged. And though but a mere boy in years, always showing the native energy and shrewdness for which he was afterward distinguished.

Eames was subsequently mustered out of the 2nd Michigan Infantry to accept a lieutenancy in another Michigan regiment. In 1863 he was honourably discharged by General Grant from this regiment, appointed aide in the Revenue Service, and served on the Mississippi River until the close of the war. He was later appointed a government detective and was instrumental in unearthing and destroying a gang of counterfeiters in Michigan.

In 1870 he married Miss Mattie Shilling of his native state, and two children were born to them.

Harry F. Landon (*150 Years of Watertown*: Watertown Daily Times, 1950) says Eames came to Watertown, New York, in 1861, "*...to install for the municipality a pumping system of his own devising. That same year, Lovett and Moses Eames (his uncle) bought Beebee Island and established a machine shop there in one of the old stone buildings which had survived the cotton factory fire.*" (It is assumed that Landon's reference to 1861 is a misprint and that Fred Eames arrived in Watertown in 1871.)

Eames received his first patent in 1874 and began to market his brake about two years later. Besides the power brake that bears his name, he invented an automatic governor and cut-off for engines of ocean steamers, a steam pump, and a multitude of other mechanical devices.

About a year after John Smith's vacuum brake went on the market, Fred Eames received his first patent (No. 153,814, of August 4, 1874). It appears that the primary difference between Eames' brake and Smith's was that the latter used a piston mounted on the car, while Eames' used diaphragms mounted separately on each truck.

He established the Eames Vacuum Brake Company on February 14, 1876, and began manufacturing the brakes in his father's machine shop on Beebee's Island. The company was capitalised at $500,000 (more than US$11 million in today's buying power) but was short of capital from the beginning.

Within three years, 29 US railroads were using Eames' brake, and by 1881, 57 railroads in nine countries had adopted it. One of its primary users was the New York Elevated Railroad (with light cars making frequent stops).

Eames traveled the world marketing his brake. To promote it in England, he purchased from Baldwin a fast locomotive built in 1880 for the Philadelphia & Reading Railroad (but not delivered), equipped it with his brake, and named it the *Lovett Eames* in honour of his father. It was shipped to England and in 1881 ran on a number of railways, but only two of the English railway companies made purchases, apparently because there was already a British-made vacuum brake available. And that was an *automatic* brake; if the train parted or the ejector failed, the brakes were applied automatically.

Eames returned home determined to create his own automatic brake; however, in June 1882, while he was away, the vice president, secretary, and directors of the company had entered into an agreement with Thomas Prosser and Son of New York, whereby they transferred and leased to the latter all the property of the company, including the patterns and the exclusive right to manufacture the brakes in the United States under Eames' patents. Evidently the Eames Company owed Prossers $47,000, a considerable sum in those days. Eames returned from Europe immediately after the transfer, and one can imagine his mortification and anger on learning of what had transpired. He immediately took action against the conspirators—his fellow directors—together with Prosser's, to have the agreement set aside. Following institution of the action, although the shop was under the charge of the sheriff, Eames—maintaining his rage—entered and proceeded to choke the officer in charge! Eames was arrested under warrant; it appears there was much bitter feeling between Eames and the shop hands.

The following account of the outcome was gleaned from a New York newspaper:

On April 20, 1883, a judgment was given which held that all the transfers were void. The Sheriff and Eames were given immediate possession of the works, however Eames had to repay Prossers the outstanding amount. Wasting no time, Eames, his brother, the sheriff and under-sheriff together with a Baptist minister who was a stockholder, went straightaway to take possession. With the Sheriff formerly turning the works over to Eames, Eames went down into the basement to close the engine down. On his way down he passed Charles Higham, a pattern maker and Superintendent of Construction. After closing down the engine, Eames made his way up to the office and caught hold of Charles Bingham, who had been in charge of the works since the commencement of the suit. With pent-up anger he apparently attempted to throw Bingham out of the building but the Sheriff intervened. Proceeding to another level, Eames went to the end of the building where a door had been put in to connect with an adjoining property. Higham had retreated behind this door. On seeing Eames steaming toward the door, Higham locked it. On coming to the locked door, Eames promptly broke a glass pane in it with his cane, put his hand through and unlocked the door. Up until this time, Eames and Higham had been the best of friends and had never had any cross words. Higham was probably terrified on seeing Eames in his seething state.

Workman said they saw Eames raise his cane above his head as if to strike him. Higham later said Eames hit him twice. In any case, Higham pulled out a .32 revolver and fired two shots at Eames who fell to the floor dead. One bullet had hit him in the lower part of the head whilst the second had entered his right breast passing through his heart. The Rev. Townley, who was following close behind, found Eames on the floor and Higham standing over him as if he intended to shoot him again. The reverend gentleman promptly took a club and beat Higham off. Higham gave himself up and put himself in the custody of the Sheriff. After going to inform his wife of what had happened he went with the Sheriff to the local gaol. It was said that during the afternoon, Higham was prostrate with grief.

Higham was 33 years of age and had worked for the company almost since its inception. Workmen said they never knew Higham carried a gun although some outsiders believe that since the trouble had begun the shop was like an 'arsenal' and that all the men had been armed. A postmortem examination was quickly arranged as was a coronial enquiry. Higham subsequently went for trial in December 1883 and was found "not guilty" on the grounds of self-defence.

Though vacuum brakes were doing well overseas, in the United States they were losing out to the Westinghouse automatic. One of the few large sales was to the narrow-gauge Denver, South Park & Pacific, which outfitted seven hundred cars in 1880. A year later, the Union Pacific (which just happened to own the DSP&P) purchased three hundred sets, also for freight cars.

In 1885, the Eames Company developed its own automatic brake, spurred on by several railroads that wanted to block the Westinghouse monopoly. At the 1886 Burlington brake trials the Eames vacuum brake came out second only to the Westinghouse pneumatic, but proved too weak for long trains at a time when one of the major goals of the railroads was to increase train length to more than fifty cars.

In 1890 the Eames Company was reorganised as the New York Air Brake Company (Landon says it was incorporated in 1894, which is not necessarily excluded by the foregoing statement.). The plant was eventually moved from Beebee's Island to 'Poole Farm' in Watertown, NY.

George Westinghouse sued the New York Air Brake Company, but his earliest patents had expired, and after years of litigation, in 1912 the companies agreed on a cross-licensing arrangement that allowed them both to do business[208].

At his funeral, Eames was eulogised thus:

Mr. Eames inherited from his father a passion and genius for invention. His insight of possible mechanisms, and his resources in mechanical devices, was phenomenal, and the difficulties in the way only stimulated his enthusiasm and augmented his energy; he was always at his best when obstacles were thickest and greatest. His conceptions of the desired end of an invention were clear and strong; his knowledge of all the conditions, comprehensive and balanced; the mutual relations of these conditions clearly seen, all possible complications stripped off, and the end sought with a persistent grappling with obstacles, and an opulence of resources which soon made him master of any problem. I know of no man who, in his peculiar field, was a better illustration of the truth that the shortest distance between two points is a straight line; simplicity and efficiency were the two objective points in all his devices. Beside all the inventions that enter into his power brakes for railroad trains; plain, automatic, and duplex, he invented an automatic governor and cut-off for engines of ocean steamers, a steam pump, and a multitude of other mechanical devices.

In perfecting and securing patents for all these inventions in the United States and foreign countries, and in seeking their application to the industrial arts, he repeatedly crossed the ocean, spent months and years in England and on the continent, and did it all handicapped by inadequate capital, financial embarrassment, and costly litigations with powerful antagonists backed by millions, with whom his inventions came into competition. As a man he was loved most by those who knew him best. His friends were ardent, while his enemies feared, sometimes hated, but never despised by him. His whole life was an intense struggle, and therefore the rugged side of his nature was toward the world, but those who shared his confidence knew that no man ever had a more ready appreciation of kindness, a more open hand, or a more open heart than Mr. Eames. It is deeply to be regretted that while defending his rights, and securing his property from those who sought to wrest it from his grasp, his useful life was suddenly terminated by a bullet from a pistol in the hand of a man whom he had befriended. He was thus killed, in the spring of 1883, at

208. Historian and Smithsonian curator, John H. White, estimated in 1978 that New York Air Brake held about a twenty-five percent market share against Westinghouse's seventy-five percent.

the early age of 39 years. What he might have done, if life had been spared, it is impossible to say; but certainly his great genius and indomitable energy gave promise of the largest results.

Eames is buried in the Mountain Home cemetery in Kalamazoo, Michigan.

D.G. Blaine

Any reader who was a member of the Air Brake Association up until the early 2000's will know of Dave Blaine and his contributions to and through that organization.

FIGURE 301: D.G. BLAINE

David George Blaine was born in New York City on May 5, 1918, the son of a medical doctor. Early in his life, the family moved to Oak Park, Illinois. The Blaine home in Oak Park was only one block from the Chicago & North Western Railroad tracks, and Dave became intrigued with trains at an early age. He believed his first train ride was probably going downtown with his mother on the old Chicago Rapid Transit.

His father had an old WWI medical corps friend who suggested he bring his family to California and become an associate in an x-ray lab. At the time, Mr. Blaine senior chose not to accept, but then in the fall of 1930, his brother passed away. He decided to bid on and was awarded the laboratory position in California, and the family moved out to Pasadena (16 km northeast of downtown Los Angeles) during the Christmas period of 1930.

Dave Blaine started railroading with the Southern Pacific at their Los Angeles Shops in 1936, and soon, at the suggestion of the general superintendent, started a pre-engineering course at UCLA ("it only cost $40 per semester."). After three and a half years, he enrolled at Purdue University in West Lafayette, Indiana, which, at that time, had a railroad mechanical engineering course with emphasis on car, locomotive, and railway shops design, air braking (the tests that proved the AB control valve were made at Purdue over the period between 1924 and 1929), and an active locomotive testing laboratory. At Purdue, Blaine was active in the local student chapter of the American Society of Mechanical Engineers.

Blaine graduated from Purdue in 1941, cum laude, with a BSME degree. That same year—having been refused for military service due to his eyesight ("in those days, everybody had to be a sharp shooter.")—started a 42-year career with the Westinghouse Air Brake Company (although he did stay on the SP seniority list until about 1946 or 1947). His initial rate was $135 per month (compared to the $75 per month offered by his former railroad employer as a tracer at the SP general mechanical headquarters in Sacramento). His initial duties at WABCO were: *...primarily to learn air brake systems and how they worked. They had a regular special apprentice class, which of course, is old railroad terminology. So we studied more damn old equipment, which I hardly ever ran into.*

He later served as a test and research engineer at Wilmerding, Pennsylvania (1941–1945):

All of the work involved a 150-car freight rack and we had a 25-car passenger test rack that had a D22, F-1864 relay valves, speed governor control and so forth. There were about 250,000 AB valves in service. From a technical standpoint the main advantage [of the AB control valve] *was, you could always get an emergency* [brake application] *if you had a rapid brake pipe reduction...* [whereas] *with the K* [triple valve], *if it had 15 pounds brake cylinder pressure or more it probably wouldn't go to emergency. The AB had the advantage of being approved* [by the Interstate Commerce Commission and AAR] *for 150-car train operation. That meant you could make an emergency at any speed and the train wouldn't automatically just derail.* [With the K triple] *They tried to run trains of 100 to 125 cars and sometimes they got away with it. But it wasn't a sure thing, primarily because of slack action.* [On the subject of sticking brakes]

The AB valve had a feature called Release Insuring and the K did not. If you had piston ring leakage that was on the high side the leakage would slip right on past the pistons and the brake would stay in the Applied position.

Many of the projects Dave Blaine was involved with during the war years: *...were for ship controls, pneumatic controls for small ships like sub chasers and so forth. They were operable from the pilothouse, where in the old days they would telegraph what they wanted from the bridge to the engine room. But I also horned in on every railroad project we had going. I wanted to make sure they knew I was there for railroads.*

After the war, he worked on the first diaphragm valve, ultimately called the AC (see Chapter 2). *It came out in the mid-50s...went over like a lead balloon. I don't think there were more than 300,000 sold. The special pipe bracket it required was its downfall. It operated fine in a unit train...* [Erie Mining, it might be noted, used the AC control valves in fully-equipped taconite iron ore trains.]

Blaine worked his way up to commercial engineer (1945–1947) and then moved to Chicago where he was field engineer (1947–1951, with opportunities to travel to the West Coast on the locomotives of the Northern Pacific RR, Union Pacific, Santa Fe, and "the Burlington"); representative (1951–1954); and then district engineer (1954–1960). As district engineer, Blaine primarily dealt with installations with much of the work being related to troubleshooting on both freight and passenger equipment: *There was a DC electro-pneumatic system on a D-22* [control valve], *and it was prone to develop all kinds of hard-to-find grounds and shorts.*

Later Blaine was promoted to manager value analysis (1960–1961); western district engineer (1962–1966); senior district engineer (1967–1973); and senior marketing representative (1973–1983). He acknowledged that the working relationships between the railroads and supply companies were different than they are now, but that: *...the customer is always right and this* [concept] *was hard for many of the engineers back at the home office to accept.*

Following his retirement from WABCO, Blaine served as a consultant to the AAR and several individual railroads. A member of ASME from his student days, he was later an active participant in the Air Brake Association and several other professional groups. The ABA elected him a life member - only the third in that organisation's ninety-plus year history.

At a valedictory dinner in his honour, he was asked to compare train operation between the time he joined WABCO and the time he retired. Blaine responded:

The AB tests hit it on the head when they said 150 cars was a practical maximum train length to operate trains. There were trains operated longer than that, particularly coal trains operated between southern Ohio, Kentucky and the Great Lakes. C&O were known for operating 250- and 270-car trains. Severe slack action was the main problem - particularly, if a heavy reduction had been made or if you had been in power for a long time.

At about the same time, pressure-maintaining was being developed which made it easier to operate 150 car trains; that is, 150 x 50-foot cars. If the cars were piggybacks or 89 feet long then the number was reduced to about 75. I believe pressure-maintaining was the one feature or invention to help railroads most to operate trains during my lifetime. Before pressure maintaining, leakage in the brake pipe would cause the brake application to increase. If the train had the maximum leakage allowed of 5 psi per minute then the total brake application would increase from a light reduction to a full Service application in a few minutes. The engineer could stall the train if he didn't closely watch it. On long descending grades, brakes would have to be cycled on and off to keep the train speed in check but this could lead to false [brake pipe pressure] *gradients, which could also result in braking problems. Pressure maintaining prevented these braking problems.*

On Southern Pacific, Santa Fe, Denver & Rio Grande Western, Northern Pacific and Great Northern most enginemen practiced "bridge braking" with the H-6 Brake Valve. This was a form of manual pressure-maintaining. Another product that made a significant impact on train operation was the composition brake shoe.

Asked about the evolution of brake systems on passenger trains over the past fifty years or so, Blaine stated:

The first commercial application of Electro-Pneumatic braking systems was on the New York City Subway in 1906. It recognised the uniformity of the response to the braking command was of immense advantage and I pushed hard for that while at the AAR and think my white paper on Electro-Pneumatic freight brake helped the ECP get put on the agenda. On the advent of "speed control," [an advanced passenger car brake control system not seen in Australasia] *speed governor control was unique with the D22 equipment. It had been tried out in various forms on the infamous* [or famous] *Wilmington tests conducted on the Pennsylvania Railroad in 1935 and 1936. Of interest, the Pennsylvania test department said, "Yeah, that is great, but we ain't going to buy any of that expensive stuff." Speed governor control overcame the friction disadvantages of the cast iron brake shoe.*

Referring to the later introduction of composition brake shoes, Blaine said:

You didn't need all that speed governor control. I remember disc brakes were first broadly applied on passenger cars I think, about 1935 or '36, courtesy of the Budd Company by C. LaVergne Eksergian, who was looking for a quiet, uniform passenger car friction brake.

On crystal-ball gazing, he said: *Well, I would like to see a universal application of automatic pressure-maintaining. I believe it* [ECP] *is going to become a standard.*

David Blaine published several books and many articles over his long career, under the auspices of the ASME and the ABA, and authored the air brake section of *Mark's Mechanical Engineer's Handbook*, 9th and 10th editions. He died on January 5, 2002.

8.

Glossary

Association of American Railroads

The AAR is an industry association dating from 1934 for North American railroads, suppliers, and associated companies to manage the collective interests of its members. Among many other functions, it prescribes certain physical specifications for equipment and prepares standards for mechanical components to facilitate interchange between member railroads. It also formulates standard operations and maintenance processes and procedures.

Accelerated application

A feature of the ABDW control valve that provides for local brake pipe reductions at each car following initial Quick-Service. This control valve activity accelerates the rate of propagation of a brake pipe reduction along the train, in turn hastening an application of vehicle brakes along the total train length. The beneficial effect of this activity is that brake application along the train is more uniform, thus reducing the potential for buff slack action. From alongside, the Accelerated Application Valve can be heard operating during a brake pipe reduction by a series of rapid-fire 'clicks' and staccato airflow noises.

Accelerated Service release

A feature of the ABD control valve that assists the brake pipe recharge at each car so as to serially transmit the release rapidly along the train. The Accelerated Service Release valve functions to supply some pressure from each car Emergency reservoir into the brake pipe locally to assist in brake pipe recharge and to speed the rate of propagation of the pneumatic 'release' signal along the brake pipe.

Accelerated Emergency release

This feature hastens release from an Emergency brake application. Without it, release and complete recharge would take more than twice as long as release and recharge from Full Service, due to the necessity to recharge *both* reservoirs as well as the brake pipe. In this case, when brake pipe pressure begins to increase (when the locomotive driver initiates a brake release), the control valve allows both the Auxiliary Reservoir *and* the control pipe (and thus also the dummy reservoir) to "back-dump" into the brake pipe at each wagon. This continues until brake pipe pressure has increased to a point within approximately 70 kPa (10 psi) of equalising with the brake pipe and Auxiliary Reservoirs (in combination). The feature greatly reduces the amount of air requiring to be supplied from the locomotive.

Australasian Railway Association

The ARA is a member-based association that represents the interests of the rail sector in Australia and New Zealand. The association assists its members by providing relevant information on a wide range of topics affecting the rail industry, including rail research, communications, safety, and infrastructure. The ARA is also active in the development of rail industry policy to ensure industry views are adequately represented.

Auxiliary reservoir

A storage volume that receives air pressure from the brake pipe as a medium for applying brakes on a car or locomotive. In most freight car equipment, it is combined with the Emergency reservoir in one divided structure.

Brake pipe

A system of piping and hose couplings by which all vehicles in a train are connected, to permit the delivery of compressed air from the locomotive for the control of air braking activity on the vehicles.

Brake pipe pressure

The pressure of air in the brake pipe expressed usually in kilopascals (kPa) in Australasia and in pounds-per-square-inch (psi) in North America. Other countries, for example, on the Indian subcontinent and in Europe, may use kilograms-per-square-centimetre (kg/cm^2, an alternative expression to kPa) as the unit of pressure.

Braking ratio

A proportional value, originally referred to as "braking power", obtained by dividing the total braking force of the rail vehicle by its weight, and usually expressed as a percentage. Early ratios were usually 100 percent at 50 psi brake cylinder pressure, this being considered an acceptable compromise between the points of "wheelslide/no-slide" on good rail. However, since—by the mid-to-late 1880s—most of the rail being produced could provide merely an <u>average</u> head condition, and because, in the United States, the new Quick-Action triple valve increased Emergency brake cylinder pressure by around 20 percent (i.e. to 60 psi), design braking ratios were eventually reduced to around 75 percent at 50 psi (equivalent to 90 percent at 60 psi). Higher ratios were possible in passenger service provided they could be reduced at lower speed by releasing and re-applying the brakes with a lesser brake cylinder pressure, to complete the stop.

The theoretical calculable figure—assuming 100 percent efficiency of the brake rigging system—is called the Gross Braking Ratio, while a similar reckoning using the <u>actual</u> shoe force (therefore taking into account the rigging efficiency) is known as the Net Braking Ratio.

Clean Cab

From the late 1960s and throughout the 1970s, there had been vociferous criticism from US transport accident investigation agency, the National Transportation Safety Board, of the structural performance of locomotives—and particularly their crew cabs—during collisions and of the perceived failure of the Federal Railroad Authority to pursue the regulation of safer locomotive cabs. This led the FRA to form the Locomotive Control Compartment Committee composed of representatives of the major railroads - the Federal Railroad Administration, NTSB, and AAR, plus craft unions, the United Transportation Union, and the Brotherhood of Locomotive Engineers. The influence of this cab committee has since resulted in many Federal safety regulations and standards, including the Clean Cab Requirements, which were adopted by the AAR in 1975 and incorporated in their Manual of Standards and Recommended Practices. Consisting largely of requirements for padding, corner-rounding, control knob recessing, and other ergonomic improvements, these requirements were developed to "clean" the diesel locomotive cab of unnecessary protuberances and to stipulate some generic design requirements for new cab equipment. They alleviated many of the minor injuries being sustained in locomotive cabs at the time and have resulted in the more user-friendly cab environments being enjoyed by locomotive crews today. Certain Clean Cab features were first applied in Australia from the 1980s by the two major Pilbara iron ore railroads.

Consist (noun)

Pronounced *CON*-sist. This is railway terminology that refers to the collection of vehicles comprising the whole, hence the locomotive ***con***sist (group of multi-coupled locomotives) or the train ***con***sist (the lineup of coupled freight or passenger vehicles comprising the train or rake). Note, in this context, a *rake* of vehicles is a group of coupled vehicles comprising either a portion of a train or a group not yet existing as part of a train (e.g., a line of coupled vehicles on a siding).

Coupler (mechanical)

A device for connecting cars and/or locomotives together. The standard North American automatic "knuckle" coupler—originally called the "Janney" (after its inventor) or "buck-eye" coupler—is required to couple automatically by impact and be capable of uncoupling without personnel going between the vehicles. These couplers are often referred to generically (but not entirely accurately) by a manufacturer's name, for example, Alliance.

Narrow-gauge railways have often used the Norwegian or "chopper" coupler consisting of a single central buffer with a mechanical hook that sits in a slot in that buffer. In overall shape, the hook is considered to resemble a meat chopper, hence the name. Because chopper couplings are not particularly strong, Australasian railways that had used them have inevitably converted to automatic couplers, which are then often (and incorrectly) referred to as "buffers."

Coupler slack

This is the degree of freedom of movement or "looseness" in the installed coupler and drawgear package of a rail vehicle. There are two kinds: *loose* (or "free") *slack* and *spring slack*. Loose slack is that existing between adjoining vehicles without any compression of draftgear assemblies, and spring slack is the additional movement offering and often occurring within the assembly when it is compressed. Spring slack also tends to react, driving the slack action in the opposite direction with the potential to increase train shocks. The total amount of coupler slack within a train is a function of the number of vehicles and the physical condition of the coupler and draftgear assemblies (see also Slack Action).

Collapsible or "by-pass" equalising discharge piston

This was a feature of some US brake equipments, whereby the triple valve piston stem was hollow and the discharge valve free to slide vertically within it but held in a normal relative position by a spring. Whenever equalising reservoir pressure slightly exceeded that in the brake pipe (e.g. at brake release on a train of any reasonable length), then owing to the small volume of the equalising reservoir as compared to the brake pipe and auxiliary and supplementary reservoirs, the excess pressure above the equalising piston forced it down (against the force of the internal spring), exposing the bypass grooves in its bushing and thus permitting equalisation of pressures across it. Having substantially the same pressures in the equalising reservoir and brake pipe *during brake pipe recharge* made it possible to make a re-application more promptly (i.e. before a full recharge had been attained) than was possible with a solid-type piston.

This feature was not otherwise required for train operations in Australia and New Zealand, although an early E-P brake used on EMU stock in both New Zealand and New South Wales did include a bypass piston. It is also notable that air brake schematic diagrams for the NSWGR 40 Class diesel-electric locomotives show a bypass piston as part of the automatic brake valve with the 6-SL schedule.

Direct Release

Describes the normal release functioning of AAR (i.e. North American-derived) freight car brake equipment, specifically the operating valve. When a brake pipe pressure increase moves the operating valve to its Release position, all brake cylinder pressure is exhausted. Passenger and locomotive brake equipments of 26-L and later types can be set to provide only *direct release,* and this applies to most such equipment used in Australia and New Zealand.

DIN edge connectors

Edge card connectors are embedded devices that mate with the edges of single- or double-sided printed circuit boards (PCBs) in order to provide an external electrical connection. Industry standards define power specifications such as voltage levels, power pin placements, and power requirements. The contacts of edge card connectors are made of metal and are plated with a thin, metallic coating of gold, copper, nickel, or silver that improves conductivity and prevents rust and corrosion. Contacts are housed in a termination assembly that is usually made of plastic. Edge card connectors are used widely in personal computers and, depending on their components, are suitable for lead-free applications.

"DIN" stands for *Deutsches Insitut für Normung*, the German standards-setting organisation. A DIN connector is one that conforms to one of the standards defined by DIN.

Double Check Valve

A non-return pneumatic valve having two input lines and seats arranged such that air flowing from one input will seal the valve at the other seat and flow to a third, common delivery connection. In other words, the delivery connection can receive air from only one of the two input lines at any time.

Draftgear

A shock-cushioning unit provided to transmit compressive and tensile forces (i.e. buff and draft) between the coupler and vehicle structure. This equipment is contained within the coupler yoke and at its inner end bears against the follower plates, which press against draft lugs attached to the vehicle structure.

Dynamic brake

An electrical means of speed retardation of a locomotive (and therefore the train) obtained by switching the locomotive traction motors to become separately-excited generators. The traction motor armatures, or (for AC machines, rotors) are then forced—by the kinetic energy of the moving train—to rotate within a strong magnetic field that tends to oppose this rotation, thus imparting a retarding force against the movement of the train. This retarding force can be used to stabilise or to slow the speed of a train, and is referred to as "braking effort." There are two forms of dynamic braking:

<u>Rheostatic</u>**:**

With diesel-electric locomotives, a heavy resistance (the D/B resistor grid - 700 amps/600 volts) is connected in the circuit to consume the electrical energy produced, by converting it to heat. The resistance of this grid is low (less than one ohm) so that a condition close to a dead short occurs, thus allowing for maximum energy to be dissipated. Cooling fans powered either by this waste electrical energy or—in some older locomotives by engine-driven blowers—then dissipate this heat. The locomotive driver controls the above-mentioned traction motor field strength—and thus the degree of retardation—by moving the Braking Lever (which is attached to a rheostat) to vary alternator excitation and thus output to the traction motor field windings. With "standard" dynamic braking, the lower limit of the usable speed range of the locomotive is about 33 km/hr (20 mph). At this speed, the maximum braking effort is produced but below this speed, the braking force drops off in a linear fashion to almost nil force at a standstill. At 16 or 17 km/hr (10 mph), only half of the maximum amount of braking force is available.

With most modern diesel locomotives, *extended-range* dynamic braking is available. With this feature, portions of the resistor grid are automatically shorted out by switch gear as speed decreases, reducing the resistance across the grid and allowing more current to flow. In this way, the resistance to traction motor rotation is maintained at a high value, permitting such locomotives to be capable of being dynamic-braked to a stand (or almost so).

Regenerative:

With straight-electric locomotives, the current generated by dynamic braking can be fed back into the overhead contact wire, assuming an approximate power use elsewhere on the network; otherwise, the energy output must also be dissipated as waste heat.

With both systems, the amount of braking current produced—actually traction motor armature current—is presented on the driving ammeter (or load meter) and must be closely monitored by the driver for good train-handling outcomes. Because dynamic braking can fail or otherwise be inoperative on any locomotive without necessarily preventing train operation, it should be considered a secondary system to air braking.

It should be noted that the popular concept of *reversal* of traction motors is a misnomer. Dynamic Braking does not in any sense involve the "reversing" of traction circuits.

ECP braking

Electronically Controlled Pneumatic braking is a recent technological development that provides for the air brake operating signal to all cars in the train to be more or less instantaneous rather than relying on pneumatic propagation through the brake pipe. This signal may be transmitted either by a radio link or by a cabled connection, with the latter system appearing (as this is being written) to be prevailing over wireless systems. A continuous, charged brake pipe is still required; however, it exists only to provide the operating "muscle" on each car—via storage volumes and the operating valve—rather than being required to transmit the operating signal. Thus, the brake pipe and therefore the auxiliary reservoirs can be kept fully charged at all times for virtually "inexhaustible" operation. Because of its instantaneous controllability, ECP braking provides for graduated release capability on even the longest of freight trains. This functionality, arguably, renders ECP as the greatest single advance in railway air braking since the advent of the triple valve.

EMD

The ***E***lectro-***M***otive ***D***ivision of General Motors Corporation, latterly known as the General Motors Locomotive Group, has manufactured diesel-electric locomotives for the US and international markets since 1941. The company's origins date back to the privately-owned Electro-Motive Engineering Company founded in Cleveland, Ohio, in 1922 and becoming the Electro-Motive Company (EMC) in 1925. By 1930, General Motors—having decided to develop the nascent diesel engine—purchased the Cleveland-based Winton Engine Company, and later its chief customer, EMC. In 1941, EMC and part of the Winton Engine Co. were merged to create GM's Electro-Motive Division. Locomotive-building plants were established in both the United States (La Grange, Illinois, nowadays a southern suburb of Chicago) and Canada (at London, Ontario—midway between Detroit and Toronto—and known as the Diesel Division of General Motors of Canada Ltd.). EMD has benefited from license-building arrangements with engineering firms in many foreign countries that have spread both domestic and export EMD locomotive models around the globe. In 2005, General Motors sold the division to a corporate partnership—Greenbriar Equity Group LLC and Berkshire Partners LLC—which formed Electro-Motive Diesel Incorporated to facilitate the purchase, thus preserving the well-known "EMD" branding. In August 2010, Progress Rail Services Corporation, a wholly-owned subsidiary of Caterpillar Incorporated, purchased Electro-Motive Diesel Inc. from Greenbriar, Berkshire, and their investment partners, retaining both the US and Canadian plants.

Nowadays the world's second-largest builder of railway locomotives, EMD claims to have the largest installed base of diesel-electric locomotives in both North America and internationally. Supporting EMD in Europe is their aftersales operation set up in 2006, Electro-Motive Services International (EMSI). This organisation provides contract and field maintenance, technical assistance, locomotive rebuilds, and after-market parts for European operators of their Class 66 locomotive.

EMERGENCY APPLICATION

A brake pipe pressure reduction at a faster rate than for a Service application, and sufficient to cause the operating valve—where so-designed—to move to its Emergency position and deliver a higher brake cylinder pressure than for a Full Service application. Operating valves that are not designed to provide a specific Emergency brake cylinder pressure (i.e. most such valves in Australasian state rail systems) will still provide a Quick-Action response to the Emergency pneumatic signal to assists its propagation along the brake pipe. As much as is possible, this will contribute to the development of brake cylinder pressure (to the equivalent of a Full Service application) as quickly as is possible.

EPROM

An ***E***rasable ***P***rogrammable ***R***ead-***O***nly ***M***emory chip is a type of processor used in computers and other electronic devices to store small amounts of data that must be saved when power is removed (as in calibration tables or for device configuration). Such chips are referred to as "non-volatile." Once programmed, an EPROM can be erased only by exposing it to strong ultraviolet light. Eprom's are often used in railway electronic applications.

EQUALISATION (PRESSURE)

The final pressure resulting when two volumes, initially at different pressures, become the same.

<u>Service equalisation</u> exists when, during a Service brake application, *brake pipe* pressure has reduced to and settled at the reduced *equalising reservoir* value. This can also be referred to as an "effective" brake application. Ideally, train brake applications should only be released after attaining this state.

<u>Full Service equalisation</u> refers to the *auxiliary reservoir* and *brake cylinder* volumes being openly connected and thus of equal pressure. This will result from an effective Full Service brake application in which the maximum Service brake pipe pressure reduction has been made and maximum Service brake cylinder pressure attained.

<u>Emergency equalisation</u> exists when both the *auxiliary and* (where provided) *emergency reservoir volumes* are connected to the *brake cylinder volume*. This will result in the Emergency brake cylinder pressure—which is greater than that achieved at Full Service—being attained. The design and operation of "non-US" operating valves used in Australia and New Zealand does not provide a true Emergency application (see Emergency Application above).

Feed valve

A valve that is an integral part of earlier driver's brake valve installations and that reduces main reservoir to a determined amount for delivery to the equalising reservoir and brake pipe.

Full service application

A brake application at the Service rate that provides the maximum Service brake cylinder pressure from the particular operating brake pipe pressure (see also Service Application).

Full service reduction

A brake pipe pressure reduction at the Service rate that is sufficient in degree to cause the development of maximum service brake cylinder pressure.

Graduated release

A feature of some locomotive and passenger brake equipment that permits brake cylinder pressure to be reduced in steps proportional to an incremental increase in brake pipe pressure. In other words, an Automatic brake application can be partly released in steps in much the same manner as it can be created. With conventional air braking, this measure of control is not normally technically feasible with trains over about five hundred metres in length. ECP braking, however, has made graduated release possible on freight trains of any length (see ECP Braking).

Interstate Commerce Commission

The ICC was a body created in the United States in 1887 to regulate railroads and later the trucking industry. Its name carries the implication that the original intention may have been that it become the "single roof" over various regulatory efforts, including communications. Conceived as a result of widespread and longstanding agitation by community groups who believed that the railroads possessed economic power that they systematically abused, it was empowered to set and publish rates as well as address civil rights issues when passengers filed complaints. It was the first independent agency in the United States, though was not ultimately considered effective. The agency was abolished in 1995, its remaining functions being transferred to the Surface Transportation Board. The Commission's five (later seven) members—who were not permitted to have economic ties to the industries they regulated—were appointed by the president with the consent of the United States Senate. The model represented by the ICC is said to have had a strong influence on the founders of Australia and the creation of an Inter-State Commission which is considered never to have assumed the role intended for it.

Inshot

A feature of operating valves that provides an effective, early response at each vehicle through a rapid development of brake cylinder pressure during the initial stage of the application. It is designed to reduce train slack compressive impacts (buff slack action) and overcome the effect of brake cylinder displacement upon cylinder pressure as the piston moves through its stroke. The inshot pressure is generally supplied to the brake cylinder via an alternate circuit additional to the normal brake cylinder supply orifice. When a specific brake cylinder pressure is achieved, the secondary circuit is automatically interrupted, terminating the inshot function, and leaving further brake cylinder pressure development to occur via the normal orifice. The usual inshot pressure in a direct release system is 70 kPa (10 psi). If the initial brake pipe reduction is greater than a Minimum Reduction, the inshot feature is bypassed. It should not be confused with Quick-Service.

LSI

In the early 1990s, the AAR Railway Electronics Task Force developed ***L**ocomotive **S**ystem **I**ntegration* standards and specifications. These are designed to promote a competitive supply market in railway electronics and to achieve a level of interchangeability for new control equipment (e.g. screen-based displays and ECP brake systems) being produced by North American locomotive and electronics manufacturers. Although the standards are used extensively, changes in technology require constant updates to permit LSI architecture to more easily accept added functionality to an LSI-equipped locomotive. Other ongoing imperatives are to reduce the cost and ease the implementation of additional LSI functionality or changes to existing functionality.

The RETF is composed of railway officers, suppliers, and other interested parties; its mission being to support the railroads in maintaining electronics standards and specifications. A configuration management plan establishes practices for all AAR-developed hardware, software, and documentation.

Magnet valve

A valvular device—actuated by an electromagnetic coil—that directs or controls airflow.

Operating valve (noun)

A generic term for the device employed to control brake activity on the car or locomotive. Variously known as a triple valve, control valve, or distributing valve, this device is designed to apply or release the vehicle brakes in response to variations in brake pipe pressure. As part of this functioning, the operating valve also; (1) controls auxiliary and emergency reservoir charging; and (2) brake cylinder exhaust. In relay brake systems, the operating valve provides a 'control' pressure to the brake cylinder relay valve to furnish a different braking ratio.

Through technical evolution, operating valves have progressed from being able to provide Quick-Action and Quick-Service features to having Retarded Recharge, Accelerated Application, Accelerated Release, and Brake Cylinder Inshot and Reduction-Ensuring (or "brake cylinder maintaining"), functionality. All of these features were designed to speed up the propagation of the pneumatic signal along the brake pipe, to get brake blocks against the wheels as quickly as possible to assist in controlling immediate slack action, to reduce the volume of recharge air that has to be supplied to the brake pipe to effect a release of brakes, and/or to assist with providing a degree of stability to the operating valve itself.

Over-reduction

A Service brake pipe reduction to a pressure lower than that at which the auxiliary reservoir and brake cylinder equalise. This is sometimes necessary when another application is required closely following a release and before auxiliary reservoirs have had time to fully recharge. In such a case, an over-reduction may be required to achieve the maximum possible brake cylinder pressure (considering the circumstances) and—with a 26C driver's brake valve—is accomplished by moving the Automatic brake valve handle into the "over-reduction zone," that is, to that portion of the quadrant between the Suppression and Handle Off positions. Should this be required, self-lapping will not occur, and the brake valve handle must be returned to the "Service Zone" to interrupt the exhaust once the required brake pipe reduction has been achieved, or the reduction will continue down to zero at the Service rate. This action requires good driver judgment and is usually only required (and effective) in passenger train operations where graduated release is not available. Appropriate outcomes using over-reduction with longer trains cannot be assured and making an over-reduction in freight train operation would generally be avoided.

Pressure-maintaining

An Automatic brake valve feature that will sustain brake pipe pressure to the equalising reservoir value against the effects of reasonable brake pipe leakage. This feature permits the degree of brake application to remain under the control of the driver. Brake pipe leakage is maintained such that the brakes do not apply to a greater degree than called-for by the driver. The requirement to release the brakes to avoid stalling of the train as a result of excessive braking due to leakage increasing the magnitude of the application is thereby eliminated. The advent of the pressure maintaining brake valve has arguably reduced the degree of skill and judgment required of the locomotive driver when braking a "direct release" train on heavy grade territory, but it has been a technological advance that few locomotive drivers would voluntarily forego.

Pulse Width Modulation

PWM is a comparatively recent technique made practical by modern electronic power switches. PWM of a signal or power source is a method of transmitting information on a series of pulses and is an efficient way of providing intermediate amounts of electrical power between the settings "fully-on" and "fully-off." The data being transmitted are encoded on the width of these pulses to either convey information over a communications channel or control the amount of power being sent to a load. PWM works well with digital controls, which, because of their on/off nature, can easily set the needed duty cycle.

Quick-action

A feature designed into air brake operating valves whereby an Emergency brake pipe reduction is passed rapidly (at approximately 320 m/sec or 960 ft/sec, regardless of operating brake pipe pressure) from car-to-car along the train with the objective of commencing the development of car brake cylinder pressure in the shortest possible time in order to prevent the onset of severe slack action.

Quick-Service

A feature designed into air brake operating valves that provides a controlled Service rate of brake pipe reduction by providing a small, initial local reduction of brake pipe pressure at each car, this action being transmitted serially from car to car. The resistance to the flow of air through a long brake pipe is relatively high due to natural laws, and, therefore, without the supplementary venting of brake pipe pressure at each car, the total brake pipe reduction would have to occur at the driver's brake valve and be significantly protracted. This extended time interval between the application of brakes at the front and rear of the train would result in excessive compressive train slack action while also increasing the stopping distance.

Note that Quick Service is available only for the initial brake brake pipe pressure reduction. Subsequent brake pipe reductions made to increase the application do not produce Quick-Service activity, the feature being reactivated only after the operating valve piston has returned to its Release position. In this respect, Direct Release has an advantage over Graduated Release in that Quick-Service activity is available much sooner after initiating the brake release. That said, it should be noted that the Quick-Service function of a Graduated Release system will propagate faster than that of a Direct Release system (at around 200 m/sec or 600 ft/sec versus 180 m/sec or 540 ft/sec). This is because—since the Graduated Release brake system lacks the Emergency brake feature of the Direct Release arrangement—it does not require a differentiation between Service and Emergency rates of brake pipe reduction.

Relay valve

A device that receives controlling pressure from an input source and functions to deliver and maintain a corresponding output pressure. If being employed as a brake cylinder relay valve, it may be located close to a supply (also called a "supplementary") reservoir and one or more brake cylinders to hasten the action of application and release. Through the use of differential diaphragms of different sizes, a lower (step-down) or higher (step-up) pressure may be delivered.

Retarded recharge

Also variously referred to as Retarded Release and Uniform Release, this is an automatic function provided by the operating valve during Release and Recharge as well as initial system charging by which the charging flow of brake pipe air to the reservoir (*auxiliary reservoir* for direct release systems and *control reservoir* for graduated release systems) is restricted and, in some cases, isolated to reduce the likelihood of overcharge on head-end cars and to provide more air volume to flow back into the brake pipe to obtain a more uniform release and recharge of brakes throughout the train. This latter advantage was and is more critical where older brake equipments designed to admit main reservoir pressure directly into the brake pipe during Release were (and are) employed. Since the advent of Accelerated Release, this feature is of reduced value where modern, maintaining brake systems are in use.

Specific to Direct Release brake systems, the Retarded Recharge feature will also provide greater brake stability against the occurrence of undesired brake applications (UDEs) caused by excessive reservoir leakage, probably its most beneficial effect in the modern-day context. By way of explanation, on a Direct Release brake system, a rapid drop in reservoir pressure on a car will create a sufficient pressure differential between brake pipe and auxiliary reservoir to cause the operating valve to move to its Retarded Recharge position. The resulting restricted charging port or groove will tend to control or limit the replacement flow of brake pipe air to the exhausting reservoir, thus preventing or at least minimising the chance of an uncommanded brake application that might otherwise result if Quick-Service activity were to occur.

The Retarded Recharge action during brake release extends only as far back into the train as approximately a 20 kPa (3 psi) greater pressure of brake pipe-over-auxiliary reservoir can be obtained - approximately 500–600 metres. This operative distance will presumably also apply from any distributed locomotives.

Rollingstock

An inclusive term for all the vehicles that move upon a railway track. It usually includes both powered and unpowered vehicles, for example locomotives, railroad cars, coaches, and wagons, however, the term is sometimes used to refer only to unpowered vehicles, specifically excluding locomotives.

There is an almost universal inconsistency in presenting the word (either as one or as two words) and there seems to be no formally-established protocol for this within or without the industry. Some dictionaries even hyphenate the word (e.g. as for 'rolling-pin'). Throughout this book I have chosen to present it as one word, adhering to the logic that *rollingstock* is to rail operations as *livestock* is to animal husbandry and *feedstock* is to industry.

SCADA

Supervisory ***C***ontrol and ***D***ata ***A***cquisition; a microprocessor-based technology for the remote control of pipeline and other energy utility operations.

Screen-based control

This is a technology nowadays considered a normal locomotive specification. In basic terms, computer screens are used to replace the conventional gauges and some switches on a locomotive control stand or desk. Integrated Function Computer and Integrated Function Display (IFC and IFD) together comprise the proprietary screen-based control and data display system produced by General Electric Transportation and supplied on their diesel-electric locomotives. The other major North American builder, EMD, has used the Rockwell-produced Integrated Cab Electronics (ICE) system and more latterly their own system, Functionally Integrated Railroad Electronics (FIRE).

With this technology, "switches" are usually touch-sensitive *membrane* switches on panels or *touch-screen* "switches" on display screens. Locomotive operating, maintenance, and fault data can be accessed and system configuration (e.g. brake equipment Cut-in and Cut-out) carried out by the locomotive driver or maintainer.

Service application

A reduction of brake pipe pressure—usually made at the driver's Automatic brake valve—at a rate and amount sufficient to cause the operating valve to move to its Service position. Note that it is not appropriate to refer to a "Service" or "Full Service" application in the context of locomotive Independent brake operation since valvular manipulation of the Independent brake and its operation is quite different and there is no comparative "Emergency" capability.

Service rate

The 'normal' rate—slower than Emergency—at which the brake pipe pressure reduces or is reduced to cause the operating valve to assume its Service position.

Slack action

The free movement—permitted by the amount of coupler slack present—that may occur between adjoining coupled rail vehicles while moving. Slack action will occur in almost all moving trains and has the potential to cause severe and damaging in-train forces. These forces result from the rate-of-change of slack from a state of tension to one of compression (referred to as "run-in", which is dynamic *buff* force) or vice versa (known as "run-out", which is dynamic *draft* force). When coupling slack runs in or out rapidly, one part of the train attains a greater or lesser speed than another, with the resultant shock coming from the stress caused by the drawgear suddenly being required to make uniform the speed of both parts of the train. The extent of the shock will depend partly upon the difference in speed of the vehicles in these portions of the train and partly on the total mass that must be suddenly altered in speed. While mass is obviously a critical factor in this equation, the rate-of-change in speed is more so, as the shock imposed varies with the square of the speed. Thus, a sudden change of speed within some part of a train of 5 km/hr will cause nine times the shock of a speed change of 2 km/hr.

The state of slack within a train—from buff to draft or vice versa—cannot be altered both gently and quickly at the same time, but slack action can be effectively controlled and its harsh effects alleviated through the proper use by drivers of appropriate train-handling practices, including correct air brake manipulation (very often as part of coordinated throttle and air brake activity). For more information, refer to chapter 6.

Slug units

A railroad slug is an accessory to a diesel-electric locomotive, to which it is connected. In essence, it is a locomotive without any power generation capability. It has bogies and traction motors but is unable to move under its own power. Instead, it is connected to its "mother" locomotive, which provides current to operate the traction motors. A slug is used to increase adhesive weight, allowing full horsepower to be applied at a lower speed and permitting a higher maximum tractive effort. Slugs are often used in low-speed operations, such as yard shunting, especially hump shunting.

Slug operation is possible because at low speeds, a diesel-electric locomotive prime mover/alternator set is capable of producing more electric power than the locomotives' traction motors can effectively use.

A slug increases the number of traction motors available to the locomotive, increasing both its pulling and braking power. As well as this, the electrical load on each traction motor is reduced, helping to prevent overheating from excess current. Slugs can be built new or converted from existing locomotives. They typically carry ballast to increase their weight and improve traction.

Soft key

A control button or control surface 'touch-point' that is flexibly programmable to invoke any of a number of functions rather than being associated with a single fixed function or set of functions. A soft key may be located alongside the display screen of a control portal (e.g. flat-screen display) or portable device such as a cellular phone, or might be a touch-point depicted graphically on a touch-sensitive display screen. The button invokes a function described by the text at that moment shown on the display adjacent to the button.

Stringlining (train slack-action)

An extremely violent slack run-out event occurring as a train negotiates a curve, especially with the existence of the heavy-duty couplers generally used by heavy-haul mineral railroads, will result in the train tending to "pull itself straight" across the chord of the curve. If the tension is not relieved by the train parting, it may derail toward the inside of the curve, also pulling the track structure laterally to the inside of the curve as it does so.

As long as there is no sudden and extreme draft slack action, and as long as no part of the curving train is derailed, then it will track relatively freely around the curve with each individual car subject and reacting to a direct draft force applied from the car immediately ahead. However, any train totally in draft tension while traversing a curve and sufficiently lengthy that it occupies all or most of the curve, will possess an inherent tendency to stringline.

Transportation Technology Center Inc./Transportation Technology Center

Formed in January 1998 as a wholly-owned subsidiary of the AAR, the Transportation Technology Center Incorporated (TTCI), is located at and manages the FRA's Transportation Technology Center (TTC) near Pueblo, Colorado. TTC operates under a care, custody, and control contract with the FRA. The 52-square mile, secure and remote site operates a large array of specialised testing facilities and railway infrastructure and provides isolated testing for all categories of freight and passenger rollingstock, vehicle and track components, and safety devices. TTCI's engineering team is available to help provide solutions to problems related to track issues, rail vehicles and components, vehicle track interaction, and railroad operations. It describes its mission as "accelerating the use of clean, safe, and efficient technologies by railways worldwide."

Vent valve

A valve or valvular portion of a rail vehicle brake system that responds to an Emergency rate of brake pipe reduction to vent the brake pipe locally at that vehicle and thereby serially propagate the Emergency application throughout the train.

UIC

The ***U***nion ***I***nternationale des **C**hemins de Fer (International Union of Railways) is a European-based, worldwide organisation for international cooperation among railways and promotion of the rail transport mode. Founded in 1922, its initial purpose was standardisation and improvement of conditions for railway construction and operations, especially in view of international traffic throughout continental Europe.

Latterly, the organisation has been revamped in order to cope with a series of new challenges - in particular railway liberalisation, increasing competition from other modes, and the growing economic constraints on railways together with the challenge of transport market globalisation. Its current stated mission is to "promote rail transport at a world level in order to meet challenges of mobility and sustainable development."

Headquartered in Paris, France, the UIC currently groups over 170 members (railways, rail operators, infrastructure managers, railway service providers, public transport companies, etc.) on all five continents.

ABOUT THE AUTHOR

FERGUS HAS WORKED in the rail transport environment for over 50 years. He grew up on a sheep farm in the New Zealand province of South Otago and began his career in 1970 at age 17 as an engine cleaner in the Dunedin steam locomotive depot of New Zealand Railways. He went on to become a locomotive engineman on passenger and freight operations with diesel and electric traction in various locations around New Zealand.

In 1981 he accepted a position as a locomotive driver with Mt Newman Mining (later to become BHP Billiton Iron Ore) in the Pilbara region of Western Australia and he and his wife moved to Port Hedland. Fergus went on to training and supervisory positions within the BHP Iron Ore Railroad Operations division. After 13 years in this environment—with BHP and later, Robe River Iron Associates—Fergus took up an operations position with National Rail Corporation (now Pacific National).

In the ensuing 20 years he has worked in a private contract capacity as a locomotive driver, operating from various locations throughout Western Australia and New South Wales, and in contract linehaul service between Adelaide and Perth. Along the way he spent four months in India training locomotive and supervisory staff in the operation of the LOCOTROL distributed motive power system (the subject of the author's second book – see below), and a year in SE Asia and two years in Queensland in technical management roles involving the establishment of luxury tourist passenger train services.

Fergus now lives in Melbourne, Australia, and has retired from a 15-year career as a transport safety investigator for a state government agency. He was a foundation member of the Melbourne Rebels Super Rugby club and continues to support the NZ All Blacks. He enjoys the V8 Supercars competition, and although the financial burdens of life have rendered him a lapsed private pilot, he still pursues an active interest in aviation and likes to explore his adopted state of Victoria at weekends. Navigation of America's Great Loop is at the top of his bucket list.

[*The Locotrol Story – THE ONLY GAME IN TOWN* - ISBN: 978-0-6485290-1-9, via IngramSpark, La Vergne, TN, USA, and Ingram Content Group, Melbourne, Australia. Available from online bookstores or from the author – femoffat@gmail.com]

www.ingramcontent.com/pod-product-compliance
Lightning Source LLC
LaVergne TN
LVHW070944210926
848482LV00002B/91